Discovering Evolution Equations
with Applications

Volume 1-Deterministic Equations

Published Titles

Advanced Differential Quadrature Methods, Zhi Zong and Yingyan Zhang
Computing with hp-ADAPTIVE FINITE ELEMENTS, Volume 1, One and Two Dimensional Elliptic and Maxwell Problems, Leszek Demkowicz
Computing with hp-ADAPTIVE FINITE ELEMENTS, Volume 2, Frontiers: Three Dimensional Elliptic and Maxwell Problems with Applications, Leszek Demkowicz, Jason Kurtz, David Pardo, Maciej Paszyński, Waldemar Rachowicz, and Adam Zdunek
CRC Standard Curves and Surfaces with Mathematica®*: Second Edition*, David H. von Seggern
Discovering Evolution Equations with Applications: Volume 1-Deterministic Equations, Mark A. McKibben
Exact Solutions and Invariant Subspaces of Nonlinear Partial Differential Equations in Mechanics and Physics, Victor A. Galaktionov and Sergey R. Svirshchevskii
Geometric Sturmian Theory of Nonlinear Parabolic Equations and Applications, Victor A. Galaktionov
Introduction to Fuzzy Systems, Guanrong Chen and Trung Tat Pham
Introduction to non-Kerr Law Optical Solitons, Anjan Biswas and Swapan Konar
Introduction to Partial Differential Equations with MATLAB®, Matthew P. Coleman
Introduction to Quantum Control and Dynamics, Domenico D'Alessandro
Mathematical Methods in Physics and Engineering with Mathematica, Ferdinand F. Cap
Mathematical Theory of Quantum Computation, Goong Chen and Zijian Diao
Mathematics of Quantum Computation and Quantum Technology, Goong Chen, Louis Kauffman, and Samuel J. Lomonaco
Mixed Boundary Value Problems, Dean G. Duffy
Multi-Resolution Methods for Modeling and Control of Dynamical Systems, Puneet Singla and John L. Junkins
Optimal Estimation of Dynamic Systems, John L. Crassidis and John L. Junkins
Quantum Computing Devices: Principles, Designs, and Analysis, Goong Chen, David A. Church, Berthold-Georg Englert, Carsten Henkel, Bernd Rohwedder, Marlan O. Scully, and M. Suhail Zubairy
A Shock-Fitting Primer, Manuel D. Salas
Stochastic Partial Differential Equations, Pao-Liu Chow

CHAPMAN & HALL/CRC APPLIED MATHEMATICS
AND NONLINEAR SCIENCE SERIES

Discovering Evolution Equations with Applications

Volume 1-Deterministic Equations

Mark A. McKibben

Goucher College

Baltimore, Maryland

CRC Press

Taylor & Francis Group

Boca Raton   London   New York

CRC Press is an imprint of the
Taylor & Francis Group, an **informa** business

A CHAPMAN & HALL BOOK

About the text cover. You might very well be wondering about the significance of the text cover. Would you believe that it embodies the main driving force behind the text? Indeed, the initial-value problem in the middle from which all arrows emanate serves as a theoretical central hub that mathematically binds models depicted by the illustrations. Each of the eight pictures illustrates a scenario described by a mathematical model (involving partial differential equations) that is studied in this text. As you will see, all of these models can be written abstractly in the form of the initial-value problem in the middle of the cover. So, despite the disparate nature of the fields in which these models arise, they can all be treated under the same theoretical umbrella. This is the power of the abstract theory developed in this text.

Reading from left to right, and top to bottom, the fields depicted in the pictures are as follows: air pollution, infectious disease epidemiology, neural networks, chemical kinetics, combustion, population dynamics, spatial pattern formation, and soil mechanics.

CRC Press
Taylor & Francis Group
6000 Broken Sound Parkway NW, Suite 300
Boca Raton, FL 33487-2742

First issued in paperback 2017

ISBN 13: 978-1-138-11778-5 (pbk)
ISBN 13: 978-1-4200-9207-3 (hbk)

Library of Congress Cataloging-in-Publication Data

McKibben, Mark A.
   Discovering evolution equations with applications : volume 1-deterministic equations / Mark A. McKibben.
     v. cm. -- (Discovering evolution equations with applications ; v. 1) (Chapman & Hall/CRC applied mathematics and nonlinear science series)
   Includes bibliographical references and index.
   ISBN 978-1-4200-9207-3 (hardcover : alk. paper)
   1. Evolution equations. I. Title. II. Series.

QA377.3.M35 2011
515'.353--dc22   2010021874

Visit the Taylor & Francis Web site at
http://www.taylorandfrancis.com

and the CRC Press Web site at
http://www.crcpress.com

Dedicated to my wife and best friend, Jodi

Contents

Preface

The mathematical modeling of complex phenomena that evolve over time relies heavily on the analysis of a variety of systems of ordinary and partial differential equations. Such models are developed in very disparate areas of study, ranging from the physical and natural sciences and population ecology to economic, neural networks, and infectious disease epidemiology. Despite the eclectic nature of the fields in which these models are formulated, various groups of them share enough common characteristics that make it possible to study them within a unified theoretical framework. Such study is an area of functional analysis commonly referred to as the theory of evolution equations.

One thread of development in this vast field is the study of evolution equations that can be written in an abstract form analogous to a system of finite-dimensional linear ordinary differential equations. The ability to represent the solution of such a finite-dimensional system by a variation of parameters formula involving the matrix exponential prompts one, by analogy, to identify the entity that plays the role of the matrix exponential in a more abstract setting. Depending on the class of equations, this entity can be interpreted as a linear C_0-semigroup, a nonlinear semigroup, a (co)sine family, etc. A general theory is then developed in each situation and applied, to the extent possible, to all models within its parlance.

The literature for the theory of evolution equations is massive. Numerous monographs and journal articles have been written, the total sum of which covers a practically insurmountable amount of ground. While there exist five-volume magna opi which provide excellent accounts of the big picture of aspects of the field (for instance, **[107, 340]**), most books written on evolution equations tend to either provide a thorough treatment of a particular class of equations in tremendous depth for a beginner or focus on presenting an assimilation of materials devoted to a very particular timely research direction (see **[39, 40, 48, 49, 61, 93, 108, 128, 129, 130, 142, 156, 171, 179, 205, 206, 207, 237, 245, 249, 266, 267, 276, 295, 300 306, 317, 318, 329, 341]**). The natural practice in such mathematics texts, given that they are written for readers trained in advanced mathematics, is to pay little attention to preliminary material or behind-the-scenes detail. Needless to say, initiating study in this field can be daunting for beginners. This begs the question, "How do newcomers obtain an overview of the field, in a reasonable amount of time, that prepares them to enter and initially navigate the research realm?" This is what prompted me to embark on writing the current volume. The purpose of this volume is to provide an engaging, accessible account of a rudimentary core of theoretical results that should be understood by anyone studying evolution equations in a way that gradually builds the reader's intuition and that culminates in a discussion of an area of active research.

To accomplish this task, I have opted to write the book using a so-called discovery approach, the ultimate goal of which is to engage you, the reader, in the actual mathematical enterprise of studying evolution equations. Some characteristics of this approach that you will encounter in the text are mentioned below.

What are the "discovery approach" features of the text?

I have tried to extract the essence of my approaches to teaching this material to newcomers to the field and conducting my own research, and incorporate these features into the actual prose of the text. For one, I pose questions of all types throughout the development of the material, from verifying details and illustrating theorems with examples to posing (and proving) conjectures of actual results and analyzing broad strokes that occur within the development of the theory itself. At times, the writing takes the form of a conversation with you, by way of providing motivation for a definition, or setting the stage for the next step of a theoretical development, or prefacing an important theorem with a plain-English explanation of it. I sometimes pose rhetorical questions to you as a lead-in to a subsequent section of the text. The inclusion of such discussion facilitates "seeing the big picture" of a theoretical development that I have found naturally connects its various stages. You are not left guessing why certain results are being developed or why a certain path is being followed. As a result, the exposition in the text, at times, may lack the "polished style" of a mathematical monograph, and the language used will be colloquial English rather than the standard mathematical language that you would encounter in a journal article. But, this style has the benefit of encouraging you to not simply passively read the text, but rather work through it, which is essential to obtaining a meaningful grasp of the material.

I deliberately begin each chapter with a discussion of models, many of which are studied in several chapters and modified along the way to motivate the particular theory to be developed in a given chapter. The intent is to illustrate how taking into account natural additional complexity gives rise to more complicated initial-boundary value problems which, in turn, are formulated using more general abstract evolution equations. This connectivity among different fields and the centrality of the theory of evolution equations to their study is illustrated on the cover of the text.

The driving force of the discussion is the substantive collection of more than 500 questions and exercises dispersed throughout the text. I have inserted questions of all types directly into the development of the chapters with the intention of having you pause and either process what has just been presented or react to a rhetorical question posed. You might be asked to supply details in an argument, verify a definition or theorem using a particular example, create a counterexample to show why an extension of a theorem from one setting to another fails, or conjecture and prove a result of your own based on previous material, etc. The questions, in essence, constitute much of the behind-the-scenes detail that goes into actually formulating the theory. In the spirit of the conversational nature of the text, I have included a section entitled Guidance for Exercises at the end of the first nine chapters that provides two layers of hints for each exercise. Layer one, labeled as "A Small Nudge in a Right

Direction" is intended to help you get started if you are stumped. The idea is that you will re-attempt the exercise using the hint. If you find this hint to be insufficient, the second layer of hints, labeled as "An Additional Thrust in a Right Direction" provides a more substantive suggestion as to how to proceed. In addition to this batch of exercises, you will encounter more than 1,000 questions or directives enclosed in parentheses throughout all parts of the text. The purpose of these less formal, yet equally important questions, is to alert you to when details are being omitted or to call your attention to a specific portion of a proof to which I want you to pay close attention. You will likely view the occurrence of these questions to be, at times, disruptive. And, this is exactly the point of including them! The tendency is to gloss over details when working through material as technical as this, but doing so too often will create gaps in understanding. It is my hope that the inclusion of the combination of the two layers of hints for the formal exercises and this frequent questioning will reduce any reluctance you might have in working through the text.

Finally, each chapter concludes with a section in which some of the models used to motivate the chapter are revisited, but are now modified in order to account for an additional complexity. The impetus is to direct your thinking toward what awaits you in the next chapter. This short, but natural, section is meant to serve as a connective link between chapters.

For whom is this book accessible?

It is my hope that anybody possessing a basic familiarity with the real numbers and at least an exposure to the most elementary of differential equations, be it a student, engineer, scientist, or mathematician specializing in a different area, can work through this text to gain an initial understanding of evolution equations, how they are used in practice, and more than 30 different areas of study to which the theory applies. Indeed, while the level of the mathematics discussed in the text is conventionally viewed as a topic that a graduate student would encounter after studying functional analysis, all of the underlying tools of functional analysis necessary to intelligently work through the text are included, chapter by chapter as they arise. This, coupled with the conversational style in which the text is written, should make the material naturally accessible to such a broad audience.

What material does this text cover, in broad strokes?

The present volume consists of eleven chapters. The text opens with a substantive chapter devoted to creating a basic analysis "toolbox," the purpose of which is to arm you with the bare essentials of analysis needed to work through the rest of the book. If you are familiar with the topics in the chapter, I suggest you peruse the chapter to get a feel for the notation and terminology prior to moving on.

Chapter 2 is devoted to the development of the theory for homogeneous finite-dimensional ODEs, which acts as a springboard into the development of its abstract counterpart in a general Banach space. The discussion proceeds to the case of linear homogeneous abstract evolution equations in Chapter 3, and subsequently in the next two chapters to the inhomogeneous and semi-linear cases. The case in which the forcing term is a functional (acting from one function space to another) is addressed in Chapter 6, followed by a discussion of two classes of implicit evolution equations in Chapter 7. Our jaunt through the linear setting concludes with a study of delay evolution equations in Chapter 8. The leap is then made into the fully nonlinear case in Chapter 9. Finally, the last two chapters are devoted to a brief discussion of accessible topics of active research.

For each class of equations, a core of theoretical results concerning the following main topics is developed: the existence and uniqueness of solutions (in a variety of senses) under various growth and compactness assumptions, continuous dependence upon initial data and parameters, convergence results regarding the initial data, and elementary stability results (in a variety of senses).

A substantive collection of mathematical models arising in areas such as heat conduction, advection, fluid flow through fissured rocks, transverse vibrations in extensible beams, thermodynamics, population ecology, pharmacokinetics, spatial pattern formation, pheromone transport, neural networks, and infectious disease epidemiology are developed in stages throughout the text. In fact, the reason for studying the class of abstract equations of a given chapter is motivated by first considering modified versions of the model(s) discussed in the previous chapter, and subsequently formulating the batch of newly-created initial-boundary value problems in the form of the abstract equation to be studied in that chapter.

Acknowledgments

Writing this text has been one of the most positive and fulfilling experiences of my professional career thus far. I found various stages of writing this book to be truly energizing and uplifting, while others required me to plumb the very depths of my patience and perseverance. I truly realize that a project like this would have never come to fruition without the constant support, encouragement, and good humor of many colleagues, students, friends, and family. While it is virtually impossible to acknowledge each and every individual who has, in some way, influenced me in a manner that helped either to steer me toward writing this book or to navigate murky (and at times very choppy) waters during the writing phase, I would like to acknowledge several at this time.

First and foremost, there are two notable women who have been indelible sources of encouragement, energy, and support for as long as they have been in my life. My wife, to whom the first volume is dedicated, prompted me for years to write this book. And, once I actually took her advice and began the process, she never abandoned my urgent and recurrent needs for technological help, editorial expertise, or idea-sounding, and has never begrudged me for momentary lack of patience, uttering of angst-provoked witty remarks, or necessary "idea jotting" at 3 a.m. And, my mother, to whom the second volume is to be dedicated, has supported all of my scholastic and professional endeavors for as long as I can remember. She relentlessly encourages me to temper hard work with balance; she always seems to know when to provide a good-humored story to lighten a stressed mood and when to politely remind me to "take a break!" This book would never have materialized if it were not for both of you; your unwavering support of my endeavors is a significant driving force!

Next, I am extremely lucky to have had two outstanding mentors during my college years. My first exposure to college-level mathematics, by way of analysis, took place nearly two decades ago in a one-on-one tutorial with Dr. David Keck. His unbridled enthusiasm for teaching and learning mathematics and well-timed witty humor have been infectious. Achieving the depth of his passion and honing my skills to mirror his innate ability to teach mathematics are goals to which I will continue to aspire for the duration of my academic career. And, my dissertation advisor, Dr. Sergiu Aizicovici, took me under his wing as a graduate student and introduced me to various facets of the study of evolution equations and the world of mathematical research. His abilities to make the area come to life, to help a newcomer navigate the practically insurmountable literature with ease, and to tolerate and provide meaningful answers to even the most rudimentary of questions (which I admittedly asked quite often!) in a way that honed my intuition surrounding the subject matter are among the many reasons why I chose to pursue this area of research. You both have left indelible imprints on my development as a mathematician and educator.

Many people have been kind enough to provide honest feedback during various stages of this project. My colleagues Dr. Robert Lewand, Dr. Scott Sibley, Dr. Bernadette Tutinas, Dr. Cynthia Young, and Dr. Jill Zimmerman all provided valuable comments on portions of the prospectus that undoubtedly led to a stronger proposal. My students Shana Lieberman, Jennifer Jordan, and Jordan Yoder endured various portions of the manuscript by way of independent study. I am proud to report that all three of them survived the experience (and seemed to enjoy it) and identified their fair share of errors in the early versions of the text! My colleague Dr. Tom Kelliher provided invaluable TEX help during early stages of the project; thank you for helping such a TEX neophyte! A special thanks to my wife, Jodi, for reading the prospectus and manuscript, and who will never let me live down the fact that she found a mathematical error in the prospectus. And, I would like to especially thank Dr. David Keck for reading the manuscript line by line and offering very detailed remarks and pointed questions that led to the correction of errors and an overall improvement of the final manuscript.

I would like to thank the entire Taylor & Francis team. To my editor, Bob Stern, thank you for approaching me about writing this text and for patiently guiding me through the process from beginning to end. To my project coordinator, Jessica Vakili, who answered my production, stylistic, and marketing questions in a very helpful and timely manner. To my copy editor, Karen Simon, for keeping the publication of this book on track. To Kevin Craig for designing the awesome cover of this book. And to Shashi Kumar, who helped me to overcome various LATEX issues throughout the typesetting process.

And, last but not least, I would like to thank you, the reader, for embarking on this journey with me through an amazingly rich field of mathematics. I hope your study is as fulfilling as mine has been thus far.

Mark A. McKibben

Chapter 1

A Basic Analysis Toolbox

Overview

The purpose of this chapter is to provide you with a succinct, hands-on introduction to elementary analysis that focuses on notation, main definitions and results, and the techniques with which you should be comfortable prior to working through this text. Additional topics will be introduced throughout the text whenever needed. Little is assumed beyond a working knowledge of the properties of real numbers, the "freshmen calculus," and a tolerance for mathematical rigor. Keep in mind that the presentation is not intended to be a complete exposition of real analysis. You are encouraged to refer to texts devoted to more comprehensive treatments of analysis (see [**16, 63, 174, 175, 195, 196, 205, 246, 283, 286, 299**]).

1.1   Some Basic Mathematical Shorthand

Symbolism is used heftily in mathematical exposition. Careful usage of some basic notation can streamline the verbiage. Some of the common symbols used are as follows.

Let P and Q be statements. (If the statement P changes depending on the value of some parameter x, we denote this dependence by writing $P(x)$.)

1.) The statement "not P," called the *negation of P*, is denoted by "$\neg P$."

2.) The statement "P or Q" is denoted by "$P \vee Q$," while the statement "P and Q" is denoted by "$P \wedge Q$."

3.) The statement "If P, then Q" is called an *implication*, and is denoted by "$P \Longrightarrow Q$" (read "P implies Q"). Here, P is called the *hypothesis* and Q is the *conclusion*.

4.) The statement "P if, and only if, Q" is denoted by "P iff Q" or "$P \Longleftrightarrow Q$." Precisely, this means "$(P \Longrightarrow Q) \wedge (Q \Longrightarrow P)$."

5.) The statement "$Q \Longrightarrow P$" is the *converse* of "$P \Longrightarrow Q$."

6.) The statement "$\neg Q \Longrightarrow \neg P$" is the *contrapositive* of "$P \Longrightarrow Q$." These two statements are equivalent.

7.) The symbol "$\exists$" is an existential quantifier and is read as "there exists" or "there is at least one."

8.) The symbol "$\forall$" is a universal quantifier and is read as "for every" or "for any."

Exercise 1.1.1. Let P, Q, R, and S be statements.
i.) Form the negation of "$P \wedge (Q \wedge R)$."
ii.) Form the negation of "$\exists\, x$ such that $P(x)$ holds."
iii.) Form the negation of "$\forall x$, $P(x)$ holds."
iv.) Form the contrapositive of "$(P \wedge Q) \Longrightarrow (\neg R \vee S)$."

Remark. Implication is a transitive relation in the sense that

$$((P \Longrightarrow Q) \wedge (Q \Longrightarrow R)) \Longrightarrow (P \Longrightarrow R).$$

For instance, a sequence of algebraic manipulations used to solve an equation is technically such a string of implications from which we conclude that the values of the variable obtained in the last step are the solutions of the original equation. Mathematical proofs are comprised of strings of implications, albeit of a somewhat more sophisticated nature.

1.2   Set Algebra

Informally, a *set* can be thought of as a collection of objects (e.g., real numbers, vectors, matrices, functions, other sets, etc.); the contents of a set are referred to as its *elements*. We usually label sets using upper case letters and their elements by lower case letters. Three sets that arise often and for whom specific notation will be reserved are:

$$\mathbb{N} = \{1, 2, 3, ...\}$$
$$\mathbb{Q} = \text{the set of all rational numbers}$$
$$\mathbb{R} = \text{the set of all real numbers}$$

If P is a certain property and A is the set of all objects having property P, we write $A = \{x : x \text{ has } P\}$ or $A = \{x | x \text{ has } P\}$. A set with no elements is *empty*, denoted by $\emptyset$.

If A is not empty and a is an element of A, we denote this fact by "$a \in A$." If a is not an element of A, a fact denoted by "$a \notin A$," where is it located? This prompts us to prescribe a universal set $\mathscr{U}$ that contains all possible objects of interest in our discussion. The following definition provides an algebra of sets.

Definition 1.2.1. Let A and B be sets.
i.) A is a *subset* of B, written $A \subset B$, whenever $x \in A \Longrightarrow x \in B$.
ii.) A *equals* B, written $A = B$, whenever $(A \subset B) \wedge (B \subset A)$.
iii.) The *complement of* A *relative to* B, written $B \setminus A$, is the set $\{x | x \in B \wedge x \notin A\}$.

Specifically, the complement relative to $\mathscr{U}$ is denoted by $\widetilde{A}$.

iv.) The *union of A and B* is the set $A \cup B = \{x | x \in A \vee x \in B\}$

v.) The *intersection of A and B* is the set $A \cap B = \{x | x \in A \wedge x \in B\}$

vi.) $A \times B = \{(a,b) | a \in A \wedge b \in B\}$

Proving set equality requires that we show two implications. Use this fact when appropriate to complete the following exercises:

Exercise 1.2.1. Let A, B, and C be sets. Prove the following:

i.) $A \subset B$ iff $\widetilde{B} \subset \widetilde{A}$.

ii.) $A = (A \cap B) \cup (A \setminus B)$

iii.) $A \cap (B \cup C) = (A \cap B) \cup (A \cap C)$ and $A \cup (B \cap C) = (A \cup B) \cap (A \cup C)$

iv.) $\widetilde{(A \cap B)} = \widetilde{A} \cup \widetilde{B}$ and $\widetilde{(A \cup B)} = \widetilde{A} \cap \widetilde{B}$

Exercise 1.2.2. Explain how you would prove $A \neq B$.

Exercise 1.2.3. Formulate an extension of Def. 1.2.1(iv)-(vi) that works for any finite number of sets.

It is often necessary to consider the union or intersection of more than two sets, possibly infinitely many. So, we need a succinct notation for unions and intersections of an arbitrary number of sets. Let $\Gamma \neq \emptyset$. (We think of the members of Γ as labels.) Suppose to each $\gamma \in \Gamma$, we associate a set A_γ. The collection of all these sets, namely $\mathscr{A} = \{A_\gamma | \gamma \in \Gamma\}$, is a *family of sets indexed by* Γ. We define

$$\bigcup_{\gamma \in \Gamma} A_\gamma = \{x | \exists \gamma \in \Gamma \text{ such that } x \in A_\gamma\} \tag{1.1}$$

$$\bigcap_{\gamma \in \Gamma} A_\gamma = \{x | \forall \gamma \in \Gamma, x \in A_\gamma\} \tag{1.2}$$

If $\Gamma = \mathbb{N}$, we write $\bigcup_{n=1}^{\infty}$ and $\bigcap_{n=1}^{\infty}$ in place of $\bigcup_{\gamma \in \Gamma}$ and $\bigcap_{\gamma \in \Gamma}$, respectively.

Exercise 1.2.4. Let A be a set and $\{A_\gamma | \gamma \in \Gamma\}$ a family of sets indexed by Γ. Prove:

i.) $A \cap \bigcup_{\gamma \in \Gamma} A_\gamma = \bigcup_{\gamma \in \Gamma} (A \cap A_\gamma)$ and $A \cup \bigcap_{\gamma \in \Gamma} A_\gamma = \bigcap_{\gamma \in \Gamma} (A \cup A_\gamma)$

ii.) $\left(\bigcup_{\gamma \in \Gamma} A_\gamma\right)^{\sim} = \bigcap_{\gamma \in \Gamma} \widetilde{A}_\gamma$ and $\left(\bigcap_{\gamma \in \Gamma} A_\gamma\right)^{\sim} = \bigcup_{\gamma \in \Gamma} \widetilde{A}_\gamma$

iii.) $A \times \bigcup_{\gamma \in \Gamma} A_\gamma = \bigcup_{\gamma \in \Gamma} (A \times A_\gamma)$ and $A \times \bigcap_{\gamma \in \Gamma} A_\gamma = \bigcap_{\gamma \in \Gamma} (A \times A_\gamma)$

iv.) $\bigcap_{\gamma \in \Gamma} A_\gamma \subset A_{\gamma_0} \subset \bigcup_{\gamma \in \Gamma} A_\gamma, \forall \gamma_0 \in \Gamma$.

1.3 Functions

The concept of a function is central to the study of mathematics.

Definition 1.3.1. Let A and B be sets.

i.) A subset $f \subset A \times B$ satisfying

    a.) $\forall x \in A, \exists y \in B$ such that $(x,y) \in f$,

    b.) $(x,y_1) \in f \wedge (x,y_2) \in f \implies y_1 = y_2$,

    is called a *function from A into B.* We say f is $B-valued$, denoted by $f : A \to B$.

ii.) The set A is called the *domain* of f, denoted dom(f).

iii.) The *range* of f, denoted by rng(f), is given by rng$(f) = \{f(x)|x \in A\}$.

Remarks.

1. <u>Notation</u>: When defining a function using an explicit formula, say $y = f(x)$, the notation $x \mapsto f(x)$ is often used to denote the function. Also, we indicate the general dependence on a variable using a dot, say $f(\cdot)$. If the function depends on two independent variables, we distinguish between them by using a different number of dots for each, say $f(\cdot, \cdot\cdot)$.

2. The term *mapping* is used synonymously with the term *function.*

3. rng$(f) \subset B$.

Exercise 1.3.1. Precisely define what it means for two functions f and g to be equal.

    The following classification plays a role in determining if a function is invertible.

Definition 1.3.2. $f : A \to B$ is called

i.) *one-to-one* if $f(x_1) = f(x_2) \implies x_1 = x_2, \forall x_1, x_2 \in A$;

ii.) *onto* whenever rng$(f) = B$.

    We sometimes wish to apply functions in succession in the following sense.

Definition 1.3.3. Suppose that $f : \text{dom}(f) \to A$ and $g : \text{dom}(g) \to B$ with rng$(g) \subset$ dom(f). The *composition of f with g*, denoted $f \circ g$, is the function $f \circ g : \text{dom}(g) \to A$ defined by $(f \circ g)(x) = f(g(x))$.

Exercise 1.3.2. Show that, in general, $f \circ g \neq g \circ f$.

Exercise 1.3.3. Let $f : \text{dom}(f) \to A$ and $g : \text{dom}(g) \to B$ be such that $f \circ g$ is defined. Prove:

i.) If f and g are onto, then $f \circ g$ is onto.

ii.) If f and g are one-to-one, then $f \circ g$ is one-to-one.

    At times, we need to compute the functional values for all members of a subset of the domain, or perhaps determine the subset of the domain whose collection of functional values is a prescribed subset of the range. These notions are made precise below.

Definition 1.3.4. Let $f : A \to B$.

i.) For $X \subset A$, the *image of X under f* is the set $f(X) = \{f(x)|x \in X\}$.

ii.) For $Y \subset B$, the *pre-image of Y under f* is the set

$$f^{-1}(Y) = \{x \in \mathscr{A} \mid \exists y \in Y \text{ such that } y = f(x)\}.$$

The following related properties are useful.

Proposition 1.3.5. *Suppose $f : A \to B$ is a function, X, X_1, X_2, and X_γ, $\gamma \in \Gamma$, are all subsets of A and Y, Y_1, Y_2, and Y_γ, $\gamma \in \Gamma$, are all subsets of B. Then,*
i.) a.) $X_1 \subset X_2 \Longrightarrow f(X_1) \subset f(X_2)$
  b.) $Y_1 \subset Y_2 \Longrightarrow f^{-1}(Y_1) \subset f^{-1}(Y_2)$
ii.) a.) $f\left(\bigcup_{\gamma \in \Gamma} X_\gamma\right) = \bigcup_{\gamma \in \Gamma} f(X_\gamma)$
  b.) $f^{-1}\left(\bigcup_{\gamma \in \Gamma} Y_\gamma\right) = \bigcup_{\gamma \in \Gamma} f^{-1}(Y_\gamma)$
iii.) a.) $f\left(\bigcap_{\gamma \in \Gamma} X_\gamma\right) \subset \bigcap_{\gamma \in \Gamma} f(X_\gamma)$
  b.) $f^{-1}\left(\bigcap_{\gamma \in \Gamma} Y_\gamma\right) = \bigcap_{\gamma \in \Gamma} f^{-1}(Y_\gamma)$
iv.) a.) $X \subset f^{-1}(f(X)))$
  b.) $f\left(f^{-1}(Y)\right) \subset Y.$

Exercise 1.3.4.
i.) Prove Prop. 1.3.5.
ii.) Impose conditions on f that would yield equality in Prop. 1.3.5(iv)(a) and (b).

We often consider functions whose domains and ranges are subsets of $\mathbb{R}$. For such functions, the notion of monotonicity is often a useful characterization.

Definition 1.3.6. Let $f : \text{dom}(f) \subset \mathbb{R} \to \mathbb{R}$ and suppose that $\varnothing \ne S \subset \text{dom}(f)$. We say that f is
i.) *nondecreasing on S* whenever $x_1, x_2 \in S$ with $x_1 < x_2 \Longrightarrow f(x_1) \le f(x_2)$;
ii.) *nonincreasing on S* whenever $x_1, x_2 \in S$ with $x_1 < x_2 \Longrightarrow f(x_1) \ge f(x_2)$.

Remark. The prefix "non" in both parts of Def. 1.3.6 is removed when the inequality is strict.

The arithmetic operations of real-valued functions are defined in the natural way. For such functions, consider the following exercise.

Exercise 1.3.5. Suppose that $f : \text{dom}(f) \subset \mathbb{R} \to \mathbb{R}$ and $g : \text{dom}(g) \subset \mathbb{R} \to \mathbb{R}$ are nondecreasing (resp. nonincreasing) functions on their domains.
i.) Which of the functions $f + g$, $f - g$, $f \cdot g$, and $\frac{f}{g}$, if any, are nondecreasing (resp. nonincreasing) on their domains?
ii.) Assuming that $f \circ g$ is defined, must it be nondecreasing (resp. nonincreasing) on its domain?

1.4   The Space $(\mathbb{R}, |\cdot|)$

1.4.1   Order Properties

The basic arithmetic and order features of the real number system are likely familiar, even if you have not worked through its formal construction. For our purposes, we shall begin with a set $\mathbb{R}$ equipped with two operations, addition and multiplication, satisfying these algebraic properties:

   (i) addition and multiplication are both commutative and associative;
   (ii) multiplication distributes over addition;
   (iii) adding zero to any real number yields the same real number;
   (iv) multiplying a real number by one yields the same real number;
   (v) every real number has a unique additive inverse; and
   (vi) every nonzero real number has a unique multiplicative inverse.

Moreover, $\mathbb{R}$ equipped with the natural "$<$" ordering is an ordered field and obeys the following properties:

Proposition 1.4.1. (Order Features of $\mathbb{R}$)
For all $x,y,z \in \mathbb{R}$, the following are true:
i.) *Exactly one of the relationships $x = y$, $x < y$, or $y < x$ holds;*
ii.) $x < y \implies x+z < y+z$;
iii.) $(x < y) \wedge (y < z) \implies x < z$;
iv.) $(x < y) \wedge (c > 0) \implies cx < cy$;
v.) $(x < y) \wedge (c < 0) \implies cx > cy$;
vi.) $(0 < x < y) \wedge (0 < w < z) \implies 0 < xw < yz$.

The following is an immediate consequence of these properties and is often the underlying principle used when verifying an inequality.

Proposition 1.4.2. *If $x,y \in \mathbb{R}$ are such that $x < y+\varepsilon$, $\forall \varepsilon > 0$, then $x \le y$.*

Proof. Suppose not; that is, $y < x$. Observe that for $\varepsilon = \frac{x-y}{2} > 0$, $y+\varepsilon = \frac{x+y}{2} < x$. (Why?) This is a contradiction. Hence, it must be the case that $x \le y$. $\square$

Remark. The above argument is a very simple example of a proof by contradiction. The strategy is to assume that the conclusion is false and then use this additional hypothesis to obtain an obviously false statement or a contradiction of another hypothesis in the claim. More information about elementary proof techniques can be found in **[299]**.

Exercise 1.4.1.
i.) Let $x,y > 0$. Prove that $xy \le \frac{x^2+y^2}{2}$.
ii.) Show that if $0 < x < y$, then $x^n < y^n$, $\forall n \in \mathbb{N}$.

1.4.2   Absolute Value

The above is a heuristic description of the familiar algebraic structure of $\mathbb{R}$. When equipped with a distance-measuring artifice, a deeper topological structure of $\mathbb{R}$ can be defined and studied. This is done with the help of the absolute value function.

Definition 1.4.3. For any $x \in \mathbb{R}$, the *absolute value* of x, denoted $|x|$, is defined by

$$|x| = \begin{cases} x, x \geq 0, \\ -x, x < 0. \end{cases}$$

This can be viewed as a measurement of distance between real numbers within the context of a number line. For instance, the solution set of the equation "$|x - 2| = 3$" is the set of real numbers x that are "3 units away from 2," namely $\{-1, 5\}$.

Exercise 1.4.2. Determine the solution set for the following equations.
i.) $|x - 3| = 0$
ii.) $|x + 6| = 2$.

Proposition 1.4.4. *These properties hold for all $x, y, z \in \mathbb{R}$ and $a \geq 0$:*
i.) $-|x| = \min\{-x, x\} \leq x \leq \max\{-x, x\} = |x|$;
ii.) $|x| \geq 0, \forall x \in \mathbb{R}$;
iii.) $|x| = 0$ *iff* $x = 0$;
iv.) $\sqrt{x^2} = |x|$;
v.) $|xy| = |x| |y|$;
vi.) $|x| \leq a$ *iff* $-a \leq x \leq a$;
vii.) $|x + y| \leq |x| + |y|$;
viii.) $|x - y| \leq |x - z| + |z - y|$;
ix.) $|\, |x| - |y| \,| \leq |x - y|$;
x.) $|x - y| < \varepsilon, \forall \varepsilon > 0 \implies x = y$.

Exercise 1.4.3. Prove Prop. 1.4.4.
Exercise 1.4.4. Let $n \in \mathbb{N}$ and $x_1, x_2, \ldots, x_n, y_1, y_2, \ldots, y_n \in \mathbb{R}$. Prove:
i.) **(Cauchy-Schwarz)** $\sum_{i=1}^{n} x_i y_i \leq \left(\sum_{i=1}^{n} x_i^2\right) \left(\sum_{i=1}^{n} y_i^2\right)$,
ii.) **(Minkowski)** $\left(\sum_{i=1}^{n} (x_i + y_i)^2\right)^{1/2} \leq \left(\sum_{i=1}^{n} x_i^2\right)^{1/2} + \left(\sum_{i=1}^{n} y_i^2\right)^{1/2}$,
iii.) $\left|\sum_{i=1}^{n} x_i\right|^M \leq \left(\sum_{i=1}^{n} |x_i|\right)^M \leq n^{M-1} \sum_{i=1}^{n} |x_i|^M$, $\forall M \in \mathbb{N}$.

1.4.3 Completeness Property of $(\mathbb{R}, |\cdot|)$

It turns out that $\mathbb{R}$ has a fundamental and essential property referred to as *completeness*, without which the study of analysis could not proceed. We introduce some terminology needed to state certain fundamental properties of $\mathbb{R}$.

Definition 1.4.5. Let $\varnothing \neq S \subset \mathbb{R}$.
i.) S is *bounded above* if $\exists u \in \mathbb{R}$ such that $x \leq u, \forall x \in S$;
ii.) $u \in \mathbb{R}$ is an *upper bound* of S (ub(S)) if $x \leq u, \forall x \in S$;
iii.) $u_0 \in \mathbb{R}$ is the *maximum* of S (max(S)) if u_0 is an ub(S) and $u_0 \in S$;
iv.) $u_0 \in \mathbb{R}$ is the *supremum* of S (sup(S)) if u_0 is an ub(S) and $u_0 \leq u$, for any other $u = $ub($S$);

The following analogous terms can be defined by reversing the inequality signs in Def. 1.4.5: *bounded below*, *lower bound* of S (lb(S)), *minimum* of S (min(S)), and *infimum of S* (inf(S)).

Exercise 1.4.5. Formulate precise definitions of the above terms.

Exercise 1.4.6. Let $\varnothing \neq S \subset \mathbb{R}$.
i.) How would you prove that $\sup(S) = \infty$?
ii.) Repeat (i) for $\inf(S) = -\infty$.

Definition 1.4.6. A set $\varnothing \neq S \subset \mathbb{R}$ is *bounded* if $\exists M > 0$ such that $|x| \leq M, \forall x \in S$.

It can be formally shown that $\mathbb{R}$ possesses the so-called *completeness property*. The importance of this concept in the present and more abstract settings cannot be overemphasized. We state it in the form of a theorem to highlight its importance. Consult **[16, 195]** for a proof.

Theorem 1.4.7. *If $\varnothing \neq S \subset \mathbb{R}$ is bounded above, then $\exists u \in \mathbb{R}$ such that $u = \sup(S)$. We say $\mathbb{R}$ is complete.*

Remark. The duality between the statements concerning sup and inf leads to the formulation of the following alternate statement of the completeness property:

$$If \varnothing \neq T \subset \mathbb{R} \text{ is bounded below, then } \exists v \in \mathbb{R} \text{ such that } v = \inf(S). \qquad (1.3)$$

Exercise 1.4.7. Prove that (1.3) is equivalent to Thrm 1.4.7.

Proposition 1.4.8. (Properties of inf and sup) *Let $\varnothing \neq S, T \subset \mathbb{R}$.*
i.) *Assume $\exists \sup(S)$. Then, $\forall \varepsilon > 0, \exists x \in S$ such that $\sup(S) - \varepsilon < x \leq \sup(S)$.*
ii.) *If $S \subset T$ and $\exists \sup(T)$, then $\exists \sup(S)$ and $\sup(S) \leq \sup(T)$.*
iii.) *Let $S + T = \{s + t | s \in S \wedge t \in T\}$. If S and T are bounded above, then $\exists \sup(S + T)$ and it equals $\sup(S) + \sup(T)$.*
iv.) *Let $c \in \mathbb{R}$ and define $cS = \{cs | s \in S\}$. If S is bounded, then $\exists \sup(cS)$ given by*

$$\sup(cS) = \begin{cases} c \cdot \sup(S), & if\, c \geq 0, \\ c \cdot \inf(S), & if\, c < 0. \end{cases} \qquad (1.4)$$

v.) *Let $\varnothing \neq S, T \subset (0, \infty)$ and define $S \cdot T = \{s \cdot t | s \in S \wedge t \in T\}$. If S and T are bounded above, then $\exists \sup(S \cdot T)$ and it equals $\sup(S) \cdot \sup(T)$.*

Proof. We prove (iii) and leave the others for you to verify as an exercise.
Since S and T are nonempty, $S + T \neq \varnothing$. Further, since

$$s + t \leq \sup(S) + \sup(T), \forall s \in S, t \in T, \qquad (1.5)$$

it follows that $\sup(S) + \sup(T)$ is an ub$(S + T)$. (Why?) Hence, $\exists \sup(S + T)$ and

$$\sup(S + T) \leq \sup(S) + \sup(T). \qquad (1.6)$$

To establish the reverse inequality, let $\varepsilon > 0$. By Prop. 1.4.8, $\exists s_0 \in S$ and $t_0 \in T$ such that

$$\sup(S) - \frac{\varepsilon}{2} < s_0 \text{ and } \sup(T) - \frac{\varepsilon}{2} < t_0. \tag{1.7}$$

Consequently,

$$\sup(S) + \sup(T) - \varepsilon < s_0 + t_0 \leq \sup(S+T). \tag{1.8}$$

Thus, we conclude from Prop. 1.4.2 that

$$\sup(S) + \sup(T) \leq \sup(S+T). \tag{1.9}$$

Claim (iii) now follows from (1.6) and (1.9). (Why?) $\qquad\square$

Exercise 1.4.8.
i.) Prove the remaining parts of Prop. 1.4.8.
ii.) Formulate statements analogous to those in Prop. 1.4.8 for infs. Indicate the changes that must be implemented in the proofs.

Remark. Prop 1.4.8(i) indicates that we can get "arbitrarily close" to $\sup(S)$ with elements of S. This is especially useful in convergence arguments.

1.4.4 Topology of $\mathbb{R}$

You have worked with open and closed intervals in the calculus, but what do the terms *open* and *closed* mean? Is there any significant difference between them? The notion of an open set is central to the construction of a so-called *topology* on $\mathbb{R}$. Interestingly, many of the theorems from calculus are formulated on closed, bounded intervals for very good reason. As we proceed with our analysis of $\mathbb{R}$, you shall see that many of these results are consequences of some fairly deep topological properties of $\mathbb{R}$ which, in turn, follow from the completeness property.

Definition 1.4.9. Let $S \subset \mathbb{R}$.
i.) x is an *interior point* (int pt) of S if $\exists \varepsilon > 0$ such that $(x-\varepsilon, x+\varepsilon) \subset S$.
ii.) x is a *limit point* (lim pt) of S if $\forall \varepsilon > 0$, $(x-\varepsilon, x+\varepsilon) \cap S$ is infinite.
iii.) x is a *boundary point* (bdry pt) of S if

$$\forall \varepsilon > 0, (x-\varepsilon, x+\varepsilon) \cap S \neq \emptyset \text{ and } (x-\varepsilon, x+\varepsilon) \cap \tilde{S} \neq \emptyset.$$

iv.) The *boundary* of S is the set $\partial S = \{x \in \mathbb{R} | x \text{ is a bdry pt of } S\}$.
v.) The *interior* of S is the set $\text{int}(S) = \{x \in \mathbb{R} | x \text{ is an int pt of } S\}$.
vi.) The *derived set* of S is the set $S' = \{x \in \mathbb{R} | x \text{ is a lim pt of } S\}$.
vii.) The *closure* of S is the set $\text{cl}_{\mathbb{R}}(S) = S \cup S'$.
viii.) S is *open* if every point of S is an int pt of S.
ix.) S is *closed* if S contains all of its lim pts.

Illustrating these concepts using a number line can facilitate your understanding of them. Do so when completing the following exercise.

Exercise 1.4.9. For each of these sets S, compute int(S), S', and $\mathrm{cl}_{\mathbb{R}}(S)$. Also, determine if S is open, closed, both, or neither.
i.) $[1,5]$
ii.) $\mathbb{Q}$
iii.) $\left\{ \frac{1}{n} \middle| n \in \mathbb{N} \right\}$
iv.) $\mathbb{R}$
v.) $\varnothing$

It is not difficult to establish the following duality between a set and its complement. It is often a useful tool when proving statements about open and closed sets.

Proposition 1.4.10. *Let $S \subset \mathbb{R}$. S is open iff $\widetilde{S}$ is closed.*

Exercise 1.4.10. Verify the following properties of open and closed sets.
i.) Let $n \in \mathbb{N}$. If $G_1, \ldots, G_n$ is a finite collection of open sets, then $\bigcap_{k=1}^{n} G_k$ is open.
ii.) Let $n \in \mathbb{N}$. If $F_1, \ldots, F_n$ is a finite collection of closed sets, then $\bigcup_{k=1}^{n} F_k$ is closed.
iii.) Let $\Gamma \neq \varnothing$. If G_γ is open, $\forall \gamma \in \Gamma$, then $\bigcup_{\gamma \in \Gamma} G_\gamma$ is open.
iv.) Let $\Gamma \neq \varnothing$. If F_γ is closed, $\forall \gamma \in \Gamma$, then $\bigcap_{\gamma \in \Gamma} F_\gamma$ is closed.
v.) If $S \subset T$, then int$(S) \subset$ int(T).
vi.) If $S \subset T$, then $\mathrm{cl}_{\mathbb{R}}(S) \subset \mathrm{cl}_{\mathbb{R}}(T)$.

Exercise 1.4.11. Let $\varnothing \neq S \subset \mathbb{R}$. Prove the following:
i.) If S is bounded above, then $\sup(S) \in \mathrm{cl}_{\mathbb{R}}(S)$.
ii.) If S is bounded above and closed, then $\max(S) \in S$.
iii.) Formulate results analogous to (i) and (ii) assuming that S is bounded below.

Intuitively, S' is the set of points to which elements of S become arbitarily close. It is natural to ask if there are proper subsets of $\mathbb{R}$ that sprawl widely enough through $\mathbb{R}$ as to be sufficiently near every real number. Precisely, consider sets of the following type:

Definition 1.4.11. A set $\varnothing \neq S \subset \mathbb{R}$ is *dense* in $\mathbb{R}$ if $\mathrm{cl}_{\mathbb{R}}(S) = \mathbb{R}$.

Exercise 1.4.12. Identify two different subsets of $\mathbb{R}$ that are dense in $\mathbb{R}$.

By way of motivation for the first major consequence of completeness, consider the following exercise.

Exercise 1.4.13. Provide examples, if possible, of sets $S \subset \mathbb{R}$ illustrating the following scenarios.
i.) S is bounded, but $S' = \varnothing$.
ii.) S is infinite, but $S' = \varnothing$.
iii.) S is bounded and infinite, but $S' = \varnothing$.

As you discovered in Exercise 1.4.13, the combination of bounded and infinite for a set S of real numbers implies the existence of a limit point of S. This is a consequence of the following theorem due to Bolzano and Weierstrass.

Theorem 1.4.12. (Bolzano-Weierstrass) *If S is a bounded, infinite subset of $\mathbb{R}$, then $S' \neq \emptyset$.*

Outline of Proof: Let $T = \{x \in \mathbb{R} | S \cap (x, \infty) \text{ is infinite}\}$. Then,

$$T \neq \emptyset. \text{ (Why?)} \tag{1.10}$$
$$T \text{ is bounded above. (Why?)} \tag{1.11}$$
$$\exists \sup(T); \text{ call it } t. \text{ (Why?)} \tag{1.12}$$
$$\forall \varepsilon > 0, S \cap (t - \varepsilon, \infty) \text{ is infinite. (Why?)} \tag{1.13}$$
$$\forall \varepsilon > 0, S \cap [t + \varepsilon, \infty) \text{ is finite. (Why?)} \tag{1.14}$$
$$\forall \varepsilon > 0, S \cap (t - \varepsilon, t + \varepsilon) \text{ is infinite. (Why?)} \tag{1.15}$$
$$t \in S'. \text{ (Why?)} \tag{1.16}$$

This completes the proof. $\square$

Exercise 1.4.14. Provide the details in the proof of Thrm. 1.4.12. Where was completeness used?

Another important concept is that of compactness. Some authors define this notion more generally using open covers (see **[16]**).

Definition 1.4.13. A set $S \subset \mathbb{R}$ is *compact* if every infinite subset of S has a limit point in S.

Remark. The "in S" portion of Def. 1.4.13 is crucial, and it distinguishes between the sets $(0, 1)$ and $[0, 1]$, for instance. (Why?) This is evident in Thrm. 1.4.14.

Exercise 1.4.15. Try to determine if the following subsets of $\mathbb{R}$ are compact.
i.) Any finite set.
ii.) $\{\frac{1}{n} | n \in \mathbb{N}\}$ versus $\{\frac{1}{n} | n \in \mathbb{N}\} \cup \{0\}$
iii.) $\mathbb{Q}$
iv.) $\mathbb{Q} \cap [0, 1]$
v.) $\mathbb{N}$
vi.) $\mathbb{R}$
vii.) $(0, 1)$ versus $[0, 1]$

Both the completeness property and finite dimensionality of $\mathbb{R}$ enter into the proof of the following characterization theorem for compact subsets of $\mathbb{R}$. The proof can be found in **[16]**.

Theorem 1.4.14. (Heine-Borel) *A set $S \subset \mathbb{R}$ is compact iff S is closed and bounded.*

Exercise 1.4.16. Revisit Exer. 1.4.15 in light of Thrm. 1.4.14.

1.5   Sequences in $(\mathbb{R}, |\cdot|)$

Sequences play a prominent role in analysis, especially in the development of numerical schemes used for approximation purposes.

1.5.1   Sequences and Subsequences

Definition 1.5.1. A *sequence* in $\mathbb{R}$ is a function $x : \mathbb{N} \to \mathbb{R}$. We often write x_n for $x(n), n \in \mathbb{N}$, called the nth*-term* of the sequence, and denote the sequence itself by $\{x_n\}$ or by enumerating the range as $x_1, x_2, x_3, \ldots$

The notions of monotonicity and boundedness given in Defs. 1.3.6 and 1.4.6 apply in particular to sequences. We formulate them in this specific setting for later reference.

Definition 1.5.2. A sequence is called
i.) *nondecreasing* whenever $x_n \leq x_{n+1}, \forall n \in \mathbb{N}$;
ii.) *increasing* whenever $x_n < x_{n+1}, \forall n \in \mathbb{N}$;
iii.) *nonincreasing* whenever $x_n \geq x_{n+1}, \forall n \in \mathbb{N}$;
iv.) *decreasing* whenever $x_n > x_{n+1}, \forall n \in \mathbb{N}$;
v.) *monotone* if any of (i) - (iv) are satisfied;
vi.) *bounded above* (resp. *below*) if $\exists M \in \mathbb{R}$ such that $x_n \leq M$ (resp. $x_n \geq M$), $\forall n \in \mathbb{N}$;
vii.) *bounded* whenever $\exists M > 0$ such that $|x_n| \leq M, \forall n \in \mathbb{N}$.

Exercise 1.5.1. Explain why a nondecreasing (resp. nonincreasing) sequence must be bounded below (resp. above).

Definition 1.5.3. If $x : \mathbb{N} \to \mathbb{R}$ is a sequence in $\mathbb{R}$ and $n : \mathbb{N} \to \mathbb{N}$ is an increasing sequence in $\mathbb{N}$, then the composition $x \diamond n : \mathbb{N} \to \mathbb{R}$ is called a *subsequence* of x in $\mathbb{R}$.

Though this is a formal definition of a subsequence, let us examine carefully what this means using more conventional notation. Suppose that the terms of Def. 1.5.3 are represented by $\{x_n\}$ and $\{n_k\}$, respectively. Since $\{n_k\}$ is increasing, we know that $n_1 < n_2 < n_3 < \ldots$. Then, the official subsequence $x \circ n$ has values $(x \circ n)(k) = x(n(k))$ which, using our notation, can be written as $x_{n_k}, \forall k \in \mathbb{N}$. Thus, the integers n_k are just the indices of those terms of the original sequence that are retained in the subsequence as k increases, and roughly speaking, the remainder of the terms are omitted.

1.5.2   Limit Theorems

We now consider the important notion of convergence.

Definition 1.5.4. A sequence $\{x_n\}$ has *limit* L whenever $\forall \varepsilon > 0, \exists N \in \mathbb{N}$ (N depending in general on ε) such that

$$n \geq N \implies |x_n - L| < \varepsilon.$$

In such case, we write $\lim_{n\to\infty} x_n = L$ or $x_n \longrightarrow L$ and say that $\{x_n\}$ *converges (or is convergent) to L.* Otherwise, we say $\{x_n\}$ *diverges.*

If we paraphrase Def. 1.5.4, it would read: $\lim_{n\to\infty} x_n = L$ whenever given <u>any</u> open interval $(L - \varepsilon, L + \varepsilon)$ around L (that is, no matter how small the positive number ε is), it is the case that $x_n \in (L - \varepsilon, L + \varepsilon)$ for all but possibly finitely many indices n. That is, the "tail" of the sequence ultimately gets into every open interval around L. Also note that, in general, the smaller the ε, the larger the index N must be used (to get deeper into the tail) since ε is an error gauge, namely how far the terms are from the target. We say N must be chosen "sufficiently large" as to ensure the tail behaves in this manner for the given ε.

Exercise 1.5.2. i.) Precisely define $\lim_{n\to\infty} x_n \neq L$.

ii.) Prove that $x_n \longrightarrow L$ iff $|x_n - L| \longrightarrow 0$.

Example. As an illustration of Def. 1.5.4, we prove that $\lim_{n\to\infty} \frac{2n^2+n+5}{n^2+1} = 2$.

Let $\varepsilon > 0$. We must argue that $\exists N \in \mathbb{N}$ such that

$$n \geq N \implies \left| \frac{2n^2 + n + 5}{n^2 + 1} - 2 \right| < \varepsilon. \tag{1.17}$$

To this end, note that $\exists N \in \mathbb{N}$ such that $N > 3$ and $N\varepsilon > 2$. (Why?) We show this N "works." Indeed, observe that $\forall n \geq N$,

$$\left| \frac{2n^2 + n + 5}{n^2 + 1} - 2 \right| = \left| \frac{2n^2 + n + 5 - 2n^2 - 2}{n^2 + 1} \right| = \frac{n+3}{n^2+1}. \tag{1.18}$$

Subsequently, by choice of N, we see that $n \geq N > 3$ and for all such n,

$$\frac{n+3}{n^2+1} < \frac{2n}{n^2+1} < \frac{2n}{n^2} = \frac{2}{n} < \frac{2}{N} < \varepsilon. \tag{1.19}$$

(Why?) Thus, by definition, it follows that $\lim_{n\to\infty} \frac{2n^2+n+5}{n^2+1} = 2$. $\square$

Exercise 1.5.3. Use Def. 1.5.4 to prove that $\lim_{n\to\infty} \frac{a}{n} = 0$, $\forall a \in \mathbb{R}$.

We now discuss the main properties of convergence. We mainly provide outlines of proofs, the details of which you are encouraged to provide.

Proposition 1.5.5. *If $\{x_n\}$ is a convergent sequence, then its limit is unique.*

Outline of Proof: Let $\lim_{n\to\infty} x_n = L_1$ and $\lim_{n\to\infty} x_n = L_2$ and suppose that, by way of contradiction, $L_1 \neq L_2$.

Let $\varepsilon = \frac{|L_1 - L_2|}{2}$. Then, $\varepsilon > 0$. (Why?)
$\exists N_1 \in \mathbb{N}$ such that $n \geq N_1 \implies |x_n - L_1| < \varepsilon$. (Why?)
$\exists N_2 \in \mathbb{N}$ such that $n \geq N_2 \implies |x_n - L_2| < \varepsilon$. (Why?)
Choose $N = \max\{N_1, N_2\}$. Then, $|x_N - L_1| < \varepsilon$ and $|x_N - L_2| < \varepsilon$. (Why?)

Consequently, $2\varepsilon = |L_1 - L_2| \leq |x_N - L_1| + |x_N - L_2| < 2\varepsilon$. (Why?)
Thus, $L_1 = L_2$. (How?)
This completes the proof. $\square$

Proposition 1.5.6. *If $\{x_n\}$ is a convergent sequence, then it is bounded.*

Outline of Proof: Assume that $\lim_{n \to \infty} x_n = L$. We must produce an $M > 0$ such that
$|x_n| \leq M, \forall n \in \mathbb{N}$. Using $\varepsilon = 1$ in Def. 1.5.4, we know that

$$\exists N \in \mathbb{N} \text{ such that } n \geq N \implies |x_n - L| < \varepsilon = 1. \tag{1.20}$$

Using Prop. 1.4.4(ix) in (1.20) then yields

$$|x_n| < |L| + 1, \forall n \geq N. \tag{1.21}$$

(Tell how.) For how many values of n does x_n possibly not satisfy (1.21)? How do
you use this fact to construct a positive real number M satisfying Def. 1.5.2(vii)? $\square$

Proposition 1.5.7. (Squeeze Theorem) *Let $\{x_n\}, \{y_n\}$, and $\{z_n\}$ be sequences such
that*

$$x_n \leq y_n \leq z_n, \forall n \in \mathbb{N}, \tag{1.22}$$

and

$$\lim_{n \to \infty} x_n = L = \lim_{n \to \infty} z_n. \tag{1.23}$$

Then, $\lim_{n \to \infty} y_n = L$.

Outline of Proof: Let $\varepsilon > 0$. From (1.23), we know that $\exists N_1, N_2 \in \mathbb{N}$ such that

$$|x_n - L| < \varepsilon, \forall n \geq N_1 \text{ and } |z_n - L| < \varepsilon, \forall n \geq N_2. \tag{1.24}$$

Specifically,

$$-\varepsilon < x_n - L, \forall n \geq N_1 \text{ and } z_n - L < \varepsilon, \forall n \geq N_2. \tag{1.25}$$

Choose $N = \max\{N_1, N_2\}$. Using (1.25), we see that

$$-\varepsilon < x_n - L, \text{ and } z_n - L < \varepsilon, \forall n \geq N. \text{ (Why?)}$$

Using this with (1.22), we can conclude that

$$n \geq N \implies -\varepsilon < y_n - L < \varepsilon. \text{ (Why?)}$$

Hence, $\lim_{n \to \infty} y_n = L$, as desired. $\square$

Remark. The conclusion of Prop. 1.5.7 holds true if we replace (1.22) by

$$\exists N_0 \in \mathbb{N} \text{ such that } x_n \leq y_n \leq z_n, \forall n \geq N_0. \tag{1.26}$$

Suitably modify the way N is chosen in the proof of Prop. 1.5.7 to account for this
more general condition. (Tell how.)

Proposition 1.5.8. *If* $\lim\limits_{n\to\infty} x_n = L$, *where* $L \neq 0$, *then* $\exists m > 0$ *and* $N \in \mathbb{N}$ *such that*

$$|x_n| > m, \forall n \geq N.$$

(In words, if a sequence has a nonzero limit, then its terms must be bounded away from zero for sufficiently large indices n.)

Outline of Proof:

Let $\varepsilon = \frac{|L|}{2}$. Then, $\varepsilon > 0$. (Why?)

$\exists N \in \mathbb{N}$ such that $|x_n - L| < \varepsilon = \frac{|L|}{2}, \forall n \geq N$. (Why?)

Thus, $||x_n| - |L|| < \frac{|L|}{2}, \forall n \geq N$. (Why?)

That is, $-\frac{|L|}{2} < |x_n| - |L| < \frac{|L|}{2}, \forall n \geq N$. (Why?)

So, $\frac{|L|}{2} < |x_n|, \forall n \geq N$.

The conclusion follows by choosing $m = \frac{|L|}{2}$. (Why?) $\square$

Proposition 1.5.9. *Suppose that* $\lim\limits_{n\to\infty} x_n = L$ *and* $\lim\limits_{n\to\infty} y_n = M$. *Then,*

i.) $\lim\limits_{n\to\infty} (x_n + y_n) = L + M$;

ii.) $\lim\limits_{n\to\infty} x_n y_n = LM$.

Outline of Proof:

<u>Proof of (i):</u> The strategy is straightforward. Since there are two sequences, we split the given error tolerance ε into two parts of size $\frac{\varepsilon}{2}$ each, apply the limit definition to each sequence with the $\frac{\varepsilon}{2}$ tolerance, and finally put the two together using the triangle inequality.

Let $\varepsilon > 0$. Then, $\frac{\varepsilon}{2} > 0$. We know that

$$\exists N_1 \in \mathbb{N} \text{ such that } |x_n - L| < \frac{\varepsilon}{2}, \forall n \geq N_1. \text{ (Why?)} \tag{1.27}$$

$$\exists N_2 \in \mathbb{N} \text{ such that } |y_n - M| < \frac{\varepsilon}{2}, \forall n \geq N_2. \text{ (Why?)} \tag{1.28}$$

How do you then select $N \in \mathbb{N}$ such that (1.27) and (1.28) hold simultaneously, for all $n \geq N$? For such an N, observe that

$$n \geq N \implies |(x_n + y_n) - (L + M)| \leq |x_n - L| + |y_n - M| < \varepsilon. \tag{1.29}$$

(Why?) Hence, we conclude that $\lim\limits_{n\to\infty} (x_n + y_n) = L + M$.

<u>Proof of (ii):</u> This time the strategy is a bit more involved. We need to show that $|x_n y_n - LM|$ can be made arbitrarily small for sufficiently large n using the hypotheses that $|x_n - L|$ and $|y_n - M|$ can each be made arbitrarily small for sufficiently large n. This requires two approximations, viz., making x_n close to L while simultaneously making y_n close to M. This suggests that we bound $|x_n y_n - LM|$ above by an expression involving $|x_n - L|$ and $|y_n - M|$. To accomplish this, we add and subtract

the same middle term in $|x_n y_n - LM|$ and apply certain absolute value properties. Precisely, observe that

$$\begin{aligned} |x_n y_n - LM| &= |x_n y_n - M x_n + M x_n - LM| \\ &= |x_n (y_n - M) + M (x_n - L)| \quad\quad\quad (1.30) \\ &\le |x_n| |y_n - M| + |M| |x_n - L|. \end{aligned}$$

(This trick is a workhorse throughout the text!) The tack now is to show that both terms on the right-side of (1.30) can be made less than $\frac{\varepsilon}{2}$ for sufficiently large n.

Let $\varepsilon > 0$. Proposition 1.5.6 implies that $\exists K > 0$ for which

$$|x_n| \le K, \forall n \in \mathbb{N}. \quad\quad\quad (1.31)$$

Also, since $\{y_n\}$ is convergent to M, $\exists N_1 \in \mathbb{N}$ such that

$$|y_n - M| < \frac{\varepsilon}{2K}, \forall n \ge N_1. \quad\quad\quad (1.32)$$

In view of (1.31) and (1.32), we obtain

$$n \ge N_1 \implies |x_n| |y_n - M| \le K |y_n - M| < K \cdot \frac{\varepsilon}{2K} = \frac{\varepsilon}{2}. \quad\quad\quad (1.33)$$

This takes care of the first term in (1.30). Next, since $\{x_n\}$ is convergent to L, $\exists N_2 \in \mathbb{N}$ such that

$$|x_n - L| < \frac{\varepsilon}{2(|M|+1)}, \forall n \ge N_2. \quad\quad\quad (1.34)$$

Now, argue in a manner similar to (1.33) to conclude that

$$n \ge N_2 \implies |M| |x_n - L| < \frac{\varepsilon}{2}. \text{ (Tell how.)} \quad\quad\quad (1.35)$$

Choose $N \in \mathbb{N}$ so that (1.33) and (1.35) hold simultaneously, $\forall n \ge N$. Then, use (1.30) - (1.35) to conclude that

$$n \ge N \implies |x_n y_n - LM| < \varepsilon. \text{ (How?)}$$

Hence, we conclude that $\lim_{n \to \infty} x_n y_n = LM$. This completes the proof. $\quad\quad\square$

Since we were unfolding the argument in somewhat reverse order for motivation, it would be better now to start with $\varepsilon > 0$ and reorganize the train of the suggested argument into a polished proof. (Do so!)

Exercise 1.5.4. Let $c \in \mathbb{R}$ and assume that $\lim_{n \to \infty} x_n = L$ and $\lim_{n \to \infty} y_n = M$. Prove that:

i.) $\lim_{n \to \infty} c x_n = cL$,

ii.) $\lim_{n \to \infty} (x_n - y_n) = L - M$.

The following lemma can be proven easily using induction. (Tell how.)

Lemma 1.5.10. *If $\{n_k\} \subset \mathbb{N}$ is an increasing sequence, then $n_k \geq k$, $\forall k \in \mathbb{N}$.*

Proposition 1.5.11. *If $\lim\limits_{n\to\infty} x_n = L$ and $\{x_{n_k}\}$ is any subsequence of $\{x_n\}$, then $\lim\limits_{k\to\infty} x_{n_k} = L$. (In words, all subsequences of a sequence convergent to L also converge to L.)*

Outline of Proof: Let $\varepsilon > 0$. There exists $N \in \mathbb{N}$ such that

$$|x_n - L| < \varepsilon, \forall n \geq N.$$

Now, fix any $K_0 \geq N$ and use Lemma 1.5.10 to infer that

$$k \geq K_0 \implies n_k > k \geq K_0 \geq N \implies |x_{n_k} - L| < \varepsilon. \text{ (Why?)}$$

The conclusion now follows. (Tell how.) $\square$

Exercise 1.5.5. Prove that if $\lim\limits_{n\to\infty} x_n = 0$ and $\{y_n\}$ is bounded, then $\lim\limits_{n\to\infty} x_n y_n = 0$.

Exercise 1.5.6.
i.) Prove that if $\lim\limits_{n\to\infty} x_n = L$, then $\lim\limits_{n\to\infty} |x_n| = |L|$.
ii.) Provide an example of a sequence $\{x_n\}$ for which $\exists \lim\limits_{n\to\infty} |x_n|$, but $\nexists \lim\limits_{n\to\infty} x_n$.

Exercise 1.5.7. Prove the following:
i.) If $\lim\limits_{n\to\infty} x_n = L$, then $\lim\limits_{n\to\infty} x_n^p = L^p$, $\forall p \in \mathbb{N}$.
ii.) If $x_n > 0$, $\forall n \in \mathbb{N}$, and $\lim\limits_{n\to\infty} x_n = L$, then $\lim\limits_{n\to\infty} \sqrt{x_n} = \sqrt{L}$.

Important connections between sequences and the derived set and closure of a set are provided in the following exercise.

Exercise 1.5.8. Let $\varnothing \neq S \subset \mathbb{R}$. Prove the following:
i.) $x \in S'$ iff $\exists \{x_n\} \subset S \setminus \{x\}$ such that $\lim\limits_{n\to\infty} x_n = x$.
ii.) $x \in \text{cl}_{\mathbb{R}}(S)$ iff $\exists \{x_n\} \subset S$ such that $\lim\limits_{n\to\infty} x_n = x$.

Proposition 1.5.12. *If $\{x_n\}$ is a bounded sequence in $\mathbb{R}$, then there exists a convergent subsequence $\{x_{n_k}\}$ of $\{x_n\}$.*

Outline of Proof: Let $\mathscr{R}_x = \{x_n | n \in \mathbb{N}\}$. We split the proof into two cases.

Case 1: $\mathscr{R}_x$ is a finite set, say $\mathscr{R}_x = \{y_1, y_2, \ldots, y_m\}$.
    It cannot be the case that the set $x^{-1}(\{y_i\}) = \{n \in \mathbb{N} | x_n = y_i\}$ is finite, for every $i \in \{1, 2, \ldots, m\}$ because $\mathbb{N} = \bigcup_{i=1}^m x^{-1}(\{y_i\})$. (Why?) As such, there is at least one $i_0 \in \{1, 2, \ldots, m\}$ such that $x^{-1}(\{y_{i_0}\})$ is infinite. Use this fact to inductively construct a sequence $n_1 < n_2 < \ldots$ in $\mathbb{N}$ such that $x_{n_k} = y_{i_0}$, $\forall k \in \mathbb{N}$. (Tell how.) Observe that $\{x_{n_k}\}$ is a convergent subsequence of $\{x_n\}$. (Why?)

Case 2: $\mathscr{R}_x$ is infinite.

Since $\{x_n\}$ is bounded, it follows from Thrm. 1.4.12 that $\mathscr{R}'_x \neq \varnothing$, say $L \in \mathscr{R}'_x$. Use the definition of limit point to inductively construct a subsequence $\{x_{n_k}\}$ of $\{x_n\}$ such that $x_{n_k} \longrightarrow L$. How does this complete the proof? $\square$

The combination of the hypotheses of monotonicity and boundedness implies convergence, as the next result suggests.

Proposition 1.5.13. *If $\{x_n\}$ is nondecreasing sequence which is bounded above, then $\{x_n\}$ converges and $\lim\limits_{n\to\infty} x_n = \sup\{x_n | n \in \mathbb{N}\}$.*

Outline of Proof: Since $\{x_n | n \in \mathbb{N}\}$ is a nonempty subset of $\mathbb{R}$ which is bounded above, $\exists \sup\{x_n | n \in \mathbb{N}\}$, call it L. (Why?) Let $\varepsilon > 0$. Then,

$$\exists N \in \mathbb{N} \text{ such that } L - \varepsilon < x_N. \text{ (Why?)}$$

Consequently,

$$n \geq \mathbb{N} \Longrightarrow L - \varepsilon < x_N \leq x_n \leq L < L + \varepsilon \Longrightarrow |x_n - L| < \varepsilon.$$

(Why?) This completes the proof. $\square$

Exercise 1.5.9. Formulate and prove a result analogous to Prop. 1.5.13 for nonincreasing sequences.

Exercise 1.5.10.
i.) Let $\{x_k\}$ be a sequence of nonnegative real numbers. For every $n \in \mathbb{N}$, define $s_n = \sum_{k=1}^{n} x_k$. Prove that the sequence $\{s_n\}$ converges iff it is bounded above.
ii.) Prove that $\left\{ \frac{a^n}{n!} \right\}$ converges, $\forall a \in \mathbb{R}$. In fact, $\lim\limits_{n\to\infty} \frac{a^n}{n!} = 0$.

Now that we know about subsequences, it is convenient to introduce a generalization of the notion of the limit of a real-valued sequence. We make the following definition.

Definition 1.5.14. Let $\{x_n\} \subset \mathbb{R}$ be a sequence.
i.) We say that $\lim\limits_{n\to\infty} x_n = \infty$ whenever $\forall r > 0$, $\exists N \in \mathbb{N}$ such that $x_n > r$, $\forall n \geq N$. (In such case, we write $x_n \to \infty$.)
ii.) For every $n \in \mathbb{N}$, let $u_n = \sup\{x_k : k \geq n\}$. We define the *limit superior* of x_n by

$$\overline{\lim_{n\to\infty}} x_n = \inf\{u_n | n \in \mathbb{N}\} = \inf_{n \in \mathbb{N}} \left(\sup_{k \geq n} x_k \right).$$

iii.) The dual notion of *limit inferior*, denoted $\underline{\lim}\limits_{n\to\infty} x_n$, is defined analogously with sup and inf interchanged in (ii), viz.,

$$\underline{\lim_{n\to\infty}} x_n = \sup_{n \in \mathbb{N}} \left(\inf_{k \geq n} x_k \right).$$

Some properties of limit superior (inferior) are gathered below. The proofs are standard and can be found in standard analysis texts (see [195]).

Proposition 1.5.15. (Properties of limit superior and inferior)
i.) $\overline{\lim\limits_{n\to\infty}} x_n = p \in \mathbb{R}$ iff $\forall \varepsilon > 0$,
   a.) *There exist only finitely many n such that $x_n > p + \varepsilon$, and*
   b.) *There exist infinitely many n such that $x_n > p - \varepsilon$;*
ii.) $\overline{\lim\limits_{n\to\infty}} x_n = p \in \mathbb{R}$ iff p is the largest limit of any subsequence of $\{x_n\}$;
iii.) $\overline{\lim\limits_{n\to\infty}} x_n = \infty$ iff $\forall r \in \mathbb{R}$, $\exists$ infinitely many n such that $x_n > r$;
iv.) If $x_n < y_n$, $\forall n \in \mathbb{N}$, then
   a.) $\overline{\lim\limits_{n\to\infty}} x_n \leq \overline{\lim\limits_{n\to\infty}} y_n$,
   b.) $\underline{\lim\limits_{n\to\infty}} x_n \leq \underline{\lim\limits_{n\to\infty}} y_n$;
v.) $\overline{\lim\limits_{n\to\infty}} (-x_n) = -\underline{\lim\limits_{n\to\infty}} x_n$;
vi.) $\underline{\lim\limits_{n\to\infty}} x_n \leq \overline{\lim\limits_{n\to\infty}} x_n$
vii.) $\lim\limits_{n\to\infty} x_n = p$ iff $\underline{\lim\limits_{n\to\infty}} x_n \leq \overline{\lim\limits_{n\to\infty}} x_n = p$;
viii.) $\underline{\lim\limits_{n\to\infty}} x_n + \underline{\lim\limits_{n\to\infty}} y_n \leq \underline{\lim\limits_{n\to\infty}} (x_n + y_n) \leq \overline{\lim\limits_{n\to\infty}} (x_n + y_n) \leq \overline{\lim\limits_{n\to\infty}} x_n + \overline{\lim\limits_{n\to\infty}} y_n$;
ix.) If $x_n \geq 0$ and $y_n \geq 0$, $\forall n \in \mathbb{N}$, then $\overline{\lim\limits_{n\to\infty}} (x_n y_n) \leq \left(\overline{\lim\limits_{n\to\infty}} x_n \right) \left(\overline{\lim\limits_{n\to\infty}} y_n \right)$, *provided the product on the right is not of the form $0 \cdot \infty$.*

1.5.3   Cauchy Sequences

Definition 1.5.16. A sequence $\{x_n\}$ is a *Cauchy sequence* if $\forall \varepsilon > 0, \exists N \in \mathbb{N}$ such that

$$n, m \geq N \implies |x_n - x_m| < \varepsilon.$$

Intuitively, the terms of a Cauchy sequence squeeze together as the index increases. Given any positive error tolerance ε, there is an index past which any two terms of the sequence, no matter how greatly their indices differ, have values within the tolerance of ε of one another. For brevity, we often write "$\{x_n\}$ is Cauchy" instead of "$\{x_n\}$ is a Cauchy sequence."

Exercise 1.5.11. Prove that the following statements are equivalent.
   (1) $\{x_n\}$ is a Cauchy sequence.
   (2) $\forall \varepsilon > 0, \exists N \in \mathbb{N}$ such that $|x_{N+p} - x_{N+q}| < \varepsilon, \forall p, q \in \mathbb{N}$.
   (3) $\forall \varepsilon > 0, \exists N \in \mathbb{N}$ such that $|x_n - x_N| < \varepsilon, \forall n \geq N$.
   (4) $\forall \varepsilon > 0, \exists N \in \mathbb{N}$ such that $|x_{N+p} - x_N| < \varepsilon, \forall p \in \mathbb{N}$.
   (5) $\lim\limits_{n\to\infty} (x_{n+p} - x_n) = 0, \forall p \in \mathbb{N}$.

We could have included the statement "$\{x_n\}$ is a convergent sequence." in the above list and asked which others imply it or are implied by it. Indeed, which of the two statements

$$\{x_n\} \text{ is a convergent sequence.}$$

and

$$\{x_n\} \text{ is a Cauchy sequence.}$$

seems stronger to you? Which implies which, if either? We will revisit this question after the following lemma.

Lemma 1.5.17. (Properties of Cauchy Sequences in $\mathbb{R}$)
i.) *A Cauchy sequence is bounded.*
ii.) *If a Cauchy sequence $\{x_n\}$ has a subsequence $\{x_{n_k}\}$ which converges to L, then $\{x_n\}$ itself converges to L.*

Outline of Proof:
Proof of (i): Let $\{x_n\}$ be a Cauchy sequence. Then, by Def. 1.5.16, $\exists N \in \mathbb{N}$ such that

$$n, m \geq N \implies |x_n - x_m| < 1.$$

In particular,

$$n \geq N \implies |x_n - x_N| < 1.$$

Starting with the last statement, argue as in Prop. 1.5.6 that $|x_n| \leq M, \forall n \in \mathbb{N}$, where

$$M = \max\{|x_1|, |x_2|, \ldots, |x_{N-1}|, |x_N| + 1\}.$$

So, $\{x_n\}$ is bounded.

Proof of (ii): Let $\varepsilon > 0$. $\exists N_1 \in \mathbb{N}$ such that

$$n, m \geq N_1 \implies |x_n - x_m| < \frac{\varepsilon}{2} \tag{1.36}$$

and $\exists N_2 \in \mathbb{N}$ such that

$$k \geq N_2 \implies |x_{n_k} - L| < \frac{\varepsilon}{2}. \tag{1.37}$$

(Why?) Now, how do you select N so that (1.36) and (1.37) hold simultaneously? Let $n > N$ and choose any $k \in \mathbb{N}$ such that $k \geq N$. Then, $n \geq N_1$ and $n_k \geq N$. (Why?) As such,

$$n \geq N \implies |x_n - L| = |x_n - x_{n_k} + x_{n_k} - L| \leq |x_n - x_{n_k}| + |x_{n_k} - L| < \varepsilon.$$

(Why?) This completes the proof. $\square$

We now shall prove that convergence and Cauchy are equivalent notions in $\mathbb{R}$.

Theorem 1.5.18. (Cauchy Criterion in $\mathbb{R}$)
$\{x_n\}$ is convergent $\iff$ $\{x_n\}$ is a Cauchy sequence.

Outline of Proof:

Proof of $\Longrightarrow$): Suppose that $\lim_{n\to\infty} x_n = L$ and let $\varepsilon > 0$. Then, $\exists N \in \mathbb{N}$ such that

$$n \geq N \Longrightarrow |x_n - L| < \frac{\varepsilon}{2}.$$

Thus,

$$n, m \geq N \Longrightarrow |x_n - x_m| \leq |x_n - L| + |x_m - L| < \frac{\varepsilon}{2} + \frac{\varepsilon}{2} = \varepsilon. \ (\text{Why?})$$

Thus, $\{x_n\}$ is a Cauchy sequence.

Proof of $\Longleftarrow$): Let $\{x_n\}$ be a Cauchy sequence. Then, $\{x_n\}$ is bounded (Why?) and so contains a convergent subsequence $\{x_{n_k}\}$. (Why?) Denote its limit by L. Then, it follows that, in fact, $\{x_n\}$ converges to L, as needed. (Why?) This completes the proof. $\square$

Remark. The proofs of the results

  i.) A bounded sequence has a convergent subsequence,

  ii.) A bounded monotone sequence converges,

  iii.) Cauchy Criterion in $\mathbb{R}$

all require the completeness property of $\mathbb{R}$, the first indirectly via Bolzano-Weierstrass which in turn uses it, the second directly, and the third via use of the first. Actually, all three statements are not only consequences of the completeness property of $\mathbb{R}$, but are equivalent to it. In fact, in more general settings in which order is no longer available (cf. Sections 1.6 and 1.7), completeness of the space is *defined* to be the property that all Cauchy sequences converge <u>in the space.</u>

Exercise 1.5.12. For every $n \in \mathbb{N}$, define $s_n = \sum_{k=1}^{n} \frac{1}{k}$. Prove that $\{s_n\}$ diverges.

1.5.4   A Brief Look at Infinite Series

Sequences defined by forming partial sums using terms of a second sequence (e.g., see Exer. 1.5.12) often arise in applied analysis. You might recognize them by the name *infinite series*. We shall provide the bare essentials of this topic below. A thorough treatment can be found in **[196]**.

Definition 1.5.19. Let $\{a_n\}$ be a sequence in $\mathbb{R}$.

i.) The sequence $\{s_n\}$ defined by $s_n = \sum_{k=1}^{n} a_k$, $n \in \mathbb{N}$ is the *sequence of partial sums of* $\{a_n\}$.

ii.) The pair $(\{a_n\}, \{s_n\})$ is called an *infinite series*, denoted by $\sum_{n=1}^{\infty} a_n$ or $\sum a_n$.

iii.) If $\lim_{n\to\infty} s_n = s$, then we say $\sum a_n$ *converges* and has *sum s*; we write $\sum a_n = s$. Otherwise, we say $\sum a_n$ *diverges*.

Remarks.

1. The sequence of partial sums can begin with an index n strictly larger than 1.

2. Suppose $\sum_{k=1}^{\infty} a_k = s$ and $s_n = \sum_{k=1}^{n} a_k$. Observe that

$$s_n + \underbrace{\sum_{k=n+1}^{\infty} a_k}_{\text{Tail}} = s. \tag{1.38}$$

Since $\lim_{n\to\infty} s_n = s$, it follows that $\lim_{n\to\infty} \sum_{k=n+1}^{\infty} a_k = 0$. (Why?)

Example. (Geometric Series)

Consider the series $\sum_{k=0}^{\infty} cx^k$, where $c, x \in \mathbb{R}$. For every $n \geq 0$, subtracting the expressions for s_n and xs_n yields

$$s_n = c\left[1 + x + x^2 + \ldots + x^n\right] \tag{1.39}$$
$$\underline{-xs_n = c\left[x + x^2 + \ldots + x^n + x^{n+1}\right]} \tag{1.40}$$
$$(1-x)s_n = c\left[1 - x^{n+1}\right]$$

Hence,

$$s_n = \begin{cases} \frac{c\left[1-x^{n+1}\right]}{1-x}, & x \neq 1, \\ c(n+1), & x = 1. \end{cases}$$

If $|x| < 1$, then $\lim_{n\to\infty} x^{n+1} = 0$, so that $\lim_{n\to\infty} s_n = \frac{c}{1-x}$. Otherwise, $\lim_{n\to\infty} s_n$ does not exist. (Tell why.)

Exercise 1.5.13. Let $p \in \mathbb{N}$. Determine the values of x for which $\sum_{n=p}^{\infty} c(5x+1)^{3n}$ converges, and for such x, determine its sum.

Proposition 1.5.20. $\sum a_n$ *converges iff* $\{s_n\}$ *is Cauchy iff* $\forall \varepsilon > 0, \exists N \in \mathbb{N}$ *such that* $n \geq N \implies \left|\sum_{k=1}^{p} a_{n+k}\right| < \varepsilon$.

Outline of Proof: The first equivalence is immediate (Why?) and the second follows from Exer. 1.5.11. (Tell how.) □

Corollary 1.5.21. (n^{th} *-term test*) *If* $\sum a_n$ *converges, then* $\lim_{n\to\infty} a_n = 0$.

Outline of Proof: Take $p = 1$ in Prop. 1.5.20. □

Exercise 1.5.14. Prove that $\sum_{n=1}^{\infty} \frac{n!}{a^n}$ diverges, $\forall a > 0$.

Proposition 1.5.22. (Comparison Test)
If $a_n, b_n \geq 0, \forall n \in \mathbb{N}$, *and* $\exists c > 0$ *and* $N \in \mathbb{N}$ *such that* $a_n \leq cb_n, \forall n \geq N$, *then*
i.) $\sum b_n$ *converges* $\implies \sum a_n$ *converges;*
ii.) $\sum a_n$ *diverges* $\implies \sum b_n$ *diverges.*

Outline of Proof: Use Prop. 1.5.13 (Tell how.) □

Example. Consider the series $\sum_{n=1}^{\infty} \frac{5n}{3^n}$. Since $\lim_{n\to\infty} \frac{5n}{3^{n/2}} = 0$, $\exists N \in \mathbb{N}$ such that

$$n \geq N \implies \frac{5n}{3^{n/2}} < 1 \implies \frac{5n}{3^n} < \frac{1}{3^{n/2}} = \left(\frac{1}{\sqrt{3}}\right)^n. \tag{1.41}$$

(Why?) But, $\sum_{n=1}^{\infty} \left(\frac{1}{\sqrt{3}}\right)^n$ is a convergent geometric series. Thus, Prop. 1.5.22 implies that $\sum_{n=1}^{\infty} \frac{5n}{3^n}$ converges.

Definition 1.5.23. A series $\sum a_n$ is *absolutely convergent* if $\sum |a_n|$ converges.

It can be shown that rearranging the terms of an absolutely convergent series does not affect convergence (see **[196]**). So, we can regroup terms at will, which is especially useful when groups of terms simplify nicely. For instance, refer to the example following (2.17).

Proposition 1.5.24. (Ratio Test)
Suppose $\sum a_n$ is a series with $a_n \neq 0$, $\forall n \in \mathbb{N}$. Let

$$r = \varliminf_{n\to\infty} \left|\frac{a_{n+1}}{a_n}\right| \text{ and } R = \varlimsup_{n\to\infty} \left|\frac{a_{n+1}}{a_n}\right|,$$

(where R could be ∞). Then,
i.) $R < 1 \implies \sum a_n$ converges absolutely;
ii.) $r > 1 \implies \sum a_n$ diverges;
iii.) If $r \leq 1 \leq R$, then the test is inconclusive.

Outline of Proof: We argue as in **[195]**.
Proof of (i): Assume $R < 1$ and choose x such that $R < x < 1$. Observe that

$$\varlimsup_{n\to\infty} \left|\frac{a_{n+1}}{a_n}\right| = R < x \implies \exists N \in \mathbb{N} \text{ such that } \left|\frac{a_{n+1}}{a_n}\right| \leq x, \forall n \geq N$$
$$\implies |a_{n+1}| \leq |a_n| x, \forall n \geq N. \tag{1.42}$$

Thus,

$$|a_{N+1}| \leq |a_N| x$$
$$|a_{N+2}| \leq |a_{N+1}| x \leq |a_N| x^2$$
$$|a_{N+3}| \leq |a_{N+2}| x \leq |a_{N+1}| x^2 \leq |a_N| x^3$$
$$\vdots$$

(Why?) What can be said about the series

$$|a_N| \left(x + x^2 + x^3 + \ldots\right)?$$

Use Prop. 1.5.22 to conclude that $\sum |a_n|$ converges.

Proof of (ii): Next, assume $1 < r$. Observe that

$$\lim_{n \to \infty} \left| \frac{a_{n+1}}{a_n} \right| = r > 1 \implies \exists N \in \mathbb{N} \text{ such that } \left| \frac{a_{n+1}}{a_n} \right| \geq 1, \forall n \geq N$$

$$\implies |a_{n+1}| \geq |a_n| \geq |a_N| > 0, \forall n \geq N. \tag{1.43}$$

(Why?) Thus, $a_n \nrightarrow 0$. (So what?)

Proof of (iii): For both $\sum \frac{1}{n}$ and $\sum \frac{1}{n^2}$, $r = R = 1$, but $\sum \frac{1}{n}$ diverges and $\sum \frac{1}{n^2}$ converges. $\square$

Exercise 1.5.15. Determine if $\sum_{n=1}^{\infty} \frac{n^n}{n!}$ converges.

Finally, we will need to occasionally multiply two series in the following sense.

Definition 1.5.25. Given two series $\sum_{n=0}^{\infty} a_n$ and $\sum_{n=0}^{\infty} b_n$, define

$$c_n = \sum_{k=0}^{n} a_k b_{n-k}, \forall n \geq 0.$$

The series $\sum_{n=0}^{\infty} c_n$ is called the *Cauchy product* of $\sum_{n=0}^{\infty} a_n$ and $\sum_{n=0}^{\infty} b_n$.

To see why this is a natural definition, consider the partial sum $\sum_{n=0}^{p} c_n$ and form a grid by writing the terms $a_0, \ldots, a_p$ as a column and $b_0, \ldots, b_p$ as a row. Multiply the terms from each row and column pairwise and observe that the sums along the diagonals (formed left to right) coincide with $c_0, \ldots, c_p$. (Check this!)

The following proposition describes a situation when such a product converges. The proof of this and other related results can be found in **[16]**.

Proposition 1.5.26. *If $\sum_{n=0}^{\infty} a_n$ and $\sum_{n=0}^{\infty} b_n$ both converge absolutely, then the Cauchy product $\sum_{n=0}^{\infty} c_n$ converges absolutely and $\sum_{n=0}^{\infty} c_n = \left(\sum_{n=0}^{\infty} a_n \right) \left(\sum_{n=0}^{\infty} b_n \right)$.*

1.6 The Spaces $\left(\mathbb{R}^N, \|\cdot\|_{\mathbb{R}^N} \right)$ and $\left(\mathbb{M}^N(\mathbb{R}), \|\cdot\|_{\mathbb{M}^N(\mathbb{R})} \right)$

We now introduce two spaces of objects with which you likely have some familiarity, namely vectors and square matrices, as a first step in formulating more abstract spaces. The key observation is that the characteristic properties of $\mathbb{R}$ carry over to these spaces, and their verification requires minimal effort. As you work through this section, use your intuition about how vectors in two and three dimensions behave to help you understand the more abstract setting.

1.6.1 The Space $\left(\mathbb{R}^N, \|\cdot\|_{\mathbb{R}^N}\right)$

Definition 1.6.1. For every $N \in \mathbb{N}$, $\mathbb{R}^N = \underbrace{\mathbb{R} \times \cdots \times \mathbb{R}}_{N \text{ times}}$ is the set of all ordered N-tuples of real numbers. This set is often loosely referred to as *N-space*.

A typical element of $\mathbb{R}^N$ (called a *vector*) is denoted by a boldface letter, say $\mathbf{x}$, representing the ordered N-tuple $\langle x_1, x_2, \dots, x_N \rangle$. (Here, x_k is the kth *component* of $\mathbf{x}$.) The *zero element* in $\mathbb{R}^N$ is the vector $\mathbf{0} = \underbrace{\langle 0, 0, \dots, 0 \rangle}_{N \text{ times}}$.

The algebraic operations defined in $\mathbb{R}$ can be applied componentwise to define the corresponding operations in $\mathbb{R}^N$. Indeed, we have:

Definition 1.6.2. (Algebraic Operations in $\mathbb{R}^N$)
Let $\mathbf{x} = \langle x_1, x_2, \dots, x_N \rangle$ and $\mathbf{y} = \langle y_1, y_2, \dots, y_N \rangle$ be elements of $\mathbb{R}^N$ and $c \in \mathbb{R}$,
i.) $\mathbf{x} = \mathbf{y}$ if and only if $x_k = y_k$, $\forall k \in \{1, \dots, N\}$,
ii.) $\mathbf{x} + \mathbf{y} = \langle x_1 + y_1, x_2 + y_2, \dots, x_N + y_N \rangle$,
iii.) $c\mathbf{x} = \langle cx_1, cx_2, \dots, cx_N \rangle$.

The usual properties of commutativity, associativity, and distributivity of scalar multiplication over addition carry over to this setting by applying the corresponding property in $\mathbb{R}$ componentwise. For instance, since $x_i + y_i = y_i + x_i$, $\forall i \in \{1, \dots, n\}$, it follows that

$$
\begin{aligned}
\mathbf{x} + \mathbf{y} &= \langle x_1 + y_1, x_2 + y_2, \dots, x_N + y_N \rangle \\
&= \langle y_1 + x_1, y_2 + x_2, \dots, y_N + x_N \rangle \\
&= \mathbf{y} + \mathbf{x}.
\end{aligned}
\tag{1.44}
$$

Exercise 1.6.1. Establish associativity of addition and distributivity of scalar multiplication over addition in $\mathbb{R}^N$.

1.6.1.1 Geometric and Topological Structure

From the viewpoint of its geometric structure, what is a natural candidate for a distance-measuring artifice for $\mathbb{R}^N$? There is more than one answer to this question, arguably the most natural of which is the Euclidean distance formula, defined below.

Definition 1.6.3. Let $\mathbf{x} \in \mathbb{R}^N$. The *(Euclidean) norm* of $\mathbf{x}$, denoted $\|\mathbf{x}\|_{\mathbb{R}^N}$, is defined by

$$
\|\mathbf{x}\|_{\mathbb{R}^N} = \sqrt{\sum_{k=1}^{N} x_k^2}.
\tag{1.45}
$$

We say that the *distance between $\mathbf{x}$ and $\mathbf{y}$ in $\mathbb{R}^N$* is given by $\|\mathbf{x} - \mathbf{y}\|_{\mathbb{R}^N}$.

Remarks.
1. When referring to the norm generically or as a function, we write $\|\cdot\|_{\mathbb{R}^N}$.

2. There are other "equivalent" ways to define a norm on $\mathbb{R}^N$ that are more convenient to use in some situations. Indeed, a useful alternative norm is given by

$$\|\mathbf{x}\|_{\mathbb{R}^N} = \max_{1 \le i \le N} |x_i|. \qquad (1.46)$$

By *equivalent*, we do not mean that the numbers produced by (1.45) and (1.46) are the same for a given $\mathbf{x} \in \mathbb{R}^N$. In fact, this is false in a big way! Rather, two norms $\|\cdot\|_1$ and $\|\cdot\|_2$ are *equivalent* if there exist constants $0 < \alpha < \beta$ such that

$$\alpha \|\mathbf{x}\|_1 \le \|\mathbf{x}\|_2 \le \beta \|\mathbf{x}\|_1, \, \forall \mathbf{x} \in \mathbb{R}. \qquad (1.47)$$

Suffice it to say that you can choose whichever norm is most convenient to work with within a given series of computations, as long as you don't decide to use a different one halfway through! **By default, we use (1.45) unless otherwise specified.**

Exercise 1.6.2. Let $\varepsilon > 0$. Provide a geometric description of these sets.
i.) $A = \left\{ \mathbf{x} \in \mathbb{R}^2 \mid \|\mathbf{x}\|_{\mathbb{R}^2} < \varepsilon \right\}$
ii.) $B = \left\{ \mathbf{y} \in \mathbb{R}^3 \mid \|\mathbf{y} - \langle 1, 0, 0 \rangle\|_{\mathbb{R}^3} \ge \varepsilon \right\}$
iii.) $C = \left\{ \mathbf{y} \in \mathbb{R}^3 \mid \|\mathbf{y} - \mathbf{x}_0\|_{\mathbb{R}^3} = 0 \right\}$, where $\mathbf{x}_0 \in \mathbb{R}^3$ is prescribed.

The $\mathbb{R}^N$-norm satisfies similar properties as $|\cdot|$ (cf. Prop. 1.4.4), summarized below.

Proposition 1.6.4. *Let* $\mathbf{x}, \mathbf{y} \in \mathbb{R}^N$ *and* $c \in \mathbb{R}$. *Then,*
i.) $\|\mathbf{x}\|_{\mathbb{R}^N} \ge 0$,
ii.) $\|c\mathbf{x}\|_{\mathbb{R}^N} = |c| \, \|\mathbf{x}\|_{\mathbb{R}^N}$,
iii.) $\|\mathbf{x} + \mathbf{y}\|_{\mathbb{R}^N} \le \|\mathbf{x}\|_{\mathbb{R}^N} + \|\mathbf{y}\|_{\mathbb{R}^N}$,
iv.) $\mathbf{x} = \mathbf{0}$ *iff* $\|\mathbf{x}\|_{\mathbb{R}^N} = 0$.

Exercise 1.6.3. Prove Prop. 1.6.4 using Def. 1.6.3. Then, redo it using (1.46).

Exercise 1.6.4. Let $M, p \in \mathbb{N}$. Prove the following string of inequalities:

$$\left\| \sum_{i=1}^{M} \mathbf{x}_i \right\|_{\mathbb{R}^N}^{p} \le \left(\sum_{i=1}^{M} \|\mathbf{x}_i\|_{\mathbb{R}^N} \right)^{p} \le M^{p-1} \sum_{i=1}^{M} \|\mathbf{x}_i\|_{\mathbb{R}^N}^{p} \qquad (1.48)$$

The space $\left(\mathbb{R}^N, \|\cdot\|_{\mathbb{R}^N} \right)$ has an even richer geometric structure since it can be equipped with a so-called *inner product* which enables us to define orthonormality (or perpendicularity) and, by extension, the notion of angle in the space. Precisely, we have:

Definition 1.6.5. Let $\mathbf{x}, \mathbf{y} \in \mathbb{R}^N$. The *inner product of $\mathbf{x}$ and $\mathbf{y}$*, denoted $\langle \mathbf{x}, \mathbf{y} \rangle_{\mathbb{R}^N}$, is defined by

$$\langle \mathbf{x}, \mathbf{y} \rangle_{\mathbb{R}^N} = \sum_{i=1}^{N} x_i y_i. \qquad (1.49)$$

Note that taking the inner product of any two elements of $\mathbb{R}^N$ produces a real number. Some of its properties are as follows:

Proposition 1.6.6. (Properties of the Inner Product on $\mathbb{R}^N$)
Let $\mathbf{x}, \mathbf{y}, \mathbf{z} \in \mathbb{R}^N$ *and* $c \in \mathbb{R}$. *Then,*
i.) $\langle c\mathbf{x}, \mathbf{y} \rangle_{\mathbb{R}^N} = \langle \mathbf{x}, c\mathbf{y} \rangle_{\mathbb{R}^N} = c \langle \mathbf{x}, \mathbf{y} \rangle_{\mathbb{R}^N}$;
ii.) $\langle \mathbf{x} + \mathbf{y}, \mathbf{z} \rangle_{\mathbb{R}^N} = \langle \mathbf{x}, \mathbf{z} \rangle_{\mathbb{R}^N} + \langle \mathbf{y}, \mathbf{z} \rangle_{\mathbb{R}^N}$;
iii.) $\langle \mathbf{x}, \mathbf{x} \rangle_{\mathbb{R}^N} \geq 0$;
iv.) $\langle \mathbf{x}, \mathbf{x} \rangle_{\mathbb{R}^N} = 0$ *iff* $\mathbf{x} = \mathbf{0}$;
v.) $\langle \mathbf{x}, \mathbf{x} \rangle_{\mathbb{R}^N} = \|\mathbf{x}\|_{\mathbb{R}^N}^2$;
vi.) $\langle \mathbf{x}, \mathbf{z} \rangle_{\mathbb{R}^N} = \langle \mathbf{y}, \mathbf{z} \rangle_{\mathbb{R}^N}$, $\forall \mathbf{z} \in \mathbb{R}^N \Longrightarrow \mathbf{x} = \mathbf{y}$.

Verifying these properties is straightforward and will be argued in a more general setting in Section 1.7. (Try proving them here!) Property (v) is of particular importance because it asserts that an inner product generates a norm.

Exercise 1.6.5. Prove Prop. 1.6.6.

The following Cauchy-Schwarz inequality is very important.

Proposition 1.6.7. (Cauchy-Schwarz Inequality)
Let $\mathbf{x}, \mathbf{y} \in \mathbb{R}^N$. *Then,*

$$|\langle \mathbf{x}, \mathbf{y} \rangle_{\mathbb{R}^N}| \leq \|\mathbf{x}\|_{\mathbb{R}^N} \|\mathbf{y}\|_{\mathbb{R}^N} \tag{1.50}$$

Outline of Proof: For any $\mathbf{y} \in \mathbb{R}^N \setminus \{\mathbf{0}\}$,

$$0 \leq \left\langle \mathbf{x} - \left(\frac{\langle \mathbf{x}, \mathbf{y} \rangle_{\mathbb{R}^N}}{\|\mathbf{y}\|_{\mathbb{R}^N}^2} \right) \mathbf{y}, \mathbf{x} - \left(\frac{\langle \mathbf{x}, \mathbf{y} \rangle_{\mathbb{R}^N}}{\|\mathbf{y}\|_{\mathbb{R}^N}^2} \right) \mathbf{y} \right\rangle_{\mathbb{R}^N}.$$

(So what?) Why does (1.50) hold for $\mathbf{y} = \mathbf{0}$? □

The inner product can be used to formulate a so-called *orthonormal basis* for $\mathbb{R}^N$. Precisely, let

$$\mathbf{e}_1 = \langle 1, 0, \ldots, 0 \rangle, \mathbf{e}_2 = \langle 0, 1, 0, \ldots, 0 \rangle, \ldots, \mathbf{e}_n = \langle 0, \ldots, 0, 1 \rangle,$$

and observe that

$$\|\mathbf{e}_i\|_{\mathbb{R}^N} = 1, \forall i \in \{1, \ldots, N\}, \tag{1.51}$$
$$\langle \mathbf{e}_i, \mathbf{e}_j \rangle_{\mathbb{R}^N} = 0, \text{ whenever } i \neq j. \tag{1.52}$$

This is useful because it yields the following unique representation for the members of $\mathbb{R}^N$ involving the inner product.

Proposition 1.6.8. *For every* $\mathbf{x} \in \mathbb{R}^N$,

$$\mathbf{x} = \sum_{i=1}^{N} \langle \mathbf{x}, \mathbf{e}_i \rangle_{\mathbb{R}^N} \mathbf{e}_i. \tag{1.53}$$

If $\mathbf{x} = \langle x_1, x_2, \ldots, x_N \rangle$, then (1.53) is a succinct way of writing

$$\mathbf{x} = \langle x_1, 0, \ldots, 0 \rangle + \langle 0, x_2, \ldots, 0 \rangle + \langle 0, 0, \ldots, x_N \rangle. \tag{1.54}$$

(Tell why.) Heuristically, this representation indicates how much to "move" in the direction of each basis vector to arrive at $\mathbf{x}$.

For any $x_0 \in \mathbb{R}$ and $\varepsilon > 0$, an open interval centered at x_0 with radius ε is defined by

$$(x_0 - \varepsilon, x_0 + \varepsilon) = \{ x \in \mathbb{R} \mid |x - x_0| < \varepsilon \}. \tag{1.55}$$

Since $\| \cdot \|_{\mathbb{R}^N}$ plays the role of $|\cdot|$ and shares its salient characteristics, it is natural to define an open N-ball centered at $\mathbf{x}_0$ with radius ε by

$$\mathscr{B}_{\mathbb{R}^N}(\mathbf{x}_0; \varepsilon) = \left\{ \mathbf{x} \in \mathbb{R}^N \mid \| \mathbf{x} - \mathbf{x}_0 \|_{\mathbb{R}^N} < \varepsilon \right\}. \tag{1.56}$$

Exercise 1.6.6. Interpret (1.56) geometrically in $\mathbb{R}^2$ and $\mathbb{R}^3$.

The terminology and results developed for $(\mathbb{R}, |\cdot|)$ in Section 1.4 can be extended to $\left(\mathbb{R}^N, \| \cdot \|_{\mathbb{R}^N} \right)$ with the only formal change being to replace $|\cdot|$ by $\| \cdot \|_{\mathbb{R}^N}$. Theorem 1.4.12 also holds, but a different approach is used to prove it because of the lack of ordering in $\mathbb{R}^N$. (See **[16]** for details.)

Exercise 1.6.7. Convince yourself of the validity of the generalization of the topological results to $\mathbb{R}^N$.

1.6.1.2   Sequences in $\mathbb{R}^N$

Definition 1.6.9. A function $\mathbf{x} : \mathbb{N} \longrightarrow \mathbb{R}^N$ is a *sequence* in $\mathbb{R}^N$.

The definitions of convergent and Cauchy sequences are essentially the same as in $\mathbb{R}$ and all the results carry over without issue, requiring only that we replace $|\cdot|$ by $\| \cdot \|_{\mathbb{R}^N}$.

Exercise 1.6.8. Convince yourself that the results from Sections 1.5.1 - 1.5.3 extend to the $\mathbb{R}^N$ setting.

Use the fact that the algebraic operations in $\mathbb{R}^N$ are performed componentwise to help you complete the following exercise.

Exercise 1.6.9.
i.) Consider the two real sequences $\{x_m\}$ and $\{y_m\}$ whose m^{th} terms are given by

$$x_m = \frac{1}{m^2}, \ y_m = \frac{2m}{4m+2}, \ m \in \mathbb{N}.$$

Show that $\lim_{m \to \infty} \langle x_m, y_m \rangle = \langle 0, \frac{1}{2} \rangle$.
ii.) Generally, if $\lim_{m \to \infty} x_m = p$ and $\lim_{m \to \infty} y_m = q$, what can you conclude about $\{ \langle x_m, y_m \rangle \}$

in $\mathbb{R}^2$?

iii.) Consider a sequence $\{\langle (x_1)_m, (x_2)_m, \ldots, (x_N)_m \rangle\}$ in $\mathbb{R}^N$. Establish a necessary and sufficient condition for this sequence to converge in $\mathbb{R}^N$.

A strategy similar to the one used in Exer. 1.6.9, coupled with Thrm. 1.5.18, is used to prove the following theorem.

Theorem 1.6.10. (Cauchy Criterion in $\mathbb{R}^N$)
$\{\mathbf{x}_n\}$ *converges in $\mathbb{R}^N$ iff $\{\mathbf{x}_n\}$ is a Cauchy sequence in $\mathbb{R}^N$. We say $\left(\mathbb{R}^N, \|\cdot\|_{\mathbb{R}^N}\right)$ is complete.*

Use the properties of sequences in $\mathbb{R}^N$ to complete the following exercises.

Exercise 1.6.10. Assume that $\{\mathbf{x}_m\}$ is a convergent sequence in $\mathbb{R}^N$. Prove that $\lim\limits_{m \to \infty} \|\mathbf{x}_m\|_{\mathbb{R}^N} = \left\| \lim\limits_{m \to \infty} \mathbf{x}_m \right\|_{\mathbb{R}^N}$. (We say that the norm $\|\cdot\|_{\mathbb{R}^N}$ is continuous.)

Exercise 1.6.11. Let $a \in \mathbb{R}$. Compute $\sup\left\{ \left\| \frac{a\mathbf{x}}{\|\mathbf{x}\|_{\mathbb{R}^N}} \right\|_{\mathbb{R}^N} : \mathbf{x} \in \mathbb{R}^N \setminus \{\mathbf{0}\} \right\}$.

Exercise 1.6.12. Let $\delta, \varepsilon > 0$, $a, b \in \mathbb{R}$, and $\mathbf{x}_0 \in \mathbb{R}^N$. Compute $\sup\{\|\mathbf{z}\|_{\mathbb{R}^N} : \mathbf{z} \in \mathscr{A}\}$, where
$$\mathscr{A} = \{a\mathbf{x} + b\mathbf{y} : \mathbf{x} \in \mathscr{B}_{\mathbb{R}^N}(\mathbf{x}_0; \varepsilon) \wedge \mathbf{y} \in \mathscr{B}_{\mathbb{R}^N}(\mathbf{x}_0; \delta)\}.$$

Exercise 1.6.13. Let $\mathbf{x} \in \mathbb{R}^N \setminus \{\mathbf{0}\}$, $a \neq 0$, and $p \in \mathbb{N}$. Must the series $\sum_{m=p}^{\infty} \left\| \frac{\mathbf{x}}{a\|\mathbf{x}\|_{\mathbb{R}^N}} \right\|_{\mathbb{R}^N}^{2m}$ converge? If so, can you determine its sum?

Exercise 1.6.14. Let $\mathbf{x} \in \mathscr{B}_{\mathbb{R}^2}\left(\mathbf{0}; \frac{1}{3}\right)$ and $p \in \mathbb{N}$. Must the series $\sum_{m=p}^{\infty} \left| \langle 2\mathbf{x}, \frac{1}{4}\mathbf{x} \rangle_{\mathbb{R}^2} \right|^{m/2}$ converge? If so, can you determine its sum?

Exercise 1.6.15. Let $\{c_n\}$ be a real sequence and $\mathbf{z} \in \mathbb{R}^N$. Assuming convergence of all series involved, prove that
$$\left\| \sum_{n=1}^{\infty} c_n \mathbf{z} \right\|_{\mathbb{R}^N} \leq \sum_{n=1}^{\infty} |c_n| \|\mathbf{z}\|_{\mathbb{R}^N}.$$

Exercise 1.6.16. Let $R > 0$ and $\langle a, b, c \rangle \in \partial\mathscr{B}_{\mathbb{R}^3}(\mathbf{0}; R)$ and define the function $\mathbf{f} : \mathbb{R} \longrightarrow \mathbb{R}^3$ by $\mathbf{f}(t) = \langle a\sin\left(\frac{t}{\pi}\right), b\cos(2t + \pi), c \rangle$. Show that $\{\|\mathbf{f}(t)\|_{\mathbb{R}^3} : t \in \mathbb{R}\} < \infty$.

1.6.2 The Space $\left(\mathbb{M}^N(\mathbb{R}), \|\cdot\|_{\mathbb{M}^N(\mathbb{R})}\right)$

A mathematical description of certain scenarios involves considering vectors whose components are themselves vectors. Indeed, consider
$$\mathbf{A} = \langle \mathbf{x}_1, \mathbf{x}_2, \ldots, \mathbf{x}_N \rangle, \tag{1.57}$$

where

$$\mathbf{x}_i = \langle x_{i1}, x_{i2}, \ldots, x_{iN} \rangle, \ 1 \le i \le N. \tag{1.58}$$

Viewing $\mathbf{x}_i$ in column form enables us to express $\mathbf{A}$ more elegantly as the $N \times N$ matrix

$$\mathbf{A} = \begin{bmatrix} x_{11} & x_{12} & \cdots & x_{1N} \\ x_{21} & x_{22} & \cdots & x_{2N} \\ \vdots & \vdots & \vdots & \vdots \\ x_{N1} & x_{N2} & \cdots & x_{NN} \end{bmatrix}. \tag{1.59}$$

It is typical to write

$$\mathbf{A} = [x_{ij}], \text{ where } 1 \le i, j \le N, \tag{1.60}$$

and refer to x_{ij} as the ij^{th} *entry* of $\mathbf{A}$.

Definition 1.6.11. Let $N \in \mathbb{N} \setminus \{1\}$. $\mathbb{M}^N(\mathbb{R})$ is the set of all $N \times N$ matrices with real entries.

The following terminology is standard in this setting.

Definition 1.6.12. Let $\mathbf{A} \in \mathbb{M}^N(\mathbb{R})$.
i.) $\mathbf{A}$ is *diagonal* if $x_{ij} = 0$, whenever $i \ne j$;
ii.) $\mathbf{A}$ is *symmetric* if $x_{ij} = x_{ji}, \forall i, j \in \{1, \ldots, N\}$;
iii.) The *trace* of $\mathbf{A}$ is the real number $\text{trace}(\mathbf{A}) = \sum_{i=1}^{N} x_{ii}$;
iv.) The *zero matrix*, denoted $\mathbf{0}$, is the unique member of $\mathbb{M}^N(\mathbb{R})$ for which $x_{ij} = 0$, $\forall 1 \le i, j \le N$.
v.) The *identity matrix*, denoted $\mathbf{I}$, is the unique diagonal matrix in $\mathbb{M}^N(\mathbb{R})$ for which $x_{ii} = 1, \forall 1 \le i \le N$.
vi.) The *transpose* of $\mathbf{A}$, denoted $\mathbf{A}^T$, is the matrix $\mathbf{A}^T = [x_{ji}]$. (That is, the ij^{th} entry of $\mathbf{A}^T$ is x_{ji}.)

We assume a modicum of familiarity with elementary matrix operations and gather some basic ones below. Note that some, but not all, of the operations are performed entry-wise.

Definition 1.6.13. (Algebraic Operations in $\mathbb{M}^N(\mathbb{R})$)
Let $\mathbf{A} = [a_{ij}]$, $\mathbf{B} = [b_{ij}]$, and $\mathbf{C} = [c_{ij}]$ be in $\mathbb{M}^N(\mathbb{R})$ and $\alpha \in \mathbb{R}$.
i.) $\mathbf{A}^0 = \mathbf{I}$,
ii.) $\alpha \mathbf{A} = [\alpha a_{ij}]$,
iii.) $\mathbf{A} + \mathbf{B} = [a_{ij} + b_{ij}]$,
iv.) $\mathbf{AB} = \left[\sum_{r=1}^{N} a_{ir} b_{rj} \right]$.

Exercise 1.6.17. Consider the operations defined in Def. 1.6.13.
i.) Does $\mathbf{A} + \mathbf{B} = \mathbf{B} + \mathbf{A}, \forall \mathbf{A}, \mathbf{B} \in \mathbb{M}^N(\mathbb{R})$?
ii.) Does $\mathbf{AB} = \mathbf{BA}, \forall \mathbf{A}, \mathbf{B} \in \mathbb{M}^N(\mathbb{R})$?
iii.) Must either $(\mathbf{A} + \mathbf{B})\mathbf{C} = \mathbf{AC} + \mathbf{BC}$ or $\mathbf{C}(\mathbf{A} + \mathbf{B}) = \mathbf{CA} + \mathbf{CB}$ hold, $\forall \mathbf{A}, \mathbf{B}, \mathbf{C} \in \mathbb{M}^N(\mathbb{R})$?
iv.) Does $(\mathbf{AB})\mathbf{C} = \mathbf{A}(\mathbf{BC}), \forall \mathbf{A}, \mathbf{B}, \mathbf{C} \in \mathbb{M}^N(\mathbb{R})$?

We assume familiarity with the basic properties of determinants of square matrices (see **[174]**). They are used to define invertibility.

Proposition 1.6.14. *For any* $\mathbf{A} \in \mathbb{M}^N(\mathbb{R})$ *for which* $\det(\mathbf{A}) \neq 0$, *there exists a unique* $\mathbf{B} \in \mathbb{M}^N(\mathbb{R})$ *such that* $\mathbf{AB} = \mathbf{BA} = \mathbf{I}$. *We say* $\mathbf{A}$ *is invertible and write* $\mathbf{B} = \mathbf{A}^{-1}$.

The notion of an *eigenvalue* arises in the study of stability theory of ordinary differential equations (ODEs). Precisely, we have

Definition 1.6.15. Let $\mathbf{A} \in \mathbb{M}^N(\mathbb{R})$.
i.) A complex number λ_0 is an *eigenvalue* of $\mathbf{A}$ if $\det(\mathbf{A} - \lambda_0 \mathbf{I}) = 0$.
ii.) An eigenvalue λ_0 has *multiplicity* M if $\det(\mathbf{A} - \lambda_0 \mathbf{I}) = p(\lambda)(\lambda - \lambda_0)^M$; that is, $(\lambda - \lambda_0)^M$ divides evenly into $\det(\mathbf{A} - \lambda_0 \mathbf{I})$.

Exercise 1.6.18. Let $\mathbf{A} = \begin{bmatrix} a & 0 \\ 0 & b \end{bmatrix}$, where $a, b \neq 0$.
i.) Compute the eigenvalues of $\mathbf{A}$.
ii.) Compute $\mathbf{A}^{-1}$ and its eigenvalues.
iii.) Generalize the computations in (i) and (ii) to the case of a diagonal $N \times N$ matrix $\mathbf{B}$ whose diagonal entries are all nonzero. Fill in the blank:

If λ is an eigenvalue of $\mathbf{B}$, then _____ is an eigenvalue of $\mathbf{B}^{-1}$.

We can equip $\mathbb{M}^N(\mathbb{R})$ with various norms in the spirit of those used in $\mathbb{R}^N$. Let $\mathbf{A} \in \mathbb{M}^N(\mathbb{R})$. Three standard choices for $\|\mathbf{A}\|_{\mathbb{M}^N}$ are

$$\|\mathbf{A}\|_{\mathbb{M}^N} = \left[\sum_{i=1}^{N}\sum_{j=1}^{N} |a_{ij}|^2\right]^{1/2}, \tag{1.61}$$

$$\|\mathbf{A}\|_{\mathbb{M}^N} = \sum_{i=1}^{N}\sum_{j=1}^{N} |a_{ij}|, \tag{1.62}$$

$$\|\mathbf{A}\|_{\mathbb{M}^N} = \max_{1 \leq i,j \leq N} |a_{ij}|. \tag{1.63}$$

It can be shown that (1.61) - (1.63) are equivalent in a sense similar to (1.47).

Exercise 1.6.19. Prove that each of (1.61) - (1.63) satisfies the properties in Prop. 1.6.4, appropriately extended to $\mathbb{M}^N(\mathbb{R})$.

Exercise 1.6.20. Let $\mathbf{A}, \mathbf{B} \in \mathbb{M}^N(\mathbb{R})$. Prove that $\|\mathbf{AB}\|_{\mathbb{M}^N} \leq \|\mathbf{A}\|_{\mathbb{M}^N}\|\mathbf{B}\|_{\mathbb{M}^N}$.

As in $\mathbb{R}^N$, a *sequence* in $\mathbb{M}^N(\mathbb{R})$ is an $\mathbb{M}^N(\mathbb{R})$-valued function whose domain is $\mathbb{N}$. If $\lim_{m\to\infty} \|\mathbf{A}_m - \mathbf{A}\|_{\mathbb{M}^N} = 0$, we say $\{\mathbf{A}_m\}$ *converges* to $\mathbf{A}$ in $\mathbb{M}^N(\mathbb{R})$ and write "$\mathbf{A}_m \longrightarrow \mathbf{A}$ in $\mathbb{M}^N(\mathbb{R})$." The similarity in the definitions of the norms used in $\mathbb{R}^N$ and $\mathbb{M}^N(\mathbb{R})$ suggests that checking convergence is performed entry-wise. The same is true of Cauchy sequences in $\mathbb{M}^N(\mathbb{R})$. (Convince yourself!) By extension of Thrm.

1.6.10, we can argue that $\left(M^N(\mathbb{R}), \|\cdot\|_{M^N(\mathbb{R})}\right)$ is complete with respect to any of the norms (1.61) - (1.63). (Tell why carefully.)

Let $\mathbf{A} \in M^N(\mathbb{R})$ and $\mathbf{x} \in \mathbb{R}^N$. From the definition of matrix multiplication, if we view $\mathbf{x}$ as a $N \times 1$ column matrix, then since $\mathbf{A}$ is an $N \times N$ matrix, we know that $\mathbf{Ax}$ is a well-defined $N \times 1$ column matrix which can be identified as a member of $\mathbb{R}^N$. As such, the function $\mathbf{f_A} : \mathbb{R}^N \longrightarrow \mathbb{R}^N$ given by $\mathbf{f_A}(\mathbf{x}) = \mathbf{Ax}$ is well-defined. Such mappings are used frequently in Chapter 2.

Exercise 1.6.21. Prove that $\left\{ \left\| \begin{bmatrix} e^{-t} & 0 \\ 0 & e^{-2t} \end{bmatrix} \mathbf{x} \right\|_{\mathbb{R}^2} : \|\mathbf{x}\|_{\mathbb{R}^2} = 1 \wedge t > 0 \right\}$ is bounded.

Exercise 1.6.22. If $\left\{ (x_{ij})_m \right\}$ is a real Cauchy sequence for each $1 \leq i, j \leq 2$, must the sequence $\{\mathbf{A}_m\}$, where $\mathbf{A}_m = \left[(x_{ij})_m \right]$, be Cauchy in $M^2(\mathbb{R})$? Prove your claim.

Exercise 1.6.23. Let $\{\mathbf{A}_m\}$ be a sequence in $M^N(\mathbb{R})$.
i.) If $\lim\limits_{m \to \infty} \mathbf{A}_m = \mathbf{0}$, what must be true about each of the N^2 sequences formed using the entries of $\mathbf{A}_m$?
ii.) If $\lim\limits_{m \to \infty} \mathbf{A}_m = \mathbf{I}$, what must be true about each of the N^2 sequences formed using the entries of $\mathbf{A}_m$?
iii.) More generally, if $\lim\limits_{m \to \infty} \mathbf{A}_m = \mathbf{B}$, what must be true about each of the N^2 sequences formed using the entries of $\mathbf{A}_m$?

Exercise 1.6.24.
i.) If $\mathbf{A}_m \longrightarrow \mathbf{A}$ in $M^N(\mathbb{R})$, must $\mathbf{A}_m\mathbf{x} \longrightarrow \mathbf{Ax}$ in $\mathbb{R}^N$, $\forall \mathbf{x} \in \mathbb{R}^N$?
ii.) If $\mathbf{x}_m \longrightarrow \mathbf{x}$ in $\mathbb{R}^N$, must $\mathbf{Ax}_m \longrightarrow \mathbf{Ax}$ in $\mathbb{R}^N$, $\forall \mathbf{A} \in M^N(\mathbb{R})$?
iii.) If $\mathbf{A}_m \longrightarrow \mathbf{A}$ in $M^N(\mathbb{R})$ and $\mathbf{x}_m \longrightarrow \mathbf{x}$ in $\mathbb{R}^N$, must $\{\mathbf{A}_m\mathbf{x}_m\}$ converge in $\mathbb{R}^N$? If so, what is its limit?

1.7 Abstract Spaces

Many other spaces possess the same salient features regarding norms, inner products, and completeness exhibited by $\left(\mathbb{R}^N, \|\cdot\|_{\mathbb{R}^N}\right)$ and $\left(M^N(\mathbb{R}), \|\cdot\|_{M^N(\mathbb{R})}\right)$. At the moment we would need to verify them for each such space that we encountered individually, which is inefficient. Rather, it would be beneficial to examine a more abstract structure possessing these characteristics and establish results directly for *them*. In turn, we would need only to verify that a space arising in an investigation had this basic structure and then invoke all concomitant results immediately. This will save us considerable work in that we will not need to reformulate all properties each time we introduce a new space. This section is devoted to the development of such abstract structures. (See **[202]** for a thorough treatment.)

1.7.1 Banach Spaces

We begin with the notion of a linear space over $\mathbb{R}$.

Definition 1.7.1. A *real linear space* X is a set equipped with addition and scalar multiplication by real numbers satisfying the following properties:
i.) $x + y = y + x$, $\forall x, y \in X$;
ii.) $x + (y + z) = (x + y) + z$, $\forall x, y, z \in X$;
iii.) There exists a unique element $0 \in X$ such that $x + 0 = 0 + x$, $\forall x \in X$;
iv.) For every $x \in X$, there exists a unique element $-x \in X$ such that

$$x + (-x) = (-x) + x = 0, \ \forall x \in X;$$

v.) $a(bx) = (ab)x$, $\forall x \in X$ and $a, b \in \mathbb{R}$;
vi.) $a(x + y) = ax + ay$, $\forall x, y \in X$ and $a \in \mathbb{R}$;
vii.) $(a + b)x = ax + bx$, $\forall x \in X$ and $a, b \in \mathbb{R}$.

Restricting attention to a subset $\mathscr{Y}$ of elements of a linear space $\mathscr{X}$ that possesses the same structure as the larger space leads to the following notion.

Definition 1.7.2. Let X be a real linear space. A subset $Y \subset X$, equipped with the same operations as X, is a *linear subspace* of X if
i.) $x, y \in Y \implies x + y \in Y$,
ii.) $x \in Y \implies ax \in Y$, $\forall a \in \mathbb{R}$.

Exercise 1.7.1.
i.) Verify that $(\mathbb{R}, |\cdot|)$, $\left(\mathbb{R}^N, \|\cdot\|_{\mathbb{R}^N}\right)$, and $\left(\mathbb{M}^N(\mathbb{R}), \|\cdot\|_{\mathbb{M}^N(\mathbb{R})}\right)$ are linear spaces.
ii.) Is $Y = \left\{ \mathbf{A} \in \mathbb{M}^N(\mathbb{R}) : \mathbf{A} \text{ is diagonal} \right\}$ a linear subspace of $\mathbb{M}^N(\mathbb{R})$?

We can enhance the structure of a real linear space by introducing a topology so that limit processes can be performed. One way to accomplish this is to equip the space with a norm in the following sense.

Definition 1.7.3. Let X be a real linear space. A real-valued function $\|\cdot\|_X : X \longrightarrow \mathbb{R}$ is a *norm* on X if $\forall x, y \in X$ and $a \in \mathbb{R}$,
i.) $\|x\|_X \geq 0$,
ii.) $\|ax\|_X = |a| \, \|x\|_X$,
iii.) $\|x + y\|_X \leq \|x\|_X + \|y\|_X$,
iv.) $x = 0$ iff $\|x\|_X = 0$.

We say that the *distance* between x and y is $\|x - y\|_X$. We use this to obtain the following richer abstract structure.

Definition 1.7.4. A real linear space X equipped with a norm $\|\cdot\|_X$ is called a *(real) normed linear space.*

We know from our work in Sections 1.4 - 1.6 that $(\mathbb{R}, |\cdot|)$, $\left(\mathbb{R}^N, \|\cdot\|_{\mathbb{R}^N}\right)$, and $\left(\mathbb{M}^N(\mathbb{R}), \|\cdot\|_{\mathbb{M}^N(\mathbb{R})}\right)$ are all normed linear spaces. Many of the normed linear spaces that we will encounter are collections of functions satisfying certain properties. Some standard function spaces (aka *Sobolev spaces*) and typical norms with which they are equipped are listed below. Momentarily, we assume an intuitive understanding of continuity, differentiability, and integrability. These notions will be defined more rigorously in Section 1.8. A detailed technical treatment of Sobolev spaces can be found in [1].

Some Common Function Spaces: Let $I \subset \mathbb{R}$ and X be a normed linear space.
1.) $\mathbb{C}(I;X) = \{f : I \longrightarrow X \,|\, f$ is continuous on $I\}$ equipped with the *sup norm*

$$\|f\|_{\mathbb{C}} = \sup_{t \in I} \|f(t)\|_X \tag{1.64}$$

2.) $\mathbb{C}^n(I;X) = \{f : I \longrightarrow X \,|\, f$ is n times continuously differentiable on $I\}$ equipped with

$$\|f\|_{\mathbb{C}^n} = \sup_{t \in I} \left[\|f(t)\|_X + \|f'(t)\|_X + \ldots + \left\| f^{(n)}(t) \right\|_X \right] \tag{1.65}$$

3.) Let $1 \leq p < \infty$. $\mathbb{L}^p(I;\mathbb{R}) = \{f : I \longrightarrow \mathbb{R} \,|\, f^p$ is integrable on $I\}$ equipped with

$$\|f\|_{\mathbb{L}^p} = \left[\int_I |f(t)|^p \, dt \right]^{\frac{1}{p}} \tag{1.66}$$

4.) Let $1 \leq p < \infty$. $\mathbb{L}^p_{\mathrm{loc}}(I;\mathbb{R}) = \{f : I \longrightarrow \mathbb{R} \,|\, f^p$ is integrable on compact subsets of $I\}$ equipped with (1.66).

5.) $\mathbb{H}^2(I;\mathbb{R}) = \{f \in \mathbb{L}^2(I;\mathbb{R}) \,|\, f', f''$ exist and $f'' \in \mathbb{L}^2(I;\mathbb{R})\}$ equipped with

$$\|f\|_{\mathbb{H}^2} = \left[\int_I |f(t)|^2 \, dt \right]^{\frac{1}{2}} \tag{1.67}$$

6.) $\mathbb{H}^1_0(a,b;\mathbb{R}) = \{f : (a,b) \longrightarrow \mathbb{R} \,|\, f'$ exists and $f(a) = f(b) = 0\}$ equipped with

$$\|f\|_{\mathbb{H}^1_0} = \left[\int_I \left(|f(t)|^2 + |f'(t)|^2 \right) dt \right]^{\frac{1}{2}} \tag{1.68}$$

7.) Let $m \in \mathbb{N}$. $\mathbb{W}^{2,m}(I;\mathbb{R}) = \left\{ f \in \mathbb{L}^2(I;\mathbb{R}) \,|\, f^{(k)} \in \mathbb{L}^2(I;\mathbb{R}), \forall k = 1,\ldots,m \right\}$ equipped with

$$\|f\|_{\mathbb{W}^{2,m}} = \left[\int_I \left(|f(t)|^2 + |f'(t)|^2 + \ldots + \left| f^{(m)}(t) \right|^2 \right) dt \right]^{\frac{1}{2}}. \tag{1.69}$$

Here, $f^{(k)}$ represents the kth- order derivative of f. (Technically, this is a generalized derivative defined in a distributional sense.)

The notions of convergent and Cauchy sequences extend to any normed linear space in the same manner as in Section 1.6. For example, we say that an X-valued sequence $\{x_n\}$ *converges to x in X* if $\lim_{n \to \infty} \|x_n - x\|_X = 0$.

Exercise 1.7.2.
i.) Interpret the statement "$\lim_{n \to \infty} x_n = x$ in X" for these specific choices of X:

   a.) (1.64)
   b.) (1.66)
   c.) (1.69)

ii.) Interpret the statement "$\{x_n\}$ is Cauchy in X" for the same choices of X.

When working with specific function spaces, knowing when Cauchy sequences in X must converge <u>in X</u> is often crucial. In other words, we need to know if a space is complete in the following sense.

Definition 1.7.5. (Completeness)
i.) A normed linear space X is *complete* if every Cauchy sequence in X converges to an element of X.
ii.) A complete normed linear space is called a *Banach space*.

We shall routinely work with sequences in $\mathbb{C}(I; \mathscr{X})$, where $\mathscr{X}$ is a Banach space. We now focus on the terminology and some particular results for this space.

Definition 1.7.6. Suppose that $\varnothing \neq S \subset D \subset \mathbb{R}$ and $f_n, f : D \longrightarrow \mathscr{X}, n \in \mathbb{N}$.
i.) $\{f_n\}$ *converges uniformly to f on S* whenever $\forall \varepsilon > 0, \exists N \in \mathbb{N}$ such that

$$n \geq N \implies \sup_{x \in S} \|f_n(x) - f(x)\|_{\mathscr{X}} < \varepsilon.$$

(We write "$f_n \longrightarrow f$ uniformly on S.")
ii.) $\{f_n\}$ *converges pointwise to f on S* whenever $\lim_{n \to \infty} f_n(x) = f(x), \forall x \in S$.
iii.) $\{f_n\}$ is *uniformly bounded on S* whenever $\exists M > 0$ such that

$$\sup_{n \in \mathbb{N}} \sup_{x \in S} \|f_n(x)\|_{\mathscr{X}} \leq M.$$

iv.) $\sum_{k=1}^{\infty} f_k(x)$ *converges uniformly to f(x) on S* whenever $s_n \longrightarrow f$ uniformly on S, where $s_n(x) = \sum_{k=1}^{n} f_k(x)$.

It can be shown that each of the function spaces listed above is complete with respect to the norm provided. Verification of this requires the use of various tools involving the behavior of sequences of functions and integrability. We consider the most straightforward one in the following exercise.

Exercise 1.7.3.
i.) Prove that $\mathbb{C}([a,b]; \mathbb{R})$ equipped with the sup norm (1.64) is complete.
ii.) How does the argument change if $\mathbb{R}$ is replaced by a Banach space $\mathscr{X}$? Can $\mathscr{X}$ be any normed linear space, or must it be complete? Explain.

Strong Cautionary Remark! We have seen that a normed linear space can be equipped with different norms. As such, we must bear in mind that completeness is norm-dependent. Indeed, equipping the spaces above with norms other than those specified by (1.64) - (1.69) could foresake completeness! For instance, $\mathbb{C}([a,b];\mathbb{R})$ equipped with (1.67) is NOT complete. (See **[146]**.)

The next Cauchy-like condition for checking the uniform convergence of a series of functions follows directly from the completeness of $\mathbb{C}(I;\mathscr{X})$.

Proposition 1.7.7. $\sum_{k=1}^{\infty} f_k(x)$ *converges uniformly on S iff* $\forall \varepsilon > 0$, $\exists N \in \mathbb{N}$ *such that*

$$n \geq N \wedge p \in \mathbb{N} \Longrightarrow \sup_{x \in S} \left\| \sum_{k=n+1}^{n+p} f_k(x) \right\|_{\mathscr{X}} < \varepsilon. \tag{1.70}$$

The following convergence result is useful in certain fixed-point arguments arising in Chapter 5.

Proposition 1.7.8. (Weierstrass M-test)
Let $\{M_k\} \subset [0,\infty)$ *such that* $\forall k \in \mathbb{N}$,

$$\sup_{x \in S} \|f_k(x)\|_{\mathscr{X}} \leq M_k.$$

If $\sum_{k=1}^{\infty} M_k$ *converges, then* $\sum_{k=1}^{\infty} f_k(x)$ *converges uniformly on S.*

Outline of Proof: Let $\varepsilon > 0$. There exists $N \in \mathbb{N}$ such that

$$n \geq N \Longrightarrow \sum_{k=n+1}^{n+p} M_k < \varepsilon.$$

(Why?) For every $n \geq N$ and $p \in \mathbb{N}$, observe that $\forall x \in S$,

$$\left\| \sum_{k=n+1}^{n+p} f_k(x) \right\|_{\mathscr{X}} \leq \sum_{k=n+1}^{n+p} \|f_k(x)\|_{\mathscr{X}} \leq \sum_{k=n+1}^{n+p} M_k < \varepsilon.$$

Now, use the completeness of $\mathbb{C}(I;\mathscr{X})$ and Prop. 1.7.7. (Tell how.) $\square$

Exercise 1.7.4. Prove Props. 1.7.7 and 1.7.8.

Remark. Taylor series representations of infinitely-differentiable functions are presented in elementary calculus. Some common examples are:

$$e^x = \lim_{N \to \infty} \sum_{n=0}^{N} \frac{x^n}{n!}, x \in \mathbb{R}, \tag{1.71}$$

$$\sin(x) = \lim_{N \to \infty} \sum_{n=0}^{N} \frac{(-1)^n x^{2n+1}}{(2n+1)!}, x \in \mathbb{R}, \tag{1.72}$$

$$\cos(x) = \lim_{N \to \infty} \sum_{n=0}^{N} \frac{(-1)^n x^{2n}}{(2n)!}, x \in \mathbb{R}. \tag{1.73}$$

It can be shown that the convergence in each case is uniform on all compact subsets of $\mathbb{R}$. The benefit of such a representation is the uniform approximation of the function on the left-side by the sequence of nicely-behaved polynomials on the right-side. Generalizations of these formulae to more abstract settings will be a key tool throughout the text.

The basic topological notions of open, closed, bounded, etc. carry over to normed linear spaces in the form of the metric topology defined using the norm $\|\cdot\|_{\mathscr{X}}$. We use the following notation:

$$\mathfrak{B}_{\mathscr{X}}(x_0;\varepsilon) = \{x \in \mathscr{X} \mid \|x - x_0\|_{\mathscr{X}} < \varepsilon\} \tag{1.74}$$
$$\text{cl}_{\mathscr{X}}(\mathscr{Z}) = \text{closure of } \mathscr{Z} \text{ (in the sense of } \|\cdot\|_{\mathscr{X}}).$$

Exercise 1.7.5. Describe the elements of the ball $\mathfrak{B}_{C([0,2];\mathbb{R})}\left(x^2;1\right)$.

Some topological results, like the Bolzano-Weierstrass and Heine-Borel theorems, do not extend to the general Banach space setting because they rely on intrinsic properties of $\mathbb{R}^N$. This will present a minor obstacle in Chapter 5, at which time we will revisit the issue.

The need to restrict our attention to a particular subspace of a function space whose elements satisfy some special characteristic arises often. But, can we be certain that we remain in the subspace upon performing limiting operations involving its elements? Put differently, must a subspace of a Banach space be complete? The answer is provided easily by the following exercise.

Exercise 1.7.6. Let Y be the subspace $((0,2];|\cdot|)$ of $\mathbb{R}$. Prove that $\left\{\frac{2}{n}\right\}$ is Cauchy in Y, but that there does not exist $y \in Y$ to which $\left\{\frac{2}{n}\right\}$ converges.

If the subspace had been closed in the topological sense, would it have made a difference? It turns out that it would have indeed, as suggested by:

Proposition 1.7.9. *A closed subspace $\mathscr{Y}$ of a Banach space $\mathscr{X}$ is complete.*

Exercise 1.7.7. Prove Prop. 1.7.9.

Exercise 1.7.8. Let $(\mathscr{X}, \|\cdot\|_{\mathscr{X}})$ and $(\mathscr{Y}, \|\cdot\|_{\mathscr{Y}})$ be real Banach spaces. Prove that $(\mathscr{X} \times \mathscr{Y}; \|\cdot\|_1)$ and $(\mathscr{X} \times \mathscr{Y}; \|\cdot\|_2)$ are also Banach spaces, where

$$\|(x,y)\|_1 = \|x\|_{\mathscr{X}} + \|y\|_{\mathscr{Y}}, \tag{1.75}$$
$$\|(x,y)\|_2 = \left(\|x\|_{\mathscr{X}}^2 + \|y\|_{\mathscr{Y}}^2\right)^{1/2}. \tag{1.76}$$

1.7.2 Hilbert Spaces

Equipping $\mathbb{R}^N$ with a dot product enhanced its structure by introducing the notion of orthogonality. This prompts us to define the general notion of an inner product on a linear space.

Definition 1.7.10. Let X be a real linear space. A real-valued function $\langle \cdot, \cdots \rangle_X : X \times X \longrightarrow \mathbb{R}$ is an *inner product on X* if $\forall x, y, z \in X$ and $a \in \mathbb{R}$,
i.) $\langle x, y \rangle_X = \langle y, x \rangle_X$,
ii.) $\langle ax, y \rangle_X = a \langle x, y \rangle_X$,
iii.) $\langle x + y, z \rangle_X = \langle x, z \rangle_X + \langle y, z \rangle_X$,
iv.) $\langle x, x \rangle_X > 0$ iff $x \neq 0$.
The pair $(X, \langle \cdot, \cdots \rangle_X)$ is called a *(real) inner product space.*

Some Common Inner Product Spaces:
1.) $\mathbb{R}^N$ equipped with (1.49).
2.) $\mathbb{C}([a,b];\mathbb{R})$ equipped with

$$\langle f, g \rangle_{\mathbb{C}} = \int_a^b f(t)g(t)dt. \tag{1.77}$$

3.) $\mathbb{L}^2(a,b;\mathbb{R})$ equipped with (1.77).
4.) $\mathbb{W}^{2,m}(a,b;\mathbb{R})$ equipped with

$$\langle f, g \rangle_{\mathbb{W}^{2,k}} = \int_a^b \left[f(t)g(t) + f'(t)g'(t) + \ldots + f^{(m)}(t)g^{(m)}(t) \right] dt. \tag{1.78}$$

Exercise 1.7.9. Verify that (1.77) and (1.78) are inner products.

An inner product on X induces a norm on X via the relationship

$$\langle x, x \rangle_X^{1/2} = \|x\|_X. \tag{1.79}$$

Exercise 1.7.10. Prove that the usual norms in $\mathbb{R}^N$, $\mathbb{C}([a,b];\mathbb{R})$, and $\mathbb{W}^{2,m}(a,b;\mathbb{R})$ can be obtained from their respective inner products (1.49), (1.77) and (1.78).

Propositions 1.6.6 and 1.6.7 actually hold for general inner products. We have

Proposition 1.7.11. *Let $(X, \langle \cdot, \cdots \rangle_X)$ be an inner product space and suppose $\|\cdot\|_X$ is given by (1.79). Then, $\forall x, y \in X$ and $a \in \mathbb{R}$,*
i.) $\langle x, ay \rangle_X = a \langle x, y \rangle_X$;
ii.) $\langle x, z \rangle_X = \langle y, z \rangle_X, \forall z \in X \implies x = y$;
iii.) $\|ax\|_X = |a| \|x\|_X$;
iv.) (Cauchy-Schwarz) $|\langle x, y \rangle_X| \leq \|x\|_X \|y\|_X$;
v.) (Minkowski) $\|x + y\|_X \leq \|x\|_X + \|y\|_X$;
vi.) *If $x_n \longrightarrow x$ and $y_n \longrightarrow y$ in X, then $\langle x_n, y_n \rangle_X \longrightarrow \langle x, y \rangle_X$.*

Exercise 1.7.11. Prove Prop. 1.7.11.

Exercise 1.7.12. Interpret Prop. 1.7.11(iv) specifically for the space $\mathbb{L}^2(a,b;\mathbb{R})$. (This is a special case of the so-called *Holder inequality*.)

Since inner product spaces come equipped with a norm, it makes sense to further characterize them using completeness.

Definition 1.7.12. A *Hilbert space* is a complete inner product space.

Both $\mathbb{R}^N$ and $\mathbb{L}^2(a, b; \mathbb{R})$ equipped with their usual norms are Hilbert spaces, while $\mathbb{C}([a, b]; \mathbb{R})$ equipped with (1.77) is not. Again, the underlying norm plays a crucial role.

The notion of a *basis* encountered in linear algebra can be made precise in the Hilbert space setting and plays a central role in formulating representation formulae for elements of the space. We begin with the following definition.

Definition 1.7.13. Let $\mathscr{H}$ be a Hilbert space and $\mathfrak{B} = \{e_n | n \in K \subset \mathbb{N}\}$.
i.) The *span of* $\mathfrak{B}$ is given by $\text{span}(\mathfrak{B}) = \{\sum_{n \in K} \alpha_n e_n | \alpha_n \in \mathbb{R}, \forall n \in K\}$;
ii.) If $\langle e_n, e_m \rangle_{\mathscr{H}} = 0$, then e_n and e_m are *orthogonal*;
iii.) The members of $\mathfrak{B}$ are *linearly independent* if

$$\sum_{n \in K} \alpha_n e_n = 0 \Longrightarrow \alpha_n = 0, \forall n \in K;$$

iv.) $\mathfrak{B}$ is an *orthonormal set* if
  a.) $\|e_n\|_{\mathscr{H}} = 1, \forall n \in K$,
  b.) $\langle e_n, e_m \rangle_{\mathscr{H}} = \begin{cases} 0, & \text{if } n \neq m, \\ 1, & \text{if } n = m. \end{cases}$
v.) $\mathfrak{B}$ is a *complete set* if $(\langle x, e_n \rangle_{\mathscr{H}} = 0, \forall n \in K) \Longrightarrow (x = 0, \forall x \in \mathscr{H})$;
vi.) A complete orthonormal subset of $\mathscr{H}$ is a *basis* for $\mathscr{H}$.

The utility of a basis $\mathfrak{B}$ of a Hilbert space $\mathscr{H}$ is that every element of $\mathscr{H}$ can be decomposed into a linear combination of the members of $\mathfrak{B}$. For general Hilbert spaces, specifically those that are not finite dimensional like $\mathbb{R}^N$, the existence of a basis is not guaranteed. There are, however, sufficiency results that indicate when a basis must exist. For instance, consider

Definition 1.7.14. An inner product space is *separable* if it contains a countable dense subset $\mathfrak{D}$.

Remark. A set $\mathfrak{D}$ is *countable* if a one-to-one function $f : \mathfrak{D} \to \mathbb{N}$ exists. In such case, the elements of $\mathfrak{D}$ can be matched in a one-to-one manner with those of $\mathbb{N}$. Intuitively, $\mathfrak{D}$ has no more elements than $\mathbb{N}$. A thorough treatment of countability can be found in **[195]**.

The proof of the following result can be found in **[202]**.

Theorem 1.7.15. *Any separable inner product space has a basis.*

Example. The set

$$\mathfrak{B} = \left\{ \frac{1}{\sqrt{2\pi}} \right\} \cup \left\{ \frac{\cos(nt)}{\sqrt{\pi}} \Big| n \in \mathbb{N} \right\} \cup \left\{ \frac{\sin(nt)}{\sqrt{\pi}} \Big| n \in \mathbb{N} \right\} \qquad (1.80)$$

is an orthonormal, dense subset of $\mathbb{L}^2(-\pi, \pi; \mathbb{R})$ equipped with inner product (1.77). (Here, $\cos(n\cdot)$ means "$\cos(nt), -\pi \leq t \leq \pi$.")

Exercise 1.7.13.
i.) Prove that $\|f\|_{\mathbb{L}^2} = 1, \forall f \in \mathfrak{B}$.
ii.) Prove that $\langle f, g \rangle_{\mathbb{L}^2} = 0, \forall f \neq g \in \mathfrak{B}$.
iii.) How would you adapt the set defined in (1.80) for $\mathbb{L}^2(a, b; \mathbb{R})$, where $a < b$, such that properties (i) and (ii) remain true?

Proposition 1.7.16. (Properties of Orthonormal Sets)
Let $\mathscr{H}$ be an inner product space and $\mathscr{Y} = \{y_1, \ldots, y_n\}$ an orthonormal set in $\mathscr{H}$. Then,
i.) $\left\| \sum_{i=1}^{n} y_i \right\|_{\mathscr{H}}^2 = \sum_{i=1}^{n} \|y_i\|_{\mathscr{H}}^2$;
ii.) *The elements of $\mathscr{Y}$ are linearly independent.*
iii.) *If $x \in \text{span}(\mathscr{Y})$, then $x = \sum_{i=1}^{n} \langle x, y_i \rangle_{\mathscr{H}} y_i$;*
iv.) *If $x \in H$, then $\langle x - \sum_{i=1}^{n} \langle x, y_i \rangle_{\mathscr{H}} y_i, y_k \rangle_{\mathscr{H}} = 0, \forall k \in \{1, \ldots, n\}$.*

Exercise 1.7.14. Prove Prop. 1.7.16.

The following result is the "big deal!"

Theorem 1.7.17. (Representation Theorem for a Hilbert Space)
Let $\mathscr{H}$ be a Hilbert space and $\mathfrak{B} = \{e_n | n \in \mathbb{N}\}$ a basis for $\mathscr{H}$. Then,
i.) *For every $x \in \mathscr{H}$, $\sum_{k=1}^{\infty} |\langle x, e_k \rangle_{\mathscr{H}}|^2 \leq \|x\|_{\mathscr{H}}^2$;*
ii.) $\lim_{N \to \infty} \left\| \sum_{k=1}^{N} \langle x, e_k \rangle_{\mathscr{H}} e_k - x \right\|_{\mathscr{H}} = 0$, *and we write $x = \sum_{k=1}^{\infty} \langle x, e_k \rangle_{\mathscr{H}} e_k$.*

Outline of Proof:
Proof of (i): Observe that $\forall N \in \mathbb{N}$,

$$0 \leq \left\| \sum_{k=1}^{n} \langle x, e_k \rangle_{\mathscr{H}} e_k - x \right\|_{\mathscr{H}}^2$$
$$= \left\langle \sum_{k=1}^{n} \langle x, e_k \rangle_{\mathscr{H}} e_k - x, \sum_{k=1}^{n} \langle x, e_k \rangle_{\mathscr{H}} e_k - x \right\rangle_H$$
$$= \|x\|_H^2 - \sum_{k=1}^{n} |\langle x, e_k \rangle_{\mathscr{H}}|^2.$$

(Why?) Thus, $\sum_{k=1}^{n} |\langle x, e_k \rangle_{\mathscr{H}}|^2 \leq \|x\|_{\mathscr{H}}^2$, $\forall n \in \mathbb{N}$. The result then follows because $\left\{ \sum_{k=1}^{n} |\langle x, e_k \rangle_{\mathscr{H}}|^2 : n \in \mathbb{N} \right\}$ is an increasing sequence bounded above. (Why and so what?)

Proof of (ii): For each $N \in \mathbb{N}$, let $S_N = \sum_{k=1}^{N} \langle x, e_k \rangle_{\mathscr{H}} e_k$. The fact that $\{S_N\}$ is a Cauchy sequence in $\mathscr{H}$ follows from Prop. 1.7.16 and part (i) of this theorem. (How?) Moreover, $\{S_N\}$ must converge since $\mathscr{H}$ is complete. The fact that the limit is x follows from the completeness of $\mathfrak{B}$. (Tell how.) $\square$

Exercise 1.7.15. Provide the details in the proof of Thrm. 1.7.17(ii).

Remark. (Fourier Series)

An important application of Thrm. 1.7.17 occurs in the study of Fourier series. A technique often used to solve elementary partial differential equations is the *method of separation of variables*. This involves identifying a (sufficiently smooth) function with a unique series representation defined using a family of sines and cosines (cf. Section 3.2). To this end, we infer from the example directly following Thrm. 1.7.15 that every $f \in \mathbb{L}^2(-\pi, \pi; \mathbb{R})$ can be expressed uniquely as

$$
\begin{aligned}
f(t) &= \sum_{n=1}^{\infty} \langle f(\cdot), e_n \rangle_{\mathbb{L}^2} \, e_n \\
&= \left\langle f(\cdot), \frac{1}{\sqrt{2\pi}} \right\rangle_{\mathbb{L}^2} \frac{1}{\sqrt{2\pi}} + \sum_{n=1}^{\infty} \left\langle f(\cdot), \frac{\cos(n\cdot)}{\sqrt{\pi}} \right\rangle_{\mathbb{L}^2} \frac{\cos(nt)}{\sqrt{\pi}} \\
&\quad + \sum_{n=1}^{\infty} \left\langle f(\cdot), \frac{\sin(n\cdot)}{\sqrt{\pi}} \right\rangle_{\mathbb{L}^2} \frac{\sin(nt)}{\sqrt{\pi}}, \quad -\pi \le t \le \pi,
\end{aligned}
\tag{1.81}
$$

where the convergence of the series is in the $\mathbb{L}^2$−sense. For brevity, let

$$
a_0 = \left\langle f(\cdot), \frac{1}{\sqrt{2\pi}} \right\rangle_{\mathbb{L}^2} = \frac{1}{\sqrt{2\pi}} \int_{-\pi}^{\pi} f(t) \, dt
\tag{1.82}
$$

$$
a_n = \left\langle f(\cdot), \frac{\cos(n\cdot)}{\sqrt{\pi}} \right\rangle_{\mathbb{L}^2} = \frac{1}{\sqrt{\pi}} \int_{-\pi}^{\pi} f(t) \cos(nt) \, dt, \, n \in \mathbb{N},
\tag{1.83}
$$

$$
b_n = \left\langle f(\cdot), \frac{\sin(n\cdot)}{\sqrt{\pi}} \right\rangle_{\mathbb{L}^2} = \frac{1}{\sqrt{\pi}} \int_{-\pi}^{\pi} f(t) \sin(nt) \, dt, \, n \in \mathbb{N}.
\tag{1.84}
$$

Then, (1.81) can be written as

$$
f(t) = \frac{a_0}{\sqrt{2\pi}} + \sum_{n=1}^{\infty} \left[a_n \frac{\cos(nt)}{\sqrt{\pi}} + b_n \frac{\sin(nt)}{\sqrt{\pi}} \right], \quad -\pi \le t \le \pi.
$$

The utility of this representation will become apparent in Chapter 3. For additional details on Fourier series (see **[121, 246]**).

1.8 Elementary Calculus in Abstract Spaces

Convergent sequences and their properties play a central role in the development of the notions of limits, continuity, the derivative, and the integral. A heuristic discussion is often what is provided in an elementary calculus course, depicting the process visually by appealing to graphs and using sentences of the form, "As x gets closer to a from left or right, quantity A gets closer to quantity B." The intuition gained from such an exposition is helpful, but it needs to be formalized for the purposes of our study.

The plan of this section is to provide the formal definitions of these notions, together with their properties and important main results. The discussion we provide is a terse outline at best, and you are strongly encouraged to review these topics carefully to fill in the gaps (see **[202, 205]**). We cut to the chase and consider <u>abstract</u> functions at the onset since the development is very similar to that of real-valued functions. Of course, the drawback is that the graphical illustrations of these concepts that permeate a presentation of the calculus of real-valued functions is not available for general Banach space-valued functions. Nevertheless, retaining the mental association to the visual interpretation of the concepts is advantageous, by way of analogy.

Throughout the remainder of this chapter, $\mathscr{X}$ and $\mathscr{Y}$ are assumed to be real Banach spaces unless otherwise specified.

1.8.1   Limits

We begin with the extension of the notion of convergence (as defined for sequences) to the function setting.

Definition 1.8.1. A function $f : \mathrm{dom}(f) \subset \mathscr{X} \to \mathscr{Y}$ has *limit L (in $\mathscr{Y}$) at $x = a \in$ $(\mathrm{dom}(f))'$* if for every sequence $\{x_n\} \subset \mathrm{dom}(f)$ for which $\lim\limits_{n \to \infty} \|x_n - a\|_{\mathscr{X}} = 0$, it is the case that $\lim\limits_{n \to \infty} \|f(x_n) - L\|_{\mathscr{Y}} = 0$.

We write $\lim\limits_{x \to a} f(x) = L$ or equivalently, "$\|f(x_n) - L\|_{\mathscr{Y}} \to 0$ as $x \to a$."

Loosely speaking, the interpretation of Def. 1.8.1 for $\mathscr{X} = \mathscr{Y} = \mathbb{R}$ is that as the inputs approach a in any manner possible (i.e., via any sequence in $\mathrm{dom}(f)$ convergent to a), the corresponding functional values approach L. The benefit of this particular definition is that the limit rules follow easily from the corresponding sequence properties.

Exercise 1.8.1. Formulate and prove extensions of Prop. 1.5.5 - Prop. 1.5.7 and Prop. 1.5.9 to the present function setting.

An alternate definition equivalent to Def. 1.8.1, which is often more convenient to work with when involving certain norm estimates in an argument, is as follows:

Definition 1.8.2. A function $f : \mathrm{dom}(f) \subset \mathscr{X} \to \mathscr{Y}$ has *limit L (in $\mathscr{Y}$) at $x = a \in$ $(\mathrm{dom}(f))'$* if $\forall \varepsilon > 0, \exists \delta > 0$ for which

$$x \in \mathrm{dom}(f) \text{ and } 0 < \|x - a\|_{\mathscr{X}} < \delta \implies \|f(x) - L\|_{\mathscr{Y}} < \varepsilon. \qquad (1.85)$$

Remark. Interpreting Def. 1.8.2 verbally, we have $\lim\limits_{x \to a} f(x) = L$ provided that a is a limit point of $\mathrm{dom}(f)$ (so that points of the domain crowd against a) and for any $\varepsilon > 0$, given any open ball $\mathfrak{B}_{\mathscr{Y}}(L; \varepsilon)$ around L, there is some sufficiently small so-called "deleted" open ball $\mathfrak{B}_{\mathscr{X}}(a; \delta) \setminus \{a\}$ around a such that all members of this ball have images $f(x) \in \mathfrak{B}_{\mathscr{Y}}(L; \varepsilon)$. That is, $\forall \varepsilon > 0, \exists \delta > 0$ such that

$$f\left(\mathrm{dom}(f) \cap [\mathfrak{B}_{\mathscr{X}}(a; \delta) \setminus \{a\}]\right) \subset \mathfrak{B}_{\mathscr{Y}}(L; \varepsilon). \qquad (1.86)$$

The special case when $\mathscr{X} = \mathscr{Y} = \mathbb{R}$ shall arise often in our discussion, as will many related situations involving infinity. In particular, we have:

Definition 1.8.3. Let $f : \mathrm{dom}(f) \subset \mathbb{R} \to \mathbb{R}$.
i.) $\lim\limits_{x \to a} f(x) = \infty$ means
    a.) $a \in (\mathrm{dom}(f))'$,
    b.) $\forall M > 0, \exists \delta > 0$ such that

$$x \in \mathrm{dom}(f) \text{ and } 0 < |x - a| < \delta \implies f(x) > M. \tag{1.87}$$

ii.) $\lim\limits_{x \to \infty} f(x) = L$ means
    a.) $\mathrm{dom}(f) \cap (M, \infty) \neq \emptyset, \forall M > 0$,
    b.) $\forall \varepsilon > 0, \exists N > 0$ such that

$$x \in \mathrm{dom}(f) \text{ and } x > N \implies |f(x) - L| < \varepsilon. \tag{1.88}$$

Exercise 1.8.2.
i.) Interpret the terms in Def. 1.8.3 geometrically.
ii.) Formulate analogous definitions when ∞ is replaced by $-\infty$.

The notion of *one-sided limits* for real-valued functions arises occasionally, especially when limits are taken as the inputs approach the endpoints of an interval. Definition 1.8.2 can be naturally modified in such case, with the only changes occurring regarding which inputs near a are considered. (Form such extensions.) We denote the *right-limit at a* by $\lim\limits_{x \to a^+} f(x)$, meaning that all inputs chosen when forming sequences that approach a are comprised of values that are greater than or equal to a. Likewise, we denote the *left-limit at a* by $\lim\limits_{x \to a^-} f(x)$.

1.8.2 Continuity

Understanding the nature of continuous functions is crucial, since much of the work in this text is performed in the space $C(I; X)$. To begin, we need only to slightly modify Defs. 1.8.1 and 1.8.2 to arrive at the following stronger notion of *(norm) continuity*.

Definition 1.8.4. A function $f : \mathrm{dom}(f) \subset \mathscr{X} \to \mathscr{Y}$ is *continuous* at $a \in \mathrm{dom}(f)$ if either of these two equivalent statements hold:
i.) For every sequence $\{x_n\} \subset \mathrm{dom}(f)$ for which $\lim\limits_{n \to \infty} \|x_n - a\|_{\mathscr{X}} = 0$, it is the case that $\lim\limits_{n \to \infty} \|f(x_n) - f(a)\|_{\mathscr{Y}} = 0$. We write $\lim\limits_{x \to a} f(x_n) = f\left(\lim\limits_{n \to \infty} x_n\right) = f(a)$.
ii.) $\forall \varepsilon > 0, \exists \delta > 0$ for which

$$x \in \mathrm{dom}(f) \wedge \|x - a\|_{\mathscr{X}} < \delta \implies \|f(x) - f(a)\|_{\mathscr{Y}} < \varepsilon. \tag{1.89}$$

We say f is *continuous on* $S \subset \mathrm{dom}(f)$ if f is continuous at every element of S.

"Continuity at a" is a strengthening of merely "having a limit at a" because the limit candidate being $f(a)$ requires that a be in the domain of f. It follows from Exer. 1.8.1 that the arithmetic combinations of continuous functions preserve continuity. (Tell how.)

Exercise 1.8.3. Prove that $f : \text{dom}(f) \subset \mathscr{X} \to \mathbb{R}$ defined by $f(x) = \|x\|_{\mathscr{X}}$ is continuous.

More complicated continuous functions can be built by forming compositions of continuous functions, as the following result indicates.

Proposition 1.8.5. *Let $(\mathscr{X}, \|\cdot\|_{\mathscr{X}})$, $(\mathscr{Y}, \|\cdot\|_{\mathscr{Y}})$, and $(\mathscr{Z}, \|\cdot\|_{Z})$ be Banach spaces and suppose that $g : \text{dom}(g) \subset \mathscr{X} \to \mathscr{Y}$ and $f : \text{dom}(f) \subset \mathscr{Y} \to \mathscr{Z}$ with $\text{rng}(g) \subset \text{dom}(f)$. If g is continuous at $a \in \text{dom}(g)$ and f is continuous at $g(a) \in \text{dom}(f)$, then $f \circ g$ is continuous at a. In such case, we write*

$$\lim_{n \to \infty} f\left(g\left(x_n\right)\right) = f\left(\lim_{n \to \infty} g\left(x_n\right)\right) = f\left(g\left(\lim_{n \to \infty} x_n\right)\right) = f(g(a)).$$

Exercise 1.8.4. Prove Prop. 1.8.5 using both formulations of continuity in Def. 1.8.4.

We will frequently consider functions defined on a product space, such as $f : \mathscr{X}_1 \times \mathscr{X}_2 \to \mathscr{Y}$. Interpreting Def. 1.8.4 for such a function requires that we use $\mathscr{X}_1 \times \mathscr{X}_2$ as the space $\mathscr{X}$. This raises the question as to what is meant by the phrases "$\|(x_1,x_2) - (a,b)\|_{\mathscr{X}_1 \times \mathscr{X}_2} < \delta$" or "$\{(x_1^n, x_2^n)\} \to (a,b)$ in $\mathscr{X}_1 \times \mathscr{X}_2$." One typical product space norm is

$$\|(x_1, x_2)\|_{\mathscr{X}_1 \times \mathscr{X}_2} = \|x_1\|_{\mathscr{X}_1} + \|x_2\|_{\mathscr{X}_2}. \tag{1.90}$$

Both conditions can be loosely interpreted by focusing on controlling each of the components of the members of the product space. (Make this precise.)

Different forms of continuity are used in practice. A weaker form of continuity is to require that the function be continuous in only a selection of the input variables and that such "section continuity" hold uniformly for all values of the remaining variables. For instance, saying that a function $f : \mathscr{X}_1 \times \mathscr{X}_2 \to \mathscr{Y}$ is "continuous in $\mathscr{X}_1$ uniformly on $\mathscr{X}_2$" means that for every fixed $x_2 \in \mathscr{X}_2$, the function $g : \mathscr{X}_1 \to \mathscr{Y}$ defined by $g(x_1) = f(x_1, x_2)$ satisfies Def. 1.8.4 with $\mathscr{X} = \mathscr{X}_1$, and that the choice of $x_2 \in \mathscr{X}_2$ does not affect the estimates or convergence of sequences arising in the continuity calculations involving g.

Exercise 1.8.5.
i.) Explain what it means for a function $f : [a,b] \times \mathscr{X} \times \mathscr{X} \to \mathscr{X}$ to be continuous on $\mathscr{X} \times \mathscr{X}$ uniformly on $[a,b]$.
ii.) Explain what it means for a mapping $\Phi : \mathbb{C}([a,b]; \mathscr{X}) \to \mathbb{C}([a,b]; \mathscr{X})$ to be continuous.
iii.) Interpret Def. 1.8.4 for functions of the form $f : [a,b] \to \mathbb{M}^N(\mathbb{R})$.

The notion of continuity for real-valued functions can be modified to give meaning to *left-* and *right-sided continuity* in a manner similar to one-sided limits. All continuity results also hold for one-sided continuity. A function that possesses both left- and right-limits at $x = a$, but for which these limits are different, is said to have a *jump discontinuity* at $x = a$.

Continuous functions enjoy interesting topological properties that lead to some rather strong results concerning boundedness and the existence of fixed-points. We list the essential results below, without proof, for later reference. (See **[16, 202]** for proofs.)

Proposition 1.8.6. (Properties of Continuous Functions)
Assume that $f : \mathrm{dom}(f) \subset \mathscr{X} \to \mathscr{Y}$ is continuous.
i.) *For every open set G in $\mathscr{Y}$, $f^{-1}(G)$ is open in $\mathscr{X}$.*
ii.) *For every compact set K in $\mathscr{X}$, $f(K)$ is compact in $\mathscr{Y}$.*
iii.) *Let $\mathscr{Y} = \mathbb{R}$. If K is a compact set in $\mathscr{X}$, then f is bounded on K and $\exists x_0, y_0 \in K$ such that $f(x_0) = \inf\{f(x) : x \in K\}$ and $f(y_0) = \sup\{f(x) : x \in K\}$.*
iv.) (Intermediate-Value Theorem) *Assume $\mathrm{dom}(f) = [a,b]$ and that $\mathscr{X} = \mathscr{Y} = \mathbb{R}$. If $f(a) \neq f(b)$, then for any z between $f(a)$ and $f(b)$, $\exists c_z \in (a,b)$ such that $f(c_z) = z$.*
v.) *If $f : \mathscr{X} \to \mathscr{Y}$ and $g : \mathscr{X} \to \mathscr{Y}$ are continuous and $f = g$ on a set $\mathscr{D}$ dense in $\mathscr{X}$, then $f = g$ on $\mathscr{X}$.*

Exercise 1.8.6.
i.) If $f : [a,b] \to [a,b]$ is continuous, prove that $\exists c \in [a,b]$ such that $f(c) = c$.
ii.) Show that the conclusion of (i) fails if $[a,b]$ is replaced by a half-open, open, or unbounded interval.

We now define a concept which is stronger than continuity in the sense that for a given $\varepsilon > 0$, there exists a <u>single</u> $\delta > 0$ which "works" for every point in the set. Precisely,

Definition 1.8.7. A function $f : S \subset \mathrm{dom}(f) \subset \mathscr{X} \to \mathscr{Y}$ is *uniformly continuous (UC) on S* provided that $\forall \varepsilon > 0, \exists \delta > 0$ for which

$$x, y \in S \wedge \|x - y\|_{\mathscr{X}} < \delta \implies \|f(x) - f(y)\|_{\mathscr{Y}} < \varepsilon.$$

Remark. The critical feature of uniform continuity on S is that the δ depends on the ε only, and not on the actual points $x, y \in S$ at which we are located. That is, given any $\varepsilon > 0$, there exists $\delta > 0$ such that $\|f(x) - f(y)\|_{\mathscr{Y}} < \varepsilon$, for any pair of points $x, y \in S$ with $\|x - y\|_{\mathscr{X}} < \delta$, no matter where they are located in S. In this sense, uniform continuity of f on S is a "global" property, whereas mere continuity at $a \in S$ is a "local" property.

Example. We claim that $f : (0,1] \to \mathbb{R}$ defined by $f(x) = \frac{1}{x}$ is not UC on $(0,1]$.

To see this, let $0 < \varepsilon < 1$ and suppose that δ is any positive real number. There exists $n \in \mathbb{N}$ such that $\frac{1}{2n} < \delta$. Let $x_0 = \frac{1}{n}$ and $y_0 = \frac{1}{2n}$. Then, x_0 and y_0 are points of $(0,1]$ such that

$$|x_0 - y_0| = \left| \frac{1}{n} - \frac{1}{2n} \right| = \frac{1}{2n} < \delta,$$

but

$$|f(x_0) - f(y_0)| = |n - 2n| = n > \varepsilon.$$

Hence, no choice of δ satisfies the definition of uniform continuity and so, f is not UC on $(0,1]$.

This example illustrates the fact that continuity on S does not imply uniform continuity. However, if S is compact, the implication does hold, as the next result suggests. (See **[16]** for a proof.)

Proposition 1.8.8. *Let* $f : \mathrm{dom}(f) \subset \mathscr{X} \to \mathscr{Y}$ *and* K *a compact subset of* $\mathrm{dom}(f)$. *If* f *is continuous on* K, *then* f *is UC on* K.

Exercise 1.8.7.
i.) If $f : \mathrm{dom}(f) \subset \mathbb{R} \to \mathbb{R}$ be UC on $S \subset \mathrm{dom}(f)$. Prove that the image under f of any Cauchy sequence in S is itself a Cauchy sequence.
ii.) Prove that if f is UC on a bounded set $S \subset \mathrm{dom}(f)$, then f is bounded on S.

The notion of absolute continuity, which involves controlling the total displacement of functional values across small intervals, arises in the definition of certain function spaces.

Definition 1.8.9. A function $f : [a,b] \to \mathbb{R}$ is *absolutely continuous (AC)* on $[a,b]$ if $\forall \varepsilon > 0, \exists \delta > 0$ such that for any finite collection $\{(a_i, b_i) : i = 1, \ldots, n\}$ of pairwise disjoint open subintervals of $[a,b]$ for which $\sum_{k=1}^{n} |b_k - a_k| < \delta$, it is the case that $\sum_{k=1}^{n} |f(b_k) - f(a_k)| < \varepsilon$.

It can be shown that the usual arithmetic combinations of AC functions are also AC, and that AC functions are necessarily continuous. (Try showing this!)

1.8.3   The Derivative

Measuring the rate of change of one quantity with respect to another is central to the formulation and analysis of many mathematical models. The concept is formalized in the real-valued setting via a limiting process of quantities that geometrically resemble slopes of secant lines. We can extend this definition to $\mathscr{X}$−valued functions by making use of the norm on $\mathscr{X}$. This leads to

Definition 1.8.10. A function $f : (a,b) \to \mathscr{X}$ is *differentiable* at $x_0 \in (a,b)$ if there exists a member of $\mathscr{X}$, denoted by $f'(x_0)$, such that

$$\lim_{h \to 0} \left\| \frac{f(x_0 + h) - f(x_0)}{h} - f'(x_0) \right\|_{\mathscr{X}} = 0. \tag{1.91}$$

The number $f'(x_0)$ is called the *derivative of* f *at* x_0. We say f is *differentiable on* S if f is differentiable at every element of S.

Exercise 1.8.8. Interpret Def. 1.8.10 using the formulation of limit given in Def. 1.8.2.

One-sided derivatives for real-valued functions $f : [a,b] \to \mathbb{R}$ are naturally defined using one-sided limits. We write $\frac{d^+}{dx} f(x)|_{x=c}$ to stand for the *right-sided derivative of f at c*, $\frac{d^-}{dx} f(x)|_{x=c}$ for the *left-sided derivative of f at c*, and $\frac{d}{dx} f(x)|_{x=c}$ the *derivative of f at c*, when they exist.

Exercise 1.8.9. Explain how you would show that $\frac{d^+}{dx} f(x)|_{x=c} = \infty$.

The notion of differentiability is more restrictive than continuity, a fact typically illustrated for real-valued functions by examining the behavior of $f(x) = |x|$ at $x = 0$. Indeed, differentiable functions are necessarily continuous, but not vice versa, in the abstract setting of $\mathscr{X}$-valued functions. Further, if the derivative of a function f is itself differentiable, we say f has a second derivative. Such a function has a "higher degree of regularity" than one that is merely differentiable. The pattern continues with each order of derivative, from which the following string of inclusions is derived, $\forall n \in \mathbb{N}$:

$$\mathbb{C}^n(I; \mathscr{X}) \subset \mathbb{C}^{n-1}(I; \mathscr{X}) \subset \ldots \subset \mathbb{C}^1(I; \mathscr{X}) \subset \mathbb{C}(I; \mathscr{X}). \qquad (1.92)$$

Here, inclusion means that a space further to the left in the string is a closed linear subspace of all those occurring to its right.

We shall often work with real-valued differentiable functions. The arithmetic combinations of differentiable functions are again differentiable, although some care must be taken when computing the derivative of a product and composition. The following result provides two especially nice features of real-valued differentiable functions, the first of which is used to establish l'Hopital's rule (see [16]).

Proposition 1.8.11. (Properties of Real-Valued Differentiable Functions)
i.) (Mean Value Theorem) *If $f : [a,b] \to \mathbb{R}$ is differentiable on (a,b) and continuous on $[a,b]$, then $\exists c \in (a,b)$ for which*

$$f(b) - f(a) = f'(c)(b-a). \qquad (1.93)$$

ii.) (Intermediate Value Theorem) *If $f : I \to \mathbb{R}$ is differentiable on $[a,b] \subset I$ and $f'(a) < f'(b)$, then $\forall z \in (f'(a), f'(b))$, $\exists c \in (a,b)$ such that $f'(c) = z$.*

Exercise 1.8.10. Assume that $f : [a,b] \to \mathbb{R}$ is continuous with $f'(x) = 0$, $\forall x \in [a,b]$. Prove that f is constant on $[a,b]$.

Remarks.
1. There are other ways of defining differentiability that are guided by different applications in which such calculations arise. For instance, there are extensions of the notion of a directional derivative, as well a weaker notion of differentiability defined using *distributions*. These topics are treated in [1].

2. We shall be interested in bounded domains $\Omega \subset \mathbb{R}^N$ with a so-called *smooth boundary* $\partial\Omega$. This boundary is necessarily a curve in $\mathbb{R}^N$ and so, by *smooth* we mean that each of the N component functions used to define the curve is differentiable in the sense of Def. 1.8.10.

1.8.4 "The" Integral

Actually, saying "the" integral is misleading since there are many different notions of the integral, all loosely based on the four-step process: (i) partitioning some set, (ii) approximating a quantity of interest, (iii) summing the approximations, and (iv) taking a limit in an appropriate sense. We assume familiarity with the development and properties of the Riemann integral, and focus on the formulation of two abstract extensions of this integral.

The first integral that we define is a natural generalization of the Riemann integral applicable to $\mathscr{X}$-valued functions. A thorough treatment reveals that this integral satisfies the same basic properties as the Riemann integral. Indeed, for a function $f : [a,b] \to \mathscr{X}$, the process used to define $\int_a^b f(x)dx$ is as follows:

<u>Step 1</u> (Partition): Let $n \in \mathbb{N}$. Divide $[a,b]$ into n subintervals using $a = x_0 < x_1 < \ldots < x_{n-1} < x_n = b$. The set $\mathscr{P} = \{x_0, x_1, \ldots, x_n\}$ is a *partition* of $[a,b]$. For convenience, let

$$\triangle x_i = x_i - x_{i-1}, \forall 1 \leq i \leq n,$$
$$\|\mathscr{P}\| = \max\{\triangle x_i : 1 \leq i \leq n\}.$$

<u>Step 2</u> (Approximation): For every $i \in \{1, \ldots, n\}$, choose $x_i^\star \in [x_{i-1}, x_i]$ and form the approximation

$$\underbrace{\underbrace{f(x_i^\star)}_{\text{in } \mathscr{X}} \underbrace{\triangle x_i}_{\text{in } \mathbb{R}}}_{\text{in } \mathscr{X}}. \tag{1.94}$$

<u>Step 3</u> (Sum): Sum the approximations in (1.94) over $i \in \{1, \ldots, n\}$ to obtain

$$\underbrace{\sum_{i=1}^{n} \underbrace{f(x_i^\star)\triangle x_i}_{\text{in } \mathscr{X}}}_{\text{in } \mathscr{X}}. \tag{1.95}$$

<u>Step 4</u> (Limit): Take the limit as $\|\mathscr{P}\| \to 0$ in (1.95) to obtain

$$\underbrace{\lim_{\|\mathscr{P}\| \to 0} \underbrace{\sum_{i=1}^{n} f(x_i^\star)\triangle x_i}_{\text{in } \mathscr{X}}}_{\text{in } \mathscr{X}}. \tag{1.96}$$

Definition 1.8.12. Let $f : [a,b] \to \mathscr{X}$. If the limit in (1.96) exists, then it is a member of $\mathscr{X}$ and we say that f is *integrable* on $[a,b]$. We denote the limiting value by $\int_a^b f(x)dx$ and call it the *integral of f on* $[a,b]$.

The last step in the above construction is delicate. We refer you to **[246]** for details and suffice it to say that the process is well-defined and the following properties hold:

Proposition 1.8.13. *Assume that $f : [a,b] \to \mathscr{X}$ and $g : [a,b] \to \mathscr{X}$ are integrable on $[a,b]$. Then,*
i.) $\int_a^b f(x)dx = -\int_b^a f(x)dx;$
ii.) (Linearity) $\int_a^b [\alpha f(x) + \beta g(x)]\, dx = \alpha \int_a^b f(x)dx + \beta \int_a^b g(x)dx, \forall \alpha, \beta \in \mathbb{R};$
iii.) (Additivity) $\int_a^b f(x)dx = \int_a^c f(x)dx + \int_c^b f(x)dx, \forall c \in [a,b];$
iv.) (Monotonicity) *If* $0 \le \|f(x)\|_{\mathscr{X}} \le \|g(x)\|_{\mathscr{X}}, \forall x \in [a,b]$, *then* $\int_a^b \|f(x)\|_{\mathscr{X}}\, dx \le \int_a^b \|g(x)\|_{\mathscr{X}}\, dx.$ *(Here, the integral is taken in either the classical Riemann sense or the more general Lebesgue sense (see below));*
v.) (Equality) *If* $f(x) = g(x), \forall x \in [a,b]$, *then* $\int_a^b f(x)dx = \int_a^b g(x)dx.$
vi.) *Assume that* $f \in \mathbb{C}([a,b];\mathscr{X})$. *Then,*
  a.) $\int_a^b f(x)dx$ *exists and*

$$\left\| \int_a^b f(x)dx \right\|_{\mathscr{X}} \le \int_a^b \|f(x)\|_{\mathscr{X}}\, dx. \tag{1.97}$$

  b.) *The function $H : [a,b] \to \mathscr{X}$ defined by $H(x) = \int_a^x f(z)dz$ is differentiable and $H'(x) = f(x), \forall x \in (a,b)$.*
vii.) (FTC) *If $f \in \mathbb{C}^1((a,b);\mathscr{X})$, then $\int_x^y f'(z)dz = f(y) - f(x), \forall [x,y] \subset (a,b)$.*

Forming estimates often involves manipulating integrals in different ways, some of which are highlighted in the following exercises.

Exercise 1.8.11. Prove Prop. 1.8.13 (vi)(b) and (vii).

Exercise 1.8.12. Suppse that both $u : [a,b] \to \mathbb{R}$ and $g : [a,b] \times \mathbb{R} \to \mathbb{R}$ are differentiable. Prove that

$$\int_a^t \frac{d}{ds} g(s, u(s))\, ds = g(t, u(t)) - g(a, u(a)).$$

Exercise 1.8.13. Assume that $f \in \mathbb{C}([a,b];[0,\infty))$. Prove that $\forall \varepsilon > 0, \exists \delta > 0$ such that $\int_a^{a+\delta} f(z)dz < \varepsilon.$

Exercise 1.8.14. Assume that $f \in \mathbb{C}([a,b];\mathbb{R})$ and $u \in \mathbb{C}^1((a,b);\mathbb{R})$.
i.) Prove that the function $H : [a,b] \to \mathbb{R}$ defined by $H(x) = \int_a^{u(x)} f(z)dz$ is differentiable on (a,b). Provide a formula for its derivative.
ii.) Does this result necessarily hold if $\mathbb{R}$ is replaced by a Banach space $\mathscr{X}$?

Exercise 1.8.15. Assume that $f, u \in C([a,b]; \mathscr{X})$ and that $g : [a,b] \times \mathscr{X} \to \mathscr{X}$ is a continuous mapping for which there exist positive real numbers M_1 and M_2 such that

$$\|g(s,z)\|_{\mathscr{X}} \le M_1 \|z\|_{\mathscr{X}} + M_2, \ \forall s \in [a,b], z \in \mathscr{X}. \tag{1.98}$$

i.) Prove that the set $\left\{ f(x) + \int_a^x g(z, u(z)) dz : x \in [a,b] \right\}$ is uniformly bounded above (with respect to $\|\cdot\|_{\mathscr{X}}$) and provide an upper bound.
ii.) Let $N \in \mathbb{N}$. Determine upper bounds for these:
  a.) $\left\| f(x) + \int_a^x g(z, u(z)) dz \right\|_{\mathscr{X}}^N$
  b.) $\int_a^b \left\| f(x) + \int_a^x g(z, u(z)) dz \right\|_{\mathscr{X}}^N dx$

The notion of an improper integral in which the interval $[a,b]$ is replaced by $[a, \infty)$ or $(-\infty, b]$ can be defined as in elementary calculus. Indeed, for $f : [a, \infty) \to \mathscr{X}$, we say that

$$\int_a^\infty f(x) dx = \lim_{z \to \infty} \int_a^z f(x) dx, \tag{1.99}$$

provided the limit exists (in $\mathscr{X}$). In such case, we say the integral *converges*. Comparison theorems similar to Prop. 1.5.22 are useful when proving the convergence of such integrals.

Exercise 1.8.16. Suppose that $f : [0, \infty) \times \mathscr{X} \to \mathscr{X}$ is such that

$$\|f(t,z)\|_{\mathscr{X}} \le e^{-t}, \ \forall z \in \mathscr{X}. \tag{1.100}$$

Must $\int_0^\infty \|f(t,z)\|_{\mathscr{X}} dt$ exist, $\forall z \in \mathscr{X}$?

Computing Riemann integrals using changes of variable and integration by parts is familiar. These two "techniques," listed below, follow easily from Prop. 1.8.13(vii).

Proposition 1.8.14. (Methods for Computing Riemann Integrals)
i.) (Integration by Parts) *If $f, g : [a,b] \to \mathbb{R}$ are such that there exist f' and g' which are continuous on $[a,b]$, then fg' and $f'g$ are integrable on (a,b) and*

$$\int_a^b f(x) g'(x) dx = f(b) g(b) - f(a) g(a) - \int_a^b f'(x) g(x) dx. \tag{1.101}$$

ii.) (Change of Variable) *Suppose that $g : [a,b] \to \mathbb{R}$ is such that there exists g' which is continuous on $[a,b]$ and f is continuous on $g([a,b])$. Then,*

$$\int_a^b f(g(x)) g'(x) dx = \int_{g(a)}^{g(b)} f(u) du. \tag{1.102}$$

Both of these results also hold for improper integrals, provided that all integrals involved converge.

Exercise 1.8.17. (Leibniz's Rule) Let $f : [a,b] \times \mathbb{R} \to \mathbb{R}$ be continuously differentiable and assume that $g_1, g_2 : \mathbb{R} \to [a,b]$ are differentiable. Prove the following:

$$\frac{\partial}{\partial x} \int_{g_1(x)}^{g_2(x)} f(t,x)dt = \int_{g_1(x)}^{g_2(x)} \frac{\partial f}{\partial x}(t,x)dt + f(g_2(x),x)\frac{dg_2}{dx} - f(g_1(x),x)\frac{dg_1}{dx}.$$

Two inequalities that are workhorses in applied analysis follow immediately from the fact that $\mathbb{L}^2(a,b;\mathbb{R})$ is a Hilbert space. We state the particular versions of these inequalities for Riemann integrable functions below.

Proposition 1.8.15. *Assume that $f,g : [a,b] \to \mathbb{R}$ are integrable.*
i.) (Holder's inequality)

$$\left| \int_a^b f(x)g(x)dx \right| \le \left[\int_a^b f^2(x)dx \right]^{1/2} \left[\int_a^b g^2(x)dx \right]^{1/2}.$$

More generally,

$$\left| \int_a^b f(x)g(x)dx \right| \le \left[\int_a^b f^p(x)dx \right]^{1/p} \left[\int_a^b g^q(x)dx \right]^{1/q},$$

where $1 \le p,q < \infty$ are such that $\frac{1}{p} + \frac{1}{q} = 1$.
ii.) (Minkowski's inequality)

$$\left[\int_a^b (f(x)+g(x))^2 dx \right]^{1/2} \le \left[\int_a^b f^2(x)dx \right]^{1/2} + \left[\int_a^b g^2(x)dx \right]^{1/2}.$$

Exercise 1.8.18. Complete the following in reference to Prop. 1.8.15.
i.) Prove Prop. 1.8.15 by appealing to the appropriate Hilbert space property as it specifically applies to $\mathbb{L}^2(a,b;\mathbb{R})$.

ii.) Prove that if f is integrable on $[0,T]$, then $\left[\int_0^T f(x)dx \right]^2 \le \sqrt{T} \left[\int_0^T f^2(x)dx \right]^{1/2}$.

We conclude this section with brief commentary on the Lebesgue integral on $\mathbb{R}$. While the Riemann integral possesses some nice characteristics, the space of all Riemann integrable functions on a given interval I is not complete. This fact has some rather unfortunate consequences that can be repaired by considering a more general notion of the integral known as the *Lebesgue integral*. The construction is measure-theoretic in nature, the details of which would rightly constitute another preliminary chapter producing minimal gain regarding the thrust of this text. As such, we are going to relegate much of this technical analysis to behind the scenes and for the purposes of the present volume shall be content with treating the integrals of real-valued functions using a selection of the properties listed in Prop. 1.8.13 with the understanding that they are often really Lebesgue integrals. Many of the convergence statements actually hold *"for almost all x"* (or *"almost everywhere"* (a.e.)) in the underlying set, meaning that the statement holds outside a set of *measure zero* (think of such a set as having total length zero). The following is an important property of Lebesgue integrals that is used frequently in convergence arguments in the text.

Proposition 1.8.16. (Lebesgue Dominated Convergence (LDC))
Let $\{f_n\}$ be a sequence of (Lebesgue) integrable real-valued functions defined on $[a,b]$. Assume that
i.) $\lim_{n\to\infty} f_n(x) = f(x)$, *for almost all* $x \in [a,b]$, *and*
ii.) *there exists a Lebesgue integrable function* $g : [a,b] \to [0,\infty)$ *(called a dominator) for which*

$$|f_n(x)| \leq g(x), \forall n \in \mathbb{N}, for\, almost\, all\, x \in [a,b]. \qquad (1.103)$$

Then, f is Lebesgue integrable and

$$\lim_{n\to\infty} \int_a^b f_n(x)dx = \int_a^b \lim_{n\to\infty} f_n(x)dx = \int_a^b f(x)dx. \qquad (1.104)$$

Refer to **[283]** for a thorough treatment of measure theory and the Lebesgue integral.

Remark. Another property of the Lebesgue integral used heavily in Section 5.6 is its absolute continuity with respect to Lebesgue measure. Intuitively, the property says that the Lebesgue integral is "small" over sets with small Lebesgue measure (think of a union of intervals the sum of whose lengths is small). We use this property in the following very specific manner: If g is Lebesgue integrable on $[0,T]$, then $\forall \varepsilon > 0, \exists t < T$ such that $\int_0^t g(s)ds < \varepsilon$.

1.9 Some Elementary ODEs

Courses on elementary ordinary differential equations (ODEs) are chock full of techniques used to solve particular types of elementary differential equations. Within this vast toolbox are three particular scenarios that play a role in this text. We recall them informally here, along with some elementary exercises, to refresh your memory.

1.9.1 Separation of Variables

An ODE of the form $\frac{dy}{dx} = f(x)g(y)$, where $y = y(x)$, is called *separable* because symbolically the terms involving y can be gathered on one side of the equality and the terms involving x can be written on the other, thereby resulting in the equivalent equation (expressed in differential form) $\frac{1}{g(y)}dy = f(x)dx$. Integrating both sides yields an implicitly-defined function $H(y) = G(x) + C$ that satisfies the original ODE on some set and is called the *general solution* of the ODE. This process can be made formal through the use of appropriate changes of variable.

Exercise 1.9.1. Determine the general solution of these ODEs:
i.) $\frac{dy}{dx} = e^{2x} \csc(\pi y)$,
ii.) $\frac{dy}{dx} = ax^n$, where $n \neq -1$ and $a \in \mathbb{R} \setminus \{0\}$,
iii.) $(1 - y^3) \frac{dy}{dx} = \sum_{i=1}^{N} a_i \sin(b_i x)$, where $a_i, b_i \in \mathbb{R}$ and $n \in \mathbb{N}$.

1.9.2 First-Order Linear ODEs

A *first-order linear ODE* is of the form

$$\frac{dy}{dx} + a(x)y = b(x), \text{ where } y = y(x). \tag{1.105}$$

We shall develop the solution of the initial-value problem (IVP) obtained by coupling (1.105) with the initial condition (IC)

$$y(x_0) = y_0 \tag{1.106}$$

using a simplified version of the so-called *variation of parameters method*.

Step 1: Solve the related homogenous equation $\frac{dy_h}{dx} + a(x)y_h = 0$.
 This equation is separable, so integrating both sides over the interval (x_0, x) yields

$$\frac{dy_h}{y_h} = -a(x)dx \implies \underbrace{\ln|y_h(x)| - \ln|y_h(x_0)|}_{=\ln\left|\frac{y_h(x)}{y_h(x_0)}\right|} = -\int_{x_0}^{x} a(s)ds$$

$$\implies y_h(x) = y_h(x_0)e^{-\int_{x_0}^{x} a(s)ds}. \tag{1.107}$$

Step 2: Determine $C(x)$ for which $y(x) = C(x)y_h(x)$ satisfies (1.105).
 Substitute this function into (1.105) to obtain

$$\frac{d}{dx}[C(x)y_h(x)] + a(x)[C(x)y_h(x)] = b(x)$$

$$C(x)\frac{dy_h}{dx} + y_h(x)\frac{dC}{dx} + a(x)C(x)y_h(x) = b(x)$$

$$C(x)\underbrace{\left[\frac{dy_h}{dx} + a(x)y_h(x)\right]}_{=0 \text{ by Step 1}} + y_h(x)\frac{dC}{dx} = b(x) \tag{1.108}$$

$$\frac{dC}{dx} = \frac{b(x)}{y_h(x)} = b(x)\left[y_h(x_0)e^{-\int_{x_0}^{x} a(s)ds}\right]^{-1}$$

$$C(x) = \int_{x_0}^{x} b(s)y_h^{-1}(x_0)e^{\int_{x_0}^{s} a(t)dt} ds + K,$$

where K is an integration constant.
Step 3: Substitute (1.108) into $y(x) = C(x)y_h(x)$ and apply (1.106) to find the general

solution of the IVP .

$$y(x) = y_h(x_0)e^{-\int_{x_0}^x a(s)ds}\left[\int_{x_0}^x b(s)y_h^{-1}(x_0)e^{\int_{x_0}^s a(t)dt}ds + K\right]$$

$$= \int_{x_0}^x b(s)e^{-\int_x^x a(t)dt}ds + \underbrace{Ky_h(x_0)}\,e^{-\int_{x_0}^x a(t)dt}. \qquad (1.109)$$
$$\text{Call this } \overline{K}$$

Step 4: Apply the IC (1.106) to determine the solution of the IVP.
   Now, apply (1.106) to see that $y(x_0) = 0 + \overline{K} = y_0$. Hence, the solution of the IVP
is

$$y(x) = y_0 e^{-\int_{x_0}^x a(t)dt} + \int_{x_0}^x b(s)e^{-\int_x^x a(t)dt}ds. \qquad (1.110)$$

This formula is called the *variation of parameters formula*.

Exercise 1.9.2. Justify all steps in the derivation of (1.110).

Exercise 1.9.3. Solve the IVP:

$$\begin{cases} \frac{dy}{dx} + \frac{1}{2}y(x) = e^{-3x}, \\ y(0) = \frac{1}{2}. \end{cases}$$

1.9.3   Higher-Order Linear ODEs

   Higher-order linear ODEs with constant coefficients of the form

$$a_n x^{(n)} + a_{n-1}x^{(n-1)} + \ldots + a_1 x' + a_0 x = 0, \qquad (1.111)$$

where $a_i \in \mathbb{R}$, $a_n \neq 0$, and $n \in \mathbb{N}$, arise in Chapters 2 - 5. A more general version of
the procedure outlined in Section 1.9.2 can be used to derive the general solution of
(1.111) (see [95, 96]). We consider the special case

$$ax''(t) + bx'(t) + cx(t) = 0, \qquad (1.112)$$

where $a \neq 0$ and $b, c \in \mathbb{R}$. Assuming that the solution of (1.112) is of the form $x(t) = e^{mt}$ yields

$$e^{mt}\left(am^2 + bm + c\right) = 0 \implies am^2 + bm + c = 0. \qquad (1.113)$$

So, the nature of the solution of (1.112) is completely determined by the values of m.
There are three distinct cases regarding the nature of the solution using (1.113):

Nature of the Roots of (1.113)	General Solution of (1.112)
$m_1 \neq m_2$ (real)	$x(t) = C_1 e^{m_1 t} + C_2 e^{m_2 t}$
$m_1 = m_2$ (real)	$x(t) = C_1 e^{m_1 t} + C_2 t e^{m_1 t}$
$m_1, m_2 = \alpha \pm i\beta$	$x(t) = C_1 e^{\alpha t}\sin(\beta t) + C_2 e^{\alpha t}\cos(\beta t)$

Exercise 1.9.4. For what values of m_1 and m_2 is it guaranteed that
i.) $\lim_{t\to\infty} x(t) = 0$, $\forall C_1, C_2 \in \mathbb{R}$?

ii.) $x(\cdot)$ is a bounded function of t for a given $C_1, C_2 \in \mathbb{R}$?

The variation of parameters method can be extended to solve higher-order nonho-mogenous linear ODEs (that is, when the right-side of (1.111) is not identically zero) as well (see **[95, 96]**).

1.10 Looking Ahead

Armed with some rudimentary tools of analysis, we are ready to embark on our journey through the world of evolution equations. We will begin by investigating some models whose mathematical description involves an extension of (1.105) to vector form. Specifically, while we know that $y(x) = Ce^{ax}$ is the general solution of the ODE $\frac{dy}{dx} = ax$, it is natural to ask what plays the role of the general solution of

the equation $\frac{d\mathbf{Y}}{dx} = \mathbf{AY}$, where $\mathbf{Y} = \begin{bmatrix} y_1 \\ \vdots \\ y_N \end{bmatrix}$ and $\mathbf{A} \in \mathbb{M}^N(\mathbb{R})$. It is tempting to write

$\mathbf{Y}(t) = Ce^{\mathbf{A}t}$, but what does this really mean? Is this defined? What are its properties? We address these questions and much more in Chapter 2.

1.11 Guidance for Exercises

1.11.1 Level 1: A Nudge in a Right Direction

1.1.1 (i) Interpret this verbally. What must happen in order for this NOT to occur?
   (ii) $\forall x$, what happens?
   (iii) $\exists x$ for which what happens?
   (iv) Apply (6) directly.
1.2.1. (i) What is the contrapositive of "$x \in A \Longrightarrow x \in B$"?
   (ii) Start with $x \in (A \cap B) \cup (A \setminus B)$ and use Def. 1.2.1 to show $x \in A$. (Now what?)
   (iii) Use the approach from (ii). If P, Q, and R are statements, then $P \wedge (Q \vee R) \equiv$ __?__.
   (iv) Use the hint for Exer. 1.1.1(i). (How?)
1.2.2. Negate Def. 1.2.1(ii).
1.2.3. (iv) $\bigcup_{i=1}^{n} A_i = \{x \mid \exists i \in \{1, \ldots n\}$ such that __?__$\}$
   (v) This is similar to (iv), but the quantifier changes. (How?)
   (vi) This set must be comprised of what kind of elements?
1.2.4. The proofs of all of these statements are similar. Start with an x in the set on the left-side and, using the defining characteristics of that set, manipulate the expressions

involved to argue that x must belong to the set on the right-side. Sometimes, these implications will reverse (thereby resulting in set equality), while for others they will not.

1.3.1. You must have $\text{dom}(f) = \text{dom}(g)$, and ...

1.3.2. Many example exist. Try $f(x) = 2x$ and $g(x) = x^2$.

1.3.3. Argue these in two stages. For (ii),

$$(f \circ g)(x_1) = (f \circ g)(x_2) \implies f(\underline{g(x_1)}) = f(\underline{g(x_2)}).$$

What must be true about the underlined quantities? (Why? Now what?)

1.3.4. (i) Use the same approach as in Exer. 1.2.4. As an example, we prove (ii)(a):

$$y \in f\left(\bigcup_{\gamma \in \Gamma} X_\gamma\right) \iff \exists x \in \bigcup_{\gamma \in \Gamma} X_\gamma \text{ such that } y = f(x)$$

$$\iff \exists \gamma \in \Gamma \text{ such that } x \in X_\gamma \text{ and } y = f(x)$$

$$\iff \exists \gamma \in \Gamma \text{ such that } y \in f(X_\gamma)$$

$$\iff y \in \bigcup_{\gamma \in \Gamma} f(X_\gamma)$$

(ii) Identify which implications do not reverse.

1.3.5. (i) The arithmetic operation and ordering must work together in order for the arithmetic combination to retain the monotonicity of the functions used to form it. The sign of the output also contributes to the result. (How?)

(ii) If f and g are nondecreasing, then so is $f \circ g$. (Why?) But, something peculiar happens when f and g are nonincreasing? (What?)

1.4.1. (i) Compute $(x+y)^2$. (So what?)

(ii) Use induction with Prop. 1.4.1(vi).

1.4.2. Apply the result: $|x - a| = b$ iff $x - a = \pm b$.

1.4.3. (i) - **(vi)** are immediate consequences of Def. 1.4.3.

(vii) It follows from (i) that $-|x| \le x \le |x|$, $\forall x \in \mathbb{R}$. (So what?)

(viii) Use (vii). (How?)

(ix) Use $|x| = |(x-y)+y| \le |x-y|+|y|$. (Now what?)

(x) Use Prop. 1.4.2.

1.4.4. (i) Note that $\sum_{i=1}^{n} (\alpha x_i - \beta y_i)^2 \ge 0$, $\forall \alpha, \beta \in \mathbb{R}$. (Now what?)

(ii) Use (i).

(iii) Apply Prop. 1.4.4(vii) for the first inequality.

1.4.5. This simply involves reversing the inequalities.

1.4.6. (i) Negate Def. 1.4.5(i).

(ii) Repeat (i) for sets bounded below.

1.4.7. Let T be a nonempty set bounded below. Apply Thrm. 1.4.7 to $S = -T$ and then appropriately reverse the inequalities to get back to T. (How?)

1.4.8. *Proof of (i)*: Argue by contradiction.

Proof of (iv): Use Prop. 1.4.1(iv) and (v) with Def. 1.4.5.

Proof of (v): Why is $\sup S + \sup T + 1 > 0$?

1.4.9. A partially completed table is as follows:

S	$\text{int}(S)$	S'	$\text{cl}_{\mathbb{R}}(S)$	Open?	Closed?
$[1,5]$	$(1,5)$	$[1,5]$	$[1,5]$	No	Yes
$\mathbb{Q}$	$\emptyset$		$\mathbb{R}$		
$\{\frac{1}{n}\|n \in \mathbb{N}\}$		$\{0\}$			No
$\mathbb{R}$	$\mathbb{R}$			Yes	Yes
$\emptyset$					Yes

1.4.10. (i) Let $x \in \bigcap_{i=1}^{n} G_i$. Show $\exists \varepsilon > 0$ such that $(x - \varepsilon, x + \varepsilon) \subset \bigcap_{i=1}^{n} G_i$.

(ii) Use Prop. 1.4.10.

(iii) Use the same approach as in (i).

(iv) Use the same approach as in (ii).

(v) $\exists \varepsilon > 0$ such that $(x - \varepsilon, x + \varepsilon) \subset S \subset T$. (So what?)

(vi) Use Def. 1.4.9(vii).

1.4.11. (i) Apply Thrm. 1.4.7 and Prop. 1.4.8(i). (How?)

(ii) This follows from (i) because $\text{cl}_{\mathbb{R}}(S) = \underline{\quad?\quad}$.

(iii) Implement standard changes involving bounded below and infs.

1.4.12. Think of Exer. 1.4.9.

1.4.13. (i) Any finite set will work. (Why?)

(ii) Think of an unbounded set with "gaps."

(iii) Keep trying...

1.4.14. These all follow from the definitions of T, $\sup T$, and limit point. Completeness is used in (1.12).

1.4.15 & 1.4.16. The only compact sets are (i), $\{\frac{1}{n}\|n \in \mathbb{N}\} \cup \{0\}$, and $[0,1]$. Why are they compact? Why are the others not?

1.5.1. If $\{x_n\|n \in \mathbb{N}\}$ is nondecreasing, then x_1 is a lb $\{x_n\|n \in \mathbb{N}\}$. (Why?) Adapt this for nonincreasing sequences.

1.5.2. (i) Negate Def. 1.5.4.

(ii) Use Def. 1.5.4 directly. (How?)

1.5.3. Let $\varepsilon > 0$. $\exists N \in \mathbb{N}$ such that $|a| < \varepsilon N$. (So what?)

1.5.4. (i) Let $\varepsilon > 0$. $\exists N \in \mathbb{N}$ such that $n \geq N \Longrightarrow |x_n - L| < \frac{\varepsilon}{|c|+1}$. (So what?)

(ii) Apply Prop. 1.5.9(i) and Exer. 1.5.4(i).

1.5.5. $\exists M > 0$ such that $|y_n| \leq M, \forall n \in \mathbb{N}$. How do you control $|x_n y_n|$?

1.5.6. (i) Use Prop. 1.4.4(ix) with Def. 1.5.4.

(ii) Use $x_n = (-1)^n, n \in \mathbb{N}$.

1.5.7. (i) Argue inductively using Prop. 1.5.9(ii).

(ii) Proceed in two cases. First, assume $L \neq 0$. Observe that

$$\sqrt{x_n} - \sqrt{L} = \left(\sqrt{x_n} - \sqrt{L}\right) \frac{\left(\sqrt{x_n} + \sqrt{L}\right)}{\left(\sqrt{x_n} + \sqrt{L}\right)} = \frac{x_n - L}{\sqrt{x_n} + \sqrt{L}}.$$

Use this with Prop. 1.5.8. (How?) If $L = 0$, then $|x_n| < \varepsilon^2 \Longrightarrow \sqrt{x_n} < \varepsilon$. (Why?)

1.5.8. (i) Proof of $(\Longrightarrow)$: $\forall N \in \mathbb{N}$, show that you can choose

$$x_N \in (S \setminus \{x_1, \ldots, x_{N-1}\}) \cap \left(x - \frac{1}{N}, x + \frac{1}{N}\right).$$

(ii) This is similar to (i), but now you can have a constant sequence. (Why? How does this alter the proof?)

1.5.9. If $\{x_n\}$ is a nonincreasing sequence bounded below, then $\{x_n\}$ converges and $\lim\limits_{n\to\infty} x_n = \inf\{x_n | n \in \mathbb{N}\}$. Now, prove it!

1.5.10. For both, apply Prop. 1.5.13 or its analogous version developed in Exer. 1.5.9.

1.5.11. The main trick is identifying the indices correctly when verifying each implication. Argue $(1) \Longrightarrow (2) \Longrightarrow (3) \Longrightarrow (4) \Longrightarrow (5) \Longrightarrow (1)$ to show all statements are equivalent.

1.5.12. Consider $s_{2m} - s_{2m-1}$, $\forall m \in \mathbb{N}$.

1.5.13. Apply the example directly.

1.5.14. Use Exer. 1.5.10(ii).

1.5.15. Apply Prop. 1.5.24.

1.6.1. Argue in a manner similar to (1.44). You will need to use associativity of addition of real numbers and distributivity of multiplication over addition.

1.6.2. Since this norm measures Euclidean distance, you should expect each of these to be related to circles or spheres, somehow.

1.6.3. (i), (ii), and (iv) follow from elementary radical properties. For (iii), apply Exer. 1.4.4(ii).

1.6.4. Argue as in Exer. 1.4.4(iii).

1.6.5. Most of these follow directly using the commutativity, associativity, and distributivity properties of $\mathbb{R}$. For (iv), use Hint 2 for Exer. 1.6.3. And for (vi), use $z = x - y$.

1.6.6. Revisit Exer. 1.6.2.

1.6.8. The computations go through without incident when replacing $|\cdot|$ by $\|\cdot\|_{\mathbb{R}^N}$.

1.6.9. $\|\mathbf{x}_n - \mathbf{L}\|_{\mathbb{R}^N}^2 = \sum_{i=1}^{N} ((x_i)_n - L_i)^2 \to 0$ as $n \to \infty$ iff what happens?

1.6.10. Argue that $\left| \|\mathbf{x}_m\|_{\mathbb{R}^N} - \left\| \lim\limits_{p\to\infty} \mathbf{x}_p \right\|_{\mathbb{R}^N} \right| \to 0$ as $m \to \infty$.

1.6.11. $\left\| \frac{a\mathbf{x}}{\|\mathbf{x}\|_{\mathbb{R}^N}} \right\|_{\mathbb{R}^N} = \frac{|a| \|\mathbf{x}\|_{\mathbb{R}^N}}{\|\mathbf{x}\|_{\mathbb{R}^N}} = |a|$, $\forall \mathbf{x} \in \mathbb{R}^N \setminus \{0\}$. (Why?)

1.6.12. $\|a\mathbf{x} + b\mathbf{y}\|_{\mathbb{R}^N} \leq |a| \|\mathbf{x}\|_{\mathbb{R}^N} + |b| \|\mathbf{y}\|_{\mathbb{R}^N} \leq \underline{\quad ? \quad}$. (Now what?)

1.6.13. $\left\| \frac{\mathbf{x}}{a\|\mathbf{x}\|_{\mathbb{R}^N}} \right\|_{\mathbb{R}^N}^2 \leq \frac{1}{a^2}$, $\forall \mathbf{x} \in \mathbb{R}^N \setminus \{0\}$. (So what?)

1.6.14. $\left| \langle 2\mathbf{x}, \frac{1}{4}\mathbf{x} \rangle_{\mathbb{R}^2} \right|^{1/2} = \frac{1}{\sqrt{2}} \|\mathbf{x}\|_{\mathbb{R}^2}$. (Why? So what?)

1.6.15. Argue that $\left\| \sum_{n=1}^{p} c_n \mathbf{z} \right\|_{\mathbb{R}^N} \leq \sum_{n=1}^{p} |c_n| \|\mathbf{z}\|_{\mathbb{R}^N}$, $\forall p \in \mathbb{N}$. (Now what?)

1.6.16. Show that $\|\mathbf{f}(t)\|_{\mathbb{R}^2} \leq R^2$, $\forall t \in \mathbb{R}$. (Now what?)

1.6.17. (i) and (iii) hold; this is easily shown because the corresponding properties in $\mathbb{R}$ can be applied entrywise. (ii) rarely holds (Why?). And, (iv) is true, but the bookkeeping is a bit more tedious. (Try showing it.)

1.6.18. **(i)** Solve for λ : $\det(\mathbf{A} - \lambda\mathbf{I}) = \det \begin{bmatrix} a - \lambda & 0 \\ 0 & b - \lambda \end{bmatrix} = 0$.

(ii) $\mathbf{A}^{-1}$ is another diagonal matrix. What are its components?

(iii) Let $\mathbf{B}$ be a diagonal $N \times N$ matrix with nonzero entries $b_{11}, b_{22}, \ldots, b_{NN}$. The eigenvalues of a diagonal matrix are precisely the diagonal entries. (Now what?)

1.6.19. The only additional hitch with which we must contend is the *double* sum, but a moment's thought reveals that this can be expressed as a single sum. (Now what?)

1.6.20. Prove that

$$\left| \sum_{i=1}^{N} \sum_{j=1}^{N} \left| \sum_{r=1}^{N} a_{ir} b_{rj} \right| \le \left(\sum_{i=1}^{N} \sum_{j=1}^{N} |a_{ij}| \right) \left(\sum_{i=1}^{N} \sum_{j=1}^{N} |b_{ij}| \right).$$

1.6.21. Observe that

$$\left\| \begin{bmatrix} e^{-t} & 0 \\ 0 & e^{-2t} \end{bmatrix} \begin{bmatrix} x_1 \\ x_2 \end{bmatrix} \right\|_{\mathbb{R}^2}^2 = \left\| \begin{bmatrix} e^{-t} x_1 \\ e^{-2t} x_2 \end{bmatrix} \right\|_{\mathbb{R}^2}^2 = \left(e^{-t} x_1 \right)^2 + \left(e^{-2t} x_2 \right)^2.$$

1.6.22. Yes, naively because this boils down to entrywise calculation. Prove this formally.

1.6.23. Convergence is entrywise for all of these. So what?

1.6.24. (i) $\|\mathbf{A}_m \mathbf{x} - \mathbf{A} \mathbf{x}\|_{\mathbb{R}^N} = \|(\mathbf{A}_m - \mathbf{A}) \mathbf{x}\|_{\mathbb{R}^N} \le \|\mathbf{A}_m - \mathbf{A}\|_{\mathbb{M}^N} \|\mathbf{x}\|_{\mathbb{R}^N}$. So what?

(ii) This is similar to (i). Tell how.

(iii) Use (i) and (ii) combined with the triangle inequality.

1.7.1. These properties are known for $(\mathbb{R}, |\cdot|)$ and applying them componentwise enables you to argue that $\left(\mathbb{R}^N, \|\cdot\|_{\mathbb{R}^N}\right)$ and $\left(\mathbb{M}^N(\mathbb{R}), \|\cdot\|_{\mathbb{M}^N(\mathbb{R})}\right)$ are linear spaces.

1.7.2. For instance, (a) reads: $\forall \varepsilon > 0, \exists N \in \mathbb{N}$ such that

$$n \ge N \implies \sup_{t \in I} \|f_n(t) - f(t)\|_{\mathscr{X}} < \varepsilon$$

1.7.3. Let $\{f_n\} \subset C([a,b];\mathbb{R})$ be Cauchy. Then, $\{f_n(x)\}$ is Cauchy in $\mathbb{R}$, $\forall x \in [a,b]$. (Why?) Hence, the function $f: [a,b] \to \mathbb{R}$ given by $f(x) = \lim_{n \to \infty} f_n(x)$ is well-defined. (Why? Now what?)

1.7.4. Prop. 1.7.7: $\{s_n\}$ converges in $\mathbb{C}$ iff $\{s_n\}$ is Cauchy in $\mathbb{C}$. (Why?) Now, use a modified version of Exer. 1.5.11 to conclude.

Prop. 1.7.8: Define $s_N = \sum_{k=1}^{N} M_k$. $\{s_N\}$ is Cauchy in $\mathbb{R}$. (Why? So what?)

1.7.5. This is the set of all $z \in C([0,2];\mathbb{R})$ such that $\sup\left\{ |z(x) - x^2| : x \in [0,2] \right\} < 1$. Interpret this geometrically.

1.7.6. Showing the sequence is Cauchy is easy. Note that $\lim_{n \to \infty} \frac{2}{n} = 0$. So, what is the issue?

1.7.7. Let $\{x_n\}$ be Cauchy in $\mathscr{Y}$. Then, $\{x_n\}$ converges in $\mathscr{X}$. (Why?) How do you prove that $\{x_n\}$ actually converges in $\mathscr{Y}$?

1.7.8. Let $\{(x_n, y_n)\}$ be Cauchy in $\mathscr{X} \times \mathscr{Y}$. Can you conclude that $\{x_n\}$ is Cauchy in $\mathscr{X}$ and $\{y_n\}$ is Cauchy in $\mathscr{Y}$? (How?)

1.7.9. Linearity of the integral is a key tool here.

1.7.10. This follows immediately.

1.7.11. (i) $\langle x, ay \rangle = \langle ay, x \rangle$. (Now what?)

(ii) The hypothesis implies that $\langle x - y, z \rangle = 0$, $\forall z \in \mathscr{X}$. Now, choose z appropriately to conclude. (How?)

(iii) $\|ax\|_{\mathscr{X}}^2 = \langle ax, ax \rangle_{\mathscr{X}}$. (So what?)

(iv) Argue as in Prop. 1.6.7.

(v) Apply the hint for (iii) with $x + y$ in place of ax.

(vi) $|\langle x_n, y_n \rangle - \langle x_n, y \rangle + \langle x_n, y \rangle - \langle x, y \rangle| \leq \ldots$ (Now what?)

1.7.12. $\left| \int_a^b f(x)g(x)dx \right| \leq \ldots$ (Now what?)

1.7.13. (i) Compute $\int_{-\pi}^{\pi} f^2(x)dx$, $\forall f \in \mathfrak{B}$, using trigonometric identities as needed.

(ii) Compute $\int_{-\pi}^{\pi} f(x)g(x)dx$ using a change of variable and trigonometric identity.

(iii) First, determine $C \in \mathbb{R}$ such that $\int_a^b C^2 dx = 1$. The other basis elements need to be replaced by

$$\mathfrak{B}^\star = \left\{ c^\star \cos\left(\frac{2n\pi t}{b-a}\right) \middle| n \in \mathbb{N} \right\} \cup \left\{ c^\star \sin\frac{2n\pi t}{b-a} \middle| n \in \mathbb{N} \right\}$$

for an appropriate choice of $c^\star \in \mathbb{R}$ that ensures $\int_a^b c^\star f^2(x)dx = 1$, $\forall f \in \mathfrak{B}^\star$.

1.7.14. (i) $\|\sum_{i=1}^n y_i\|_{\mathscr{H}}^2 = \langle \sum_{i=1}^n y_i, \sum_{i=1}^n y_i \rangle_{\mathscr{H}}$ (Now what?)

(ii) Take the inner product with y_i on both sides of $\alpha_1 y_1 + \ldots + \alpha_n y_n = 0$.

(iii) Take the inner product with y_i on both sides of $\alpha_1 y_1 + \ldots + \alpha_n y_n = x$.

(iv) Simplify using the properties of inner product to get $\langle x, y_i \rangle - \langle x, y_i \rangle = 0$. (Tell how.)

1.7.15. Proof of (i): Use Prop. 1.5.13.

Proof of (ii): Let $\varepsilon > 0$. $\exists N \in \mathbb{N}$ such that

$$m, n \geq N \implies \sum_{k=m}^n |\langle x, e_k \rangle|^2 < \varepsilon.$$

(So what?)

1.8.1. Prop. 1.5.5: If $\exists \lim_{x \to a} f(x)$, then it is unique.

Prop. 1.5.6: If $\exists \lim_{x \to a} f(x)$, then $\exists M > 0$ and $\delta > 0$ such that

$$|f(x)| \leq M, \forall x \in \text{dom}(f) \cap (a - \delta, a + \delta).$$

Prop. 1.5.7: If $f(x) \leq g(x) \leq h(x)$, for all "appropriate x near a" and $\exists \lim_{x \to a} f(x) = \lim_{x \to a} h(x) = L$, then $\exists \lim_{x \to a} g(x) = L$. (Make precise the phrase in quotes!)

Prop. 1.5.9: If $\exists \lim_{x \to a} f(x) = L$ and $\lim_{x \to a} g(x) = M$, then

(i) $\exists \lim_{x \to a} (f(x) + g(x)) = L + M$;

(ii) $\exists \lim_{x \to a} (f(x) \cdot g(x)) = L \cdot M$.

1.8.2. (i) These are formal ways of defining asymptotes. Which is which?

(ii) Certain inequalities will change since the inputs of interest are different. (How?)

1.8.3. Argue as in Exer. 1.6.10.

1.8.4. The proof using Def. 1.8.4(i) is suggested by the string of equalities in the statement of Prop. 1.8.5. Alternatively, using Def. 1.8.4(ii), let $\varepsilon > 0$. Find $\delta > 0$ such that

$$\|x - a\|_{\mathscr{X}} < \delta \implies \|f(g(x)) - f(g(a))\|_{\mathscr{X}} < \varepsilon.$$

1.8.5. (i) Mimic the statement in the paragraph preceding Exer. 1.8.5 with $\mathscr{X}_2 = [a, b]$ and $\mathscr{X}_1 = \mathscr{X} \times \mathscr{X}$.

(ii) It is more intuitive to use Def. 1.8.4(i). (Do so.)

(iii) Interpret this entrywise as suggested by Exer. 1.6.22.

1.8.6. (i) Consider the function $g(x) = f(x) - x$.

(ii) Align the function so that the fixed-point would occur at one of the endpoints that you are now excluding, or use asymptotes to your advantage.

1.8.7. (i) Let $\{x_n\}$ be a Cauchy sequence in S and $\varepsilon > 0$. Prove that $\{f(x_n)\}$ is a Cauchy sequence in $\mathbb{R}$.

(ii) If f is not bounded on S, then $\forall n \in \mathbb{N}$, $\exists x_n \in S$ such that $|f(x_n)| > n$. (Now what?)

1.8.8. For any real sequence $\{x_n\}$ such that $x_n \to 0$, ...

1.8.9. How do you show that a subset of $\mathbb{R}$ is unbounded? Adapt this.

1.8.10. Use Prop. 1.8.11(i) to show that $f(x) = f(a)$, $\forall x \in [a,b]$.

1.8.11. Proof of (vi)(b): $\left\| \frac{\int_a^{x+h} f(z)dz - \int_a^x f(z)dz}{h} - f(x) \right\|_{\mathscr{X}} = \underline{\ ?\ }$

Proof of (vii): We infer from (vi)(b) that $\frac{d}{dx}\left[\int_c^x f'(s)ds - f(x)\right] = 0$, $\forall x$. (So what?)

1.8.12. Use the function $F(s) = g(s, u(s))$ and a previous result.

1.8.13. If f is continuous on $[a,b]$, then f is bounded. Use Prop. 1.8.13(iv). (How?)

1.8.14. (i) Define H as a composition of two functions. Then, use the chain rule.

(ii) Remember, the final expression must be defined in the space.

1.8.15. (i) f is continuous on a compact set. (So what?) Also, use Prop. 1.8.13(iv) and (vi)(a). (How?)

(ii) Use the upper bound from (i).

(iii) Use the upper bound from (ii).

1.8.16. Compute $\int_0^\infty e^{-t}dt$. (Now what?)

1.8.17. Use integration by parts.

1.8.18. (i) Use the Cauchy-Schwarz inequality for (i) and triangle inequality for (ii).

(ii) Apply Prop. 1.8.15(i) with $g(x) = 1$.

1.9.1. (i) $\sin(\pi y)\,dy = e^{2x}dx$ ... Now, integrate.

(ii) $dy = ax^n dx$ ... Now, integrate.

(iii) $\left(1 - y^3\right)dy = \sum_{i=1}^n a_i \sin(b_i x)\,dx$ ... Now, integrate.

1.9.2. Use linearity and additivity of the integral.

1.9.3. Apply (1.110) directly with $y_0 = \frac{1}{2}$, $x_0 = 0$, $a(x) = \frac{1}{2}$, and $b(x) = e^{-3x}$.

1.9.4. (i) All exponential terms must go to zero.

(ii) Consider the case in which the roots are complex.

1.11.2 Level 2: An Additional Thrust in a Right Direction

1.1.1. (i) Use the fact that $\neg(P \wedge Q)$ is equivalent to $(\neg P) \vee (\neg Q)$. Interpret this.

(iv) Use (i) and an equivalent form of $\neg(P \vee Q)$ similar to (i).

1.2.1. (i) Use the contrapositive for both implications.

(ii) For the reverse inclusion, begin with $x \in A$. To which of the two sets on the right-side of the equality must x belong? (So what?)

(iii) $\ldots (P \wedge Q) \vee (P \wedge R)$. Similar reasoning applied to $P \vee (Q \wedge R)$ can be used to verify the related distributive law.

(iv) Negate $P \lor Q$.

1.2.2. This boils down to arguing that either A is not a subset of B, or vice versa.

1.2.3. (iv) Fill in the blank with "$x \in A_i$."

(v) $\bigcap_{i=1}^{n} A_i = \{x \mid x \in A_i, \forall i \in \{1, \ldots, n\}\}$

(vi) $\{(x_1, \ldots, x_n) \mid x_i \in A_i, \forall i \in \{1, \ldots, n\}\}$

1.2.4. As an example, we prove (ii):

$$
\begin{aligned}
x \in \left(\bigcup_{\gamma \in \Gamma} A_\gamma \right)^{\sim} &\iff x \notin \bigcup_{\gamma \in \Gamma} A_\gamma \\
&\iff \neg \left(\exists \gamma \in \Gamma \text{ such that } x \in A_\gamma \right) \\
&\iff \forall \gamma \in \Gamma, x \notin A_\gamma \\
&\iff \forall \gamma \in \Gamma, x \in \widetilde{A_\gamma} \\
&\iff x \in \bigcap_{\gamma \in \Gamma} \widetilde{A_\gamma}.
\end{aligned}
$$

1.3.1. ... and $f(x) = g(x), \forall x \in \text{dom}(f) = \text{dom}(g)$.

1.3.3. Continuing, we have $g(x_1) = g(x_2) \implies x_1 = x_2$, where the facts that f and g are one-to-one were used (in that order). The proof of (i) is similar.

1.3.4. (ii) Try using one-to-one for one of them, and onto for the other.

1.3.5. (i) The sum is the only one for which this holds. The product would have worked if the range were restricted to $(0, \infty)$. (Why?) Why don't the others work?

(ii) If f and g are both nonincreasing, then

$$
x_1 < x_2 \implies g(x_1) > g(x_2) \implies f(g(x_1)) < f(g(x_2)).
$$

(So what?)

1.4.1. (i) Note that $(x+y)^2 \geq 0$. Expand the left-side.

(ii) At the inductive step, use $(x^n < y^n) \land (0 < x < y) \implies x^n x < y^n y$. (Now what?)

1.4.3. (vii) Apply this to both x and y and add. (Now what?) Alternatively, expand $(x+y)^2$ and apply (i), (v), and (vi).

(ix) Apply this to y also and subtract. (Now what?)

1.4.4. (i) Expand the expression and choose α, β appropriately. (How?)

(iii) For the second inequality, use Exer. 1.4.1(i) with

$$
(a+b)^N = \sum_{k=0}^{N} \binom{N}{k} a^k b^{N-k}.
$$

1.4.6. (i) Show that $\forall M > 0, \exists x \in S$ such that $x > M$.

(ii) Adapt (i) appropriately.

1.4.7. Now, argue similarly to show (1.3)$\implies$ Thrm. 1.4.7.

1.4.8. Proof of (i): If the conclusion does not hold, then $\sup(S) - \varepsilon$ is an ub(S). Why is this a contradiction?

Proof of (v): Now use Prop. 1.4.8(i) with the number $\zeta = \frac{\varepsilon}{\sup(S) + \sup(T) + 1}$ and mimic the argument of Prop. 1.4.8(iii). (Tell how.)

Regarding the proofs of the corresponding INF statements, all changes are straight-forward and primarily involve inequality reversals and the appropriate modification of Prop. 1.4.8(i). (Supply the details.)

1.4.9. Make certain to supply the details.

S	$\mathrm{int}(S)$	S'	$\mathrm{cl}_{\mathbb{R}}(S)$	Open?	Closed?
$[1,5]$	$(1,5)$	$[1,5]$	$[1,5]$	No	Yes
$\mathbb{Q}$	$\varnothing$	$\mathbb{R}$	$\mathbb{R}$	No	No
$\left\{\frac{1}{n} \mid n \in \mathbb{N}\right\}$	$\varnothing$	$\{0\}$	$\left\{\frac{1}{n} \mid n \in \mathbb{N}\right\} \cup \{0\}$	No	No
$\mathbb{R}$	$\mathbb{R}$	$\mathbb{R}$	$\mathbb{R}$	Yes	Yes
$\varnothing$	$\varnothing$	$\varnothing$	$\varnothing$	Yes	Yes

1.4.10. (i) For every $i \in \{1,\ldots,n\}$, $\exists \varepsilon_i > 0$ such that $(x - \varepsilon_i, x + \varepsilon_i) \subset G_i$. (So what?)

(**ii**) Since $\tilde{F}_1, \ldots, \tilde{F}_n$ are open, (i) $\Longrightarrow \bigcap_{i=1}^{n} \tilde{F}_i$ is open. Now apply Exer. 1.2.4 (ii).

(**iii**) $\exists \gamma_0 \in \Gamma$ such that $x \in G_{\gamma_0}$, and since G_{γ_0} is open, $\exists \varepsilon > 0$ such that $(x - \varepsilon, x + \varepsilon) \subset G_{\gamma_0}$. Now, use Exer. 1.2.4 (iv).

1.4.11. (i) Note that $\forall \varepsilon > 0$, $(\sup(S) - \varepsilon, \sup(S) + \varepsilon) \cap S \neq \varnothing$. Also, $\sup(S)$ need not be in S, but it must be in S'? (Why?)

(**ii**) $\mathrm{cl}_{\mathbb{R}}(S) = S$

1.4.12. Some possibilities are $\mathbb{Q}$, $\widetilde{\mathbb{Q}}$, $\mathbb{Q} \setminus A$, and $\widetilde{\mathbb{Q}} \setminus A$, where A is any finite subset of $\mathbb{R}$.

1.4.13. (ii) $\mathbb{N}$

(**iii**) This is not possible.

1.4.15 & 1.4.16. Compute the closures of these sets and appeal to Thrm. 1.4.14.

1.5.2. (i) $\exists \varepsilon > 0$ such that no matter what $N \in \mathbb{N}$ is chosen, $\exists n \geq N$ for which $|x_n - L| \geq \varepsilon$.

(**ii**) $|x_n - L| = ||x_n - L| - 0|$. (So what?)

1.5.4. (i) $|cx_n - cL| = |c| |x_n - L| < \frac{\varepsilon |c|}{|c| + 1}$. (So what?)

1.5.5. Let $\varepsilon > 0$. $\exists N \in \mathbb{N}$ such that $n \geq N \Longrightarrow |x_n - 0| < \frac{\varepsilon}{M+1}$. Now, argue as in Exer. 1.5.4(i).

1.5.6. (i) Let $\varepsilon > 0$. $\exists N \in \mathbb{N}$ such that $n \geq N \Longrightarrow ||x_n| - |L|| \leq |x_n - L| < \varepsilon$. (So what?)

1.5.7. (i) $\left(x_n^p \to L^p\right) \wedge \left(x_n \to L\right) \Longrightarrow x_n^p x_n \to L^p L$. (Why?)

(**ii**) Let $\varepsilon > 0$. $\exists M > 0$ and $N_1 \in \mathbb{N}$ such that $n \geq N_1 \Longrightarrow x_n > M$. Also, $\exists N_2 \in \mathbb{N}$ such that $n \geq N_2 \Longrightarrow |x_n - L| < \left(M + \sqrt{L}\right) \varepsilon$. So,

$$n \geq \max\{N_1, N_2\} \Longrightarrow \left| \frac{x_n - L}{\sqrt{x_n} + \sqrt{L}} \right| \leq \left(\frac{1}{M + \sqrt{L}} \right) |x_n - L| < \varepsilon.$$

1.5.8. (i) Proof of ($\Longleftarrow$) : Let $\varepsilon > 0$. $\exists N \in \mathbb{N}$ such that $n \geq N \Longrightarrow x_n \in (x - \varepsilon, x + \varepsilon)$. (So what?)

1.5.9. The proof is very similar to the proof of Prop. 1.5.13, but you need to use the fact that $\exists N \in \mathbb{N}$ such that $x_N < L + \varepsilon$. (Now what?)

1.5.10. The fact that $\lim\limits_{n \to \infty} \frac{a^n}{n!} = 0$ readily follows by applying Prop. 1.5.24 in conjunction with Cor. 1.5.21. (Revisit this when you reach this point.)

1.5.12. Use Thrm. 1.5.18. Find a real number ζ_0 for which $|s_{2^m} - s_{2^{m-1}}| \geq \zeta_0, \forall m \in \mathbb{N}$. Then, how do you conclude?

1.5.13. This series converges for any x such that $|5x+1|^3 < 1$. For such x, the sum is $\frac{c(5x+1)^{3p}}{1-(5x+1)^3}$.

1.5.14. If $x_n > 0$ and $x_n \to 0$, what can you say about $\left\{\frac{1}{x_n}\right\}$? Use this with Cor. 1.5.21.

1.5.15. $\lim\limits_{n \to \infty} \frac{(n+1)^{n+1}}{(n+1)!} \cdot \frac{n!}{n^n} = \lim\limits_{n \to \infty} \left(1 + \frac{1}{n}\right)^n = e.$ (So what?)

1.6.1. For instance,

$$
\begin{aligned}
c\,(\mathbf{x}+\mathbf{y}) &= c\,\langle x_1 + y_1, x_2 + y_2, \ldots, x_N + y_N \rangle \\
&= \langle c\,(x_1 + y_1), c\,(x_2 + y_2), \ldots, c\,(x_N + y_N) \rangle \\
&= \langle cx_1 + cy_1, cx_2 + cy_2, \ldots, cx_N + cy_N \rangle \\
&= \langle cx_1, cx_2, \ldots, cx_N \rangle + \langle cy_1, cy_2, \ldots, cy_N \rangle \\
&= c\mathbf{x} + c\mathbf{y}
\end{aligned}
$$

1.6.2. (i) Open circle with radius ε centered at $(0,0)$.
   (ii) Complement of an open sphere with radius ε centered at $(1,0,0)$.
   (iii) The singleton set $\{\mathbf{x}_0\}$.
1.6.3. Use $\sqrt{z^2} = |z|$ and $\sum_{i=1}^{N} a_i^2 = 0 \Longleftrightarrow a_i = 0, \forall i \in \{1, \ldots, N\}$.
1.6.4. Adapt the hints provided for Exer. 1.4.4(iii).
1.6.5. For instance, the proof of (ii) is:

$$
\begin{aligned}
\langle \mathbf{x}+\mathbf{y}, \mathbf{z} \rangle_{\mathbb{R}^N} &= \sum_{i=1}^{N} (x_i + y_i)\, z_i = \sum_{i=1}^{N} (x_i z_i + y_i z_i) \\
&= \sum_{i=1}^{N} x_i z_i + \sum_{i=1}^{N} y_i z_i = \langle \mathbf{x}, \mathbf{z} \rangle_{\mathbb{R}^N} + \langle \mathbf{y}, \mathbf{z} \rangle_{\mathbb{R}^N}.
\end{aligned}
$$

For (vi), expand $\langle \mathbf{x}, \mathbf{x}-\mathbf{y} \rangle_{\mathbb{R}^N} = \langle \mathbf{y}, \mathbf{x}-\mathbf{y} \rangle_{\mathbb{R}^N}$ to arrive at $\sum_{i=1}^{N} (x_i - y_i)^2 = 0$. (Now what?)

1.6.6. Open circle (or sphere) with radius ε centered at $\mathbf{x}_0$.
1.6.9. $(x_i)_n \to L_i$ as $n \to \infty, \forall i \in \{1, \ldots, N\}$.
1.6.10. Using Prop. 1.6.4(iii) yields

$$
0 \leq \left| \|\mathbf{x}_m - \mathbf{L} + \mathbf{L}\|_{\mathbb{R}^N} - \left\| \lim_{p \to \infty} \mathbf{x}_p \right\|_{\mathbb{R}^N} \right|
$$

$$
\leq \|\mathbf{x}_m - \mathbf{L}\|_{\mathbb{R}^N} + \underbrace{\left| \|\mathbf{L}\|_{\mathbb{R}^N} - \left\| \lim_{p \to \infty} \mathbf{x}_p \right\|_{\mathbb{R}^N} \right|}_{=0} \to 0.
$$

1.6.11. What is the supremum of a singleton set?
1.6.12. Continuing, we conclude that $\eta = |a|\,(\|\mathbf{x}_0\|_{\mathbb{R}^N} + \varepsilon) + |b|\,(\|\mathbf{x}_0\|_{\mathbb{R}^N} + \delta)$ is an

ub($\mathscr{A}$). Completeness ensures $\exists \sup(\mathscr{A})$. In order to prove that $\eta = \sup(\mathscr{A})$, let $0 < \zeta < \eta$. Produce $\mathbf{x} \in \mathcal{B}_{\mathbb{R}^N}(\mathbf{x}_0; \varepsilon)$ and $\mathbf{y} \in \mathcal{B}_{\mathbb{R}^N}(\mathbf{x}_0; \delta)$ such that $\zeta = \|a\mathbf{x} + b\mathbf{y}\|_{\mathbb{R}^N}$.

1.6.13. This is dominated by a geometric series that converges iff $|a| > 1$. (Why?)

1.6.14. $\left|\langle 2\mathbf{x}, \frac{1}{4}\mathbf{x}\rangle_{\mathbb{R}^2}\right|^{1/2} < 1, \forall \mathbf{x} \in \mathcal{B}_{\mathbb{R}^2}\left(\mathbf{0}; \frac{1}{3}\right)$. Use Prop. 1.5.22 to conclude. (How?)

1.6.15. Now, since $|c_n| \|\mathbf{z}\|_{\mathbb{R}^N}, \forall n \in \mathbb{N}$, we see that

$$\left\| \sum_{n=1}^{p} c_n \mathbf{z} \right\|_{\mathbb{R}^N} \leq \sum_{n=1}^{\infty} |c_n| \|\mathbf{z}\|_{\mathbb{R}^N}, \forall p \in \mathbb{N},$$

from which the conclusion follows. (How?)

1.6.16. Thus, R is an ub $\left(\{\|\mathbf{f}(t)\|_{\mathbb{R}^3} : t \in \mathbb{R}\}\right)$. In fact, R is the sup of this set; this can be shown using the continuity of the components of $\mathbf{f}(t)$.

1.6.18. (i) $\lambda = a, b$

(ii) $\mathbf{A}^{-1} = \begin{bmatrix} \frac{1}{a} & 0 \\ 0 & \frac{1}{b} \end{bmatrix}$. Compute the eigenvalues in the same manner as in (i).

(iii) The reciprocals of $b_{11}, b_{22}, \ldots, b_{NN}$ are the eigenvalues of $\mathbf{B}^{-1}$.

1.6.20. Expand both sides and compare the terms. Replace some terms on the left by larger terms on the right to arrive at the right-side provided.

1.6.21. Continuing, we see that $\forall t > 0$,

$$\left(e^{-t}x_1\right)^2 + \left(e^{-2t}x_2\right)^2 = e^{-2t}x_1^2 + e^{-4t}x_2^2$$
$$\leq e^{-2t}\left(x_1^2 + x_2^2\right) + e^{-4t}\left(x_1^2 + x_2^2\right)$$
$$\leq e^{-2t} + e^{-4t}$$
$$\leq 2.$$

(Note that the upper bound you end with depends on which $\mathbb{R}^N$ norm you use.)

1.6.22. Let $\varepsilon > 0$. $\forall i, j \in \{1, 2\}, \exists M_{ij} \in \mathbb{N}$ such that

$$n, m \geq M_{ij} \implies \left|(x_{ij})_n - (x_{ij})_m\right| < \frac{\varepsilon}{4}.$$

How do you use this to argue that $\{\mathbf{A}_m\}$ is a Cauchy sequence in $\mathbb{M}^2(\mathbb{R})$?

1.6.23. If $\mathbf{A}_m \to \mathbf{B}$ in $\mathbb{M}^N(\mathbb{R})$, then $\forall i, j \in \{1, \ldots, N\}$, $(a_{ij})_m \to b_{ij}$ as $m \to \infty$. Apply this to all parts of the exercise.

1.6.24. (i) Use the Squeeze Theorem with the inequality to conclude.

(ii) $\|\mathbf{A}\mathbf{x}_m - \mathbf{A}\mathbf{x}\|_{\mathbb{R}^N} \leq \|\mathbf{A}\|_{\mathbb{M}^N} \|\mathbf{x}_m - \mathbf{x}\|_{\mathbb{R}^N}$. (So what?)

(iii) $\|\mathbf{A}_m\mathbf{x}_m - \mathbf{A}\mathbf{x}\|_{\mathbb{R}^N} \leq \|\mathbf{A}_m\|_{\mathbb{M}^N} \|\mathbf{x}_m - \mathbf{x}\|_{\mathbb{R}^N} + \|\mathbf{A}_m - \mathbf{A}\|_{\mathbb{M}^N} \|\mathbf{x}\|_{\mathbb{R}^N}$. (So what?)

1.7.2. (b) reads: $\forall \varepsilon > 0, \exists N \in \mathbb{N}$ such that

$$n \geq N \implies \int_I |f_n(t) - f(t)|^p \, dt < \varepsilon^p.$$

(c) is formulated similarly. The modifications for Cauchy are obvious.

1.7.3. Observe that $|f_n(x) - f_m(x)| \leq \|f_n - f_m\|_{\mathbb{C}} \to 0$ as $n, m \to \infty$. Argue that $\|f_n - f\|_{\mathbb{C}} \to 0$ as $n \to \infty$.

(ii) The space $\mathscr{X}$ must be complete; otherwise, f would not be well-defined. The argument remains unchanged.

1.7.4. Prop. 1.7.7: Compute $\sup_{x \in S} \left\| s_{N+p}(x) - s_N(x) \right\|_{\mathscr{X}}$. (Now what?)

Prop. 1.7.8: $\sum_{k=n+1}^{n+p} M_k \to 0$ as $n \to \infty$. So, $\forall \varepsilon > 0$, $\exists N \in \mathbb{N}$ such that $n \geq N \implies \sum_{k=n+1}^{n+p} M_k < \varepsilon$. Now, apply Prop. 1.7.7.

1.7.5. Construct a tube centered at the graph of $f(x) = x^2$ on the interval $[0,2]$ by translating copies of the graph of f vertically up 1 unit and down 1 unit to form its boundaries. Any continuous function that remains strictly inside this tube is a member of this ball.

1.7.6. $\left\{ \frac{2}{n} \right\}$ does not converge in the space $\mathscr{Y}$. So, $\mathscr{Y}$ is not complete.

1.7.7. Use an appropriately modified version of Exer. 1.5.8. (How?)

1.7.8. Yes, and in fact, the completeness of the respective spaces implies that $\{x_n\}$ converges in $\mathscr{X}$ and $\{y_n\}$ converges in $\mathscr{Y}$. So, $\{(x_n, y_n)\}$ converges in $\mathscr{X} \times \mathscr{Y}$ using either norm.

1.7.11. (i) Now use Def. 1.7.10(ii), then (i).

    (ii) Choose $z = x - y$ and use Def. 1.7.10(iv).

    (iii) $\langle ax, ax \rangle_{\mathscr{X}} = a \langle x, ax \rangle_{\mathscr{X}} = a \langle ax, x \rangle_{\mathscr{X}} = a^2 \langle x, x \rangle_{\mathscr{X}}$. (Now what?)

    (v) $\ldots = \|x\|_{\mathscr{X}}^2 + 2 |\langle x, y \rangle| + \|y\|_{\mathscr{X}}^2 \leq (\|x\|_{\mathscr{X}} + \|y\|_{\mathscr{X}})^2$. (Now what?)

    (vi) $\ldots = |\langle y_n - y, x_n \rangle + \langle x_n - x, y \rangle| \leq \|y_n - y\|_{\mathscr{X}} \|x_n\|_{\mathscr{X}} + \|x_n - x\|_{\mathscr{X}} \|y\|_{\mathscr{X}}$ (Now what?)

1.7.12. $\ldots \leq \left(\int_a^b f^2(x) dx \right)^{1/2} \left(\int_a^b g^2(x) dx \right)^{1/2}$.

1.7.13. (i) Use a double-angle formula.

    (ii) Use a product-to-sum formula.

    (iii) Use $C = \frac{1}{\sqrt{b-a}}$ and use a change of variable to find $c^\star$.

1.7.14. (i) Use properties of inner product to arrive at $\sum_{i=1}^{n} \sum_{j=1}^{n} \langle y_i, y_j \rangle_{\mathscr{H}}$. (Now what?)

    (ii) Conclude that $\alpha_i = 0$, $\forall i \in \{1, \ldots, n\}$. (How? So what?)

    (iii) Now, use $\langle x, y_i \rangle = \alpha_i$ in the definition of span$(\mathscr{Y})$ to conclude.

1.7.15. Proof of (ii): To see why the limit is x, use

$$\left\langle x - \sum_{k=1}^{\infty} \langle x, e_k \rangle_{\mathscr{H}} e_k, e_j \right\rangle_{\mathscr{H}} = 0, \forall j \in \mathbb{N}.$$

(So what?)

1.8.1. The proofs mirror those of the corresponding results in the sequence setting with δ playing the role of N and the "tail of the sequence" corresponding to the "deleted neighborhood of a." Keep in mind that if there are several conditions involving different δ neighborhoods of a, then in order to ensure they hold simultaneously, take the MIN of the $\delta's$. (Why?)

1.8.2. (i) Def. 1.8.3(i) means that the graph of f has a vertical asymptote at a, while (ii) implies the existence of a horizontal asymptote.

    (ii) Alternatively, use $|\|x\|_{\mathscr{X}} - \|a\|_{\mathscr{X}}| \leq \|x - a\|_{\mathscr{X}}$ with Def. 1.8.4(ii).

1.8.4. Tackle the implication in two stages. Let $\varepsilon > 0$. First, find $\delta_1 > 0$ such that

$$\|y - g(a)\|_{\mathscr{X}} < \delta_1 \implies \|f(y) - f(g(a))\|_{\mathscr{X}} < \varepsilon.$$

Then, find $\delta_2 > 0$ such that $\|x - a\|_{\mathscr{X}} < \delta_2 \implies \|g(x) - g(a)\|_{\mathscr{X}} < \delta_1$. (Now what?)

1.8.6. (i) Show that at least one of these holds: the sign of $g(x)$ changes at some point within the interval $[a, b]$, $g(a) = 0$, or $g(b) = 0$.

1.8.7. (i) Use Def. 1.8.7 carefully to link control between the two Cauchy sequences.

(ii) Since $\{x_n\}$ is bounded, it contains a convergent subsequence $\{x_{n_k}\}$. Now use part (i). (How? So what?)

1.8.8. Substitute x_n in for h and interpret.

1.8.9. Show that $\forall N \in \mathbb{N}$, $\exists h_n \in \mathbb{R}$ such that $\frac{f(x_o + h_n) - f(x_0)}{h_n} > N$.

1.8.10. Apply the Mean Value Theorem on $[a, x]$, $\forall x \in [a, b]$.

1.8.11. Proof of (vi)(b): Observe that

$$\left\| \frac{\int_a^{x+h} f(z)dz - \int_a^x f(z)dz}{h} - f(x) \right\|_{\mathscr{X}} \leq \max \{ \|f(x) - f(z)\|_{\mathscr{X}} : |x - z| \leq h \}.$$

Also, note that $f(x) = \frac{1}{h} \int_x^{x+h} f(x)ds$.

Proof of (vii): $\int_c^x f'(s)ds = f(x) + K$, where K is a constant. Use $x = c, d$ to conclude.

1.8.12. Can you use Prop. 1.8.13(vii)?

1.8.13. If $|f(x)| \leq M$, $\forall x \in [a, b]$, choose $\delta > 0$ such that $\int_a^{a+\delta} M dx < \varepsilon$. (How?)

1.8.14. (i) $H'(x) = f(u(x)) u'(x)$

(ii) Explain why the formula in (i) is not necessarily valid in a general Banach space $\mathscr{X}$.

1.8.15. (i) $\exists M^\star > 0$ such that $\|f(x)\|_{\mathscr{X}} \leq M^\star$, $\forall x \in [a, b]$. One natural upper bound for the given set is

$$\zeta = M^\star + (b - a)(M_1 M^{\star\star} + M_2),$$

where $M^{\star\star} = \sup \{ \|u(z)\|_{\mathscr{X}} : z \in [a, b] \}$. (Verify this.)

(ii) ζ^N (Why?)

(iii) $\zeta^N (b - a)$ (Why?)

1.8.16. Since $\int_0^\infty \|f(t, z)\|_{\mathscr{X}} \, dt \leq 1$. So, the integral converges by a comparison test.

1.9.1. The integration is standard. Note that the solutions of (i) and (ii) can be solved explicitly for y.

1.9.3. $y(x) = \frac{1}{2} e^{-\frac{1}{2}x} + \int_0^x e^{-3x} e^{-\frac{1}{2}(s-x)} ds$. Now, simplify.

1.9.4. (i) m_1 and m_2 are negative.

(ii) $\alpha \pm i\beta = \pm i\beta$, $\forall \beta > 0$.

Chapter 2

Homogenous Linear Evolution Equations in $\mathbb{R}^N$

Overview

The question of how phenomena evolve over time is central to a broad range of fields within the social, natural, and physical sciences. The behavior of such phenomena is governed by established laws in the underlying field that typically describe the rates at which it and related quantities evolve over time. A precise mathematical description involves the formulation of so-called *evolution equations* whose complexity depends to a large extent on the realism of the model. We focus in this chapter on models in which the evolution equation is generated by a system of ordinary differential equations. We are in search of an abstract paradigm into which all of these models are subsumed as special cases. Once established, we can study the rudimentary properties of the abstract paradigm and subsequently apply the results to each model.

2.1 Motivation by Models

We motivate the theoretical development presented in this chapter with a discussion of some elementary models.

Model I.1 Chemical Kinetics

Chemical substances are transformed via a sequence of reactions into other products. The use of differential equations facilitates the understanding the chemical kinetics of such reactions (see [57, 81, 126, 225]). For instance, consider a first-order reaction $Y \xrightarrow{\alpha_1} Z$ in which a substance Y whose concentration at time t is denoted by $[Y](t)$, is transformed into a product Z at a rate α_1. The rate at which $[Y](t)$ reacts and is converted into Z is described by the initial-value problem (IVP):

$$\begin{cases} \frac{d[Y](t)}{dt} = -\alpha_1 [Y](t), \ t > 0, \\ \frac{d[Z](t)}{dt} = \alpha_1 [Z](t), \ t > 0, \\ [Y](0) = [Y]_0, \ [Z](0) = 0. \end{cases} \qquad (2.1)$$

Equation (2.1) is easily converted into the equivalent matrix form:

$$\begin{cases} \begin{bmatrix} [Y](t) \\ [Z](t) \end{bmatrix}' = \begin{bmatrix} -\alpha_1 & 0 \\ 0 & \alpha_1 \end{bmatrix} \begin{bmatrix} [Y](t) \\ [Z](t) \end{bmatrix}, t > 0, \\ \begin{bmatrix} [Y](0) \\ [Z](0) \end{bmatrix} = \begin{bmatrix} [Y]_0 \\ 0 \end{bmatrix}. \end{cases} \qquad (2.2)$$

A related reversible reaction $Y \xrightarrow{\alpha_1} Z$, $Y \xleftarrow{\alpha_2} Z$ in which part of the product resulting from the forward reaction is converted back into the original substance at a certain rate is described by the IVP

$$\begin{cases} \frac{d[Y](t)}{dt} = -\alpha_1 [Y](t) + \alpha_2 [Z](t), \ t > 0, \\ \frac{d[Z](t)}{dt} = \alpha_1 [Y](t) - \alpha_2 [Z](t), \ t > 0, \\ [Y](0) = [Y]_0, \ [Z](0) = [Z]_0. \end{cases} \qquad (2.3)$$

What are the concentrations of both substances at any time t, and do they approach an equilibrium state as $t \to \infty$?

Model II.1 Pharmacokinetics

The field of pharmacokinetics is concerned with studying the evolution of a substance (e.g., drugs, toxins, nutrients) administered to a living organism (by consumption, inhalation, absorption, etc.) and its effects on the organism. Models can be formed by partitioning portions of the organism into compartments, each of which is treated as a single unit; improvements of these models can be made by refining the compartments in various ways. We consider a rudimentary model motivated by the work discussed in **[120, 252]**. Let

$$y(t) = \text{concentration of the drug in the GI tract,}$$
$$z(t) = \text{concentration of the drug in the blood,}$$
$$a = \text{absorption rate into the bloodstream,}$$
$$b = \text{rate at which the drug is eliminated from the blood.}$$

Note that $a, b > 0$. The following system is obtained based on an elementary "rate in minus rate out" model:

$$\begin{cases} \frac{dy}{dt} = -ay(t), \ t > 0, \\ \frac{dz}{dt} = ay(t) - bz(t), \ t > 0, \\ y(0) = y_0, \ z(0) = 0. \end{cases} \qquad (2.4)$$

The matrix form of system (2.4) is:

$$
\begin{cases}
\begin{bmatrix} y(t) \\ z(t) \end{bmatrix}' = \begin{bmatrix} -a & 0 \\ a & -b \end{bmatrix} \begin{bmatrix} y(t) \\ z(t) \end{bmatrix}, t > 0, \\
\begin{bmatrix} y(0) \\ z(0) \end{bmatrix} = \begin{bmatrix} y_0 \\ 0 \end{bmatrix}.
\end{cases}
\tag{2.5}
$$

At what time is a certain level of the substance reached within each compartment in the body? When is the substance concentration among different compartments in equilibrium? What happens if our measurements of the parameters and/or the initial condition y_0 are a little off; are the resulting solutions drastically different?

Model III.1 Spring-Mass Systems

A second-order ordinary differential equation (ODE) describing the position $x(t)$ at time t of a mass m attached to an oscillating spring with spring constant k with respect to an equilibrium position can be derived using Newton's second law. Initially, let us assume that the movement of the mass is one-dimensional and that there is no damping factor or external driving force. In such case, Hooke's law describes the force acting on the spring, resulting in the IVP:

$$
\begin{cases}
x''(t) + \omega^2 x(t) = 0, \ t > 0, \\
x(0) = x_0, \ x'(0) = x_1,
\end{cases}
\tag{2.6}
$$

where $\omega^2 = \frac{k}{m}$, x_0 is the initial position of the mass with respect to the equilibrium and x_1 is the initial speed. This is referred to as a *harmonic oscillator* (see [137, 152]). This IVP can be converted into a system of first-order ODEs by way of the change of variable

$$
\begin{cases}
y = x, \\
z = x'
\end{cases}
\text{so that}
\begin{cases}
y' = x' = z \\
z' = x'' = -\omega^2 x = -\omega^2 y.
\end{cases}
\tag{2.7}
$$

Then, (2.6) can be written in the equivalent matrix form

$$
\begin{cases}
\begin{bmatrix} y(t) \\ z(t) \end{bmatrix}' = \begin{bmatrix} 0 & 1 \\ -\omega^2 & 0 \end{bmatrix} \begin{bmatrix} y(t) \\ z(t) \end{bmatrix}, t > 0, \\
\begin{bmatrix} y(0) \\ z(0) \end{bmatrix} = \begin{bmatrix} x_0 \\ x_1 \end{bmatrix}.
\end{cases}
\tag{2.8}
$$

Certainly, $z(t)$ is redundant (Why?), but this transformation is useful because it enables us to consider a second-order ODE in the same form as the previous two models.

Next, consider the scenario in which there are two springs, one to which a mass m_A is affixed and the other to which a mass m_B is attached. One of these springs is fastened to the mass of the other to form a system of coupled springs. The IVP governing the positions $x_A(t)$ and $x_B(t)$ of the masses m_A and m_B, respectively, with respect to the equilibrium state is given by

$$\begin{cases} m_A x_A''(t) + (k_A + k_B) x_A(t) + k_A x_B(t) = 0, \ t > 0, \\ m_B x_B''(t) - k_B x_A(t) + k_B x_B(t) = 0, \ t > 0, \\ x_A(0) = x_{0,A}, \ x_A'(0) = x_{1,A}, \\ x_B(0) = x_{0,B}, \ x_B'(0) = x_{1,B}. \end{cases} \tag{2.9}$$

Does the spring-mass system exhibit periodic behavior in any of these cases? Do the values of the initial data have any bearing on this?

Exercise 2.1.1. Convert (2.9) into matrix form.

Common Theme: Although these applications arise in vastly different contexts, the nature of the IVPs used to describe the scenarios and the questions posed are strikingly similar. Specfically, the IVPs can be written compactly in the vector form

$$\begin{cases} \mathbf{U}'(t) = \mathbf{A}\mathbf{U}(t), t > 0, \\ \mathbf{U}(0) = \mathbf{U}_0, \end{cases} \tag{2.10}$$

where $\mathbf{U} : [0, \infty) \to \mathbb{R}^N$ is the unknown function, $\mathbf{A}$ is an $N \times N$ matrix, and $\mathbf{U}_0 \in \mathbb{R}^N$ is the vector containing the initial conditions. (Tell how.) It would be efficient to initially focus our attention on the abstract IVP (2.10) and answer as many rudimentary questions as possible regarding existence and uniqueness of a solution, stability of the solutions with respect to initial data, etc. We could, in turn, apply those results to any model that could be viewed as a special case of (2.10). How do we proceed rigorously?

   Initial attempts at describing population dynamics, determining the half-life of a radioactive substance, and studying other phenomena governed by exponential growth or decay all involve the first-order IVP

$$\begin{cases} u'(t) = au(t), \ t > 0, \\ u(0) = u_0. \end{cases} \tag{2.11}$$

It is easy to verify that $u(t) = e^{at} u_0$ satisfies (2.11) $\forall t \geq 0$ (cf. Section 1.9). Moreover, all information needed to describe the dynamics of $u(t)$ is "contained within" $\{e^{at} : t \geq 0\}$ and its action on u_0. Precisely, if $a < 0$, then $|u(t)| = |e^{at} u_0| \to 0$ as $t \to \infty$, while if $a > 0$, $|u(t)| = |e^{at} u_0| \to \infty$ as $t \to \infty$. If $a = 0$, then $u(t) = u_0$, $\forall t \geq 0$. These are the only options. (Why?)

   The similarity between (2.10) and (2.11) suggests that $\mathbf{U}(t) = e^{\mathbf{A}t} \mathbf{U}_0$ is a likely candidate for a solution of (2.10). This certainly feels right, but at the moment we have done nothing more than symbol matching. Indeed, while the two IVPs "look" the same, the solutions u and $\mathbf{U}$ take values in different spaces ($\mathbb{R}$ and $\mathbb{R}^N$, respectively).

This might seem like a minor difference, but when we compare the two IVPs more closely, we are immediately faced with basic structural questions regarding $e^{\mathbf{A}t}$ since $\mathbf{A}$ is a matrix: Is $e^{\mathbf{A}t}$ itself a matrix, and if so, how do we compute with it? Does the usual calculus apply? Answering such questions requires that we define the quantity $e^{\mathbf{A}t}$ in a manner for which the end result is comparable to e^{at} in that they share the same rudimentary properties, but for which there is no reliance on the geometry inherent to $\mathbb{R}$ (e.g., a graph). Doing so would render (2.11) a special case of (2.10). We make this discussion precise in the next section. Some standard references upon which the subsequent material is based include [**14, 59, 172, 206, 220, 234, 270, 319, 320, 321**].

2.2   The Matrix Exponential

A natural definition of e^{at} that is independent of a geometric context is the Taylor representation $e^{at} = \sum_{k=0}^{\infty} \frac{(at)^k}{k!}$, which converges $\forall t \in \mathbb{R}$. But, can we define $e^{\mathbf{A}t}$ in a similar manner?

Exercise 2.2.1. Explain why for any $N \times N$ matrix $\mathbf{A}$, the finite sum $\sum_{k=0}^{m} \frac{(\mathbf{A}t)^k}{k!}$ is a well-defined $N \times N$ matrix, $\forall m \in \mathbb{N}$.

This is reason for optimism, but we must ensure that $\lim_{m \to \infty} \sum_{k=0}^{m} \frac{(\mathbf{A}t)^k}{k!}$ exists in order for a Taylor series-type definition to be meaningful. This prompts us to appeal to the Banach space $\mathbb{M}^N(\mathbb{R})$ (cf. Section 1.6.2).

Exercise 2.2.2. How would you use the fact that $\mathbb{M}^N(\mathbb{R})$ is a Banach space to conclude that such series are well-defined?

As mentioned in Section 1.8, the single-variable calculus notions of limit, derivative, and integral can be conveniently extended to matrix-valued functions $f : I \subset \mathbb{R} \to \mathbb{M}^N(\mathbb{R})$ "componentwise," meaning that if $f(t) = [a_{ij}(t)], \forall t \in I$, then

$$\lim_{t \to c} f(t) = \left[\lim_{t \to c} a_{ij}(t) \right],$$
$$f'(t) = \left[a'_{ij}(t) \right],$$
$$\int_c^d f(t)dt = \left[\int_c^d a_{ij}(t)dt \right].$$

Loosely speaking, the limit, derivative, and integral of a matrix-valued function exists if and only if the limit, derivative, and integral of each of its entries exists in the one-variable sense. So, we simply apply the one-variable results to each component and apply the matrix norm at the appropriate moment.

The following result enables us to formally define $e^{\mathbf{A}t}$.

Proposition 2.2.1. *For any* $\mathbf{A} \in \mathbb{M}^N(\mathbb{R})$ *and* $t \in \mathbb{R}$, $\lim\limits_{m\to\infty} \sum_{k=0}^{m} \frac{(\mathbf{A}t)^k}{k!}$ *exists.*

Proof. Let $\mathbf{A} \in \mathbb{M}^N(\mathbb{R})$ and $t > 0$ be given. Due to the completeness of $\mathbb{M}^N(\mathbb{R})$, it suffices to show that the sequence $\left\{ \sum_{k=0}^{m} \frac{(\mathbf{A}t)^k}{k!} : m = 0, 1, 2, \ldots \right\}$ is Cauchy in $\mathbb{M}^N(\mathbb{R})$.

Let $\varepsilon > 0$. For any $k \in \mathbb{N}$, $\frac{\left(\|\mathbf{A}\|_{\mathbb{M}^N} t\right)^k}{k!}$ is a real number and

$$\lim_{m\to\infty} \sum_{k=0}^{m} \frac{\left(\|\mathbf{A}\|_{\mathbb{M}^N} t\right)^k}{k!} = e^{\|\mathbf{A}\|_{\mathbb{M}^N} t}.$$

(Why?) Hence, $\exists M \in \mathbb{N}$ such that

$$k, l \geq M \Longrightarrow \sum_{j=k+1}^{l} \frac{\left(\|\mathbf{A}\|_{\mathbb{M}^N} t\right)^j}{j!} < \varepsilon.$$

For such k and l,

$$\left\| \sum_{j=0}^{l} \frac{(\mathbf{A}t)^j}{j!} - \sum_{j=0}^{k} \frac{(\mathbf{A}t)^j}{j!} \right\|_{\mathbb{M}^N} \leq \sum_{j=k+1}^{l} \left\| \frac{(\mathbf{A}t)^j}{j!} \right\|_{\mathbb{M}^N} \leq \sum_{j=k+1}^{l} \frac{\left(\|\mathbf{A}\|_{\mathbb{M}^N} t\right)^j}{j!} < \varepsilon,$$

as desired. $\qquad\square$

Consequently, the following definition is meaningful.

Definition 2.2.2. For any $\mathbf{A} \in \mathbb{M}^N(\mathbb{R})$ and $t \in \mathbb{R}$, the *matrix exponential* $e^{\mathbf{A}t}$ is the unique member of $\mathbb{M}^N(\mathbb{R})$ defined by

$$e^{\mathbf{A}t} = \sum_{k=0}^{\infty} \frac{(\mathbf{A}t)^k}{k!}.$$

Before studying the properties of $e^{\mathbf{A}t}$, we consider some elementary examples illustrating how to compute it. The close relationship between $\mathbf{A}$ and $e^{\mathbf{A}t}$ suggests that it is easier to determine the components of $e^{\mathbf{A}t}$ for matrices $\mathbf{A}$ possessing a simpler structure.

Example. Let $\mathbf{A} = \begin{bmatrix} a & 0 \\ 0 & b \end{bmatrix}$, where $a, b \in \mathbb{R}$. Observe that

$$e^{\begin{bmatrix} a & 0 \\ 0 & b \end{bmatrix} t} = \sum_{k=0}^{\infty} \frac{\left(\begin{bmatrix} a & 0 \\ 0 & b \end{bmatrix} t\right)^k}{k!} = \sum_{k=0}^{\infty} \frac{\begin{bmatrix} at & 0 \\ 0 & bt \end{bmatrix}^k}{k!} = \sum_{k=0}^{\infty} \frac{\begin{bmatrix} (at)^k & 0 \\ 0 & (bt)^k \end{bmatrix}}{k!}$$

$$= \sum_{k=0}^{\infty} \begin{bmatrix} \frac{(at)^k}{k!} & 0 \\ 0 & \frac{(bt)^k}{k!} \end{bmatrix} = \begin{bmatrix} \sum_{k=0}^{\infty} \frac{(at)^k}{k!} & 0 \\ 0 & \sum_{k=0}^{\infty} \frac{(bt)^k}{k!} \end{bmatrix} = \begin{bmatrix} e^{at} & 0 \\ 0 & e^{bt} \end{bmatrix}.$$

Exercise 2.2.3. Compute $e^{\mathbf{A}t}$ where $\mathbf{A}$ is a diagonal $N \times N$ matrix with real diagonal entries $a_{11}, a_{22}, \ldots, a_{NN}$. Justify all steps.

Of course, not every member of $\mathbb{M}^N(\mathbb{R})$ is diagonal. But, many are "close" in the following sense.

Exercise 2.2.4. Let $\mathbf{A} \in \mathbb{M}^N(\mathbb{R})$ and assume that $\exists \mathbf{D}, \mathbf{P} \in \mathbb{M}^N(\mathbb{R})$ such that $\mathbf{D}$ is diagonal, $\mathbf{P}$ is invertible, and $\mathbf{A} = \mathbf{P}^{-1}\mathbf{D}\mathbf{P}$. (In such case, $\mathbf{A}$ is said to be *diagonalizable*.) Show that $\forall t \in \mathbb{R}, e^{\mathbf{A}t} = \mathbf{P}^{-1}e^{\mathbf{D}t}\mathbf{P}$.

Such matrices often arise in practice. For instance, if either $\mathbf{A}$ has n distinct real eigenvalues or $\mathbf{A}$ is symmetric, then $\mathbf{A}$ is diagonalizable. For nondiagonalizable matrices $\mathbf{A}$, even though $e^{\mathbf{A}t}$ exists, computing it can be tedious. In such case, the following *Putzer algorithm* is a useful tool.

Proposition 2.2.3. (Putzer Algorithm)
Let $\mathbf{A} \in \mathbb{M}^N(\mathbb{R})$ *with eigenvalues* $\lambda_1, \lambda_2, \ldots, \lambda_N$ *(including multiplicity). Then,*

$$e^{\mathbf{A}t} = \sum_{k=0}^{N-1} r_{k+1}(t)\mathbf{P}_k,$$

where

$$\begin{cases} \mathbf{P}_0 = \mathbf{I} \\ \mathbf{P}_j = (\mathbf{A} - \lambda_1 \mathbf{I})(\mathbf{A} - \lambda_2 \mathbf{I}) \cdots (\mathbf{A} - \lambda_j \mathbf{I}), \quad j = 1, \ldots, N, \end{cases}$$

and $r_i(t)$ *is the unique solution of the IVP*

$$\begin{cases} r_1'(t) = \lambda_1 r_1(t), \ r_1(0) = 1, \\ r_i'(t) = \lambda_i r_i(t) + r_{i-1}(t), \ r_i(0) = 0, \quad i = 2, \ldots, N. \end{cases} \tag{2.12}$$

The proof of this theorem relies on the Cayley-Hamilton theorem and can be found in **[202]**. The usual variation of parameters method (cf. Section 1.9.3) can be used to determine the solutions of the IVPs in (2.12). We illustrate how to use the algorithm in a simple example.

Example. Let $\mathbf{B} = \begin{bmatrix} \lambda & 1 \\ 0 & \lambda \end{bmatrix}$, where $\lambda \neq 0$. The eigenvalues of $\mathbf{B}$ are $\lambda = \lambda_1 = \lambda_2$. From Prop. 2.2.3, we know that

$$\mathbf{P}_1 = \mathbf{B} - \lambda \mathbf{I} = \begin{bmatrix} 0 & 1 \\ 0 & 0 \end{bmatrix}$$

and

$$e^{\mathbf{B}t} = r_1(t)\begin{bmatrix} 1 & 0 \\ 0 & 1 \end{bmatrix} + r_2(t)\begin{bmatrix} 0 & 1 \\ 0 & 0 \end{bmatrix}. \tag{2.13}$$

We must solve the following two IVPs, in the order presented:

$$\begin{cases} r_1'(t) = \lambda r_1(t), \\ r_1(0) = 1, \end{cases} \tag{2.14}$$

and

$$\begin{cases} r_2'(t) = \lambda r_2(t) + r_1(t), \\ r_2(0) = 0. \end{cases} \tag{2.15}$$

The solution of (2.14) is $r_1(t) = e^{\lambda t}$. Now, substitute this expression into the right-side of (2.15) and solve the resulting equation to obtain $r_2(t) = t e^{\lambda t}$. Finally, substitute both of these into (2.13) to conclude that

$$e^{\mathbf{B}t} = e^{\lambda t}\begin{bmatrix} 1 & t \\ 0 & 1 \end{bmatrix} = \begin{bmatrix} e^{\lambda t} & t e^{\lambda t} \\ 0 & e^{\lambda t} \end{bmatrix}.$$

Exercise 2.2.5. Use Prop. 2.2.3 to calculate $e^{\mathbf{B}t}$ for these choices of $\mathbf{B}$:

i.) $\begin{bmatrix} \alpha & \beta \\ -\beta & \alpha \end{bmatrix}$, where $\alpha \in \mathbb{R}$ and $\beta \neq 0$,

ii.) $\begin{bmatrix} \alpha & 0 \\ \beta & \alpha \end{bmatrix}$, where $\alpha \in \mathbb{R}$ and $\beta \neq 0$.

The following two propositions lay the groundwork for a rigorous development of the properties of $e^{\mathbf{A}t}$.

Proposition 2.2.4. *Let* $\mathbf{A} \in \mathbb{M}^N(\mathbb{R})$ *and define* $\mathscr{B} : \mathrm{dom}(\mathscr{B}) \subset \mathbb{R}^N \to \mathbb{R}^N$ *by* $\mathscr{B}\mathbf{x} = \mathbf{A}\mathbf{x}$.
i.) $\mathrm{dom}(\mathscr{B}) = \mathbb{R}^N$;
ii.) *For every* $\alpha, \beta \in \mathbb{R}$ *and* $\mathbf{x}, \mathbf{y} \in \mathbb{R}^N$, $\mathscr{B}(\alpha\mathbf{x} + \beta\mathbf{y}) = \alpha\mathscr{B}\mathbf{x} + \beta\mathscr{B}\mathbf{y}$;
iii.) *For every* $\mathbf{x} \in \mathbb{R}^N$, $\|\mathscr{B}\mathbf{x}\|_{\mathbb{R}^N} \leq \|\mathbf{A}\|_{\mathbb{M}^N} \|\mathbf{x}\|_{\mathbb{R}^N}$.
(We say that $\mathscr{B}$ *is a bounded linear operator.)*

Proof. (i) holds since the product $\mathbf{A}\mathbf{x}$ is defined, $\forall \mathbf{x} \in \mathbb{R}^N$, and (ii) follows from the properties of matrix multiplication. As for (iii), for simplicity we prove the result for $N = 2$. Let $\mathbf{x} = \begin{bmatrix} x_1 \\ x_2 \end{bmatrix}$. Observe that

$$\begin{aligned} \|\mathbf{A}\mathbf{x}\|_{\mathbb{R}^2} &= \left\| \begin{bmatrix} a_{11} & a_{12} \\ a_{21} & a_{22} \end{bmatrix} \cdot \begin{bmatrix} x_1 \\ x_2 \end{bmatrix} \right\|_{\mathbb{R}^2} = \left\| \begin{bmatrix} a_{11}x_1 + a_{12}x_2 \\ a_{21}x_1 + a_{22}x_2 \end{bmatrix} \right\|_{\mathbb{R}^2} \\ &= |a_{11}x_1 + a_{12}x_2| + |a_{21}x_1 + a_{22}x_2| \\ &\leq (|a_{11}| + |a_{21}|)|x_1| + (|a_{12}| + |a_{22}|)|x_2| \\ &\leq \left(\sum_{i=1}^{2}\sum_{j=1}^{2} |a_{ij}| \right)|x_1| + \left(\sum_{i=1}^{2}\sum_{j=1}^{2} |a_{ij}| \right)|x_2| \\ &= \left(\sum_{i=1}^{2}\sum_{j=1}^{2} |a_{ij}| \right)(|x_1| + |x_2|) \\ &= \|\mathbf{A}\|_{\mathbb{M}^2} \|\mathbf{x}\|_{\mathbb{R}^2}. \end{aligned}$$

(Justify the steps and try proving the general case.) $\square$

In practice, $\mathbf{A}$ and $\mathscr{B}$ are used interchangeably and $\mathbf{A}$ itself is often referred to as a bounded linear operator without going through the formality of defining $\mathscr{B}$. We adopt this convention henceforth.

Exercise 2.2.6. Let $\{\mathbf{x}_k\} \subset \mathbb{R}^N$ be such that $\lim\limits_{k\to\infty} \mathbf{x}_k = \mathbf{x}$ and $\lim\limits_{k\to\infty} \mathbf{A}\mathbf{x}_k = \mathbf{y}$. Prove that $\mathbf{y} = \mathscr{B}\mathbf{x}$.

Proposition 2.2.5. *Let* $\mathbf{A} \in \mathbb{M}^N(\mathbb{R})$ *and* $\forall t \in \mathbb{R}$, *define* $S_t : \mathrm{dom}(S_t) \subset \mathbb{R}^N \to \mathbb{R}^N$ *by*
$S_t\mathbf{x} = e^{\mathbf{A}t}\mathbf{x}$.
i.) $\mathrm{dom}(S_t) = \mathbb{R}^N$;
ii.) *For every* $\alpha, \beta \in \mathbb{R}$ *and* $\mathbf{x}, \mathbf{y} \in \mathbb{R}^N$, $S_t(\alpha\mathbf{x} + \beta\mathbf{y}) = \alpha S_t\mathbf{x} + \beta S_t\mathbf{y}$;
iii.) *For every* $\mathbf{x} \in \mathbb{R}^N$, $\|S_t\mathbf{x}\|_{\mathbb{R}^N} = \left\|e^{\mathbf{A}t}\mathbf{x}\right\|_{\mathbb{R}^N} \le e^{t\|\mathbf{A}\|_{\mathbb{M}^N}}\|\mathbf{x}\|_{\mathbb{R}^N}$.

Proof. (i) and (ii) follow since $e^{\mathbf{A}t} \in \mathbb{M}^N(\mathbb{R})$. As for (iii), Prop. 2.2.4 implies that

$$\left\|e^{\mathbf{A}t}\mathbf{x}\right\|_{\mathbb{R}^N} \le \left\|e^{\mathbf{A}t}\right\|_{\mathbb{M}^N}\|\mathbf{x}\|_{\mathbb{R}^N}. \tag{2.16}$$

Continuing, we see

$$\left\|e^{\mathbf{A}t}\right\|_{\mathbb{M}^N} = \left\|\sum_{k=0}^{\infty}\frac{\mathbf{A}^k t^k}{k!}\right\|_{\mathbb{M}^N} = \left\|\lim_{m\to\infty}\sum_{k=0}^{m}\frac{\mathbf{A}^k t^k}{k!}\right\|_{\mathbb{M}^N} = \lim_{m\to\infty}\left\|\sum_{k=0}^{m}\frac{\mathbf{A}^k t^k}{k!}\right\|_{\mathbb{M}^N}$$

$$\le \lim_{m\to\infty}\sum_{k=0}^{m}\frac{\|\mathbf{A}\|_{\mathbb{M}^N}^k |t|^k}{k!} = \sum_{k=0}^{\infty}\frac{(\|\mathbf{A}\|_{\mathbb{M}^N}|t|)^k}{k!} = e^{|t|\|\mathbf{A}\|_{\mathbb{M}^N}}.$$

(Justify the steps.) Using this estimate in (2.16) yields the result. $\square$

Exercise 2.2.7. Let $\mathbf{A} \in \mathbb{M}^N(\mathbb{R})$ and $t \ge 0$. The family of operators $\{e^{\mathbf{A}t} : t \ge 0\}$ is *contractive* if $\left\|e^{\mathbf{A}t}\mathbf{x}\right\|_{\mathbb{R}^N} \le \|\mathbf{x}\|_{\mathbb{R}^N}, \forall \mathbf{x} \in \mathbb{R}^N$ and $t \ge 0$.
i.) Give an example of $\mathbf{A} \in \mathbb{M}^N(\mathbb{R})$ for which $\{e^{\mathbf{A}t} : t \ge 0\}$ is not contractive.
ii.) Determine a sufficient condition that could be imposed on a diagonal matrix $\mathbf{A} \in \mathbb{M}^N(\mathbb{R})$ to ensure that $\{e^{\mathbf{A}t} : t \ge 0\}$ is contractive.
iii.) Show that a diagonal matrix $\mathbf{A} \in \mathbb{M}^N(\mathbb{R})$ with all negative eigenvalues is such that $\langle \mathbf{A}\mathbf{x}, \mathbf{x}\rangle_{\mathbb{R}^N} \le 0, \forall \mathbf{x} \in \mathbb{R}^N$. Such a matrix operator is called *dissipative*. (More generally, it can be shown that $\{e^{\mathbf{A}t} : t \ge 0\}$ is contractive if and only if $\langle \mathbf{A}\mathbf{x}, \mathbf{x}\rangle_{\mathbb{R}^N} \le 0$, $\forall \mathbf{x} \in \mathbb{R}^N$. We will revisit this notion in Chapter 3.)

Exercise 2.2.8. Let $\alpha \in \mathbb{R}$ and $\mathbf{A} \in \mathbb{M}^N(\mathbb{R})$. Prove that $\forall t \ge 0$,

$$e^{(\alpha\mathbf{A})t}e^{\mathbf{A}t} = e^{\mathbf{A}t}e^{(\alpha\mathbf{A})t}.$$

The matrix exponential obeys the following familiar exponential rules:

Proposition 2.2.6. *Let* $\mathbf{A} \in \mathbb{M}^N(\mathbb{R})$ *and* $t \geq 0$.
i.) $e^{\mathbf{0}} = \mathbf{I}$, *where* $\mathbf{0}$ *is the zero matrix and* $\mathbf{I}$ *is the identity matrix in* $\mathbb{M}^N(\mathbb{R})$;
ii.) *For every* $t, s \geq 0$, $e^{\mathbf{A}(t+s)} = e^{\mathbf{A}t} e^{\mathbf{A}s} = e^{\mathbf{A}s} e^{\mathbf{A}t}$;
iii.) *For every* $\mathbf{A} \in \mathbb{M}^N(\mathbb{R})$ *and* $t \geq 0$, $e^{\mathbf{A}t}$ *is invertible and* $\left(e^{\mathbf{A}t}\right)^{-1} = e^{-\mathbf{A}t}$.

Proof. (i) is easily verified. (ii) holds because

$$
\begin{aligned}
e^{\mathbf{A}(t+s)} &= \sum_{n=0}^{\infty} \frac{\mathbf{A}^n (t+s)^n}{n!} = \sum_{n=0}^{\infty} \frac{\mathbf{A}^n}{n!} \sum_{k=0}^{n} \binom{n}{k} t^k s^{n-k} \\
&= \sum_{n=0}^{\infty} \sum_{k=0}^{n} \frac{\mathbf{A}^n}{n!} \frac{n!}{k!(n-k)!} t^k s^{n-k} = \sum_{n=0}^{\infty} \sum_{k=0}^{n} \frac{\mathbf{A}^k \mathbf{A}^{n-k}}{k!(n-k)!} t^k s^{n-k} \\
&= \sum_{k=0}^{\infty} \sum_{n=k}^{\infty} \frac{\mathbf{A}^k t^k}{k!} \frac{\mathbf{A}^{n-k} s^{n-k}}{(n-k)!} = \left(\sum_{k=0}^{\infty} \frac{\mathbf{A}^k t^k}{k!} \right) \left(\sum_{n=k}^{\infty} \frac{\mathbf{A}^{n-k} s^{n-k}}{(n-k)!} \right) \\
&= \left(\sum_{k=0}^{\infty} \frac{\mathbf{A}^k t^k}{k!} \right) \left(\sum_{n=0}^{\infty} \frac{\mathbf{A}^n s^n}{n!} \right) = e^{\mathbf{A}t} e^{\mathbf{A}s}.
\end{aligned}
$$

(Justify the steps.) Finally, since

$$
\mathbf{I} = e^{\mathbf{0}} = e^{\mathbf{A}t + (-\mathbf{A}t)} = e^{\mathbf{A}t} e^{-\mathbf{A}t} = e^{-\mathbf{A}t} e^{\mathbf{A}t},
$$

we conclude that (iii) holds.   □

Before pressing on with the abstract development, note that the exponential matrix $e^{\mathbf{A}t}$ has been defined and shown to satisfy the usual exponential properties. Keep in mind, though, that our goal from the viewpoint of applications is to define what is meant by a solution of IVP (2.10) in a manner that describes the evolution of the state $\mathbf{U}(t)$. Thinking of the phenomena as processes set in motion at time $t = 0$, we want the behavior of the state $\mathbf{U}(t)$ to be completely governed by the action of the matrices $\{e^{\mathbf{A}t} : t \geq 0\}$ on the initial state $\mathbf{U}_0$ at any time; that is, $\mathbf{U}(t) = e^{\mathbf{A}t} \mathbf{U}_0$. In order for this to be meaningful from a "dynamics" perspective, the family $\{e^{\mathbf{A}t} \mathbf{U}_0 : t \geq 0\}$ should possess the following characteristics:

Dynamical Characteristics:
1. $e^{\mathbf{A}(0)}$ should leave any initial condition unchanged (that is, $e^{\mathbf{A}(0)} \mathbf{U}_0 = \mathbf{U}_0$) since the process has not yet been set into motion.
2. The dynamics of (2.10) should be time invariant in the sense that computing the solution at time $t + s$ in the following two ways should yield the same result. One way, naturally, is simply to compute $\mathbf{U}(t+s) = e^{\mathbf{A}(t+s)} \mathbf{U}_0$. The alternative way is to first determine the state at time t, namely $\mathbf{U}(t) = e^{\mathbf{A}t} \mathbf{U}_0$, and subsequently, using *that* location as a new starting point, determine the state of the system s units of time into

the future. This amounts to computing $e^{\mathbf{A}s}\left(e^{\mathbf{A}t}\mathbf{U}_0\right)$. We summarize this as:

$$
e^{\mathbf{A}s}\left(\underbrace{e^{\mathbf{A}t}\;\underbrace{\mathbf{U}_0}_{\text{original state}}}_{\text{state after }t\text{ units of time}}\right) = e^{\mathbf{A}(t+s)}\;\underbrace{\mathbf{U}_0}_{\text{original state}}
$$
$$
\underbrace{}_{\text{state after an additional }s\text{ units of time}}\qquad \underbrace{}_{\text{state after }t+s\text{ units of time}}
$$

3. Loosely speaking, $\left\{e^{\mathbf{A}t}:t\geq 0\right\}$ should exhibit a certan degree of *continuous dependence* in the sense that for any initial condition $\mathbf{U}_0$, the state $e^{\mathbf{A}(t+\triangle t)}\mathbf{U}_0$ should be relatively close to the state $e^{\mathbf{A}t}\mathbf{U}_0$, provided that the time change $\triangle t$ is sufficiently small.

We have already verified that characteristics (1) and (2) hold. The following proposition addresses (3).

Proposition 2.2.7. (Continuity properties of $e^{\mathbf{A}t}$) *Let $\mathbf{A}\in \mathbb{M}^N(\mathbb{R})$.*
i.) $\lim\limits_{t\to 0^+}\left\|e^{\mathbf{A}t}-\mathbf{I}\right\|_{\mathbb{M}^N}=0;$
ii.) *For every $\mathbf{x}_0\in\mathbb{R}^N$, $\mathbf{g}:[0,\infty)\to\mathbb{R}^N$ defined by $\mathbf{g}(t)=e^{\mathbf{A}t}\mathbf{x}_0$ is continuous;*
iii.) *For every $t_0\geq 0$, $\mathbf{h}:\mathbb{R}^N\to\mathbb{R}^N$ defined by $\mathbf{h}(\mathbf{x})=e^{\mathbf{A}t_0}\mathbf{x}$ is continuous.*

Proof. We prove (i) and leave the verification of (ii) and (iii) as an exercise. Observe that $\forall t\geq 0$,

$$
0\leq\left\|e^{\mathbf{A}t}-\mathbf{I}\right\|_{\mathbb{M}^N}=\left\|\sum_{n=1}^{\infty}\frac{\mathbf{A}^n t^n}{n!}\right\|_{\mathbb{M}^N}\leq\sum_{n=1}^{\infty}\frac{(\|\mathbf{A}\|_{\mathbb{M}^N}t)^n}{n!}=e^{t\|\mathbf{A}\|_{\mathbb{M}^N}}-1.
$$

(Why?) The result follows from Prop. 1.5.7 (as extended to the function setting - cf. Exer. 1.8.1) because $\lim\limits_{t\to 0^+}e^{t\|\mathbf{A}\|_{\mathbb{M}^N}}=1$. $\qquad\square$

Exercise 2.2.9. Which of the three conditions in Prop. 2.2.7 seems to sufficiently address the continuous dependence characteristic (3)?

Exercise 2.2.10. Let $\mathbf{A}\in\mathbb{M}^N(\mathbb{R}),\mathbf{x}\in\mathbb{R}^N$, and $t_0\geq 0$.
i.) Prove that $\lim\limits_{h\to 0^+}\frac{1}{h}\int_{t_0}^{t_0+h}e^{\mathbf{A}s}\mathbf{x}ds=e^{\mathbf{A}t_0}\mathbf{x}$.
ii.) Prove that if $g:[0,\infty)\to\mathbb{R}$ is continuous, then

$$
\lim_{h\to 0^+}\frac{1}{h}\int_{t_0}^{t_0+h}g(s)e^{\mathbf{A}s}\mathbf{x}ds=g(t_0)e^{\mathbf{A}t_0}\mathbf{x}.
$$

The operators $\mathbf{A}$ and $e^{\mathbf{A}t}$ are closely related, as illustrated below.

Exercise 2.2.11. Let $\mathbf{A}=\begin{bmatrix}a & 0\\ 0 & b\end{bmatrix}$, where $a,b\in\mathbb{R}$. Verify that the following hold:

i.) For every $\mathbf{x}_0 \in \mathbb{R}^2$, $\lim_{t \to 0^+} \frac{(e^{\mathbf{A}t} - \mathbf{I})\mathbf{x}_0}{t} = \mathbf{A}\mathbf{x}_0$;

ii.) For every $t \geq 0$, $\mathbf{A}e^{\mathbf{A}t} = e^{\mathbf{A}t}\mathbf{A}$.

In fact, these properties hold $\forall \mathbf{A} \in \mathbb{M}^n(\mathbb{R})$. Indeed, we have:

Proposition 2.2.8. (Connection between A and $e^{\mathbf{A}t}$) *Let* $\mathbf{A} \in \mathbb{M}^N(\mathbb{R})$. *Then,*

i.) *For every* $\mathbf{z} \in \mathbb{R}^N$, $\lim_{h \to 0^+} \frac{(e^{\mathbf{A}h} - \mathbf{I})\mathbf{z}}{h} = \mathbf{A}\mathbf{z}$;

ii.) *For every* $t \geq 0$, $\mathbf{A}e^{\mathbf{A}t} = e^{\mathbf{A}t}\mathbf{A}$.

Proof. (ii) follows from linearity since $\mathbf{A}$ commutes with scalar multiples of $\mathbf{A}$. For (i), let $\mathbf{z} \in \mathbb{R}^N$. Observe that

$$0 \leq \frac{1}{h}\left\|\left(e^{\mathbf{A}h} - \mathbf{I}\right)\mathbf{z} - \mathbf{A}h\mathbf{z}\right\|_{\mathbb{R}^N} = \frac{1}{h}\left\|\sum_{n=2}^{\infty} \frac{\mathbf{A}^n h^n}{n!}\mathbf{z}\right\|_{\mathbb{R}^N}$$

$$\leq \frac{1}{h}\sum_{n=2}^{\infty} \frac{(\|\mathbf{A}\|_{\mathbb{M}^N} h)^n}{n!}\|\mathbf{z}\|_{\mathbb{R}^N} \leq \left[\frac{e^{h\|\mathbf{A}\|_{\mathbb{M}^N}} - 1}{h} - \|\mathbf{A}\|_{\mathbb{M}^N}\right]\|\mathbf{z}\|_{\mathbb{R}^N}$$

(Why?) An application of l'Hopital's rule shows that the right-side goes to zero as $h \to 0^+$, so that the result follows from Prop. 1.5.7 (as extended to the function setting). $\square$

For reasons that will become apparent in Chapter 3, we introduce the following definition.

Definition 2.2.9. Let $\mathbf{A} \in \mathbb{M}^N(\mathbb{R})$. The function $\mathscr{B} : \mathbb{R}^N \to \mathbb{R}^N$ defined by

$$\mathscr{B}\mathbf{z} = \lim_{h \to 0^+} \frac{(e^{\mathbf{A}h} - \mathbf{I})}{h}\mathbf{z}$$

is the *generator* of the family $\{e^{\mathbf{A}t} : t \geq 0\}$.

By Prop. 2.2.8, the operator $\mathscr{B}$ introduced in Prop. 2.2.4 generates $\{e^{\mathbf{A}t} : t \geq 0\}$. For brevity, we say "$\mathbf{A}$ generates $\{e^{\mathbf{A}t} : t \geq 0\}$." As such, to every $\mathbf{A} \in \mathbb{M}^N(\mathbb{R})$ there is associated at least one exponential family. The following theoretical questions naturally present themselves:

1.) Can the operator $\mathscr{B}$ in Prop. 2.2.4 generate more than one exponential family?
2.) Must every exponential family be generated by some $\mathbf{A} \in \mathbb{M}^N(\mathbb{R})$?
3.) Can a given exponential family arise from more than one generator?

Intuition suggests "NO" to (1) and (3) and "YES" to (2), which turns out to be true *in this setting*. But, some surprises lie ahead in more elaborate settings. We shall affirm our intuition regarding (1) and (3) in Chapter 3 in a somewhat more general setting. For now, we simply mention that the generator is important because it provides a direct link to the Cauchy problem (2.10). Making this connection precise requires the following properties.

Proposition 2.2.10. *Let* $\mathbf{A} \in \mathbb{M}^N(\mathbb{R})$ *and* $\mathbf{x} \in \mathbb{R}^N$.
i.) *For every* $\mathbf{x}_0 \in \mathbb{R}^N$, $\mathbf{g}: (0, \infty) \to \mathbb{R}^n$ *defined by* $\mathbf{g}(t) = e^{\mathbf{A}t}\mathbf{x}_0$ *is in* $\mathbb{C}^1\left((0, \infty); \mathbb{R}^N\right)$
and

$$\frac{d}{dt}e^{\mathbf{A}t}\mathbf{x}_0 = \mathbf{A}e^{\mathbf{A}t}\mathbf{x}_0 = e^{\mathbf{A}t}\mathbf{A}\mathbf{x}_0;$$

ii.) *For every* $t_0 \geq 0$ *and* $\mathbf{x}_0 \in \mathbb{R}^N$,

$$\mathbf{A}\int_0^{t_0} e^{\mathbf{A}s}\mathbf{x}_0\,ds = e^{\mathbf{A}t_0}\mathbf{x}_0 - \mathbf{x}_0;$$

iii.) *For every* $0 < s \leq t < \infty$ *and* $\mathbf{x}_0 \in \mathbb{R}^N$,

$$e^{\mathbf{A}t}\mathbf{x}_0 - e^{\mathbf{A}s}\mathbf{x}_0 = \int_s^t e^{\mathbf{A}u}\mathbf{A}\mathbf{x}_0\,du = \int_s^t \mathbf{A}e^{\mathbf{A}u}\mathbf{x}_0\,du.$$

Proof. Let $t_0 > 0$ and consider $0 < h < t_0$ such that $t_0 + h > 0$. Since

$$\frac{e^{\mathbf{A}(t_0+h)}\mathbf{x}_0 - e^{\mathbf{A}t_0}\mathbf{x}_0}{h} = \left(\frac{e^{\mathbf{A}h} - \mathbf{I}}{h}\right)e^{\mathbf{A}t_0}\mathbf{x}_0 = e^{\mathbf{A}t_0}\left(\frac{e^{\mathbf{A}h} - \mathbf{I}}{h}\right)\mathbf{x}_0,$$

we see that

$$\lim_{h\to 0^+}\left\|\frac{e^{\mathbf{A}h}\left(e^{\mathbf{A}t_0}\mathbf{x}_0\right) - \left(e^{\mathbf{A}t_0}\mathbf{x}_0\right)}{h} - e^{\mathbf{A}t_0}\left(\mathbf{A}\mathbf{x}_0\right)\right\|_{\mathbb{R}^N} = \lim_{h\to 0^+}\left\|e^{\mathbf{A}t_0}\left[\frac{e^{\mathbf{A}h}\mathbf{x}_0 - \mathbf{x}_0}{h} - \mathbf{A}\mathbf{x}_0\right]\right\|_{\mathbb{R}^N}$$
$$= 0.$$

Hence, $\lim_{h\to 0^+}\frac{e^{\mathbf{A}h}\left(e^{\mathbf{A}t_0}\mathbf{x}_0\right) - \left(e^{\mathbf{A}t_0}\mathbf{x}_0\right)}{h}$ exists and equals both $\mathbf{A}e^{\mathbf{A}t_0}\mathbf{x}_0$ (by Def. 2.2.9) and
$e^{\mathbf{A}t_0}\mathbf{A}\mathbf{x}_0$ (by Prop. 2.2.8). (Argue similarly for $h < 0$. Tell how.) This establishes (i).
Next, observe that

$$\left(\frac{e^{\mathbf{A}h} - \mathbf{I}}{h}\right)\int_0^{t_0} e^{\mathbf{A}s}\mathbf{x}_0\,ds = \frac{1}{h}\int_0^{t_0}\left[e^{\mathbf{A}(s+h)}\mathbf{x}_0 - e^{\mathbf{A}s}\mathbf{x}_0\right]ds$$
$$= \frac{1}{h}\int_h^{t_0+h} e^{\mathbf{A}u}\mathbf{x}_0\,du - \frac{1}{h}\int_0^{t_0} e^{\mathbf{A}s}\mathbf{x}_0\,ds$$
$$= \frac{1}{h}\int_{t_0}^{t_0+h} e^{\mathbf{A}s}\mathbf{x}_0\,ds - \frac{1}{h}\int_0^h e^{\mathbf{A}s}\mathbf{x}_0\,ds.$$

(Tell why.) Thus, Exer. 2.2.10 implies that

$$\lim_{h\to 0}\left(\frac{e^{\mathbf{A}h} - \mathbf{I}}{h}\right)\int_0^{t_0} e^{\mathbf{A}s}\mathbf{x}_0\,ds = e^{\mathbf{A}t_0}\mathbf{x}_0 - \mathbf{x}_0,$$

thereby proving (ii). Finally, we conclude from (i) that

$$\frac{d}{du}e^{\mathbf{A}u}\mathbf{x}_0 = \mathbf{A}e^{\mathbf{A}u}\mathbf{x}_0 = e^{\mathbf{A}u}\mathbf{A}\mathbf{x}_0,$$

so that integrating from s to t and applying Prop. 1.8.13(vii) yields (iii). $\square$

Exercise 2.2.12. For illustration purposes, verify the results of Prop. 2.2.10 assuming that $\mathbf{A} \in \mathbb{M}^N(\mathbb{R})$ is a diagonal matrix.

We are now in a position to study IVP (2.10).

2.3   The Homogenous Cauchy Problem: Well-Posedness

The IVP (2.10) is referred to as the *homogenous Cauchy problem in* $\mathbb{R}^N$. We seek a solution in the following sense.

Definition 2.3.1. A *classical solution* of (2.10) is a function $\mathbf{U} \in \mathbb{C}\left([0,\infty);\mathbb{R}^N\right) \cap \mathbb{C}^1\left((0,\infty);\mathbb{R}^N\right)$ that satisfies the ODE and IC in (2.10).

Since $\mathbb{C}\left((0,\infty);\mathbb{R}^N\right) \subset \mathbb{C}^1\left((0,\infty);\mathbb{R}^N\right)$, the requirement that $\mathbf{U} \in \mathbb{C}\left([0,\infty);\mathbb{R}^N\right)$ reduces to demanding that $\mathbf{U}$ be right continuous at $t = 0$.

Exercise 2.3.1. Why do you suppose this particular requirement is important?

The properties of $e^{\mathbf{A}t}$ established in Section 2.2 enable us to establish the following result:

Theorem 2.3.2. *For every* $\mathbf{U}_0 \in \mathbb{R}^N$, *the IVP (2.10) has a unique classical solution given by* $\mathbf{U}(t) = e^{\mathbf{A}t}\mathbf{U}_0$.

Proof. Let $\mathbf{U}_0 \in \mathbb{R}^N$.
<u>Existence:</u> We must verify that $\mathbf{U}(t) = e^{\mathbf{A}t}\mathbf{U}_0$ satsifies Def. 2.3.1. To this end, note that Prop. 2.2.7(ii) ensures that $\mathbf{U} \in \mathbb{C}\left([0,\infty);\mathbb{R}^N\right)$, Prop. 2.2.10(i) ensures that $\mathbf{U} \in \mathbb{C}^1\left((0,\infty);\mathbb{R}^N\right)$, and $\mathbf{U}$ satisfies the first equation in (2.10). Finally, Prop. 2.2.6(i) guarantees that $\mathbf{U}$ satisfies the initial condition $\mathbf{U}(0) = \mathbf{U}_0$. This establishes existence.
<u>Uniqueness:</u> Let $\mathbf{V} \colon [0,\infty) \to \mathbb{R}^N$ be a solution to the IVP

$$\begin{cases} \mathbf{V}'(t) = \mathbf{A}\mathbf{V}(t), t > 0, \\ \mathbf{V}(0) = \mathbf{U}_0. \end{cases}$$

Define the function $\Psi \colon [0,\infty) \to \mathbb{R}^N$ by $\Psi(s) = e^{\mathbf{A}(t-s)}\mathbf{V}(s)$. Prop. 2.2.10 implies that Ψ is differentiable. (Why?) Let $t_0 > 0$. Then, $\forall s \in [0,t_0]$,

$$\frac{d\Psi}{ds}(s) = e^{\mathbf{A}(t_0-s)}\mathbf{V}'(s) + \frac{d}{ds}e^{\mathbf{A}(t_0-s)}\mathbf{V}(s)$$

$$= e^{\mathbf{A}(t_0-s)}\left(\mathbf{A}\mathbf{V}(s)\right) - \mathbf{A}e^{\mathbf{A}(t_0-s)}\mathbf{V}(s) = 0.$$

(Why?) Hence, $\forall t_0 > 0$, Ψ is a constant on $[0,t_0]$. In particular, $\forall t_0 > 0$, $\Psi(0) = \Psi(t_0)$. Consequently, $\forall t_0 > 0$,

$$\mathbf{U}(t_0) = e^{\mathbf{A}t_0}\mathbf{U}_0 = \Psi(0) = \Psi(t_0) = \mathbf{V}(t_0),$$

thereby showing uniqueness. This completes the proof.   □

What if reformulating the system of ODEs in the abstract form (2.10) requires that the initial condition be computed at some time t_0 other than zero? Technically, the results we have established do not apply. But, thanks to the following corollary, a simple change of variable remedies the situation. Hence, without loss of generality, the initial conditions for all IVPs under consideration will be specified at $t = 0$.

Corollary 2.3.3. *For every* $t_0 > 0$*, the IVP*

$$\begin{cases} \mathbf{U}'(t) = \mathbf{A}\mathbf{U}(t), \, t > t_0, \\ \mathbf{U}(t_0) = \mathbf{U}_0, \end{cases} \tag{2.17}$$

has a unique classical solution $\mathbf{U}: [t_0, \infty) \to \mathbb{R}^N$ *given by* $\mathbf{U}(t) = e^{\mathbf{A}(t-t_0)}\mathbf{U}_0$.

Proof. Let $t_0 > 0$ and suppose that $\mathbf{U}: [t_0, \infty) \to \mathbb{R}^N$ satisfies (2.17). Define the function $\mathbf{W}: [0, \infty) \to \mathbb{R}^N$ by $\mathbf{W}(t) = \mathbf{U}(t + t_0)$. Then

$$\mathbf{W}'(t) = \mathbf{U}'(t + t_0) = \mathbf{A}\mathbf{U}(t + t_0) \text{ and } \mathbf{W}(0) = \mathbf{U}(t_0) = \mathbf{U}_0.$$

Thus, $\mathbf{W}$ satisfies (2.10) and so, $\mathbf{W}(t) = e^{\mathbf{A}t}\mathbf{U}_0, \forall t > 0$. As such, the classical solution of (2.17) is given by $\mathbf{U}(t) = \mathbf{W}(t - t_0) = e^{\mathbf{A}(t-t_0)}\mathbf{U}_0$, for $t > t_0$, as desired.   □

A useful consequence of the combination of these existence results and the Putzer algorithm is that an explicit formula for the classical solution is available for such IVPs. In particular, every IVP arising in Section 2.1 has a unique classical solution expressed in terms of the appropriate matrix exponential. Consider the following example.

Example. Theorem 2.3.2 implies that IVP (2.8) describing an elementary spring-mass system has a unique classical solution given by

$$\begin{bmatrix} x(t) \\ x'(t) \end{bmatrix} = e^{\begin{bmatrix} 0 & 1 \\ -\omega^2 & 0 \end{bmatrix} t} \begin{bmatrix} x_0 \\ x_1 \end{bmatrix}, \, t > 0.$$

Certainly, the Putzer algorithm can be used to compute the matrix exponential (Try it!), but we proceed in a different manner. We begin with Def. 2.2.2 to obtain:

$$e^{\begin{bmatrix} 0 & 1 \\ -\omega^2 & 0 \end{bmatrix} t} = \sum_{n=0}^{\infty} \begin{bmatrix} 0 & 1 \\ -\omega^2 & 0 \end{bmatrix}^n \frac{t^n}{n!}$$

$$= \mathbf{I} + \begin{bmatrix} 0 & 1 \\ -\omega^2 & 0 \end{bmatrix} t - \omega^2 \mathbf{I} \frac{t^2}{2!} - \omega^2 \begin{bmatrix} 0 & 1 \\ -\omega^2 & 0 \end{bmatrix} \frac{t^3}{3!}$$

$$+ \omega^4 \mathbf{I} \frac{t^4}{4!} + \omega^4 \begin{bmatrix} 0 & 1 \\ -\omega^2 & 0 \end{bmatrix} \frac{t^5}{5!} + \dots$$

Continuing in this manner definitely suggests a pattern. The fact that we can re-arrange the order of the terms of an absolutely convergent series (cf. Def. 1.5.23) enables us to continue this equality to obtain

$$= \sum_{n=0}^{\infty} (-1)^n \omega^{2n} \frac{t^{2n}}{(2n)!} \mathbf{I} + \sum_{n=0}^{\infty} (-1)^n \omega^{2n} \frac{t^{2n+1}}{(2n+1)!} \begin{bmatrix} 0 & 1 \\ -\omega^2 & 0 \end{bmatrix}$$

$$= \sum_{n=0}^{\infty} (-1)^n \frac{(\omega t)^{2n}}{(2n)!} \mathbf{I} + \sum_{n=0}^{\infty} (-1)^n \frac{(\omega t)^{2n+1}}{\omega(2n+1)!} \begin{bmatrix} 0 & 1 \\ -\omega^2 & 0 \end{bmatrix}$$

$$= \cos(\omega t) \begin{bmatrix} 1 & 0 \\ 0 & 1 \end{bmatrix} + \frac{1}{\omega} \sin(\omega t) \begin{bmatrix} 0 & 1 \\ -\omega^2 & 0 \end{bmatrix}$$

$$= \begin{bmatrix} \cos(\omega t) & \frac{1}{\omega}\sin(\omega t) \\ -\omega\sin(\omega t) & \cos(\omega t) \end{bmatrix},$$

where the Taylor series (1.72) and (1.73) have been used. Hence, the classical solution of (2.8) can be expressed equivalently as

$$\begin{bmatrix} x(t) \\ x'(t) \end{bmatrix} = \begin{bmatrix} \cos(\omega t) & \frac{1}{\omega}\sin(\omega t) \\ -\omega\sin(\omega t) & \cos(\omega t) \end{bmatrix} \begin{bmatrix} x_0 \\ x_1 \end{bmatrix}.$$

Of interest is the first row, namely $x(t) = (\cos(\omega t)) x_0 + (\sin(\omega t)) \left(\frac{x_1}{\omega}\right)$. Note that $x(t)$ is periodic with period $\frac{2\pi}{\omega}$.

Exercise 2.3.2. (Model III.1 revisited)
i.) The parameters k and m in Model III.1 are subject to measurement error. In real-ity, they are within a small tolerance of the true measurements $\hat{k}$ and $\hat{m}$; that is, $k \in (\hat{k} - \varepsilon_1, \hat{k} + \varepsilon_1)$ and $m \in (\hat{m} - \varepsilon_2, \hat{m} + \varepsilon_2)$, for some $\varepsilon_1, \varepsilon_2 > 0$ such that $\hat{m} - \varepsilon_2 > 0$.
    a.) What is the resulting range of possible values for $\hat{\omega}^2$?
    b.) Let $\hat{x}(\cdot)$ denote the classical solution of (2.8) for a specified value of $\hat{\omega}^2$ in (a). To what function does $\hat{x}(\cdot)$ become uniformly closer as $\varepsilon_1, \varepsilon_2 \to 0$?
ii.) Suppose that for a given spring (with spring constant k), we successively attach a sequence of bobs whose masses m_n approach the number m^*. For every $n \in \mathbb{N}$, find a formula for the classical solution $\mathbf{x}_n$ of (2.8). Does there exist $\mathbf{x}^*$ for which $\mathbf{x}_n \longrightarrow \mathbf{x}^*$ uniformly as $n \to \infty$?

Exercise 2.3.3. (Model II.1 revisited)
i.) Determine an explicit formula for the classical solution of IVP (2.5).
ii.) Consider a sequence $\{a_n : n \in \mathbb{N}\} \subset \mathbb{R}$ for which $\lim_{n \to \infty} a_n = a$. For every $n \in \mathbb{N}$, find a formula for the classical solution $\mathbf{U}_n$ of (2.5). Does there exist $\mathbf{U}^*$ for which $\mathbf{U}_n \longrightarrow \mathbf{U}^*$ uniformly as $n \to \infty$?
iii.) Error can occur in the measurement of any parameter, including the initial con-ditions. Consider two versions of IVP (2.5), one for which the initial data are $y(0) = y_0, z(0) = 0$ and the other for which $y(0) = \overline{y_0}, z(0) = \overline{z_0}$. Denote the classical so-lutions by $\mathbf{U}(t)$ and $\overline{\mathbf{U}}(t)$, respectively, and assume that $|y_0 - \overline{y_0}| < \delta_1$ and $|\overline{z_0}| < \delta_2$. For any $T > 0$, determine $\sup\left\{\left\|\mathbf{U}(t) - \overline{\mathbf{U}}(t)\right\|_{\mathbb{R}^N} : 0 \le t \le T\right\}$.

The above two exercises address what is commonly referred to as *continuous dependence on parameters*. Such topics are explored more extensively in the next section.

2.4 Perturbation and Convergence Results

The act of measuring of parameters is inevitably imprecise, since it is done at the mercy of imperfect environmental conditions, faulty equipment, and human error and inconsistency. The continuity properties of Prop. 2.2.7 are useful when dealing with small pertubations in such measurements. For simplicity, we shall consider a 2×2 system, although the discussion applies for any such $N \times N$ system. Suppose that a certain phenomenon is described by the IVP

$$\begin{cases} x'(t) = \alpha x(t) + \beta y(t), \\ y'(t) = \overline{\alpha} x(t) + \overline{\beta} y(t), \\ x(0) = x_0, \ y(0) = y_0. \end{cases} \tag{2.18}$$

In reality, due to measurement error, a more precise description would be

$$\begin{cases} x'(t) = (\alpha + \varepsilon_1) x(t) + (\beta + \varepsilon_2) y(t), \\ y'(t) = (\overline{\alpha} + \overline{\varepsilon_1}) x(t) + \left(\overline{\beta} + \overline{\varepsilon_2}\right) y(t), \\ x(0) = x_0, \ y(0) = y_0. \end{cases} \tag{2.19}$$

where $\varepsilon_i, \overline{\varepsilon_i} \in \mathbb{R} \ (i = 1, 2)$. Letting $\mathbf{A} = \begin{bmatrix} \alpha & \beta \\ \overline{\alpha} & \overline{\beta} \end{bmatrix}$ and $\mathbf{B}_\varepsilon = \begin{bmatrix} \varepsilon_1 & \varepsilon_2 \\ \overline{\varepsilon_1} & \overline{\varepsilon_2} \end{bmatrix}$, we can rewrite (2.18) and (2.19) as

$$\begin{cases} \mathbf{U}'(t) = \mathbf{A}\mathbf{U}(t), t > 0, \\ \mathbf{U}(0) = \mathbf{U}_0, \end{cases} \tag{2.20}$$

and

$$\begin{cases} \mathbf{U}'_\varepsilon(t) = (\mathbf{A} + \mathbf{B}_\varepsilon) \mathbf{U}_\varepsilon(t), t > 0, \\ \mathbf{U}_\varepsilon(0) = \mathbf{U}_0, \end{cases} \tag{2.21}$$

respectively. Since $\mathbf{A} + \mathbf{B}_\varepsilon \in \mathbb{M}^2(\mathbb{R})$, the classical solution of (2.21) is $\mathbf{U}_\varepsilon(t) = e^{(\mathbf{A}+\mathbf{B}_\varepsilon)t}\mathbf{U}_0$. Naturally, we want to know how this differs from $\mathbf{U}(t) = e^{\mathbf{A}t}\mathbf{U}_0$. To this end, note that $\forall t > 0$,

$$\|\mathbf{U}_\varepsilon(t) - \mathbf{U}(t)\|_{\mathbb{R}^N} \leq \left\|e^{(\mathbf{A}+\mathbf{B}_\varepsilon)t} - e^{\mathbf{A}t}\right\|_{\mathbb{M}^N} \|\mathbf{U}_0\|_{\mathbb{R}^N}. \tag{2.22}$$

We would like to further say that

$$\left\|e^{(\mathbf{A}+\mathbf{B}_\varepsilon)t} - e^{\mathbf{A}t}\right\|_{\mathbb{M}^N} = \left\|e^{\mathbf{A}t} e^{\mathbf{B}_\varepsilon t} - e^{\mathbf{A}t}\right\|_{\mathbb{M}^N} = \left\|e^{\mathbf{A}t} \left(e^{\mathbf{B}_\varepsilon t} - \mathbf{I}\right)\right\|_{\mathbb{M}^N} \tag{2.23}$$

so that the closer $\mathbf{B}_\varepsilon$ is to $\mathbf{0}$, the closer $\mathbf{U}_\varepsilon$ is to $\mathbf{U}$ (each in the sense of the $\mathbb{M}^N$norm). But, this hinges on the equality $e^{(\mathbf{A}+\mathbf{B}_\varepsilon)t} = e^{\mathbf{A}t}e^{\mathbf{B}_\varepsilon t}$. Does this hold, in general? An initial glance might suggest an unfettered yes, but consider the following exercise.

Exercise 2.4.1.
i.) Let $T > 0$. Show that if $e^{(\mathbf{A}+\mathbf{B}_\varepsilon)t} = e^{\mathbf{A}t}e^{\mathbf{B}_\varepsilon t}$, $\forall 0 \leq t \leq T$, then $\|\mathbf{U}_\varepsilon - \mathbf{U}\|_{\mathbb{C}([0,T];\mathbb{R}^N)} \to 0$ as $\mathbf{B}_\varepsilon \to \mathbf{0}$.
ii.) Prove that if $\mathbf{A}$ and $\mathbf{B}$ are $N \times N$ diagonal matrices, then $e^{\mathbf{A}+\mathbf{B}} = e^{\mathbf{A}}e^{\mathbf{B}}$.
iii.) Let $\mathbf{A} = \begin{bmatrix} 2 & 1 \\ 0 & 2 \end{bmatrix}$ and $\mathbf{B} = \begin{bmatrix} \frac{1}{2} & 0 \\ 0 & \frac{1}{3} \end{bmatrix}$. Show that $\mathbf{AB} \neq \mathbf{BA}$ and that $e^{\mathbf{A}}e^{\mathbf{B}} \neq e^{\mathbf{A}+\mathbf{B}}$.
Hence, the desired equality (2.23) does not hold.

Lack of commutativity of these matrices presents an obstacle when trying to verify (2.23), even for a reasonably tame perturbation. However, if this hurdle is removed, then we can verify (2.23), as suggested by the following exercise.

Exercise 2.4.2.
i.) Prove that if $\mathbf{AB} = \mathbf{BA}$, then $e^{\mathbf{B}}e^{\mathbf{A}} = e^{\mathbf{A}}e^{\mathbf{B}} = e^{\mathbf{A}+\mathbf{B}}$.
ii.) Explain how (i) helps when verifying that $\|\mathbf{U}_\varepsilon - \mathbf{U}\|_{\mathbb{C}([0,T];\mathbb{R}^N)} \to 0$ as $\varepsilon \to 0^+$ in (2.22).

A related notion is the approximation of a given IVP by a sequence of IVPs whose parameters are successively better approximations of those in the original IVP. Precisely, let $\{\mathbf{A}_k\} \subset \mathbb{M}^N$ and $\{\mathbf{U}_0^k\} \subset \mathbb{R}^N$. For each $k \in \mathbb{N}$, consider the IVP

$$\begin{cases} \mathbf{U}'_k(t) = \mathbf{A}_k\mathbf{U}_k(t), t > 0, \\ \mathbf{U}_k(0) = \mathbf{U}_0^k. \end{cases} \tag{2.24}$$

Proposition 2.4.1. *Let $\mathbf{A} \in \mathbb{M}^N$ and $\mathbf{U}_0 \in \mathbb{R}^N$. Assume that*
(H2.1) $\{e^{\mathbf{A}t} : t \geq 0\}$ *and* $\{e^{\mathbf{A}_k t} : t \geq 0\}$ *are contractive, $\forall k \in \mathbb{N}$;*
(H2.2) $\mathbf{A}_k e^{\mathbf{A}t} = e^{\mathbf{A}t}\mathbf{A}_k$, *$\forall k \in \mathbb{N}$ and $t \geq 0$;*
(H2.3) *For every $\mathbf{z} \in \mathbb{R}^N$, $\lim\limits_{k \to \infty} \|\mathbf{A}_k\mathbf{z} - \mathbf{A}\mathbf{z}\|_{\mathbb{R}^N} = 0$ and $\lim\limits_{k \to \infty} \|\mathbf{U}_0^k - \mathbf{U}_0\|_{\mathbb{R}^N} = 0$.*
Then, $\forall T > 0$ and $\mathbf{z} \in \mathbb{R}^N$,

$$\left\| e^{\mathbf{A}_k t}\mathbf{z} - e^{\mathbf{A}t}\mathbf{z} \right\|_{\mathbb{R}^N} \longrightarrow 0 \; uniformly \; on \; [0,T] \; as \; k \to \infty$$

and

$$\lim_{k \to \infty} \|\mathbf{U}_k - \mathbf{U}\|_{\mathbb{C}([0,T];\mathbb{R}^N)} = 0,$$

where $\mathbf{U}$ is the classical solution of (2.10).

Proof. Let $\mathbf{z} \in \mathbb{R}^N$, $0 \le s \le t \le T$, and $\varepsilon > 0$. Then, $\forall k \in \mathbb{N}$,

$$e^{\mathbf{A}_k t}\mathbf{z} - e^{\mathbf{A}t}\mathbf{z} = -\int_0^t \frac{d}{ds}\left[e^{\mathbf{A}_k(t-s)}e^{\mathbf{A}s}\mathbf{z}\right]ds$$

$$= \int_0^t e^{\mathbf{A}_k(t-s)}(\mathbf{A}_k - \mathbf{A})e^{\mathbf{A}s}\mathbf{z}\,ds$$

$$= \int_0^t e^{\mathbf{A}_k(t-s)}e^{\mathbf{A}s}(\mathbf{A}_k\mathbf{z} - \mathbf{A}\mathbf{z})\,ds.$$

(Why?) Hence,

$$\left\|e^{\mathbf{A}_k t}\mathbf{z} - e^{\mathbf{A}t}\mathbf{z}\right\|_{\mathbb{R}^N} \le t\,\|\mathbf{A}_k\mathbf{z} - \mathbf{A}\mathbf{z}\|_{\mathbb{R}^N}.$$

It follows from **(H2.3)** that $\exists K \in \mathbb{N}$ such that

$$k \ge K \Rightarrow \|\mathbf{A}_k\mathbf{z} - \mathbf{A}\mathbf{z}\|_{\mathbb{R}^N} < \frac{\varepsilon}{T}.$$

The conclusion now follows. (Tell how.) □

Exercise 2.4.3. Fill in the missing details in the proof of Prop. 2.4.1.

Exercise 2.4.4. Recall that if $a, x \in \mathbb{R}$, then $e^{ax} = \lim_{n \to \infty}\left(1 - \frac{ax}{n}\right)^{-n}$. So, for sufficiently large $n \in \mathbb{N}$, $\left(1 - \frac{ax}{n}\right)^{-n}$ is a good approximation of e^{ax}. We are interested in the extent to which this generalizes to the matrix setting. To this end, let $\alpha, \beta \in \mathbb{R}$, $\mathbf{x} \in \mathbb{R}^2$, and consider $\mathbf{A} = \begin{bmatrix} \alpha & 0 \\ 0 & \beta \end{bmatrix}$. Calculate $\lim_{n \to \infty}\left(\mathbf{I} - \frac{t\mathbf{A}}{n}\right)^{-n}\mathbf{x}$. Does it equal $e^{\mathbf{A}t}\mathbf{x}$?

2.5   A Glimpse at Long-Term Behavior

We know that $\mathbf{U}(t) = e^{\mathbf{A}t}\mathbf{U}_0$ is the unique classical solution of (2.10) on $[0, \infty)$. We now investigate the behavior of this solution as $t \to \infty$. Some natural questions concerning the so-called *long-term behavior* (a.k.a., *asymptotic behavior*) of the classical solution $\mathbf{U}(t)$ are:

1. Does there exist $\mathbf{U}^\star \in \mathbb{R}^N$ for which $\lim_{t \to \infty}\|\mathbf{U}(t) - \mathbf{U}^\star\|_{\mathbb{R}^N} = 0$?

2. Is $\mathbf{U}(t)$ *time-periodic*, meaning $\exists p > 0$ such that $\mathbf{U}(t + p) = \mathbf{U}(t)$, $\forall t > 0$?

3. Is $\mathbf{U}(t)$ *globally bounded*, meaning $\exists M > 0$ such that $\|\mathbf{U}(t)\|_{\mathbb{R}^N} \le M$, $\forall t > 0$?

4. Can $\mathbf{U}(t)$ *blow-up in finite time*, meaning $\exists T^\star > 0$ such that

$$\lim_{t \to (T^\star)^-}\|\mathbf{U}(t)\|_{\mathbb{R}^N} = \infty?$$

Since the classical solution is $\mathbf{U}(t) = e^{\mathbf{A}t}\mathbf{U}_0$, these questions ultimately concern the nature of $\{e^{\mathbf{A}t} : t \ge 0\}$, which in turn is directly linked to the matrix $\mathbf{A}$. How, specifically, does the relationship between $\mathbf{A}$ and $\{e^{\mathbf{A}t} : t \geq 0\}$ translate into different long-term behavior? The following exercises shed some light on this situation.

Exercise 2.5.1. Let $\mathbf{A} = \begin{bmatrix} \alpha & 0 \\ 0 & \beta \end{bmatrix}$, where $\alpha, \beta \in \mathbb{R}$. Answer the following three questions for each description of α and β to follow:

(I) Let $t \geq 0$. Compute $e^{\mathbf{A}t}$ and determine an upper bound for $\left\| e^{\mathbf{A}t} \right\|_{\mathbb{M}^N}$.

(II) Does $\exists \lim\limits_{t \to \infty} \left\| e^{\mathbf{A}t} \mathbf{x} \right\|_{\mathbb{R}^N}$? Does it depend on the choice of $\mathbf{x} \in \mathbb{R}^N$?

(III) Address each of the four questions posed at the beginning of this section for the classical solution of the corresponding IVP.

i.) $\alpha, \beta < 0$;
ii.) $\alpha, \beta > 0$;
iii.) Exactly one of α, β is equal to zero and the other is strictly positive;
iv.) Exactly one of α, β is equal to zero and the other is strictly negative;
v.) $\alpha < 0 < \beta$.

Exercise 2.5.2. Let $\mathbf{A} = \begin{bmatrix} 0 & \beta \\ -\beta & 0 \end{bmatrix}$, where $\beta \neq 0$.

i.) Compute $e^{\mathbf{A}t}$ and describe the long-term behavior of $e^{\mathbf{A}t} \mathbf{U}_0$ for a given $\mathbf{U}_0 \in \mathbb{R}^2$.

ii.) More generally, consider $\mathbf{B} = \begin{bmatrix} \alpha & \beta \\ -\beta & \alpha \end{bmatrix}$, where $\alpha, \beta \neq 0$. Answer questions (I) - (III) from Exer. 2.5.1, first assuming $\alpha < 0$ and then for $\alpha > 0$.

The eigenvalues of the matrix $\mathbf{A}$ significantly impact the structure of $e^{\mathbf{A}t}$ (cf. Prop. 2.2.3). Every $N \times N$ matrix has N complex eigenvalues, including multiplicity. As such, it is reasonable to expect a connection between them and the long-term behavior of $\{e^{\mathbf{A}t} : t \geq 0\}$. This is apparent in the following simple scenario.

Exercise 2.5.3. Consider the matrix $\mathbf{A} = \begin{bmatrix} \alpha & 0 \\ 0 & \beta \end{bmatrix}$, whose eigenvalues are α and β. For each case listed in Exer. 2.5.1, associate the nature of the eigenvalues with the long-term behavior of the classical solution $\mathbf{U}(t)$ of (2.10). (For instance, if $\alpha, \beta < 0$, then the eigenvalues of $\mathbf{A}$ are both negative and in such case, $\lim\limits_{t \to \infty} \|\mathbf{U}(t)\|_{\mathbb{R}^2} = 0$.)

The nature of the eigenvalues characterize the long-term behavior of $\{e^{\mathbf{A}t} : t \geq 0\}$, and in turn the classical solution of (2.10). Of course, we have not proved this; we have merely affirmed its truth for particular types of matrices. What if two different matrices $\mathbf{A}$ and $\mathbf{B}$ have precisely the same eigenvalues; must $\{e^{\mathbf{A}t} : t \geq 0\}$ and $\{e^{\mathbf{B}t} : t \geq 0\}$ behave in the same manner? The answer is yes, but this requires proof. The idea is to define an equivalence relation (via matrix similarity) on $\mathbb{M}^N(\mathbb{R})$ and characterize the classes by distinct combinations of types of eigenvalues (real vs. imaginary, positive/zero/negative, etc.). A "nice" representative with which it is easy to work and which determines the behavior of the corresponding matrix exponential is chosen from each class; the matrix exponential of any member of the class behaves in exactly the same manner. Formally doing this requires the use of Jordan forms, discussed in [172].

In light of the above discussion, the following representation theorem is very useful. (See [202] for a proof.)

Proposition 2.5.1. *Let* $\mathbf{A} \in \mathbb{M}^N(\mathbb{R})$ *with eigenvalues* $\lambda_1, \ldots, \lambda_N$ *(including multiplicity) and suppose that* $\{\mathbf{e}_k : k = 1, \ldots, N\}$ *is a basis for* $\mathbb{R}^N$. *If* $\mathbf{A}\mathbf{z} = \sum_{k=1}^N \lambda_k \langle \mathbf{z}, \mathbf{e}_k \rangle_{\mathbb{R}^N} \mathbf{e}_k$, $\forall \mathbf{z} \in \mathbb{R}^N$, *then* $e^{\mathbf{A}t}\mathbf{z} = \sum_{k=1}^N e^{\lambda_k t} \langle \mathbf{z}, \mathbf{e}_k \rangle_{\mathbb{R}^N} \mathbf{e}_k$.

Example. Let $\mathbf{A} = \begin{bmatrix} \alpha & 0 \\ 0 & \beta \end{bmatrix}$, whose eigenvalues are α and β. Assume that $\mathbf{e}_1 = \begin{bmatrix} 1 \\ 0 \end{bmatrix}$ and $\mathbf{e}_2 = \begin{bmatrix} 0 \\ 1 \end{bmatrix}$. Let $\mathbf{z} = \begin{bmatrix} z_1 \\ z_2 \end{bmatrix} \in \mathbb{R}^2$ and observe that

$$\mathbf{A}\mathbf{z} = \begin{bmatrix} \alpha & 0 \\ 0 & \beta \end{bmatrix} \begin{bmatrix} z_1 \\ z_2 \end{bmatrix} = \begin{bmatrix} \alpha z_1 \\ \beta z_2 \end{bmatrix} = \alpha \langle \mathbf{z}, \mathbf{e}_1 \rangle_{\mathbb{R}^2} \mathbf{e}_1 + \beta \langle \mathbf{z}, \mathbf{e}_2 \rangle_{\mathbb{R}^2} \mathbf{e}_2.$$

Hence,

$$e^{\mathbf{A}t}\mathbf{z} = \begin{bmatrix} e^{\alpha t} & 0 \\ 0 & e^{\beta t} \end{bmatrix} \begin{bmatrix} z_1 \\ z_2 \end{bmatrix} = \begin{bmatrix} e^{\alpha t} z_1 \\ e^{\beta t} z_2 \end{bmatrix} = e^{\alpha t} \langle \mathbf{z}, \mathbf{e}_1 \rangle_{\mathbb{R}^2} \mathbf{e}_1 + e^{\beta t} \langle \mathbf{z}, \mathbf{e}_2 \rangle_{\mathbb{R}^2} \mathbf{e}_2.$$

Different notions of long-term behavior are characterized by the type of convergence used when computing the limit as $t \to \infty$. Some of these are provided in the following definition.

Definition 2.5.2. Let $\mathbf{A} \in \mathbb{M}^N(\mathbb{R})$. We say $\{e^{\mathbf{A}t} : t \geq 0\}$ is
i.) *uniformly stable* if $\lim_{t \to \infty} \|e^{\mathbf{A}t}\|_{\mathbb{M}^N} = 0$;
ii.) *exponentially stable* if $\exists M \in \mathbb{N}$ and $\alpha > 0$ such that $\|e^{\mathbf{A}t}\|_{\mathbb{M}^N} \leq M e^{-\alpha t}, \forall t \geq 0$;
iii.) *strongly stable* if $\lim_{t \to \infty} \|e^{\mathbf{A}t}\mathbf{z}\|_{\mathbb{R}^N} = 0, \forall \mathbf{z} \in \mathbb{R}^N$.

For certain cases, Prop. 2.5.1 enables us to reduce the question of the stability of $\{e^{\mathbf{A}t} : t \geq 0\}$ to examining the nature of the eigenvalues of $\mathbf{A}$. This is explored in the following exercises.

Exercise 2.5.4. Let $\mathbf{A} \in \mathbb{M}^N(\mathbb{R})$ be diagonal and assume that all eigenvalues $\lambda_1, \ldots, \lambda_N$ of $\mathbf{A}$ are negative real numbers.
i.) Prove that $\{e^{\mathbf{A}t} : t \geq 0\}$ is uniformly stable.
ii.) What happens if at least one eigenvalue is positive? Does a similar conclusion hold if some, but not all, of the eigenvalues are zero and the remaining ones are negative? Explain.

Exercise 2.5.5. **(Model I.1 revisited)**
Does $\exists \alpha_1 \in \mathbb{R} \setminus \{0\}$ for which the classical solution of IVP (2.2) is exponentially stable? How about uniformly stable?

Exercise 2.5.6. Consider the three notions of stability in Def. 2.5.2. Which seems to be the strongest? Do any imply the others? Try to formally justify your assertions.

Much more can be said about the above notions and so-called *Lyapunov stability*. Consult the discussion in [**59, 65, 220, 319, 320**].

2.6   Looking Ahead

The content of this chapter establishes the well-posedness of the IVP (2.10) (that is, the existence and uniqueness of a classical solution that depends continuously on the initial data) for any $N \times N$ matrix $\mathbf{A}$ and any initial condition $\mathbf{U}_0$. This theory can be used in the description of numerous applications. However, not all phenomena can be described using a linear system of ODEs. In fact, as soon as the description depends on more than one variable (e.g., both time *and* position), the use of *partial differential equations* (PDEs) is required. For instance, the following initial-boundary value problem (IBVP) is a classical model of heat conduction in a one-dimensional rod of length L:

$$\begin{cases} \frac{\partial u}{\partial t}(z,t) = k\frac{\partial^2 u}{\partial z^2}(z,t), \ 0 \le z \le L, t > 0, \\ u(z,0) = u_0(z), \ 0 \le z \le L, \\ u(0,t) = u(L,t) = 0, \ t > 0. \end{cases} \tag{2.25}$$

Here, the constant k is a physical parameter involving the density of the rod and thermal conductivity, and $u(z,t)$ represents the temperature at position z along the rod at time t. A loose comparison of (2.25) to (2.10) suggests that we can identify $\mathbf{U}$ with the unknown $u : [0,L] \times [0,\infty) \to \mathbb{R}$, the left-side with $\frac{\partial u}{\partial t}$, and the right-side somehow with the differential operator $A = \frac{\partial^2}{\partial z^2}$. But, this certainly does not fall under the parlance of (2.10) since $\frac{\partial^2}{\partial z^2}$ cannot be identified with any member of $\mathbb{M}^N(\mathbb{R})$ (as we shall see in Chapter 3). Still, expressing (2.25) in an abstract form similar to (2.10) is not unreasonable, although this clearly requires some tweaking and redefining of terms. The following suggests that a solution of (2.25) exhibits some exponential-like properties.

Exercise 2.6.1.
i.) Verify that the function $u : [0,L] \times [0,\infty) \to \mathbb{R}$ defined below satisfies all portions of (2.25):

$$u(z,t) = \sum_{m=1}^{\infty} e^{-\left(\frac{m\pi}{L}\right)^2 kt} \left\langle u_0(\cdot), \sin\left(\frac{m\pi}{L}\cdot\right)\right\rangle_{\mathbb{L}^2(0,L;\mathbb{R})} \sin\left(\frac{m\pi z}{L}\right). \tag{2.26}$$

ii.) Note that the right-side of (2.26) really has <u>three</u> inputs, namely $t, z,$ and $u_0(\cdot)$. To emphasize these distinct dependencies, we write $u(z,t) = S(t)[u_0](z)$. Verify the following:

a.) $S(0)[u_0](z) = u_0(z)$,

b.) $S(t_1 + t_2)[u_0](z) = S(t_1)(S(t_2)[u_0])(z)$.

These resemble the exponential properties in Prop. 2.2.5, which suggests that there ought to be *some* entity that plays the role of $\{e^{At} : t \geq 0\}$ in this scenario. But, what *exactly* is it, and to what extent does the theory developed in this chapter generalize to this setting? This leg of the journey awaits us in Chapter 3.

2.7 Guidance for Exercises

2.7.1 Level 1: A Nudge in a Right Direction

2.1.1. Apply the trick in (2.7) twice, once each for x_A and x_B. Identify the dimensions of the vectors and matrices involved.

2.2.1. Explain why $\mathbf{A}t$, $(\mathbf{A}t)^k$, and $\frac{(\mathbf{A}t)^k}{k!}$ are meaningful, $\forall t \geq 0$ and $k \in \mathbb{N}$ using properties of matrix multiplication. (Then what?)

2.2.2. The fact that we are dealing with limits of sequences of elements in $\mathbb{M}^N(\mathbb{R})$ suggests that we must somehow appeal to completeness. (How?)

2.2.3. Mimic the steps used to compute $e^{\begin{bmatrix} a & 0 \\ 0 & b \end{bmatrix} t}$. Compute $\mathbf{A}t$, $(\mathbf{A}t)^k$, and $\frac{(\mathbf{A}t)^k}{k!}$, where

$$\mathbf{A} = \begin{bmatrix} a_{11} & 0 & 0 & \cdots & 0 \\ 0 & a_{22} & 0 & \cdots & 0 \\ \vdots & \vdots & \ddots & & \vdots \\ 0 & 0 & & \ddots & 0 \\ 0 & 0 & 0 & \cdots & a_{NN} \end{bmatrix} \tag{2.27}$$

(Now what?)

2.2.4. Show $\mathbf{P}^{-1}(\mathbf{A}+\mathbf{B})\mathbf{P} = \mathbf{P}^{-1}\mathbf{A}\mathbf{P} + \mathbf{P}^{-1}\mathbf{B}\mathbf{P}$ and $(\mathbf{P}^{-1}\mathbf{D}\mathbf{P})^k = \mathbf{P}^{-1}\mathbf{D}^k\mathbf{P}$, $\forall k \in \mathbb{N}$.

2.2.5. Compute the eigenvalues in each case and form the IVPs. Be certain to solve them in the order suggested in the example following Prop. 2.2.3.

2.2.6. Prove that $\lim_{k \to \infty} \|\mathbf{A}x_k - \mathbf{A}x\|_{\mathbb{R}^N} = 0$. (Now what?)

2.2.7. i.) Use (2.16) to guide your thinking.

ii.) Use Exer. 2.2.3.

iii.) Let $\mathbf{A}$ be given by (2.27), where $a_{ii} < 0$, $\forall i \in \{1, \ldots, N\}$. Compute $\|\mathbf{A}\|_{\mathbb{M}^N}$.

2.2.8. Use Def. 2.2.2 directly for both exponentials and consider the partial sums. (Now what?)

2.2.9. (i) and (iii). Which seems stronger?

2.2.10. i.) Let $\varepsilon > 0$. Prove that $\exists \delta > 0$ such that $0 < |h| < \delta \implies$

$$\left\| \frac{1}{h} \int_{t_0}^{t_0+h} e^{As}\mathbf{x}ds - e^{At_0}\mathbf{x} \right\|_{\mathbb{R}^N} = \left\| \frac{1}{h} \int_{t_0}^{t_0+h} \left[e^{As} - e^{At_0} \right] \mathbf{x}ds \right\|_{\mathbb{R}^N} < \varepsilon.$$

ii.) This is similar to (i), but with $\left[e^{As} - e^{At_0} \right]$ replaced by

$$\left[g(s)e^{As} - g(t_0)e^{At_0} \right] = \left[g(s)e^{As} - g(s)e^{At_0} + g(s)e^{At_0} - g(t_0)e^{At_0} \right].$$

2.2.11. i.) Let $\mathbf{x}_0 = \begin{bmatrix} x_1 \\ x_2 \end{bmatrix}$ and prove that

$$\left\| \begin{bmatrix} \frac{e^{at}-1}{t} & 0 \\ 0 & \frac{e^{bt}-1}{t} \end{bmatrix} \begin{bmatrix} x_1 \\ x_2 \end{bmatrix} - \begin{bmatrix} ax_1 \\ bx_2 \end{bmatrix} \right\|_{\mathbb{R}^2} \longrightarrow 0$$

ii.) Compute both quantities directly and compare.
2.2.12. Use Exer. 2.2.3 and apply all calculus operations entrywise to verify this.
2.3.1. What could happen if $\mathbf{U}$ was not right continuous at $t = 0$?
2.3.2. i.) a.) Recall that $\hat{\omega}^2 = \frac{\hat{k}}{\hat{m}}$.
   b.) Let $\hat{x}(t) = (\cos(\hat{\omega}t))x_0 + (\sin(\hat{\omega}t))\left(\frac{x_1}{\hat{\omega}}\right)$. What happens to the range of $\hat{\omega}^2$ as $\varepsilon_1, \varepsilon_2 \to 0$.
   ii.) Observe that $\lim_{n\to\infty} \omega_n^2 = \lim_{n\to\infty} \frac{k}{m_n}$. Use this in conjunction with (i)(b).
2.3.3. i.) $\mathbf{U}(t) = e^{\begin{bmatrix} -a & 0 \\ a & -b \end{bmatrix}t} \begin{bmatrix} y_0 \\ 0 \end{bmatrix}$. Now compute the exponential matrix.
   ii.) Compute $e^{\begin{bmatrix} -a_n & 0 \\ a_n & -b \end{bmatrix}t}$ and then $\lim_{n\to\infty} e^{\begin{bmatrix} -a_n & 0 \\ a_n & -b \end{bmatrix}t}$.
   iii.) Note that $\|\mathbf{U}(t) - \overline{\mathbf{U}}(t)\|_{\mathbb{R}^N} \le e^{At} \|\mathbf{U}_0 - \overline{\mathbf{U}}_0\|_{\mathbb{R}^N}$. (So what?)
2.4.1. i.) Note that $\|e^{\mathbf{B}_\varepsilon t} - \mathbf{I}\|_{\mathbb{M}^N} \longrightarrow 0$ as $\mathbf{B}_\varepsilon \to \mathbf{0}$. (So what?)
ii.) Let $\mathbf{A}$ and $\mathbf{B}$ be diagonal matrices with diagonal entries $a_{11}, a_{22}, \dots, a_{NN}$ and $b_{11}, b_{22}, \dots, b_{NN}$, respectively. Use the rule $e^{x+y} = e^x e^y$, $\forall x, y \in \mathbb{R}$, when comparing $e^{\mathbf{A}+\mathbf{B}}$ to $e^{\mathbf{A}}e^{\mathbf{B}}$.
iii.) Use the Putzer algorithm.
2.4.2. i.) Use the binomial theorem to compute $(a+b)^n$.
   ii.) Use the hint for Exer. 2.4.1(i).
2.4.3. Use Prop. 2.2.10 and (H2.2) for the first string of equalities. (Now what?)
2.4.4. Observe that

$$\left(\mathbf{I} - \frac{t\mathbf{A}}{n}\right)^{-n} = \begin{bmatrix} 1 - \frac{\alpha t}{n} & 0 \\ 0 & 1 - \frac{\beta t}{n} \end{bmatrix}^{-n} = \begin{bmatrix} \left(1 - \frac{\alpha t}{n}\right)^{-n} & 0 \\ 0 & \left(1 - \frac{\beta t}{n}\right)^{-n} \end{bmatrix}.$$

2.5.1. I) $e^{\mathbf{A}t}$ has the same form for (i) - (v), and $\|e^{\mathbf{A}t}\|_{\mathbb{M}^N} \le e^{t\|\mathbf{A}\|}_{\mathbb{M}^N}$, $\forall t$. (So what?)
II) Compute the limits entrywise. When does the term e^{at} have a limit as $t \to \infty$?
III) (1) Only for those for which the limit in (II) exists. (Why?)

(2) Does $\exists p > 0$ such that $e^{a(t+p)} = e^{at}$, $\forall t \geq 0$?

(3) For which values of $a \in \mathbb{R}$ is the set $\{e^{at} | t \geq 0\}$ bounded?

(4) What must be true about the graph of $y = e^{at}$ in order for this to occur?

2.5.2. i.) Use Exer. 2.2.5.

ii.) Use Exer. 2.2.5. The hints for Exer. 2.5.1 also apply here.

2.5.3. The eigenvalues of $\mathbf{A}$ are α and β. Now, the conclusions follow immediately from Exer. 2.5.1. (How?)

2.5.4. i.) Use $\left\| e^{\mathbf{A}t} \right\|_{\mathbb{M}^N} \leq e^{t\|\mathbf{A}\|_{\mathbb{M}^N}}$, $\forall t$.

ii.) The scenario of having one positive eigenvalue is different from the one in which some, but not all, of the eigenvalues are zero and the remaining ones are negative?

2.5.5. $\|\mathbf{A}\|_{\mathbb{M}^N} \geq 0$. (So what?)

2.5.6. Use (2.16). Under what conditions does $\lim\limits_{t \to \infty} \left\| e^{\mathbf{A}t} \right\|_{\mathbb{M}^N} = 0$.

2.6.1. i.) Note that $\left\langle u_0(\cdot), \sin\left(\frac{m\pi}{L}\cdot\right) \right\rangle_{\mathbb{L}^2(0,L;\mathbb{R})}$ is independent of z.

ii.) a.) Use the Fourier representation for $u_0(\cdot)$ (cf. Section 1.7.2).

b.) Use the rule $e^{x+y} = e^x e^y$, $\forall x, y \in \mathbb{R}$. (How?)

2.7.2   Level 2: An Additional Thrust in a Right Direction

2.1.1. Use the change of variable $v_1 = x_A$, $v_2 = x'_A$, $v_3 = x_B$, $v_4 = x'_B$. Use the first two equations in (2.9) to arrive at a matrix equation for which the unknown vector has 4 components and the coefficient matrix on the right-side is 4×4.

2.2.1. Note that the dimensions of $\frac{(\mathbf{A}t)^k}{k!}$ are the same $\forall k \in \mathbb{N}$. Can you add such matrices?

2.2.2. Look ahead to Prop. 2.2.1.

2.2.3. If $\mathbf{A}$ is given by (2.27), then

$$e^{\mathbf{A}t} = \begin{bmatrix} e^{a_{11}t} & 0 & 0 & \cdots & 0 \\ 0 & e^{a_{22}t} & 0 & \cdots & 0 \\ \vdots & \vdots & \ddots & & \vdots \\ 0 & 0 & & \ddots & 0 \\ 0 & 0 & 0 & \cdots & e^{a_{NN}t} \end{bmatrix}$$

2.2.4. Now apply the Taylor representation formula.

2.2.5. i.) $e^{\mathbf{B}t} = e^{\alpha t} \begin{bmatrix} \cos(\beta t) & \sin(\beta t) \\ -\sin(\beta t) & \cos(\beta t) \end{bmatrix}$

ii.) $e^{\mathbf{B}t} = e^{\alpha t} \begin{bmatrix} 1 & 0 \\ \beta t & 1 \end{bmatrix}$

2.2.6. Use the uniqueness of limits to conclude.

2.2.7. i.) Choose $\mathbf{A}$ such that $\left\| e^{\mathbf{A}t} \right\|_{\mathbb{M}^N} > 1$.

ii.) Calculate $\|\mathbf{A}\|_{\mathbb{M}^N}$ using (1.63). What must be true about the eigenvalues?

iii.) Using the Cauchy-Schwarz inequality, we see that

$$|\langle \mathbf{A}\mathbf{x}, \mathbf{x} \rangle| \leq \|\mathbf{A}\mathbf{x}\|_{\mathbb{R}^N} \|\mathbf{x}\|_{\mathbb{R}^N} \leq \|\mathbf{A}\|_{\mathbb{M}^N} \|\mathbf{x}\|_{\mathbb{R}^N}^2.$$

2.2.8. For any $m \in \mathbb{N}$, the partial sums corresponding to $e^{(\alpha \mathbf{A})t}$ and $e^{\mathbf{A}t}$ are equal since $(\alpha \mathbf{A})t$ and $\mathbf{A}t$ commute. (Why? So what?)

2.2.9. (i) is stronger, but is (iii) sufficient?

2.2.10. i.) Use Prop. 1.8.13(vi)(a) with the estimate

$$\left\| \left(e^{\mathbf{A}s} - e^{\mathbf{A}t_0} \right) \mathbf{x} \right\|_{\mathbb{R}^N} = \left\| \left(e^{\mathbf{A}(s-t_0)} e^{\mathbf{A}t_0} - e^{\mathbf{A}t_0} \right) \mathbf{x} \right\|_{\mathbb{R}^N}$$
$$\leq \left\| e^{\mathbf{A}(s-t_0)} - \mathbf{I} \right\|_{\mathbb{M}^N} \left\| e^{\mathbf{A}t_0} \right\|_{\mathbb{M}^N} \|\mathbf{x}\|_{\mathbb{R}^N}.$$

ii.) Argue as in (i), but now also take into account the continuity of g in different ways to obtain the desired estimate.

2.2.11. i.) Compute the limit componentwise using l'Hopital's rule. (Now what?)

ii.) Both equal the matrix $\begin{bmatrix} ae^{at} & 0 \\ 0 & be^{bt} \end{bmatrix}$.

2.3.1. A lack of such continuity could obscure the continuous dependence on the initial data. (How?)

2.3.2. i.) a.) $\hat{\omega}^2 \in \left(\frac{\hat{k}-\varepsilon_1}{\hat{m}+e_2}, \frac{\hat{k}+\varepsilon_1}{\hat{m}-e_2} \right)$.

b.) $\hat{x}(\cdot)$ gets closer to the solution $x(\cdot)$ of (2.8) corresponding to $\hat{\omega}^2 = \frac{\hat{k}}{\hat{m}}$.

ii.) Yes. $x(\cdot)$ is the solution of (2.8) corresponding to $\omega_* = \frac{k}{m^*}$.

2.3.3. i.) Assume that $a, b > 0$. Use the Putzer algorithm to compute $e^{\begin{bmatrix} -a & 0 \\ a & -b \end{bmatrix} t}$.

ii.) Determine if $\lim_{n \to \infty} e^{\begin{bmatrix} -a_n & 0 \\ a_n & -b \end{bmatrix} t} = e^{\begin{bmatrix} -a & 0 \\ a & -b \end{bmatrix} t}$. What can you conclude from this observation?

iii.) $\sup \left\{ \|\mathbf{U}(t) - \overline{\mathbf{U}}(t)\|_{\mathbb{R}^N} : 0 \leq t \leq T \right\} \leq e^{\|\mathbf{A}\|_{\mathbb{M}^N} t} \max \{\delta_1, \delta_2\}$.

2.4.1. i.) Now use $\|\mathbf{U}_\varepsilon - \mathbf{U}\|_{\mathbb{C}([0,T];\mathbb{R}^N)} \leq \left\| e^{\mathbf{A}t} \left(e^{\mathbf{B}\varepsilon t} - \mathbf{I} \right) \right\|_{\mathbb{M}^N} \leq \left\| e^{\mathbf{A}t} \right\|_{\mathbb{M}^N} \left\| e^{\mathbf{B}\varepsilon t} - \mathbf{I} \right\|_{\mathbb{M}^N}$.

ii.) Note that

$$e^{\mathbf{A}+\mathbf{B}} = \begin{bmatrix} e^{a_{11}+b_{11}} & 0 & 0 \cdots & 0 \\ 0 & e^{a_{22}+b_{22}} & 0 \cdots & 0 \\ \vdots & \vdots & \ddots & \vdots \\ 0 & 0 & \ddots & 0 \\ 0 & 0 & 0 \cdots & e^{a_{NN}+b_{NN}} \end{bmatrix} = \ldots = e^{\mathbf{A}} e^{\mathbf{B}}$$

2.4.2. i.) Observe that $\forall M \in \mathbb{N}$,

$$(\mathbf{A}+\mathbf{B})^M = \sum_{k=0}^{M} \binom{M}{k} \mathbf{A}^k \mathbf{B}^{M-k} = M! \sum_{k=0}^{M} \frac{\mathbf{A}^k}{k!} \cdot \frac{\mathbf{B}^{M-k}}{(M-k)!}.$$

As such, $e^{\mathbf{A}+\mathbf{B}} = \sum_{M=0}^{\infty} \frac{(\mathbf{A}+\mathbf{B})^M}{M!}$. Now, use Prop. 1.5.26. (How?)

ii.) Use (2.23) in (2.22) and take the sup over $[0, T]$.

2.4.3. Use (H2.1) for the next inequality. As for the conclusion, observe that $\forall t$,

$$\left\|\mathbf{U}_k(t) - \mathbf{U}(t)\right\|_{\mathbb{R}^N} \leq \left\|e^{\mathbf{A}_k t}\mathbf{U}_0^k - e^{\mathbf{A}t}\mathbf{U}_0\right\|_{\mathbb{R}^N}$$

$$\leq \left\|e^{\mathbf{A}_k t}\right\|_{\mathbb{M}^N} \left\|\mathbf{U}_0^k - \mathbf{U}_0\right\|_{\mathbb{R}^N} + \left\|e^{\mathbf{A}_k t} - e^{\mathbf{A}t}\right\|_{\mathbb{M}^N} \left\|\mathbf{U}_0\right\|_{\mathbb{R}^N}.$$

The right-side goes to zero as $k \to \infty$.

2.4.4. Let $\mathbf{x} = \begin{bmatrix} x_1 \\ x_2 \end{bmatrix}$ and observe that

$$\lim_{n \to \infty} \left(\mathbf{I} - \frac{t\mathbf{A}}{n}\right)^{-n} \mathbf{x} = \begin{bmatrix} \lim_{n \to \infty} \left(1 - \frac{\alpha t}{n}\right)^{-n} x_1 \\ \lim_{n \to \infty} \left(1 - \frac{\beta t}{n}\right)^{-n} x_2 \end{bmatrix} = \begin{bmatrix} e^{\alpha t} x_1 \\ e^{\beta t} x_2 \end{bmatrix} = \ldots = e^{\mathbf{A}t}\mathbf{x}$$

2.5.1. I) See the example following Def. 2.2.2 for $e^{\mathbf{A}t}$. Also, $\left\|e^{\mathbf{A}t}\right\|_{\mathbb{M}^N} \leq e^{\max\{\alpha,\beta\}t}$, $\forall t$.
II) There is a uniform limit of $\mathbf{0}$ in (i) and (iv); the others do not have limit functions.
III) (2) None are periodic. (Can you use Exer. 1.9.4 to determine conditions under which such a solution *would* be periodic?)
  (3) (i) and (iv) only.
  (4) None exhibit this behavior because no component has a vertical asymptote.
2.5.2. i.) Let $\mathbf{U}_0 = \begin{bmatrix} x_0 \\ y_0 \end{bmatrix}$ and observe that $e^{\mathbf{A}t}\mathbf{U}_0 = \begin{bmatrix} x_0 \cos(\beta t) + y_0 \sin(\beta t) \\ -x_0 \sin(\beta t) + y_0 \cos(\beta t) \end{bmatrix}$.
Both components have period $\frac{2\pi}{\beta}$.

  ii.) Let $\mathbf{U}_0 = \begin{bmatrix} x_0 \\ y_0 \end{bmatrix}$ and observe that $e^{\mathbf{B}t}\mathbf{U}_0 = e^{\alpha t}e^{\mathbf{A}t}\mathbf{U}_0$. If $\alpha < 0$, then $\left\|e^{\mathbf{B}t}\mathbf{U}_0\right\|_{\mathbb{R}^2} \to$
0 as $t \to \infty$, while it approaches ∞ if $\alpha > 0$. The presence of the term $e^{\alpha t}$ prevents any of these from being periodic.
2.5.4. i.) $\|\mathbf{A}\|_{\mathbb{M}^N} = \max\{\lambda_i : i = 1,\ldots,N\} < 0$. Now conclude using Def. 2.5.2(i).
(How?)
  ii.) Use the hint for (i) and note that $\max\{\lambda_i : i = 1,\ldots,N\} < 0$ iff $\lim_{t \to \infty} \left\|e^{\mathbf{A}t}\right\|_{\mathbb{M}^N} = 0$. (So what?)
2.5.5. No to both due to the presence of a positive exponential entry in $e^{\mathbf{A}t}$.
2.5.6. The implications Def. 2.5.2(i)$\Longrightarrow$(iii) and (ii)$\Longrightarrow$(iii) follow easily. All eigenvalues of $\mathbf{A}$ must have negative real parts in order for $\lim_{t \to \infty} \left\|e^{\mathbf{A}t}\right\|_{\mathbb{M}^N} = 0$. (This is a not-so-obvious consequence of Prop. 2.2.3.) The other implications are not immediate (see **[59, 220]** for a detailed discussion).

Chapter 3

Abstract Homogenous Linear Evolution Equations

Overview

Partial differential equations are often an important component of the mathematical description of phenomena. Guided by our study of (2.10) in Chapter 2, it is natural to ask whether or not certain classes of initial-boundary value problems could also be subsumed under some theoretical umbrella in the spirit of (2.10), albeit in a more elaborate sense. The quick answer is yes, provided we interpret the pieces correctly. Our work in this chapter focuses on extending the theoretical framework from Chapter 2 to a more general setting.

3.1 Linear Operators

Up to now, we have dealt only with mappings from $\mathbb{R}^N$ to $\mathbb{R}^M$, where $N, M \in \mathbb{N}$. Extending the theory from Chapter 2 to more elaborate settings will require the use of more general mappings between Banach spaces $\mathscr{X}$ and $\mathscr{Y}$. We begin with a preliminary discussion of linear operators. A thorough treatment of the topics discussed in this section can be found in **[118, 146, 185, 202, 210, 340]**.

3.1.1 Bounded versus Unbounded Operators

Definition 3.1.1. Let $(\mathscr{X}, \|\cdot\|_{\mathscr{X}})$ and $(\mathscr{Y}, \|\cdot\|_{\mathscr{Y}})$ be real Banach spaces.
i.) A *bounded linear operator* from $(\mathscr{X}, \|\cdot\|_{\mathscr{X}})$ into $(\mathscr{Y}, \|\cdot\|_{\mathscr{Y}})$ is a mapping $\mathscr{F} : \mathscr{X} \to \mathscr{Y}$ such that

    a.) (linear) $\mathscr{F}(\alpha x + \beta y) = \alpha \mathscr{F}(x) + \beta \mathscr{F}(y)$, $\forall \alpha, \beta \in \mathbb{R}$ and $x, y \in \mathscr{X}$,

    b.) (bounded) There exists $m \geq 0$ such that $\|\mathscr{F}(x)\|_{\mathscr{Y}} \leq m \|x\|_{\mathscr{X}}$, $\forall x \in \mathscr{X}$.
We denote the set of all such operators by $\mathbb{B}(\mathscr{X}, \mathscr{Y})$. If $\mathscr{X} = \mathscr{Y}$, we write $\mathbb{B}(\mathscr{X})$ and refer to its members as "bounded linear operators on $\mathscr{X}$."
ii.) Let $\mathscr{F} \in \mathbb{B}(\mathscr{X}, \mathscr{Y})$. The *operator norm* of $\mathscr{F}$, denoted by $\|\mathscr{F}\|_{\mathbb{B}(\mathscr{X}, \mathscr{Y})}$ or more

succinctly as $\|\mathscr{F}\|_{\mathbb{B}}$, is defined by

$$\|\mathscr{F}\|_{\mathbb{B}} = \inf\{m : m > 0 \wedge \|\mathscr{F}(x)\|_{\mathscr{Y}} \leq m\|x\|_{\mathscr{X}}, \forall x \in \mathscr{X}\}$$
$$= \sup\{\|\mathscr{F}(x)\|_{\mathscr{Y}} : \|x\|_{\mathscr{X}} = 1\}.$$

iii.) If there does not exist $m \geq 0$ such that $\|\mathscr{F}(x)\|_{\mathscr{Y}} \leq m\|x\|_{\mathscr{X}}$, $\forall x \in \mathscr{X}$, we say that $\mathscr{F}$ is *unbounded*.

The terminology "bounded operator" may seem to be somewhat of a misnomer in comparison to the notion of a bounded real-valued function, but the name arose because such operators map norm-bounded subsets of $\mathscr{X}$ into norm-bounded subsets of $\mathscr{Y}$. We must simply contend with this nomenclature issue on a contextual basis. Also, regarding the notation, the quantity $\mathscr{F}(x)$ is often written more succinctly as $\mathscr{F}x$ (with parentheses suppressed) as in the context of matrix multiplication.

Exercise 3.1.1.
i.) Prove that $\|\cdot\|_{\mathbb{B}}$ is a norm on $\mathbb{B}(\mathscr{X}, \mathscr{Y})$.
ii.) Let $\mathscr{F} \in \mathbb{B}(\mathscr{X}, \mathscr{Y})$. Prove that $\|\mathscr{F}x\|_{\mathscr{Y}} \leq \|\mathscr{F}\|_{\mathbb{B}}\|x\|_{\mathscr{X}}$, $\forall x \in \mathscr{X}$.

Exercise 3.1.2. Explain how to prove that an operator $\mathscr{F} : \mathscr{X} \to \mathscr{Y}$ is unbounded.

Some Examples.
1. The identity operator $I : \mathscr{X} \to \mathscr{X}$ is in $\mathbb{B}(\mathscr{X})$ with $\|I\|_{\mathbb{B}} = 1$.
2. Let $\mathbf{A} \in \mathbb{M}^N(\mathbb{R})$. The operator $\mathfrak{B} : \mathbb{R}^N \to \mathbb{R}^N$ defined by $\mathfrak{B}\mathbf{x} = \mathbf{A}\mathbf{x}$ is in $\mathbb{B}(\mathbb{R}^N)$ with $\|\mathfrak{B}\|_{\mathbb{B}} = \|\mathbf{A}\|_{\mathbb{M}^N}$ (cf. Prop. 2.2.4).
3. Assume that $g : [a,b] \times [a,b] \to \mathbb{R}$ is continuous. Let $x \in \mathbb{C}([a,b];\mathbb{R})$ and define $y : [a,b] \to \mathbb{R}$ by $y(t) = \int_a^b g(t,s)x(s)\,ds$. The operator $\mathscr{F} : \mathbb{C}([a,b];\mathbb{R}) \to \mathbb{C}([a,b];\mathbb{R})$ defined by $\mathscr{F}(x) = y$ is in $\mathbb{B}(\mathbb{C}([a,b];\mathbb{R}))$, where $\mathbb{C}([a,b];\mathbb{R})$ is equipped with the sup norm.

Exercise 3.1.3.
i.) Provide the details in Example 3 and identify an upper bound for $\|\mathscr{F}\|_{\mathbb{B}}$.
ii.) Assume that $x \in \mathbb{C}([a,b];\mathbb{R})$. If $\mathscr{F}$ is, instead, viewed as an operator on $\mathbb{L}^2(a,b;\mathbb{R})$, is $\mathscr{F} \in \mathbb{B}(\mathbb{L}^2(a,b;\mathbb{R}))$?

Exercise 3.1.4. Assume that $g : [a,b] \times [a,b] \times \mathbb{R} \to \mathbb{R}$ is a continuous mapping for which $\exists m_g > 0$ such that

$$|g(x,y,z)| \leq m_g|z|, \forall x, y \in [a,b] \text{ and } z \in \mathbb{R}. \tag{3.1}$$

For every $x \in [a,b]$, define the operator $\mathscr{F} : \mathbb{C}([a,b];\mathbb{R}) \to \mathbb{C}([a,b];\mathbb{R})$ by

$$\mathscr{F}(z)[x] = \int_a^b g(x,y,z(y))\,dy. \tag{3.2}$$

Is $\mathscr{F}$ linear? bounded?

Exercise 3.1.5. Define $\mathscr{F} : \mathbb{C}^1\left((0,a);\mathbb{R}\right) \to \mathbb{C}\left([0,a];\mathbb{R}\right)$ by $\mathscr{F}(g) = g'$.
i.) Certainly, $\mathscr{F}$ is linear. Show that $\mathscr{F}$ is unbounded if $\mathbb{C}\left([0,a];\mathbb{R}\right)$ is equipped with the sup norm.
ii.) If $\mathbb{C}\left([0,a];\mathbb{R}\right)$ is equipped with the $\mathbb{L}^2-$norm, show that $\mathscr{F}$ is in $\mathbb{B}(\mathbb{L}^2\left(0,a;\mathbb{R}\right))$ with $\|\mathscr{F}\|_{\mathbb{B}}^2 \leq a$.

Important Note. Exercise 3.1.5 illustrates the fact that changing the underlying norm (not to mention the function space) can drastically alter the nature of the operator. This has ramifications in the theoretical development, and to an even greater extent the application of the theory to actual IBVPs. Often, choosing the correct closed subspace equipped with the right norm is critical in establishing the existence of a solution to an IBVP. We will revisit this issue frequently in what is to come.

Exercise 3.1.6. Define $\mathscr{F} : \mathbb{C}^2\left((0,a);\mathbb{R}\right) \to \mathbb{C}\left([0,a];\mathbb{R}\right)$ by $\mathscr{F}(g) = g''$. Certainly, $\mathscr{F}$ is linear. Show that $\mathscr{F}$ is unbounded if $\mathbb{C}\left([0,a];\mathbb{R}\right)$ is equipped with the sup norm.

Exercise 3.1.7. Let $\mathscr{F} \in \mathbb{B}(\mathscr{X},\mathscr{Y})$ and take $\{x_n\} \subset \mathscr{X}$ such that $\lim\limits_{n\to\infty} \|x_n - x\|_{\mathscr{X}} = 0$.
i.) Show that $\lim\limits_{n\to\infty} \|\mathscr{F}x_n - \mathscr{F}x\|_{\mathscr{Y}} = 0$. (Thus, bounded linear operators are continuous.)
ii.) Explain why $\mathscr{X}$ need not be complete in order for (ii) to hold.

The notions of domain and range are the same for any mapping. One nice feature of an operator $\mathscr{F} \in \mathbb{B}(\mathscr{X},\mathscr{Y})$ is that both $\mathrm{dom}(\mathscr{F})$ and $\mathrm{rng}(\mathscr{F})$ are vector subspaces of $\mathscr{X}$ and $\mathscr{Y}$, respectively. The need to compare two operators and to consider the restriction of a given operator to a subset of its domain arise often. These notions are made precise below.

Definition 3.1.2. Let $\mathscr{F},\mathscr{G} \in \mathbb{B}(\mathscr{X},\mathscr{Y})$.
i.) We say $\mathscr{F}$ *equals* $\mathscr{G}$, written $\mathscr{F} = \mathscr{G}$, if
    a.) $\mathrm{dom}(\mathscr{F}) = \mathrm{dom}(\mathscr{G})$,
    b.) $\mathscr{F}x = \mathscr{G}x$, for all x in the common domain.
ii.) Let $\mathscr{Z} \subset \mathrm{dom}(\mathscr{F})$. The operator $\mathscr{F}|_{\mathscr{Z}} : \mathscr{Z} \subset \mathscr{X} \to \mathscr{Y}$ defined by $\mathscr{F}|_{\mathscr{Z}}(z) = \mathscr{F}(z)$, $\forall z \in \mathscr{Z}$, is called the *restriction* of $\mathscr{F}$ to $\mathscr{Z}$.

The following claim says that if two bounded linear operators "agree often enough", then they agree everywhere. This is not difficult to prove. (Compare this to Prop. 1.8.6(v).)

Proposition 3.1.3. *Let* $\mathscr{F},\mathscr{G} \in \mathbb{B}(\mathscr{X})$ *and* $\mathscr{D}$ *a dense subset of* $\mathscr{X}$. *If* $\mathscr{F}x = \mathscr{G}x$, $\forall x \in \mathscr{D}$, *then* $\mathscr{F} = \mathscr{G}$.

All operators arising in Chapter 2 were members of the Banach space $\mathbb{M}^N(\mathbb{R})$. The structure inherent to a Banach space was essential to ensure that sums of $N \times N$ matrices and limits of convergent sequences of $N \times N$ matrices were well-defined. The very act of forming a more general theory suggests that spaces playing a comparable role in the present setting will need to possess a similar structure. While many of the operators arising in our models will be unbounded, the underlying theory relies heavily on the space $\mathbb{B}(\mathscr{X},\mathscr{Y})$. In accordance with intuition, it turns out that if $\mathscr{X}$ and $\mathscr{Y}$ are sufficiently nice, then $\mathbb{B}(\mathscr{X},\mathscr{Y})$ is also. Precisely, we have

Proposition 3.1.4. *If $\mathscr{X}$ and $\mathscr{Y}$ are Banach spaces, then $\mathbb{B}(\mathscr{X},\mathscr{Y})$ equipped with the norm $\|\cdot\|_{\mathbb{B}}$ is a Banach space.*

Proof. We break the proof into two main steps.
Step 1: Show that $\mathbb{B}(\mathscr{X},\mathscr{Y})$ is a linear space.
Let $\alpha,\beta \in \mathbb{R}$ and $\mathscr{F},\mathscr{G} \in \mathbb{B}(\mathscr{X},\mathscr{Y})$. Certainly, the operator $\alpha\mathscr{F} + \beta\mathscr{G}$ is linear (Why?) and thanks to Exercise 3.1.1 we have

$$\|(\alpha\mathscr{F} + \beta\mathscr{G})x\|_{\mathscr{Y}} \leq (|\alpha|\,\|\mathscr{F}\|_{\mathbb{B}} + |\beta|\,\|\mathscr{G}\|_{\mathbb{B}})\,\|x\|_{\mathscr{X}}\,,$$

$\forall x \in \mathscr{X}$. Hence, $\alpha\mathscr{F} + \beta\mathscr{G} \in \mathbb{B}(\mathscr{X},\mathscr{Y})$ with $\|\alpha\mathscr{F} + \beta\mathscr{G}\|_{\mathbb{B}} \leq |\alpha|\,\|\mathscr{F}\|_{\mathbb{B}} + |\beta|\,\|\mathscr{G}\|_{\mathbb{B}}$.
Step 2: Show that $\mathbb{B}(\mathscr{X},\mathscr{Y})$ is complete.
    Let $\{\mathscr{F}_n\}$ be a Cauchy sequence in $\mathbb{B}(\mathscr{X},\mathscr{Y})$. We must show that $\exists\mathscr{F} \in \mathbb{B}(\mathscr{X},\mathscr{Y})$ such that $\lim_{n\to\infty} \|\mathscr{F}_n - \mathscr{F}\|_{\mathbb{B}} = 0$. We begin by identifying a candidate for $\mathscr{F}$. To this end, let $x \in \mathscr{X}$ and $\varepsilon > 0$. There exists $P \in \mathbb{N}$ such that

$$n,m \geq P \implies \|\mathscr{F}_n - \mathscr{F}_m\|_{\mathbb{B}} < \frac{\varepsilon}{\|x\|_{\mathscr{X}} + 1} < \varepsilon. \tag{3.3}$$

Using (3.3) subsequently yields

$$n,m \geq P \implies \|\mathscr{F}_n x - \mathscr{F}_m x\|_{\mathscr{Y}} \leq \|\mathscr{F}_n - \mathscr{F}_m\|_{\mathbb{B}}\,\|x\|_{\mathscr{X}} < \varepsilon. \tag{3.4}$$

Hence, $\{\mathscr{F}_n x\}$ is a Cauchy sequence in $\mathscr{Y}$ and so it is convergent to some element of $\mathscr{Y}$. This is true $\forall x \in \mathscr{X}$. This prompts us to define the limit candidate $\mathscr{F} : \mathscr{X} \to \mathscr{Y}$ by $\mathscr{F}x = \lim_{n\to\infty} \mathscr{F}_n x$.
    We now prove that $\mathscr{F}$ "works." First, we show that $\mathscr{F} \in \mathbb{B}(\mathscr{X},\mathscr{Y})$. The linearity of $\mathscr{F}$ follows since the operators $\mathscr{F}_n$ and the limit operation are all linear. Next, observe that

$$\|\mathscr{F}x\|_{\mathscr{Y}} = \left\|\lim_{n\to\infty} \mathscr{F}_n x\right\|_{\mathscr{Y}} = \lim_{n\to\infty} \|\mathscr{F}_n x\|_{\mathscr{Y}} \leq \sup_{n\in\mathbb{N}} \|\mathscr{F}_n x\|_{\mathscr{Y}} \leq \left(\sup_{n\in\mathbb{N}} \|\mathscr{F}_n\|_{\mathbb{B}}\right)\|x\|_{\mathscr{X}}\,.$$

Since $\sup_{n\in\mathbb{N}} \|\mathscr{F}_n\|_{\mathbb{B}} < \infty$ (Why?), we conclude that $\mathscr{F}$ is bounded, as needed.
    Finally, we argue that

$$\lim_{n\to\infty} \|\mathscr{F}_n - \mathscr{F}\|_{\mathbb{B}} = \lim_{n\to\infty} \left(\sup\left\{\|\mathscr{F}_n x - \mathscr{F}x\|_{\mathscr{Y}} : \|x\|_{\mathscr{X}} = 1\right\}\right) = 0.$$

Since $\{\mathscr{F}_m x\}$ converges in $\mathscr{Y}$, taking the limit as $m \to \infty$ in (3.4) yields

$$n \geq N \implies \|(\mathscr{F}_n - \mathscr{F})(x)\|_{\mathscr{Y}} = \left\|\mathscr{F}_n x - \lim_{m\to\infty} \mathscr{F}_m x\right\|_{\mathscr{Y}}$$
$$= \lim_{m\to\infty} \|\mathscr{F}_n x - \mathscr{F}_m x\|_{\mathscr{Y}}$$
$$< \varepsilon\,\|x\|_{\mathscr{X}}\,,$$

$\forall x \in \mathscr{X}$. Hence,

$$n \geq N \Longrightarrow \|\mathscr{F}_n - \mathscr{F}\|_{\mathbb{B}} = \sup\left\{\|(\mathscr{F}_n - F)(x)\|_{\mathscr{Y}} : \|x\|_{\mathscr{X}} = 1\right\} < \varepsilon.$$

Thus, $\lim_{n \to \infty} \mathscr{F}_n = \mathscr{F}$. This completes the proof. $\qquad\square$

We frequently need to apply two operators in succession in the following sense.

Proposition 3.1.5. *Let $\mathscr{F}, \mathscr{G} \in \mathbb{B}(\mathscr{X})$ where $\mathrm{rng}(\mathscr{G}) \subset \mathrm{dom}(\mathscr{F})$. The composition operator $\mathscr{F}\mathscr{G} : \mathscr{X} \to \mathscr{X}$ defined by $(\mathscr{F}\mathscr{G})(x) = \mathscr{F}(\mathscr{G}(x))$ is in $\mathbb{B}(\mathscr{X})$ and satisfies $\|\mathscr{F}\mathscr{G}\|_{\mathbb{B}} \leq \|\mathscr{F}\|_{\mathbb{B}}\|\mathscr{G}\|_{\mathbb{B}}$. (If $\mathscr{F} = \mathscr{G}$, then $\mathscr{F}^2$ is written in place of $\mathscr{F}\mathscr{G}$.)*

Exercise 3.1.8. Prove Prop. 3.1.5.

Convergence properties such as Prop. 2.4.1 are important, especially when establishing numerical results for computational purposes. Different types of convergence can be defined by equipping the underlying spaces with different topologies. Presently, we focus only on the norm-topology.

Definition 3.1.6. Let $\{\mathscr{F}_n\} \subset \mathbb{B}(\mathscr{X}, \mathscr{Y})$. We say that $\{\mathscr{F}_n\}$ is
i.) *uniformly convergent* to $\mathscr{F}$ in $\mathbb{B}(\mathscr{X}, \mathscr{Y})$ if $\lim_{n \to \infty} \|\mathscr{F}_n - \mathscr{F}\|_{\mathbb{B}} = 0$.
ii.) *strongly convergent* to $\mathscr{F}$ if $\lim_{n \to \infty} \|\mathscr{F}_n x - \mathscr{F} x\|_{\mathscr{Y}} = 0$, $\forall x \in \mathscr{X}$.
We write $\mathscr{F}_n \xrightarrow{\text{uni}} \mathscr{F}$ and $\mathscr{F}_n \xrightarrow{\text{s}} \mathscr{F}$, respectively.

Exercise 3.1.9. Show (i) $\Longrightarrow$ (ii) in Def. 3.1.6.

Convergence issues will arise when defining different types of solutions of an IVP. In general, the characteristic properties of the limit operator are enhanced when the type of convergence used is strengthened. We will revisit this notion as the need arises. Note that we do not assume $\mathscr{F} \in \mathbb{B}(\mathscr{X}, \mathscr{Y})$ in Def. 3.1.6(ii). It actually turns out to be true, but not without some work. To show this, we make use of the following powerhouse theorem.

Theorem 3.1.7. (Principle of Uniform Boundedness)
Let $I \subset \mathbb{R}$ be a (possibly uncountable) index set and let $\{\mathscr{F}_i\} \subset \mathbb{B}(\mathscr{X}, \mathscr{Y})$ be such that $\forall x \in \mathscr{X}$, $\exists m_x > 0$ for which $\sup\{\|\mathscr{F}_i x\|_{\mathscr{Y}} : i \in I\} \leq m_x$. Then, $\exists m^\star > 0$ such that $\sup\left\{\|\mathscr{F}_i\|_{\mathbb{B}(\mathscr{X}, \mathscr{Y})} : i \in I\right\} \leq m^\star$.

Consequently, we have

Corollary 3.1.8. *If $\{\mathscr{F}_n : n \in \mathbb{N}\} \subset \mathbb{B}(\mathscr{X}, \mathscr{Y})$ and $\mathscr{F}_n \xrightarrow{\text{s}} \mathscr{F}$ as $n \longrightarrow \infty$, then $\mathscr{F} \in \mathbb{B}(\mathscr{X}, \mathscr{Y})$.*

Proof. Since $\mathscr{F}_n \xrightarrow{\text{s}} \mathscr{F}$, it follows that $\{\mathscr{F}_n x : n \in \mathbb{N}\}$ is a bounded subset of $\mathscr{Y}$, $\forall x \in \mathscr{X}$. So, by Thrm. 3.1.7, $\exists m^\star > 0$ such that $\sup\{\|\mathscr{F}_n\|_{\mathbb{B}} : n \in \mathbb{N}\} \leq m^\star$. Hence, $\forall x \in \mathscr{X}$,

$$\|\mathscr{F}_n x\|_{\mathscr{Y}} \leq \|\mathscr{F}_n\|_{\mathbb{B}}\|x\|_{\mathscr{X}} \leq m^\star \|x\|_{\mathscr{X}},$$

and so

$$\|\mathscr{F}x\|_{\mathscr{Y}} = \left\|\lim_{n\to\infty}\mathscr{F}_n x\right\|_{\mathscr{Y}} = \lim_{n\to\infty}\|\mathscr{F}_n x\|_{\mathscr{Y}} \le m^\star \|x\|_{\mathscr{X}}.$$

$\square$

3.1.2   Invertible Operators

The notions of one-to-one and onto for operators, and their relationships to invertibility, coincide with the usual elementary notions for real-valued functions.

Definition 3.1.9. **i.)** A linear operator $\mathscr{F} : \mathrm{dom}(\mathscr{F}) \subset \mathscr{X} \to \mathscr{Y}$ is
a.) *one-to-one* if $\mathscr{F}x = \mathscr{F}y \Longrightarrow x = y$, $\forall x, y \in \mathrm{dom}(\mathscr{F})$.
b.) *onto* if $\mathrm{rng}(\mathscr{F}) = \mathscr{Y}$.
ii.) An operator $\mathscr{G} : \mathrm{rng}(\mathscr{F}) \to \mathrm{dom}(\mathscr{F})$ that satisfies

$$\mathscr{G}(\mathscr{F}(x)) = x, \ \forall x \in \mathrm{dom}(\mathscr{F}),$$
$$\mathscr{F}(\mathscr{G}(y)) = y, \ \forall y \in \mathrm{rng}(\mathscr{F}),$$

is an *inverse* of $\mathscr{F}$ and is denoted by $\mathscr{G} = \mathscr{F}^{-1}$. If $\mathscr{G}$ exists, we say that $\mathscr{F}$ is *invertible*.

As with real-valued functions, $\mathscr{F} \in \mathbb{B}(\mathscr{X}, \mathscr{Y})$ is invertible iff $\mathscr{F} : \mathrm{dom}(\mathscr{F}) \to \mathrm{rng}(\mathscr{F})$ is one-to-one. (Why?) The following characterization is useful.

Proposition 3.1.10. *Let* $\mathscr{F} \in \mathbb{B}(\mathscr{X}, \mathscr{Y})$.
i.) $\mathscr{F}$ *is invertible iff* $\mathscr{F}x = 0 \Longrightarrow x = 0$.
ii.) *If* $\mathscr{F}$ *is invertible, then* $\mathscr{F}^{-1}$ *is a linear operator.*

Exercise 3.1.10. Prove Prop. 3.1.10.

The following notion of a convergent series in a Banach space is needed in the next result.

Definition 3.1.11. Let $\mathscr{X}$ be a Banach space and $\{x_n\} \subset \mathscr{X}$. We say that $\sum_{n=1}^{\infty} x_n$ *converges absolutely* if $\sum_{n=1}^{\infty} \|x_n\|_{\mathscr{X}}$ converges (in the sense of Def. 1.5.19).

Exercise 3.1.11. Let $\mathscr{X}$ be a Banach space and $\{x_n\} \subset \mathscr{X}$. For each $n \in \mathbb{N}$, define $S_n = \sum_{k=1}^{n} x_k$. Prove that if $\sum_{n=1}^{\infty} \|x_n\|_{\mathscr{X}}$ converges, then $\{S_n\}$ is a strongly convergent sequence in $\mathscr{X}$.

The identity operator I on $\mathscr{X}$ is clearly invertible, as are bounded linear operators that are "sufficiently close to I" in the following sense:

Proposition 3.1.12. *Let* $\mathscr{A} \in \mathbb{B}(\mathscr{X})$. *If* $\|\mathscr{A}\|_{\mathbb{B}} < 1$, *then*
i.) $I - \mathscr{A}$ *is invertible with inverse* $(I - \mathscr{A})^{-1}$.
ii.) $(I - \mathscr{A})^{-1}$ *is bounded.*
iii.) $\left\|(I - \mathscr{A})^{-1}\right\|_{\mathbb{B}} \le \frac{1}{1 - \|\mathscr{A}\|_{\mathbb{B}}}$.

Proof. Since $\|\mathscr{A}\|_{\mathbb{B}} < 1$, $\sum_{n=0}^{\infty} \|\mathscr{A}\|_{\mathbb{B}}^{n}$ is a convergent geometric series with sum $\frac{1}{1-\|\mathscr{A}\|_{\mathbb{B}}}$. By Exer. 3.1.11, $\{\sum_{k=0}^{n} \mathscr{A}^{k} : n \in \mathbb{N}\}$ is strongly convergent in $\mathbb{B}(\mathscr{X})$ with limit $\sum_{k=0}^{\infty} \mathscr{A}^{k}$. (Why?) Further, since

$$(I - \mathscr{A}) \left(\sum_{k=0}^{\infty} \mathscr{A}^{k} \right) = I = \left(\sum_{k=0}^{\infty} \mathscr{A}^{k} \right) (I - \mathscr{A}),$$

it follows that $I - \mathscr{A}$ is invertible with inverse $(I - \mathscr{A})^{-1} = \sum_{k=0}^{\infty} \mathscr{A}^{k}$ in $\mathbb{B}(\mathscr{X})$. Finally,

$$\left\| (I - \mathscr{A})^{-1} \right\|_{\mathbb{B}} \leq \left\| \sum_{k=0}^{\infty} \mathscr{A}^{k} \right\|_{\mathbb{B}} \leq \sum_{k=0}^{\infty} \|\mathscr{A}\|_{\mathbb{B}}^{k} = \frac{1}{1 - \|\mathscr{A}\|_{\mathbb{B}}}.$$

This concludes the proof. $\qquad\qquad\qquad\qquad\qquad\qquad\qquad\qquad\qquad\qquad\square$

Exercise 3.1.12. Let $\mathscr{A} \in \mathbb{B}(\mathscr{X})$. Show that if $\mathscr{A}$ is invertible, then $\alpha\mathscr{A}$ is invertible, $\forall \alpha \neq 0$.

3.1.3 Closed Operators

Recall what it means for a set $\mathscr{D}$ to be closed in $\mathscr{X}$, assuming that $\mathscr{X}$ is equipped with the norm topology. This notion can be extended to operators in $\mathbb{B}(\mathscr{X}, \mathscr{Y})$, but we must first make precise the notions of a product space and the graph of an operator.

Proposition 3.1.13. *Let* $(\mathscr{X}, \|\cdot\|_{\mathscr{X}})$ *and* $(\mathscr{Y}, \|\cdot\|_{\mathscr{Y}})$ *be real Banach spaces. The product space*

$$\mathscr{X} \times \mathscr{Y} = \{(x,y) : x \in \mathscr{X} \wedge y \in \mathscr{Y}\}$$

equipped with the operations

$$(x_1, y_1) + (x_2, y_2) = (x_1 + x_2, y_1 + y_2), \ \forall (x_1, y_1), (x_2, y_2) \in \mathscr{X} \times \mathscr{Y},$$
$$\alpha(x, y) = (\alpha x, \alpha y), \ \forall \alpha \in \mathbb{R}, x \in \mathscr{X}, y \in \mathscr{Y},$$

and the so-called graph norm

$$\|(x,y)\|_{\mathscr{X} \times \mathscr{Y}} = \|x\|_{\mathscr{X}} + \|y\|_{\mathscr{Y}}, \tag{3.5}$$

is a Banach space.

Exercise 3.1.13. Prove Prop. 3.1.13.

Definition 3.1.14. $\mathscr{F} : \mathscr{X} \to \mathscr{Y}$ is a *closed operator* if its graph, defined by

$$\text{graph}(\mathscr{F}) = \{(x, \mathscr{F}x) : x \in \text{dom}(\mathscr{F})\}$$

is a closed set in $\mathscr{X} \times \mathscr{Y}$.

Proving that an operator $\mathscr{F} : \mathscr{X} \to \mathscr{Y}$ is closed requires that $\forall \{x_n\} \subset \mathrm{dom}(\mathscr{F})$,

$$(\|x_n - x^\star\|_{\mathscr{X}} \to 0 \,\wedge\, \|\mathscr{F}x_n - y^\star\|_{\mathscr{Y}} \to 0) \Longrightarrow (x^\star \in \mathrm{dom}(\mathscr{F}) \wedge y^\star = \mathscr{F}x^\star.)$$

We have already encountered one example of a closed linear operator in Exer. 2.2.6. So, we know that the set of closed operators intersects $\mathbb{B}(\mathscr{X}, \mathscr{Y})$. However, neither set is contained within the other. Indeed, let $\mathscr{X}$ be a Banach space and $\mathscr{V}$ a linear subspace of $\mathscr{X}$ which is not closed. Then, the operator $I|_{\mathscr{V}}$ is bounded but not closed. (Why?) Next, consider the operator $\mathscr{F} : \mathbb{C}^1((a,b);\mathbb{R}) \to \mathbb{C}([a,b];\mathbb{R})$ defined by $\mathscr{F}(g) = g'$ (cf. Exer. 3.1.5). We assert that $\mathscr{F}$ is a closed operator. To see this, let $\{f_n\} \subset \mathbb{C}^1((a,b);\mathbb{R})$ be such that $f_n \to g$ in $\mathbb{C}^1((a,b);\mathbb{R})$ and $\mathscr{F}(f_n) = f_n' \to h$ in $\mathbb{C}([a,b];\mathbb{R})$. We must argue that $g \in \mathrm{dom}(\mathscr{F})$ and $g' = h$. Indeed, $\forall t \in [a,b]$,

$$\int_a^t h(s)\,ds = \int_a^t \lim_{n \to \infty} f_n'(s)\,ds = \lim_{n \to \infty} \int_a^t f_n'(s)\,ds$$
$$= \lim_{n \to \infty} [f_n(t) - f_n(a)] = g(t) - g(a). \tag{3.6}$$

Since h is continuous, Prop. 1.8.13(vii) implies that g is differentiable and $\frac{d}{dt}\int_a^t h(s)\,ds = h(t)$, which equals $g'(t)$ by (3.6), as desired.

Exercise 3.1.14. Define $\mathscr{B} : \mathrm{dom}(\mathscr{B}) \subset \mathbb{C}^2((a,b);\mathbb{R}) \to \mathbb{C}([a,b];\mathbb{R})$ by $\mathscr{B}(g) = -g''$, where $\mathrm{dom}(\mathscr{B})$ is given by

$$\{g \in \mathbb{C}^2((a,b);\mathbb{R}) \,|\, g, g'' \in \mathbb{C}([a,b];\mathbb{R}) \wedge g(a) = g(b) = g''(a) = g''(b) = 0\}. \tag{3.7}$$

Prove that $\mathscr{B}$ is a closed operator.

Exercise 3.1.15. Prove that if $\mathscr{B} : \mathrm{dom}(\mathscr{B}) \subset \mathscr{X} \to \mathscr{Y}$ is a closed invertible operator, then $\mathscr{B}^{-1}$ is a closed operator.

The next theorem is another useful powerhouse result.

Theorem 3.1.15. (Closed Graph Theorem)
Let $\mathscr{F} : \mathrm{dom}(\mathscr{F}) \subset \mathscr{X} \to \mathscr{Y}$ be a closed linear operator. If $\mathrm{dom}(\mathscr{F})$ is closed in $\mathscr{X}$, then $\mathscr{F} \in \mathbb{B}(\mathscr{X}, \mathscr{Y})$.

Remark. In particular, if $\mathrm{dom}(\mathscr{F}) = \mathscr{X}$, then Thrm. 3.1.15 reduces to saying "if $\mathscr{F}$ is a closed, linear operator, then $\mathscr{F}$ must be bounded."

3.1.4   Densely-Defined operators

Recall that a set $\mathscr{D} \subset \mathscr{X}$ is dense in $\mathscr{X}$ if $\mathrm{cl}_{\mathscr{X}}(\mathscr{D}) = \mathscr{X}$. Intuitively, we can get arbitrarily close (in the sense of the $\mathscr{X}$−norm) to any member of $\mathscr{X}$ with elements of $\mathscr{D}$. This notion can be used to define a so-called *densely-defined operator*, as follows:

Definition 3.1.16. $\mathscr{F} : \mathrm{dom}(\mathscr{F}) \subset \mathscr{X} \to \mathscr{Y}$ is *densely-defined* if $\mathrm{cl}_{\mathscr{X}}(\mathrm{dom}(\mathscr{F})) = \mathscr{X}$.

Exercise 3.1.16. Explain why, for normed spaces, it is sufficient to produce a set $\mathscr{D} \subset \mathrm{dom}(\mathscr{F})$ for which $\mathrm{cl}_{\mathscr{X}}(\mathscr{D}) = \mathscr{X}$ in order to prove that $\mathscr{F}$ is densely-defined.

We will encounter this notion exclusively when $\mathscr{X}$ and $\mathscr{Y}$ are function spaces. The underlying details can sometimes be involved, but we will primarily focus on tame examples. For instance, consider the following.

Proposition 3.1.17. *The operator $\mathscr{B}$ with domain $\mathbb{C}^1((a,b);\mathbb{R}) \cap \mathbb{C}([a,b];\mathbb{R})$, defined in Exer. 3.1.14 is linear, closed, and densely-defined.*

Proof. The linearity of $\mathscr{B}$ is clear and the fact that $\mathscr{B}$ is closed can be shown as in Exer. 3.1.14. In order to argue that $\mathscr{B}$ is densely-defined, we produce a set $\mathscr{D} \subset \mathrm{dom}(\mathscr{B})$ for which $\mathrm{cl}_{\mathbb{C}([a,b];\mathbb{R})}(\mathscr{D}) = \mathbb{C}([a,b];\mathbb{R})$. This is easy since every $h \in \mathbb{C}^1((a,b);\mathbb{R}) \cap \mathbb{C}([a,b];\mathbb{R})$ has a unique Fourier representation given by

$$h(x) = \sum_{n=0}^{\infty} \lambda_n \cos\left(\frac{n\pi x}{2(b-a)}\right), x \in [a,b],$$

where $\{\lambda_n\} \subset \mathbb{R}$ are the Fourier coefficients of h (cf. Section 1.7.2). Since the set

$$\left\{ \sum_{n=0}^{N} \lambda_n \cos\left(\frac{n\pi x}{2(b-a)}\right) \,\middle|\, \{\lambda_n\} \subset \mathbb{R}, N \in \mathbb{N} \right\}$$

is dense in $\mathbb{C}^1((a,b);\mathbb{R}) \cap \mathbb{C}([a,b];\mathbb{R})$, we conclude that $\mathscr{B}$ is densely-defined. This completes the proof. □

Exercise 3.1.17. Consider the operator $\mathscr{B} : \mathrm{dom}(\mathscr{B}) \subset \mathbb{L}^2((a,b);\mathbb{R}) \to \mathbb{L}^2((a,b);\mathbb{R})$ defined by $\mathscr{B}(g) = -g''$, where

$$\mathrm{dom}(\mathscr{B}) = \left\{ g \in \mathbb{W}_2^2((a,b);\mathbb{R}) \,\middle|\, g(a) = g(b) = 0 \right\}. \tag{3.8}$$

Is $\mathscr{B}$ densely-defined on $\mathbb{L}^2((a,b);\mathbb{R})$? Compare this to Prop. 3.1.17.

3.2 Motivation by Models

We now consider some elementary models often discussed in an introductory PDEs course. An interesting recent account of the state of the study of PDEs is provided in **[62]**. As we progress through the text, we will encounter increasingly more complex versions of these models and many others.

Model IV.1 Advection Equation - Pollution and Traffic Flow

Suppose we wish to study the concentration levels of a certain air pollutant over time at every point z in some enclosed region $\mathscr{D}$ of space. Let $c(z,t)$ denote the concentration of this pollutant at position z in $\mathscr{D}$ at time $t > 0$, and assume that the initial distribution of pollutant concentration throughout $\mathscr{D}$ is described by the function $c(z,0) = c_0(z)$. Assume momentarily that we only account for the effect of the wind $v(z,t)$ on the concentration levels throughout $\mathscr{D}$ over time, and for now ignore any effects due to diffusion or other atmospheric, chemical, or physical factors. Then, intuitively it would seem that as time goes on, the wind would simply "push" the initial profile $c_0(z)$ through $\mathscr{D}$ without changing its shape. How do we formally describe this phenomenon?

For simplicity, we reduce the above scenario to the one-dimensional case and take $\mathscr{D} = [0,\infty)$. Assume that the concentration is zero along the boundary of $\mathscr{D}$ (which is $\{0\}$) and, for ease of computation, assume that the wind is represented by the constant V, $\forall (z,t) \in \mathscr{D} \times [0,\infty)$. This scenario can be described by the IBVP

$$\begin{cases} \frac{\partial}{\partial t}c(z,t) & = V\frac{\partial}{\partial z}c(z,t),\ z > 0, t > 0, \\ c(z,0) & = c_0(z),\ z > 0, \\ c(0,t) & = 0,\ t > 0. \end{cases} \tag{3.9}$$

As expected, the solution $c : \mathscr{D} \times [0,\infty) \to \mathbb{R}$ is given by

$$c(z,t) = c_0(z + Vt), \text{ where } z \in \mathscr{D}, t \geq 0. \tag{3.10}$$

Exercise 3.2.1. Verify that (3.10) satisfies (3.9) using the multivariable chain rule. Why is this solution sensible based on the underlying assumptions?

Interestingly, a model of the form (3.9) can also be used as an elementary description of one-dimensional uni-directional traffic flow. Indeed, suppose we instead interpret $c(z,t)$ as the number of cars along a single-lane road at position z and at time t, and $c_0(z)$ as the initial dispersion of cars along the road. The wind field then corresponds to a velocity field $V(z,t)$ describing the speed at which cars move along the road at time t. It is reasonable to assume that at any point along the road, $V(z,t)$ depends only on the traffic density ρ, so that $V(z,t) = \rho(c(z,t))$. Using a "conservation of cars" argument involving a Taylor approximation leads to the following approximate linearized model of traffic flow:

$$\begin{cases} \frac{\partial}{\partial t}c(z,t) & = \frac{\partial}{\partial z}(V(z,t)c(z,t)),\ z > 0, t > 0, \\ c(z,0) & = c_0(z),\ z > 0, \\ c(0,t) & = 0,\ t > 0. \end{cases}$$

This IBVP reduces to (3.9) when ρ is constant. (Why?) For more discussion, see [**42, 152, 253**].

Our present goal is to reformulate the IBVP (3.9) as an abstract IVP (called an *abstract evolution equation*) of the form

$$\begin{cases} \frac{d}{dt}(u(t)) = A(u(t)),\ t > 0, \\ u(0) = u_0, \end{cases} \tag{3.11}$$

for some operator A in an appropriate Banach space $\mathscr{X}$. We begin with the following naive pairing of terms between (3.9) and (3.11). We will then analyze each identification in turn.

	IBVP (**3.9**)	Abstract IVP (**3.11**)	
Solution	$c : \mathscr{D} \times [0,\infty) \to \mathbb{R}$ given by $c(z,t)$	$u : [0,\infty) \to \mathscr{X}$ given by $u(t)$	(3.12)
Initial Condition	$c_0(z)$	u_0	(3.13)
Left Side	$\frac{\partial}{\partial t} \underbrace{(\ \cdot\ )}_{\text{function of } z,t}$	$\frac{d}{dt} \underbrace{(\ \cdot\ )}_{\text{function of } t}$	(3.14)
Right Side	$V\frac{\partial}{\partial z} \underbrace{(\ \cdot\ )}_{\text{function of } z,t}$	$A \underbrace{(\ \cdot\ )}_{\text{function of } t}$	(3.15)
Boundary Condition	$c(0,t) = 0$	None	(3.16)

First, since we are attempting to reformulate a *partial* differential equation (whose solution by its very nature depends upon time and at least one other variable) as an abstract *ordinary* differential equation (whose solution depends only on time), identification (3.12) suggests that for each time $t_0 \geq 0$, the term $u(t_0)$ must "contain" the information for the entire trajectory $\{c(z,t_0) : z > 0\}$. As such, $u(t_0)$ must itself be a function of z. We write

$$u(t_0)[z] = c(z,t_0), \ \forall z > 0. \tag{3.17}$$

Remark. Brackets are often used to enclose the input z while **parentheses** are used to enclose t_0 to emphasize the distinction between the function $u(t_0)$ and *its* input z. Later in the text, once we have acquired a reasonable level of comfort with these notions, the brackets will be suppressed.

Note that (3.13) follows from (3.17) because

$$u_0[z] = u(0)[z] = c(z,0) = c_0(z), \ \forall z > 0. \tag{3.18}$$

It follows from (3.17) and (3.18) that the space $\mathscr{X}$ mentioned in (3.12) must be a space of functions. But, *which* space exactly? This is a critical issue since our choice of the space $\mathscr{X}$ directly impacts the smoothness (also called *regularity*) of the solution $u(t)$. For instance, taking $\mathscr{X}$ to be $\mathbb{C}([0,\infty);\mathbb{R})$ instead of $\mathbb{C}^2([0,\infty);\mathbb{R})$ imposes a less restrictive condition on u since the latter requires the solution to be twice continuously differentiable. As such, we could end up with a less smooth solution. On the other hand, using $\mathbb{C}^2([0,\infty);\mathbb{R})$ imposes a degree of smoothness that might be unrealistic to expect of the underlying model (3.10), for physical reasons or otherwise. The point is that there is a choice to be made. We will not presently linger on the subtleties involved in making this choice because using the underlying characteristics inherent to model (physical, ecological, economical, etc.) inevitably enter into making the appropriate choice in a nontrivial manner. But, we will definitely revisit this question repeatedly. When information specific to the model is unavailable, we choose a convenient space for $\mathscr{X}$ that ensures all identifications involved in expressing the IBVP in the abstract form (3.11) are meaningful. For the current problem, we use

$$\mathscr{X} = \{f \in \mathbb{C}([0,\infty);\mathbb{R}) \,|\, f(0) = 0\}. \tag{3.19}$$

Based on the discussion leading to (3.17) and (3.18), it is quite natural that the *partial* derivative (in t) for the real-valued function c should be transformed into an *ordinary* derivative (also in t) for the $\mathscr{X}$-valued function u. So, (3.14) is reasonable (Why?), and the derivative is also an $\mathscr{X}$-valued function. (Why?)

Finally, we must handle (3.15) and (3.16). Judging from (3.17), the boundary condition

$$c(0,t) = 0, \ \forall t > 0, \tag{3.20}$$

is easily transformed into

$$u(t)[0] = 0, \ \forall t > 0. \tag{3.21}$$

However, the expression (3.21) is nowhere to be found in (3.11), yet it must be accounted for in the transformation of (3.9) into (3.11). Consequently, we must define the operator A in a manner which satisfies

$$A\left(u(t)\right)[\cdot] = V\frac{\partial}{\partial z}c(\cdot,t), \ \forall t > 0, \tag{3.22}$$

<u>and</u> accounts for (3.20). It is apparent from (3.22) that the inputs of A and the corresponding outputs Au are functions in the space $\mathscr{X}$. As such, A must be an operator from $\mathscr{X}$ into $\mathscr{X}$. But, not *all* functions in $\mathscr{X}$ should be included in dom(A) because there exist continuous functions f such that $f(0) \neq 0$, and such functions are not helpful in our search for a solution of (3.11).

Exercise 3.2.2. Why do we exclude such functions from consideration?

Therefore, it makes sense to restrict the domain of A to include only those functions for which (3.20) holds <u>and</u> (3.22) is defined, namely

$$A[f] = V\frac{d}{dz}[f],$$
$$\mathrm{dom}(A) = \left\{ f \in \mathscr{X} \,\middle|\, \frac{df}{dz} \in \mathscr{X} \wedge f(0) = 0 \right\}. \tag{3.23}$$

Remarks.
1. The choice of dom(A) in (3.23) seems perfectly reasonable for our purposes and turns out to be a viable choice. But, for reasons addressed later in this chapter, we must also make certain that the underlying space $\mathscr{X}$ and dom(A) are mutually chosen in a manner that ensures that dom(A) is a dense linear subspace of $\mathscr{X}$. (How would you show this?)

2. The flexibility we have when choosing $\mathscr{X}$ to account for additional information originating from the context of the model is also inherent in how $\mathrm{dom}(A)$ is chosen. This is not surprising since not only are the choices of $\mathscr{X}$ and $\mathrm{dom}(A)$ linked, but also we have already incorporated additional information (from the boundary condition) into the definition of $\mathrm{dom}(A)$. If the initial state $c_0(\cdot)$ is known *a priori* to be twice differentiable, for instance, then judging from (3.10) we would expect the solution to exhibit a higher degree of regularity. We expect the solution to be twice differentiable in the present scenario. Such information can be used to further refine the operator A. Indeed, we would refine the domain in (3.23) to be

$$\mathrm{dom}(A) = \left\{ f \in \mathscr{X} \,\middle|\, \frac{df}{dz}, \frac{d^2f}{dz^2} \in \mathscr{X} \wedge f(0) = 0 \right\}. \tag{3.24}$$

3. In light of (2), we emphasize that the identification of A in (3.15) must include its domain as part of the definition of A. Indeed, the operator A coupled with the domain listed in (3.23) is NOT the same as the pairing of the operator A with the domain listed in (3.24) because they are acting on different function spaces.

Using (3.17) - (3.23) enables us to successfully reformulate (3.9) as the abstract evolution equation (3.11) in the space $\mathscr{X} = \{ f \in \mathbb{C}([0,\infty);\mathbb{R}) \,|\, f(0) = 0 \}$.

Model V.1 The Many Faces of Diffusion - Heat Conduction

Classical diffusion theory originated in 1855 with the work of the physiologist Adolf Fick. The premise is simply that a diffusion substance (e.g., heat, gas, virus) will move from areas of high level of concentration toward areas of lower concentration. As such, in the absence of other factors (like advection or external forcing terms), we expect that if the substance diffuses only over a bounded region, its concentration would, over time, become uniformly distributed throughout the region.

This phenomenon arises in many disparate settings. Some common areas include intersymbol distortion of a pulse transmitted along a cable [227, 330], pheromone transport (emitted by certain species to identify mates) [51], migratory patterns of moving herds [134, 201], the spread of infectious disease through populated areas [178, 339], and the dispersion of salt through water [209]. We will consider different interpretations of diffusion (with added complexity) as the opportunity arises. We begin with a well-known classical model of heat conduction in one and two dimensions.

Consider a one-dimensional rod of length a with uniform properties and cross-sections. Assuming that no heat is generated and the surface is insulated, the homogenous heat equation describes the evolution of temperature throughout the rod over time. This equation, coupled with the initial profile, yields the IVP

$$\begin{cases} \frac{\partial}{\partial t} z(x,t) = k \frac{\partial^2}{\partial x^2} z(x,t), \ 0 < x < a, t > 0, \\ z(x,0) = z_0(x), \ 0 < x < a, \end{cases} \tag{3.25}$$

where $z(x,t)$ represents the temperature at position x along the rod at time t and k is a proportionality constant depending on the thermal conductivity and material density. A very readable account of the derivation of this equation from basic physical principles can be found in [273].

A complete description of this phenomenon requires that we prescribe what happens to the temperature on the boundary of the rod. This can be done in many naturally-occurring ways, some of which are described below:

1. Temperature is held constant along the boundary of the rod:

$$z(0,t) = C_1 \text{ and } z(a,t) = C_2, \ \forall t > 0. \tag{3.26}$$

2. Temperature is controlled along the boundary of the rod, but changes with time:

$$z(0,t) = C_1(t) \text{ and } z(a,t) = C_2(t), \ \forall t > 0. \tag{3.27}$$

3. Heat flow rate is controlled along the boundary of the rod:

$$\frac{\partial z}{\partial x}(0,t) = C_1(t) \text{ and } \frac{\partial z}{\partial x}(a,t) = C_2(t), \ \forall t > 0. \tag{3.28}$$

4. Convection (governed by Newton's law of heating and cooling):

$$C_1 z(0,t) + C_2 \frac{\partial z}{\partial x}(0,t) = C_3(t), \ \forall t > 0,$$

$$\overline{C_1} z(a,t) + \overline{C_2} \frac{\partial z}{\partial x}(a,t) = \overline{C_3}(t), \ \forall t > 0. \tag{3.29}$$

Boundary conditions (BCs) of the forms (3.26) and (3.27) are called *Dirichlet* BCs, while those of type (3.28) are called *Neumann* BCs. If the constants/functions C_i are zero, the BCs are called *homogenous*; otherwise, they are *nonhomogenous*. We can use a mixture of the types of BCs in the formulation of an IBVP. For instance, a homogenous Dirichlet BC can be imposed at one end of the rod and a nonhomogenous Neumann BC at the other end.

We first consider the IBVP formed by coupling (3.25) with the homogenous Dirichlet BCs

$$z(0,t) = z(a,t) = 0, \ \forall t > 0. \tag{3.30}$$

The solution can be constructed using the standard *separation of variables method* involving Fourier series (cf. Section 1.7.2 and [109, 125, 135, 242, 273]). To this end, assume the solution $z(x,t)$ is of the form

$$z(x,t) = X(x)T(t). \tag{3.31}$$

Assume that $X(x) \neq 0$, for all x. (As you will see, this is not a restrictive assumption.) Substituting (3.31) into the first equation in (3.25) yields

$$\underbrace{\frac{X''(x)}{X(x)}}_{\text{Function of only } x} = \lambda = \underbrace{\overbrace{\frac{T'(t)}{kT(t)}}^{T'(t)-k\lambda T(t)=0}}_{\text{Function of only } t}$$

$$\underbrace{\phantom{\frac{X''(x)}{X(x)}}}_{X''(x)-\lambda X(x)=0}$$

where λ is a constant. The general solution of $T'(t) - k\lambda T(t) = 0$ is

$$T(t) = C_\lambda e^{\lambda kt}, \ t > 0, \tag{3.32}$$

where C_λ is a constant. We seek all values of λ (referred to as *eigenvalues*) for which the resulting BVP

$$\begin{cases} X''(x) - \lambda X(x) = 0 \\ X(0) = X(a) = 0, \end{cases} \tag{3.33}$$

has a nonzero solution. It is easily shown that when $\lambda \geq 0$, the only solution of (3.33) is the zero solution. (Do so!) But, when $\lambda < 0$, which we denote as $\lambda = -n^2$ for convenience, the general solutions of (3.33) are

$$X(x) = C_1 \cos(nx) + C_2 \sin(nx). \tag{3.34}$$

(cf. Section 1.9.3) Applying the BCs given in (3.33) to (3.34) yields the two conditions

$$C_1 = 0 \text{ and } C_2 \sin(na) = 0. \tag{3.35}$$

Since choosing $C_2 = 0$ would result in the zero solution in (3.34) and hence would not contribute meaningfully to the construction of the general solution of the IBVP, we assume instead that $\sin(na) = 0$, which is satisfied when $n = \frac{m\pi}{a}, \forall m \in \mathbb{N}$. As such, the desired eigenvalues are given by

$$\lambda_m = -\left(\frac{m\pi}{a}\right)^2, \forall m \in \mathbb{N}. \tag{3.36}$$

For each $m \in \mathbb{N}$, the function

$$z_m(x,t) = b_m X_m(x) T_m(t) = b_m e^{-\left(\frac{m\pi}{a}\right)^2 kt} \sin\left(\frac{m\pi}{a}x\right), \tag{3.37}$$

where b_m is an appropriate constant, satisfies both the heat equation and the BCs. Hence, $\forall N \in \mathbb{N}$, the finite sum

$$\sum_{m=1}^{N} b_m e^{-\left(\frac{m\pi}{a}\right)^2 kt} \sin\left(\frac{m\pi}{a}x\right)$$

satisfies them (Why?) and more importantly, the function

$$z(x,t) = \sum_{m=1}^{\infty} b_m e^{-\left(\frac{m\pi}{a}\right)^2 kt} \sin\left(\frac{m\pi}{a}x\right) \tag{3.38}$$

satisfies them. (Tell why, carefully.) Applying the initial condition (IC) in (3.25) to determine the constants $\{b_m\}$ yields

$$z_0(x) = z(x,0) = \sum_{m=1}^{\infty} b_m \sin\left(\frac{m\pi}{a}x\right). \tag{3.39}$$

Assuming that $z_0(\cdot)$ is sufficiently smooth, b_m is the Fourier coefficient given by

$$b_m = \frac{2}{a} \int_0^a z_0(w) \sin\left(\frac{m\pi}{a} w\right) dw, \ \forall m \in \mathbb{N}. \tag{3.40}$$

(cf. Section 1.7.2) Finally, substituting (3.40) into (3.38) renders the solution of the IBVP as

$$z(x,t) = \sum_{m=1}^{\infty} \left(\frac{2}{a} \int_0^a z_0(w) \sin\left(\frac{m\pi}{a}w\right) dw \right) e^{-\left(\frac{m\pi}{a}\right)^2 kt} \sin\left(\frac{m\pi}{a}x\right) \tag{3.41}$$

$$= \sum_{m=1}^{\infty} \frac{2}{a} e^{-\left(\frac{m\pi}{a}\right)^2 kt} \left\langle z_0(\cdot), \sin\left(\frac{m\pi}{a}\cdot\right) \right\rangle_{\mathbb{L}^2(0,a;\mathbb{R})} \sin\left(\frac{m\pi}{a}x\right),$$

where $0 < x < a$ and $t > 0$.

We now transform this IBVP into an abstract evolution equation of the form (3.11). To this end, let $\mathscr{X} = \mathbb{L}^2(0,a;\mathbb{R})$, assume that $z_0(\cdot) \in \mathscr{X}$, and identify the solution and IC, respectively, by

$$u(t)[x] = z(x,t), \ 0 < x < a, t > 0, \tag{3.42}$$

$$u_0[x] = u(0)[x] = z(x,0) = z_0(x), \ 0 < x < a. \tag{3.43}$$

Define the operator $A : \mathrm{dom}(A) \subset \mathbb{L}^2(0,a;\mathbb{R}) \to \mathbb{L}^2(0,a;\mathbb{R})$ by

$$A[f] = k\frac{d^2}{dx^2}[f], \tag{3.44}$$

$$\mathrm{dom}(A) = \left\{ f \in \mathscr{X} \ \middle| \ \exists \frac{df}{dx}, \frac{d^2f}{dx^2}, \frac{d^2f}{dx^2} \in \mathscr{X}, \text{ and } f(0) = f(a) = 0 \right\}.$$

Identifying the time derivatives in the same manner as in Model IV.I, we see that using (3.42) - (3.44) yields a reformulation of the given IBVP into the form (3.11) in the Banach space $\mathbb{L}^2(0,a;\mathbb{R})$.

Remark. The domain specified in (3.44) is often written more succinctly using the Sobolev space $\mathbb{H}^2(0,a)$ (cf. (1.67)); indeed, it can be expressed equivalently as

$$\mathrm{dom}(A) = \left\{ f \in \mathbb{H}^2(0,a;\mathbb{R}) \mid f(0) = f(a) = 0 \right\}.$$

Exercise 3.2.3.
i.) Carefully explain why the choice of $\mathscr{X}$ and the identifications (3.42) - (3.44) used to transform the IBVP into the abstract form (3.11) are natural and justifiable.
ii.) To which space must z_0 belong to ensure that the transformation into (3.11) is meaningful? Is assuming $z_0 \in \mathbb{L}^1(0,a;\mathbb{R})$ sufficient?

Exercise 3.2.4.
i.) Use the separation of variables method to show that the solution of the IBVP obtained by coupling (3.25) instead with the homogenous Neumann BCs

$$\frac{\partial z}{\partial x}(0,t) = \frac{\partial z}{\partial x}(a,t) = 0, \ \forall t > 0, \tag{3.45}$$

is given by

$$z(x,t) = \sum_{m=0}^{\infty} \left(\frac{2}{a} \int_0^a z_0(w) \cos\left(\frac{m\pi}{a}w\right) dw \right) e^{-\left(\frac{m\pi}{a}\right)^2 kt} \cos\left(\frac{m\pi}{a}x\right). \tag{3.46}$$

ii.) Simplify (3.46) when $z_0(x) = x$.

iii.) Transform the IBVP described in (i) into an abstract evolution equation of the form (3.11). Clearly define all identifications.

Next, we consider a similar model for heat conduction in a two-dimensional rectangular plate composed of an isotropic, uniform material. Assuming that the temperature is zero along the boundary of the rectangle, the IBVP describing the transient temperature at every point on the plate over time is given by

$$\begin{cases} \frac{\partial z}{\partial t}(x,y,t) = k\left(\frac{\partial^2 z}{\partial x^2}(x,y,t) + \frac{\partial^2 z}{\partial y^2}(x,y,t)\right), 0 < x < a, 0 < y < b, t > 0, \\ z(x,y,0) = z_0(x,y), \ 0 < x < a, 0 < y < b, \\ z(x,0,t) = 0 = z(x,b,t), \ 0 < x < a, t > 0, \\ z(0,y,t) = 0 = z(a,y,t), \ 0 < y < b, t > 0, \end{cases} \tag{3.47}$$

where $z(x,y,t)$ represents the temperature at the point (x,y) on the plate at time t. We apply a suitably-modified version of the separation of variables method to solve (3.47). Assume that the solution $z(x,y,t)$ is of the form

$$z(x,y,t) = X(x)Y(y)T(t). \tag{3.48}$$

Substituting (3.48) into the first equation in (3.47) leads to the system of ODEs

$$\begin{cases} X''(x) + \lambda_x^2 X(x) = 0, X(0) = X(a) = 0, \\ Y''(y) + \lambda_y^2 Y(y) = 0, Y(0) = Y(b) = 0, \\ T'(t) + k\lambda^2 T(t) = 0, t > 0, \end{cases} \tag{3.49}$$

where the eigenvalues λ_x^2 and λ_y^2 are related by $\lambda^2 = \lambda_x^2 + \lambda_y^2$. (Tell why.) Proceeding as before, the following functions satisfy (3.49), $\forall m, n \in \mathbb{N}$ and constants C_{mn} :

$$\begin{cases} X_m(x) = \sin\left(\frac{m\pi}{a}x\right), \ 0 < x < a, \\ Y_n(y) = \sin\left(\frac{n\pi}{b}y\right), \ 0 < y < b, \\ T_{mn}(t) = C_{mn} e^{-\left(\left(\frac{m\pi}{a}\right)^2 + \left(\frac{n\pi}{b}\right)^2\right)kt}, t > 0. \end{cases} \tag{3.50}$$

(Verify this.) For each $m, n \in \mathbb{N}$ and constants b_{mn} , the function

$$z_{mn}(x,y,t) = b_{mn} X_m(x) Y_n(y) T_{mn}(t) \tag{3.51}$$

satisfies the heat equation and the BCs in (3.47). (Why?) It can be shown that

$$z(x,y,t) = \sum_{m=1}^{\infty} \sum_{n=1}^{\infty} b_{mn} X_m(x) Y_n(y) T_{mn}(t) \tag{3.52}$$

also satisfies them. (Why?) Applying the ICs in (3.47) to determine the constants $\{b_{mn}\}$ yields

$$z_0(x,y) = z(x,y,0) = \sum_{m=1}^{\infty} \sum_{n=1}^{\infty} b_{mn} X_m(x) Y_n(y), \ 0 < x < a, 0 < y < b. \qquad (3.53)$$

Assuming that $z_0(\cdot)$ is sufficiently smooth, we conclude that

$$b_{mn} = \frac{4}{ab} \int_0^b \int_0^a z_0(w,v) \sin\left(\frac{m\pi}{a} w\right) \sin\left(\frac{n\pi}{b} v\right) dw dv, \ m, n \in \mathbb{N}, \qquad (3.54)$$

which are double Fourier series coefficients. Finally, substituting (3.54) into (3.52) and using (3.50) yields the solution of the IBVP.

Exercise 3.2.5. Justify the details in the above discussion.

Exercise 3.2.6. Formulate (3.47) as an abstract evolution equation. Proceed by making suitable modifications to the approach used in the one-dimensional case.

Exercise 3.2.7. Replace the BCs in (3.47) by the homogenous Neumann BCs

$$\frac{\partial z}{\partial x}(0,y,t) = \frac{\partial z}{\partial x}(a,y,t) = 0, 0 < y < b, t > 0,$$
$$\frac{\partial z}{\partial y}(x,0,t) = \frac{\partial z}{\partial y}(x,b,t) = 0, 0 < x < a, t > 0. \qquad (3.55)$$

i.) Solve the resulting IBVP using the separation of variables method.
ii.) Formulate the IBVP as an abstract evolution equation.

Exercise 3.2.8. Construct an IBVP for heat conduction on an n-dimensional rectangular plate $[0,a_1] \times \ldots \times [0,a_n]$ equipped with homogenous Neumann BCs. Without going through all of the computations, conjecture a form of the solution. How would you formulate this IBVP as an abstract evolution equation?

Remark. The operators A used to formulate all of the above heat conduction IBVPs abstractly are forms of the *Laplacian operator* and are often denoted using the symbol $\triangle$.

Model VI.1 Fluid Flow Through Porous Media

The following model is a special case of a so-called *Sobolev-type* IBVP arising in the study of thermodynamics [186], fluid flow through fissured rocks [41], soil mechanics [149, 197], and consolidation of clay [251, 303, 327, 330]. We shall investigate such models more thoroughly in Chapter 7. For now, we consider

$$\begin{cases} \frac{\partial}{\partial t}\left(z(x,t) - \frac{\partial^2}{\partial x^2} z(x,t)\right) = \frac{\partial^2}{\partial x^2} z(x,t), \ 0 < x < \pi, t > 0, \\ z(x,0) = z_0(x), \ 0 < x < \pi, \\ z(0,t) = z(\pi,t) = 0, \ t > 0. \end{cases} \qquad (3.56)$$

The main difference between (3.56) and (3.25) is the presence of the term $-\frac{\partial^2 z}{\partial x^2}(x,t)$, which initially hinders our effort to transform (3.56) into the abstract form (3.11). As before, let $\mathscr{X} = \mathbb{L}^2(0,\pi;\mathbb{R})$ and define the operators $A : \text{dom}(A) \subset \mathscr{X} \to \mathscr{X}$ and $B : \text{dom}(B) \subset \mathscr{X} \to \mathscr{X}$ as follows:

$$A[f] = f'', \ \text{dom}(A) = \left\{ f \in \mathbb{H}^2(0,\pi;\mathbb{R}) \,|\, f(0) = f(\pi) = 0 \right\},$$
$$B[f] = f - f'', \ \text{dom}(B) = \text{dom}(A). \tag{3.57}$$

Making the identification $u(t)[x] = z(x,t)$ enables us to reformulate (3.56) as the following abstract evolution equation in $\mathbb{L}^2(0,\pi;\mathbb{R})$:

$$\begin{cases} (Bu)'(t) = Au(t), \ t > 0, \\ u(0) = u_0. \end{cases} \tag{3.58}$$

Exercise 3.2.9. Intuitively, what would be the natural thing to *try* to do in order to further express (3.58) in the form (3.11)? What conditions are needed to justify such a transformation?

We will study the intricacies of such problems in Chapter 7. For the moment, let us just say A and B must be compatible in order to facilitate the further transition to the form (3.11). Moreover, the solution of (3.56) is given by

$$z(x,t) = \sum_{m=1}^{\infty} e^{-\left(\frac{m^2}{m^2+1}\right)^2 t} \frac{-m^4}{m^2+1} \left\langle z_0(\cdot), \sqrt{\frac{2}{\pi}} \sin(m\cdot) \right\rangle_{\mathbb{L}^2(0,\pi;\mathbb{R})} \sqrt{\frac{2}{\pi}} \sin(mx),$$
$$\tag{3.59}$$

where $0 < x < \pi, t > 0$.

Exercise 3.2.10. Verify directly that (3.59) satisfies (3.56).

Common Theme: We now recap and make some observations that motivate our theoretical study of (3.11).

First, we have shown that various IBVPs can be written in the form (3.11) by suitably choosing the space $\mathscr{X}$, defining the operator A, and identifying the solution of (3.11) with the solution of the IBVP. Each of the operators $A : \text{dom}(A) \subset \mathscr{X} \to \mathscr{X}$ arising in these models is linear, closed, densely-defined and unbounded on $\mathscr{X}$. It remains to be seen if it is crucial that an operator A satisfy all of these conditions in order for the transformation of the IBVP into the abstract evolution equation (3.11) to be meaningful.

Second, now that we have written each of these IBVPs in the common abstract form (3.11), we need to develop a theory that

1.) precisely defines what is meant by a solution of (3.11);
2.) ensures (3.11) has such a solution; and
3.) enables us to solve (3.11) in a manner that coincides with the solution of the IBVP

in an appropriate sense.

The similarity of (3.11) to its finite-dimensional counterpart suggests that we search for the entity that plays the role of $\{e^{At} : t \geq 0\}$. Once identified, it is conceivable that a solution of (3.11) could be expressed as $u(t) = e^{At}u_0$, just as before.

3.3   Introducing Semigroups

For the remainder of the chapter, $\mathscr{X}$ denotes a real Banach space with norm $\|\cdot\|_{\mathscr{X}}$. Much of the material developed in the remainder of the chapter is based on the contents of the standard references **[48 49, 93, 103, 107, 108, 123, 129, 130, 142, 205, 245, 246, 267, 300, 318, 340, 341]**.

3.3.1   Motivation

Our present goal is to develop a theory analogous to the one formulated in Chapter 2 for (3.11) in a general Banach space $\mathscr{X}$. In the finite-dimensional setting, this amounted to defining the family of operators $\{e^{At} : t \geq 0\}$ and studying the evolution of its action on the initial state $\mathbf{U}_0$ over time. Familiarity with the Taylor series representation for the real-valued exponential function prompted us to define the matrix exponential as

$$e^{\mathbf{A}t} = \sum_{n=0}^{\infty} \frac{(\mathbf{A}t)^n}{n!}, \, t \geq 0. \tag{3.60}$$

This turned out to be crucial to the development of the theory.

Based on how well this definition worked in handling the extension from the one-dimensional case to $\mathbb{R}^N$, it is natural to ask if the same definition works when the matrix $\mathbf{A}$ is replaced by an unbounded operator A. However, a careful examination of our discussion in Chapter 2 reveals that the boundedness of the operator $\mathbf{A}$ played a key role in critical stages of the development.

Exercise 3.3.1. Identify the steps where boundedness was used in Chapter 2. Try to determine if this assumption was necessary in each case.

Even if we can somehow circumvent this hurdle, any hope that the generalization of the theory in Chapter 2 to the present setting will go through hitch-free is short-lived since, as seen in Section 3.2, $\text{dom}(A)$ need not equal $\mathscr{X}$. So, there could exist some $x_0 \in \mathscr{X}$ for which $Ax_0 \notin \text{dom}(A)$. In such case, the second term of the series (3.60) would be undefined, thereby rendering (3.60) meaningless.

Exercise 3.3.2. Explain why this cannot happen when $\mathbf{A} \in \mathbb{M}^N(\mathbb{R})$.

Nevertheless, we would like to handle the infinite-dimensional evolution equation (3.11) in a manner analogous to our theoretical treatment of (2.10). Specifically, we seek to define a family of linear operators $\{S(t) : t \geq 0\}$ possessing the same three dynamical characteristics of the matrix exponential (cf. discussion following Prop. 2.2.6) without relying on the use of a convenient representation formula like (3.60). Moreover, we want the solution of (3.11) to be completely governed by the action of these operators on the initial state u_0 over time; that is, $u(t) = S(t)[u_0]$, for any time $t \geq 0$. Motivated by our desire for the candidate for the new "matrix exponential" to exhibit these salient characteristics, we introduce the following definition.

Definition 3.3.1. A family of operators $\{S(t) : t \geq 0\}$ satisfying
i.) $S(t) \in \mathbb{B}(\mathscr{X})$, $\forall t \geq 0$,
ii.) $S(0) = I$, where I is the identity operator on $\mathscr{X}$, and
iii.) $S(t + s) = S(t)S(s)$, $\forall t, s \geq 0$ (called the *semigroup property*),
is called a *semigroup of bounded linear operators on* $\mathscr{X}$.
(For brevity, we say that $\{S(t) : t \geq 0\}$ is a *linear semigroup on* $\mathscr{X}$.)

Naturally, the question of how to compute e^{At} arises. When A is an $N \times N$ matrix, the Putzer algorithm yields a nice formula. This fails in the present more general setting. And, given that an unbounded linear operator A is a more complicated entity than is an $N \times N$ matrix, is it even reasonable to expect that such an operator could be expressed using a formula like (3.60)? In fact, must there even be a semigroup $\{e^{At} : t \geq 0\}$ associated to every unbounded linear operator A? The answers to these questions are forthcoming. For now, consider the following example that illustrates the existence of a family of operators satisfying Def. 3.3.1.

Example. Consider the solution of (3.25) equipped with homogenous Neumann BCs (cf. Exer. 3.2.4). For each $t \geq 0$, define the operator $S(t) : \mathbb{L}^2(0, a; \mathbb{R}) \to \mathbb{L}^2(0, a; \mathbb{R})$ by

$$S(t)[f][x] = \sum_{m=0}^{\infty} \left(\frac{2}{a} \int_0^a f(w) \cos\left(\frac{m\pi}{a} w\right) dw \right) e^{-\left(\frac{m\pi}{a}\right)^2 kt} \cos\left(\frac{m\pi}{a} x\right).$$

We claim that $\{S(t) : t \geq 0\}$ is a linear semigroup on $\mathbb{L}^2(0, a; \mathbb{R})$.
First, we show that $S(t) \in \mathbb{B}(\mathbb{L}^2(0, a; \mathbb{R}))$. Let $t \geq 0$. For any $f, g \in \mathbb{L}^2(0, a; \mathbb{R})$, applying the linearity of the integral and convergent series immediately yields

$$S(t)[f + g][x] = S(t)[f][x] + S(t)[g][x], \ 0 < x < a.$$

This proves linearity. As for boundedness, let $f \in \mathbb{L}^2(0, a; \mathbb{R})$. Using standard inequalities from Section 1.8 and properties of convergent series with the fact that $\sup\left\{ \left| \cos\left(\frac{m\pi}{a} w\right) \right| : w \in [0, a] \right\} \leq 1$ yields

$$\|S(t)[f]\|^2_{\mathbb{L}^2(0,a;\mathbb{R})} \leq 2 \int_0^a \left[\frac{2}{a} \|f\|^2_{\mathbb{L}^2(0,a;\mathbb{R})} + \left(\frac{2}{a}\right)^2 \|f\|^2_{\mathbb{L}^2(0,a;\mathbb{R})} M \sum_{m=1}^{\infty} e^{-\left(\frac{m\pi}{a}\right)^2 2kt} \right] dx$$

$$\leq \overline{M} \|f\|^2_{\mathbb{L}^2(0,a;\mathbb{R})} < \infty, \tag{3.61}$$

for some positive constants M and $\overline{M}$ depending on a, k, t, and the convergent series $\sum_{m=1}^{\infty} e^{-\left(\frac{m\pi}{a}\right)^2 2kt}$. This proves boundedness, so that Def. 3.3.1(i) has been shown.

Next, let $f \in \mathbb{L}^2(0, a; \mathbb{R})$. We use the fact that $\left\{\sqrt{\frac{2}{a}} \cos\left(\frac{m\pi}{a} \cdot\right) : m \in \mathbb{N} \cup \{0\}\right\}$ is an orthonormal basis of $\mathbb{L}^2(0, a; \mathbb{R})$ to see that

$$S(0)[f][x] = \sum_{m=0}^{\infty} \left(\frac{2}{a} \int_0^a f(w) \cos\left(\frac{m\pi}{a} w\right) dw\right) \cos\left(\frac{m\pi}{a} x\right)$$

$$= \sum_{m=0}^{\infty} \left\langle f(\cdot), \sqrt{\frac{2}{a}} \cos\left(\frac{m\pi}{a} \cdot\right)\right\rangle_{\mathbb{L}^2(0,a;\mathbb{R})} \sqrt{\frac{2}{a}} \cos\left(\frac{m\pi}{a} x\right)$$

$$= f(x).$$

Hence, $S(0) = I$, where I is the identity operator on $\mathbb{L}^2(0, a; \mathbb{R})$.

Finally, using the fact that $e^{-\left(\frac{m\pi}{a}\right)^2 k(t_1 + t_2)} = e^{-\left(\frac{m\pi}{a}\right)^2 kt_1} \cdot e^{-\left(\frac{m\pi}{a}\right)^2 kt_2}$, it follows immediately that $S(t_1 + t_2)[f][x] = S(t_2)(S(t_1)[f][x]), \forall f \in \mathbb{L}^2(0, a; \mathbb{R})$. This estabishes the semigroup property. Thus, we have shown that $\{S(t) : t \geq 0\}$ is a linear semigroup on $\mathbb{L}^2(0, a; \mathbb{R})$. $\qquad\square$

Exercise 3.3.3. Carefully provide the missing details in (3.61).

Exercise 3.3.4.
i.) Verify that the solution of (3.25) equipped with Dirichlet BCs (cf. Model V.1) defines a linear semigroup on $\mathbb{L}^2(0, a; \mathbb{R})$.
ii.) Show that the solution of the two-dimensional homogenous heat equation on $[0, a] \times [0, b]$, coupled with homogenous Neumann BCs, defines a linear semigroup on $\mathbb{L}^2((0, a) \times (0, b); \mathbb{R})$.
iii.) Based on your conjecture in Exer. 3.2.8, what would be the natural candidate for the semigroup and underlying space $\mathscr{X}$ for the n dimensional version of (ii)?

Exercise 3.3.5. Consider the solution of (3.56). For each $t \geq 0$, define the operator $S(t) : \mathbb{L}^2(0, \pi; \mathbb{R}) \to \mathbb{L}^2(0, \pi; \mathbb{R})$ by

$$S(t)[f][x] = \sum_{m=1}^{\infty} e^{-\left(\frac{m^2}{m^2+1}\right)^2 t} \left\langle f(\cdot), \sqrt{\frac{2}{\pi}} \sin(m \cdot)\right\rangle_{\mathbb{L}^2(0,\pi;\mathbb{R})} \sqrt{\frac{2}{\pi}} \sin(mx).$$

Prove that $\{S(t) : t \geq 0\}$ is a linear semigroup on $\mathbb{L}^2(0, \pi; \mathbb{R})$.

Exercise 3.3.6. For each $t \geq 0$, define the operator $S(t) : \mathbb{C}([0, \infty); \mathbb{R}) \to \mathbb{C}([0, \infty); \mathbb{R})$ by $S(t)[f][x] = f(x + vt)$. Show that $\{S(t) : t \geq 0\}$ is a linear semigroup on $\mathbb{C}([0, \infty); \mathbb{R})$.

A linear semigroup $\{S(t) : t \geq 0\}$ as defined in Def. 3.3.1 bears a striking resemblance to the matrix exponential $\{e^{\mathbf{A}t} : t \geq 0\}$ studied in Chapter 2. This is no accident, but what plays the role of the generator $\mathbf{A}$ in the present setting? Guided by Def. 2.2.9, we proceed as follows.

Let $\{S(t) : t \geq 0\}$ be a linear semigroup on $\mathscr{X}$ and define

$$\mathscr{D} = \left\{ x \in \mathscr{X} \,\middle|\, \exists \lim_{h \to 0^+} \frac{(S(h) - I)x}{h} \right\}. \tag{3.62}$$

It follows directly from the linearity of the semigroup $\{S(t) : t \geq 0\}$ that $\mathscr{D}$ is a linear subspace of $\mathscr{X}$. (Tell why.) Now, define the operator $A : \mathscr{D} \subset \mathscr{X} \to \mathscr{X}$ by

$$Ax = \lim_{h \to 0^+} \frac{(S(h) - I)x}{h}. \tag{3.63}$$

Observe that A is linear on $\mathscr{D}$. (Why?) Note that this is precisely how the generator of $\{e^{\mathbf{A}t} : t \geq 0\}$ was defined when $\mathbf{A} \in \mathbb{M}^N(\mathbb{R})$, with the notable difference being that in the finite-dimensional setting $\mathscr{D} = \mathbb{R}^N = \mathscr{X}$, whereas in the infinite-dimensional setting we will find that $\mathscr{D}$ need only be a dense linear subspace of $\mathscr{X}$ and, in general, is not the entire space. Henceforth, we write $\mathrm{dom}(A)$ in place of $\mathscr{D}$.

Formally, we have the following definition.

Definition 3.3.2. A linear operator $A : \mathrm{dom}(A) \subset \mathscr{X} \to \mathscr{X}$ defined by

$$Ax = \lim_{h \to 0^+} \frac{(S(h) - I)x}{h},$$

$$\mathrm{dom}(A) = \left\{ x \in \mathscr{X} \,\middle|\, \exists \lim_{h \to 0^+} \frac{(S(h) - I)x}{h} \right\}$$

is called an *(infinitesimal) generator* of $\{S(t) : t \geq 0\}$. We say A *generates* $\{S(t) : t \geq 0\}$.

The similarity between the above discussion and the analogous one in Chapter 2 prompts us to introduce the notation $S(t) \equiv e^{At}$ for a linear semigroup on $\mathscr{X}$ whose generator is A. We anticipate many of the same properties, like $\frac{d}{dt} S(t) = A e^{At}$, to carry over to the present setting.

Remark. Note that we have intentionally used a subtle notational change when referring to a linear semigroup (A is now italicized) in contrast to the matrix exponential in Def. 2.2.2 (where $\mathbf{A}$ is bold-faced). We distinguish between the two simply to highlight when we are dealing with a finite-dimensional setting. But, keep in mind that we intend for Def. 2.2.2 to be subsumed as a special case of Def. 3.3.1, which renders this notational difference moot.

Exercise 3.3.7. Based on your experience from Chapter 2, try to identify the operators that generate each of the semigroups in Exercises 3.3.4 - 3.3.6.

All linear semigroups arising in the next several chapters will satisfy Def. 3.3.1. We can further characterize them based on the strength of the convergence used to ensure that dynamical characteristic 3 (regarding continuous dependence, see the discussion following Prop. 2.2.6) holds. We consider two such natural distinctions below.

3.3.2   Uniformly Continuous Semigroups

The continuous dependence property suggests that a linear semigroup $\{e^{\mathbf{A}t} : t \geq 0\}$ must be continuous in t, in *some* sense. We proved in Prop. 2.2.7 that

$$\text{For any } \mathbf{A} \in \mathbb{M}^N(\mathbb{R}), \; \lim_{t \to 0^+} \left\| e^{\mathbf{A}t} - I \right\|_{\mathbb{B}(\mathbb{R}^N)} = 0.$$

This is a strong result concerning the (uniform) continuity of $\{e^{\mathbf{A}t} : t \geq 0\}$ and certainly implies the continuous dependence described in dynamical characteristic 3. (Why?) Using this idea in the present setting motivates our first characterization of linear semigroups.

Definition 3.3.3. The linear semigroup $\{e^{\mathbf{A}t} : t \geq 0\}$ on $\mathscr{X}$ is called *uniformly continuous* (U.C.) if $\lim_{t \to 0^+} \left\| e^{\mathbf{A}t} - I \right\|_{\mathbb{B}(\mathscr{X})} = 0$.

Exercise 3.3.8. Assume that $\{e^{\mathbf{A}t} : t \geq 0\}$ satisfies Def. 3.3.3. Prove that $\forall t > 0$, $\exists \lim_{h \to 0} \left\| e^{\mathbf{A}(t+h)} - e^{\mathbf{A}t} \right\|_{\mathbb{B}(\mathscr{X})} = 0$. (Hence, the mapping $t \mapsto \left\| e^{\mathbf{A}t} \right\|_{\mathbb{B}(\mathscr{X})}$ is continuous.)

Which linear semigroups, if any, actually satisfy this rather strong condition? More pointedly, will those arising in our models satisfy it? The following characterization result directly answers this question.

Theorem 3.3.4. (Only bounded linear operators can generate U.C. semigroups!)
A linear operator $A : \mathrm{dom}(A) \subset \mathscr{X} \to \mathscr{X}$ generates a U.C. semigroup on $\mathscr{X}$ if and only if $A \in \mathbb{B}(\mathscr{X})$ and $\mathrm{dom}(A) = \mathscr{X}$. Moreover, this semigroup is unique.

Proof. ($\Longleftarrow$) Let $A \in \mathbb{B}(\mathscr{X})$ with $\mathrm{dom}(A) = \mathscr{X}$. For each $t \geq 0$, consider the operator $S(t) : \mathscr{X} \to \mathscr{X}$ defined by

$$S(t) = \sum_{n=0}^{\infty} \frac{(At)^n}{n!}. \tag{3.64}$$

One can argue as in the proof of Prop. 2.2.6 (with the $\mathscr{X}$-norm in place of the $\mathbb{R}^N$-norm) to verify that $\{S(t) : t \geq 0\}$ satisfies Def. 3.3.1. The uniform continuity holds since

$$0 \leq \|S(t) - I\|_{\mathbb{B}(\mathscr{X})} = \left\| \sum_{n=1}^{\infty} \frac{(At)^n}{n!} \right\|_{\mathbb{B}(\mathscr{X})} \leq \sum_{n=1}^{\infty} \frac{t^n \|A\|_{\mathbb{B}(\mathscr{X})}^n}{n!} \leq e^{t\|A\|_{\mathbb{B}(\mathscr{X})}} - 1, \tag{3.65}$$

where the right-side goes to zero as $t \to 0^+$. (Why?) The squeeze theorem then implies that $\lim_{t \to 0^+} \|S(t) - I\|_{\mathbb{B}(\mathscr{X})} = 0$, so that $\{S(t) : t \geq 0\}$ is a U.C. semigroup.

The fact that A is the generator of $\{S(t) : t \geq 0\}$ follows easily since

$$0 \leq \left\| \frac{S(t) - I}{t} - A \right\|_{\mathbb{B}(\mathscr{X})} \leq \left\| tA^2 \sum_{n=2}^{\infty} \frac{A^{n-2}t^{n-2}}{n!} \right\|_{\mathbb{B}(\mathscr{X})} \tag{3.66}$$

$$\leq t \|A\|_{\mathbb{B}(\mathscr{X})}^2 e^{t\|A\|_{\mathbb{B}(\mathscr{X})}},$$

and the right-side of (3.66) goes to zero as $t \to 0^+$. (Why?) Thus, A satisfies Def. 3.3.2, as needed.

($\Longrightarrow$)Suppose that A generates a U.C. semigroup $\{T(t) : t \geq 0\}$. Applying elementary integral properties (cf. Section 1.8.4) with the semigroup property yields

$$\left(\frac{T(h) - I}{h}\right) \int_0^\delta T(t)dt = \frac{1}{h}\left[\int_0^\delta T(h)T(t)dt - \int_0^\delta T(t)dt\right]$$

$$= \frac{1}{h}\left[\int_\delta^{\delta+h} T(u)du - \int_0^h T(t)dt\right]. \qquad (3.67)$$

There exists $\delta > 0$ such that

$$\left\|\left(\frac{1}{\delta}\int_0^\delta T(t)dt\right) - I\right\|_{\mathbb{B}(\mathcal{X})} < 1. \qquad (3.68)$$

(Why?) Hence, $\int_0^\delta T(t)dt$ is an invertible operator by Prop. 3.1.12, so that (3.67) reduces to

$$\left(\frac{T(h) - I}{h}\right) = \left(\int_0^\delta T(t)dt\right)^{-1}\left[\frac{1}{h}\int_\delta^{\delta+h} T(u)du - \frac{1}{h}\int_0^h T(t)dt\right]. \qquad (3.69)$$

Subsequently, taking the limit as $h \to 0^+$ in (3.69) yields

$$\lim_{h \to 0^+}\left(\frac{T(h) - I}{h}\right) = \left(\int_0^\delta T(t)dt\right)^{-1}[T(\delta) - I]. \qquad (3.70)$$

As such, Def. 3.3.2 enables us to conclude that $\forall x \in \mathcal{X}$,

$$Ax = \lim_{h \to 0^+}\frac{T(h)x - x}{h} = \left(\int_0^\delta T(t)dt\right)^{-1}[T(\delta) - I]x.$$

Hence, $A = \left(\int_0^\delta T(t)dt\right)^{-1}[T(\delta) - I] \in \mathbb{B}(\mathcal{X})$, as needed.

It remains to prove the uniqueness of this semigroup. Suppose $\{S(t) : t \geq 0\}$ and $\{T(t) : t \geq 0\}$ are both U.C. semigroups on $\mathcal{X}$ for which the operator A is expressed by

$$A = \lim_{h \to 0^+}\frac{S(h) - I}{h} = \lim_{h \to 0^+}\frac{T(h) - I}{h}. \qquad (3.71)$$

For any $b > 0$, Exer. 3.3.8 guarantees that $\exists M > 0$ such that

$$\sup_{0 \leq t \leq b} \|T(t)\|_{\mathbb{B}(\mathcal{X})} \leq M \text{ and } \sup_{0 \leq t \leq b} \|S(t)\|_{\mathbb{B}(\mathcal{X})} \leq M. \qquad (3.72)$$

Let $\varepsilon > 0$. We claim that $\sup\limits_{0 \le t \le b} \|T(t) - S(t)\|_{\mathbb{B}(\mathscr{X})} < \varepsilon$. Indeed, note that (3.71) guarantees the existence of $\delta > 0$ such that $\forall 0 < h < \delta < b$,

$$\frac{1}{h}\|T(h) - S(h)\|_{\mathbb{B}(\mathscr{X})} = \left\|\frac{T(h) - I}{h} - \frac{S(h) - I}{h}\right\|_{\mathbb{B}(\mathscr{X})} < \frac{\varepsilon}{2bM^{\Sigma_{k=1}^{p}k}},$$

so that

$$\|T(h) - S(h)\|_{\mathbb{B}(\mathscr{X})} < \left(\frac{\varepsilon}{2bM^{\Sigma_{k=1}^{p}k}}\right)h. \tag{3.73}$$

(Tell why.) Also, $\exists p \in \mathbb{N}$ sufficiently large such that when $N = 2^p$, it is the case that $\frac{t}{N} < \delta$, $\forall 0 \le t \le b$. (Why?) As such, (3.73) applies to the set $\{\frac{t}{N} : 0 \le t \le b\}$. An application of the semigroup property yields

$$\|T(t) - S(t)\|_{\mathbb{B}(\mathscr{X})} = \left\|T\left(N \cdot \frac{t}{N}\right) - S\left(N \cdot \frac{t}{N}\right)\right\|_{\mathbb{B}(\mathscr{X})}$$

$$= \left\|\left[T\left(\frac{t}{N}\right)\right]^N - \left[S\left(\frac{t}{N}\right)\right]^N\right\|_{\mathbb{B}(\mathscr{X})}$$

Factoring the last expression subsequently yields, with the help of standard norm properties and (3.73),

$$\|T(t) - S(t)\|_{\mathbb{B}(\mathscr{X})} = \left\|T\left(\frac{t}{N}\right) - S\left(\frac{t}{N}\right)\right\|_{\mathbb{B}(\mathscr{X})} \prod_{k=1}^{p}\left\|\left[T\left(\frac{t}{N}\right)\right]^k - \left[S\left(\frac{t}{N}\right)\right]^k\right\|_{\mathbb{B}(\mathscr{X})}$$

$$\le \left(\frac{\varepsilon}{2bM^{\Sigma_{k=1}^{p}k}}\right)\frac{t}{N} \cdot \prod_{k=1}^{p}\left[\left\|T\left(\frac{t}{N}\right)\right\|_{\mathbb{B}(\mathscr{X})}^k + \left\|S\left(\frac{t}{N}\right)\right\|_{\mathbb{B}(\mathscr{X})}^k\right]$$

$$\le \left(\frac{\varepsilon}{2bM^{\Sigma_{k=1}^{p}k}}\right)\frac{t}{N} \cdot \prod_{k=1}^{p}2M^k \tag{3.74}$$

$$= \left(\frac{\varepsilon}{2bM^{\Sigma_{k=1}^{p}k}}\right)\frac{t}{N} \cdot 2M^{\Sigma_{k=1}^{p}k}$$

$$< \varepsilon$$

Since $\varepsilon > 0$ was arbitrary, we conclude that $S(t) = T(t)$, $\forall 0 \le t \le b$. Moreover, since this is true for all $b > 0$, uniqueness follows. This completes the proof. $\quad\square$

Important Remark! We have established that for such an operator A, the semi-group $S(t) = e^{tA}$ by uniqueness and satisfies the inequality $\|e^{tA}\|_{\mathbb{B}(\mathscr{X})} \le e^{t\|A\|_{\mathbb{B}(\mathscr{X})}}$, $\forall t \ge 0$. This will be particularly useful in the proof of Prop. 3.5.3.

Exercise 3.3.9. Fill in the following details in the proof of Thrm. 3.3.4.
i.) Verify the last inequalities in (3.65) and (3.66).
ii.) Where was the fact that $\text{dom}(A) = \mathscr{X}$ used in the proof of $(\Longleftarrow)$?
iii.) Carefully verify (3.67).
iv.) Carefully verify (3.74).

Theorem 3.3.4 has an important implication concerning the extension of the general theory developed in Chapter 2 to the present setting. Indeed, each of the operators A used to reformulate the IBVPs in Section 3.2 as the abstract evolution equation (3.11) in a Banach space $\mathscr{X}$ is unbounded. And, Thrm. 3.3.4 states in no uncertain terms that such operators cannot generate U.C. semigroups on $\mathscr{X}$. Consequently, Def. 3.3.3 falls woefully short of what we need for our present purposes. Now what?

3.3.3 Strongly Continuous Semigroups

A moment's thought suggests that the restrictive nature of Thrm. 3.3.4 is not surprising since the uniform continuity of Def. 3.3.3 was rather severe. The question is whether we can weaken this somewhat while still preserving a reasonable level of continuous dependence so that all three dynamical properties are preserved.

Intuitively, strong continuity, which is weaker than uniform continuity, should be sufficient, since for a given fixed initial state u_0, we simply need to make certain that small changes in t do not produce radically different outputs. There is no reason to demand from the very beginning that this dependence be uniform for an entire set of initial states. We are cautiously optimistic that weakening *uniform* continuity to *strong* continuity, defined below, will facilitate the extension of the theory to accommodate unbounded linear operators.

Definition 3.3.5. The linear semigroup $\{e^{At} : t \geq 0\}$ on $\mathscr{X}$ is called *strongly continuous* (S.C.) if $\lim\limits_{t \to 0^+} \left\| e^{At}x - x \right\|_{\mathscr{X}} = 0, \forall x \in \mathscr{X}$. In such case, $\{e^{At} : t \geq 0\}$ is referred to as a C_0−*semigroup on* $\mathscr{X}$.

Exercise 3.3.10. Why must a U.C. semigroup also be a C_0−semigroup?

Exercise 3.3.11. Show that the family of operators defined by the solution to the advection equation in Model IV.1 is a C_0−semigroup on $\mathbb{C}([0,\infty);\mathbb{R})$.

The semigroups arising in the models in Section 3.2 are, in fact, C_0−semigroups, but we postpone our verification of this fact until Section 3.5, at which time we will also formally identify their generators.

Henceforth, for simplicity, we focus almost exclusively on C_0−semigroups. Exercise 3.3.10 ensures that all properties of C_0−semigroups are inherited by U.C. semigroups. We gather several properties of C_0−semigroups and their generators in the following theorem, many of which were previously explored in the finite-dimensional setting. Most of those arguments work here with a simple change of norm.

Theorem 3.3.6. (Properties of C_0−semigroups and their Generators)
Let $\{e^{At} : t \geq 0\}$ *be a* C_0−*semigroup with generator* $A : \mathrm{dom}(A) \subset \mathscr{X} \to \mathscr{X}$. *Then,*
i.) $\mathrm{dom}(A)$ *is a linear subspace of* $\mathscr{X}$.
ii.) *The mapping* $t \mapsto \left\| e^{tA} \right\|_{\mathbb{B}(\mathscr{X})}$ *is bounded on bounded subsets of* $[0,\infty)$. *Moreover,*

there exist constants $\omega \in \mathbb{R}$ *and* $M \geq 1$ *such that*

$$\left\| e^{tA} \right\|_{\mathbb{B}(\mathscr{X})} \leq M e^{\omega t}, \ \forall t \geq 0. \tag{3.75}$$

iii.) *For every* $x_0 \in \mathscr{X}$, *the function* $g \colon [0,\infty) \to \mathscr{X}$ *defined by* $g(t) = e^{At}x_0$ *is continuous.*

iv.) *For every* $t_0 \geq 0$, *the function* $h \colon \mathscr{X} \to \mathscr{X}$ *defined by* $h(x) = e^{At_0}x$ *is continuous.*

v.) *For every* $x_0 \in \mathscr{X}$ *and* $t_0 \geq 0$, $\lim\limits_{h \to 0^+} \frac{1}{h} \int_{t_0}^{t_0 + h} e^{As}x_0 ds = e^{At_0}x_0$.

vi.) *For every* $x \in \mathrm{dom}(A)$ *and* $t \geq 0$, $e^{At}x \in \mathrm{dom}(A)$. *Moreover,* $\forall x \in \mathrm{dom}(A)$, *the function* $g \colon (0,\infty) \to \mathscr{X}$ *defined by* $g(t) = e^{At}x_0$ *is in* $\mathbb{C}^1 \left((0,\infty) ; \mathrm{dom}(A) \right)$ *and*

$$\frac{d}{dt} e^{At}x_0 = \underbrace{A e^{At}x_0 = e^{At}A x_0}_{\text{i.e., } A \text{ and } e^{At} \text{ commute}}. \tag{3.76}$$

vii.) *For every* $x_0 \in \mathscr{X}$ *and* $t_0 \geq 0$,

$$A \int_0^{t_0} e^{As}x_0 ds = e^{At_0}x_0 - x_0. \tag{3.77}$$

viii.) *For every* $x_0 \in \mathrm{dom}(A)$ *and* $0 < s \leq t < \infty$,

$$e^{At}x_0 - e^{As}x_0 = \int_s^t e^{Au}A x_0 du = \int_s^t A e^{Au}x_0 du. \tag{3.78}$$

ix.) *A is a densely-defined operator.*

x.) *A is a closed operator.*

xi.) *If A generates two* C_0-*semigroups* $\{S(t) : t \geq 0\}$ *and* $\{T(t) : t \geq 0\}$ *on* $\mathscr{X}$, *then* $S(t) = T(t), \ \forall t \geq 0.$

Proof. Properties (iii) - (viii) are verified in the same manner as in Chapter 2. (Tell how.) We prove the others.

Property (i) follows directly from the linearity of A and $\left\{ e^{At} : t \geq 0 \right\}$. (Why?)

In order to prove (ii), we first show there exist $\delta_0 > 0$ and $M > 1$ such that

$$\sup_{0 \leq t \leq \delta_0} \left\| e^{At} \right\|_{\mathbb{B}(\mathscr{X})} \leq M. \tag{3.79}$$

Suppose, by way of contradiction, that (3.79) does not hold. Then, $\forall n \in \mathbb{N}$, $\exists t_n \in \left[0, \frac{1}{n} \right]$ such that $\left\| e^{At_n} \right\|_{\mathbb{B}(\mathscr{X})} \geq n$. (Why?) Theorem 3.1.7 then ensures the existence of at least one $z_0 \in \mathscr{X}$ for which the sequence $\left\{ \left\| e^{At_n}z_0 \right\|_{\mathscr{X}} : n \in \mathbb{N} \right\}$ is unbounded. This results in a contradiction since the strong continuity of $\left\{ e^{At} : t \geq 0 \right\}$ implies that $\lim\limits_{n \to \infty} e^{At_n}z_0 = z_0$. (Tell how.)

Now, let $m \in \mathbb{N}$, and $0 \leq \tau \leq \delta_0$. Define $t = m\delta_0 + \tau$. Observe that

$$\left\| e^{At} \right\|_{\mathbb{B}(\mathscr{X})} = \left\| e^{A(m\delta_0 + \tau)} \right\|_{\mathbb{B}(\mathscr{X})} = \left\| \underbrace{e^{A\delta_0} \cdots e^{A\delta_0}}_{m \text{ times}} e^{A\tau} \right\|_{\mathbb{B}(\mathscr{X})}$$

$$\leq \left\| e^{A\delta_0} \right\|_{\mathbb{B}(\mathscr{X})}^m \left\| e^{A\tau} \right\|_{\mathbb{B}(\mathscr{X})} \leq M^{m+1}. \tag{3.80}$$

Since $t \geq m\delta_0$,

$$M^{m+1} \leq M \cdot M^{\frac{t}{\delta_0}} = Me^{\omega t}, \tag{3.81}$$

where $\omega = \delta_0^{-1} \log M$. (Why?)

Next, we prove (ix). Let $x_0 \in \mathscr{X}$. For each $n \in \mathbb{N}$, define $y_n = n \int_0^{\frac{1}{n}} e^{As} x_0 ds$. It follows from (vii) that $\{y_n\} \subset \text{dom}(A)$ and from (v) that $\lim_{n \to \infty} y_n = x_0$. (Tell how.) Hence, $x_0 \in \text{cl}_{\mathscr{X}}(\text{dom}(A))$. This proves (ix).

As for (x), let $\{x_n\} \subset \text{dom}(A)$ be such that

$$\lim_{n \to \infty} x_n = x \text{ and } \lim_{n \to \infty} Ax_n = y. \tag{3.82}$$

From (viii), it follows that

$$e^{Ah} x_n - x_n = \int_0^h e^{Au} Ax_n du, \tag{3.83}$$

so that letting $n \to \infty$ yields (by LDC)

$$\begin{aligned} e^{Ah} x - x &= \lim_{n \to \infty} \left[e^{Ah} x_n - x_n \right] = \lim_{n \to \infty} \int_0^h e^{Au} Ax_n du \\ &= \int_0^h e^{Au} \left(\lim_{n \to \infty} Ax_n \right) du = \int_0^h e^{Au} y du. \end{aligned} \tag{3.84}$$

Since (v) implies that $\lim_{h \to 0^+} \frac{1}{h} \int_0^h e^{Au} Ax du = y$, using (3.83) and (3.84) together yields

$$Ax = \lim_{h \to 0^+} \frac{\left(e^{Ah} - I \right) x}{h} = \lim_{h \to 0^+} \frac{1}{h} \int_0^h e^{Au} Ax du = y.$$

Thus, $x \in \text{dom}(A)$ and $y = Ax$, thereby proving that A is a closed operator.

Finally, we verify (xi). Let $t > 0$. It suffices to show that

$$S(t)x = T(t)x, \ \forall x \in \text{dom}(A). \tag{3.85}$$

(Why?) To this end, let $x \in \text{dom}(A)$ and define the mapping $\psi : [0, t] \to \mathscr{X}$ by

$$\psi(s) = S(t - s)T(s)x, \ 0 \leq s \leq t. \tag{3.86}$$

Using (v), we see that ψ is differentiable and $\forall 0 \leq s \leq t$,

$$\begin{aligned} \frac{d\psi}{ds}(s) &= S(t - s)\frac{d}{ds}(T(s)x) + \frac{d}{ds}(S(t - s)x)T(s) \\ &= S(t - s)AT(s)x - AS(t - s)xT(s) = 0. \end{aligned}$$

Hence, ψ is constant on $[0, t]$, so that $\psi(0) = \psi(t)$. Thus, we conclude that (3.85) holds. (Why?) This completes the proof. $\square$

Definition 3.3.7. A C_0-semigroup is called *contractive* on $\mathscr{X}$ if $M = 1$ and $\omega \leq 0$ in Thrm. 3.3.6(ii).

We shall focus on this particular subclass of C_0-semigroups when developing many of the major theorems later in the chapter since there are fewer technical details than in the case of a general C_0-semigroup.

Exercise 3.3.12. Provide the following details in the proof of Thrm. 3.3.6.
i.) Verify the arguments of (iii) - (vii) when $h < 0$ and/or $t_0 = 0$.
ii.) Justify the string of equalities in (3.84).
iii.) Why does showing (3.85) suffice to prove (xi)?

Exercise 3.3.13. Let $\left\{ e^{At} : t \geq 0 \right\}$ be a C_0-semigroup on $\mathscr{X}$ with generator $A :$ $\mathrm{dom}(A) \subset \mathscr{X} \to \mathscr{X}$. Prove that if $f : [0, \infty) \to \mathbb{R}$ is continuous, then $\forall x_0 \in \mathscr{X}$ and $t_0 > 0$,

$$\lim_{h \to 0} \frac{1}{h} \int_{t_0}^{t_0+h} f(s) e^{As} x_0 ds = f(t_0) e^{At_0} x_0.$$

Exercise 3.3.14. Let $\alpha \in \mathbb{R}$ and assume that $A : \mathrm{dom}(A) \subset \mathscr{X} \to \mathscr{X}$ generates a C_0-semigroup $\left\{ e^{At} : t \geq 0 \right\}$ on $\mathscr{X}$. Prove that $\left\{ e^{\alpha t} e^{At} : t \geq 0 \right\}$ is a C_0-semigroup on $\mathscr{X}$ with generator $\alpha I - A$.

Exercise 3.3.15. Prove that the semigroup arising in the example directly following Def. 3.3.1 is contractive.

Exercise 3.3.16. Prove that if $A : \mathrm{dom}(A) \subset \mathscr{X} \to \mathscr{X}$ generates a C_0-semigroup $\left\{ e^{At} : t \geq 0 \right\}$ on $\mathscr{X}$, then $\forall \alpha > 0$, $\frac{1}{\alpha} A$ generates the C_0-semigroup $\left\{ e^{A\left(\frac{t}{\alpha}\right)} : t \geq 0 \right\}$ on $\mathscr{X}$. Is the condition that $\alpha > 0$ necessary?

3.4 The Abstract Homogenous Cauchy Problem

The abstract IVP

$$\begin{cases} u'(t) = Au(t), \ t > 0, \\ u(0) = u_0, \end{cases} \tag{3.87}$$

is referred to as an *abstract homogenous Cauchy problem in* $\mathscr{X}$. When does (3.87) have a solution? In fact, what do we even mean by a solution? We make precise one notion of a solution below.

Definition 3.4.1. A function $u : [0, \infty) \to \mathscr{X}$ is a *classical solution* of (3.87) if
i.) $u \in \mathbb{C}([0, \infty); \mathscr{X})$,
ii.) $u(t) \in \mathrm{dom}(A), \forall t > 0$,

iii.) $u \in \mathbb{C}^1((0,\infty);\mathrm{dom}(A))$,
iv.) (3.87) is satisfied.

Exercise 3.4.1. Why must we impose condition (ii) in Def. 3.4.1? Why was this not specified in Def. 2.3.1 when $\mathscr{X} = \mathbb{R}^N$?

The properties of Thrm. 3.3.6 enable us to establish the existence and uniqueness of a classical solution of (3.87) just as we did in Chapter 2. This is summarized in the next theorem.

Theorem 3.4.2. *If $A : \mathrm{dom}(A) \subset \mathscr{X} \to \mathscr{X}$ generates a C_0-semigroup $\{e^{At} : t \geq 0\}$ on $\mathscr{X}$, then $\forall u_0 \in \mathrm{dom}(A)$, (3.87) has a unique classical solution given by $u(t) = e^{At}u_0$.*

Exercise 3.4.2. Prove Thrm. 3.4.2.

It is necessary to assume that $u_0 \in \mathrm{dom}(A)$ in Thrm. 3.4.2. Indeed, the function $u(t) = e^{At}u_0$ might not be differentiable if $u_0 \in \mathscr{X} \setminus \mathrm{dom}(A)$. In such case, u could not formally satisfy (3.87) in the sense of Def. 3.4.1. This is explored in the following two exercises.

Exercise 3.4.3. Let $\mathscr{X} = \mathbb{C}_B((0,\infty);\mathbb{R})$ denote the space of continuous real-valued functions that remain bounded on $(0,\infty)$. This is a Banach space when equipped with the usual sup norm. Define $A : \mathrm{dom}(A) \subset \mathscr{X} \to \mathscr{X}$ by

$$Af = f'$$
$$\mathrm{dom}(A) = \left\{ f \in \mathbb{C}_B((0,\infty);\mathbb{R}) \,\middle|\, f' \in \mathbb{C}_B((0,\infty);\mathbb{R}) \right\}.$$

Choose $u_0 \in \mathbb{C}_B((0,\infty);\mathbb{R})$ for which u_0' does not exist for at least some $t_0 \in (0,\infty)$. (Why is such a choice possible?) Show that for such a choice of u_0, (3.87) does not have a classical solution in the sense of Def. 3.4.1.

Exercise 3.4.4. Why do we not encounter the difficulty arising in Exer. 3.4.3 when $A \in \mathbb{B}(\mathscr{X})$?

The fact that we have been using $t = 0$ as the starting time of IVP (3.87) is not essential, but it does suffice, as illustrated by the following corollary (which is proven in the same manner as Cor. 2.3.3). As a result, we assume henceforth that all abstract IVPs are equipped with ICs evaluated at $t = 0$.

Corollary 3.4.3. *Assume the conditions of Theorem 3.4.2. For any $t_0 > 0$, the IVP*

$$\begin{cases} u'(t) = Au(t), \ t > t_0, \\ u(t_0) = u_0, \end{cases} \tag{3.88}$$

has a unique classical solution $u : [t_0,\infty) \to \mathscr{X}$ given by $u(t) = e^{A(t-t_0)}u_0$.

From the viewpoint of applications, we must account for the possibility that although we seek a classical solution of the IVPs of the form (3.87) for a precisely prescribed IC, some degree of error in measurement is inherently present. Consequently, in all likelihood, the actual IC is not u_0 but rather $u_0 + \varepsilon$ for some small $\varepsilon \neq 0$. Of course, some of these measurements are more likely to occur than others, and accounting for such probabilistic features requires that we view (3.87) stochastically, a notion explored in Volume 2. For now, we seek a classical solution of the IVP

$$\begin{cases} u'(t) & = Au(t),\, t > 0, \\ u(0) & = u_0 + \varepsilon. \end{cases} \tag{3.89}$$

Intuitively, unless there are specific reasons to suggest otherwise, we expect small changes in the initial data to produce only small changes in the resulting solution on any *bounded* time interval $[0,T]$. This would be an unrealistic expectation on an unbounded time interval since small changes, over an infinite amount of time, can naturally lead to drastically different outcomes. (Why?)

One way to establish such continuous dependence is to verify the convergence of the classical solutions to a suitable approximating sequence of IVPs to the classical solution of the original IVP. Indeed, suppose $A : \text{dom}(A) \subset \mathscr{X} \to \mathscr{X}$ generates a C_0–semigroup $\{e^{At} : t \geq 0\}$ on $\mathscr{X}$ and let $u_0 \in \text{dom}(A)$. Let $\{u_0^n\} \subset \text{dom}(A)$ be a sequence such that $\lim\limits_{n\to\infty} \left\| u_0^n - u_0 \right\|_{\mathscr{X}} = 0$. For each $n \in \mathbb{N}$, consider the IVP

$$\begin{cases} u_n'(t) = Au_n(t),\, t > 0, \\ u_n(0) = u_0^n. \end{cases} \tag{3.90}$$

Theorem 3.4.2 guarantees that $\forall n \in \mathbb{N}$, (3.90) has a unique classical solution $u_n : [0,\infty) \to \mathscr{X}$. The relationship between $\{u_n\}$ and u is made precise in the following theorem.

Theorem 3.4.4. (Continuous Dependence on Initial Data)
Assuming the above context, let $u_n : [0,\infty) \to \mathscr{X}$ denote the unique classical solution of (3.87). Then, $\forall T > 0, \lim\limits_{n\to\infty} \left\| u_n - u \right\|_{\mathbb{C}([0,T];\mathscr{X})} = 0$.

Exercise 3.4.5. Prove Thrm. 3.4.4.

Exercise 3.4.6. Provide a sufficient condition that would guarantee the stronger outcome

$$\lim\limits_{n\to\infty} \left\| u_n - u \right\|_{\mathbb{C}([0,\infty);\mathscr{X})} = 0.$$

The dependence of the classical solution in the advection model on the initial state was evident, since it was simply a translation of the initial profile (cf. (3.10)). As such, if the initial state were n times continuously differentiable, the same would be true of the classical solution. (Why?) As we encounter more general evolution equations, the strength of the connection between the regularity of the IC and that of the classical solution may weaken, but it will still exist to some degree. For the class of evolution equations presently under investigation, this connection is very strong, as we shall see shortly. But first, we make precise the notion of an integer power of an operator.

Definition 3.4.5. Let $A : \text{dom}(A) \subset \mathscr{X} \to \mathscr{X}$ be a linear operator. For each $n \in \mathbb{N}$, the nth *power of* A, denoted A^n, is defined by

$$A^n = AA^{n-1}, \text{ where } A^0 = I,$$
$$\text{dom}(A^n) = \left\{ x \in \text{dom}\left(A^{n-1}\right) \middle| A^{n-1}x \in \text{dom}(A) \right\}. \tag{3.91}$$

Exercise 3.4.7. Establish that the following hold $\forall n \in \mathbb{N}$.

i.) $\text{dom}(A^n) \subset \text{dom}\left(A^{n-1}\right)$.

ii.) $\frac{d^n}{dt^n}\left(e^{At}z\right) = A^n e^{At} z = e^{At} A^n z, \forall z \in \text{dom}(A^n)$.

iii.) $\left(\text{dom}(A^n), \|\cdot\|_{\text{dom}(A^n)}\right)$ is a Banach space when equipped with the norm

$$\|z\|_{\text{dom}(A^n)}^2 \equiv \|z\|_{\mathscr{X}}^2 + \|Az\|_{\mathscr{X}}^2 + \dots + \left\|A^{n-1}z\right\|_{\mathscr{X}}^2.$$

The following lemma is needed in order to verify the regularity result that immediately follows. The proof of part (i) is somewhat technical and can be found in **[341]**. The remaining parts are not difficult to establish. (Try it!)

Lemma 3.4.6. *Assume that* $A : \text{dom}(A) \subset \mathscr{X} \to \mathscr{X}$ *generates a* C_0−*semigroup* $\left\{e^{At} : t \geq 0\right\}$ *on* $\mathscr{X}$.

i.) $\text{cl}_{\mathscr{X}}\left(\bigcap_{n=1}^{\infty} \text{dom}(A^n)\right) = \mathscr{X}$.

ii.) $\text{dom}(A^n)$ *is dense in* $\mathscr{X}$, $\forall n \in \mathbb{N}$.

iii.) *The operator* $B = A\big|_{\text{dom}(A^n)}$ *generates the* C_0−*semigroup* $\left\{e^{Bt} : t \geq 0\right\}$ *on* $\mathscr{X}$.

Proposition 3.4.7. *Assume that* $A : \text{dom}(A) \subset \mathscr{X} \to \mathscr{X}$ *generates a* C_0−*semigroup* $\left\{e^{At} : t \geq 0\right\}$ *on* $\mathscr{X}$ *and let* $n \in \mathbb{N}$. *If* $u_0 \in \text{dom}(A^n)$, *then the classical solution of* (3.87) *belongs to the space* $\bigcap_{k=0}^{n} C^{n-k}\left([0,\infty); \text{dom}(A^k)\right)$.

Exercise 3.4.8.

i.) Prove parts (ii) and (iii) of Lemma 3.4.6.

ii.) Use induction to prove Prop. 3.4.7.

Exercise 3.4.9.

i.) Consider the IBVP (3.25) equipped with homogenous Neumann BCs. Let $n \in \mathbb{N}$. If $z_0 \in C^{2n}\left((0,a); \mathbb{R}\right)$, what can you conclude about the classical solution z ?

ii.) What level of regularity is sufficient to impose on the initial data in order to guarantee that the classical solution of (3.47) belongs to $C^3\left((0,a) \times (0,b); \text{dom}(A^2)\right)$?

We have shown that each of the IBVPs considered in Section 3.2 has a unique classical solution whose regularity is directly linked to that of the IC. Of course, we already *knew* this, having applied other means from which we were able to determine an explicit formula for the solution. However, in other applications, finding an explicit formula for the solution may not be feasible, or even possible. And, in such cases, we cannot simply work backwards to deduce whether or not the natural candidate for the operator A generates a C_0−semigroup. What do we do?

If we can successfully reformulate the IBVP as the abstract evolution equation (3.87), a logical next step would be to determine whether or not the operator A generates a C_0−semigroup on our choice of space $\mathscr{X}$, without relying on an expression for the semigroup itself. Certainly, if $A \in \mathbb{B}(\mathscr{X})$ with dom$(A) = \mathscr{X}$, then we know automatically that it generates a U.C. semigroup on $\mathscr{X}$. However, if A is an unbounded linear operator, the situation is more delicate. If we can somehow manage to conclude that A *does* generate a C_0−semigroup, then Thrm. 3.4.2 and Prop. 3.4.7 ensure the existence and uniqueness of a suitably regular classical solution of (3.87).

As such, our next task is to establish criteria that ensure an unbounded linear operator A generates a C_0−semigroup. This requires some fairly deep theorems developed in the next section.

3.5   Generation Theorems

Must an unbounded linear operator $A : \text{dom}(A) \subset \mathscr{X} \rightarrow \mathscr{X}$ generate a C_0−semigroup on $\mathscr{X}$? The evidence gathered thus far suggests no, but can we establish criteria that ensure when such an operator must be a generator? We investigate this important question in this section, beginning with the following notions.

Definition 3.5.1. Let $A : \text{dom}(A) \subset \mathscr{X} \rightarrow \mathscr{X}$ be a linear operator.
i.) The *resolvent set* of A, denoted $\rho(A)$, is the set of all complex numbers λ for which there exists an operator $R_\lambda(A) \in \mathbb{B}(\mathscr{X})$ such that

   a.) For every $\forall y \in \mathscr{X}$, $R_\lambda(A)y \in \text{dom}(A)$ and $(\lambda I - A) R_\lambda(A)y = y$, and

   b.) for every $x \in \text{dom}(A)$, $R_\lambda(A)(\lambda I - A)x = x$.
ii.) For any $\lambda \in \rho(A)$, $R_\lambda(A) = (\lambda I - A)^{-1}$ is the *resolvent operator of A*.

The reason why these operators are helpful will become apparent soon. For the moment, in order to gain familiarity with them, complete the following exercise in the special case when $\mathbf{A} \in \mathbb{M}^2(\mathbb{R})$.

Exercise 3.5.1. Let $\mathbf{A} = \begin{bmatrix} 1 & 0 \\ 0 & 2 \end{bmatrix}$.
i.) For what values of λ does $R_\lambda(\mathbf{A})$ exist?
ii.) Compute $R_{-1}(\mathbf{A})$ and $R_0(\mathbf{A})$.
iii.) Show that $\forall x \in \mathbb{R}^2$, $\lim\limits_{\lambda \to \infty} \lambda R_\lambda(\mathbf{A})x = x$.

iv.) Repeat (i) - (iii) for $\mathbf{A} = \begin{bmatrix} 0 & \beta \\ -\beta & 0 \end{bmatrix}$, where $\beta > 0$.

3.5.1   Hille-Yosida and Feller–Miyadera–Phillips Theorems

Thinking ahead toward formulating a characterization result for C_0-semigroups, a reasonable strategy might be to approximate the operator A by a sequence of "nicer" operators known to generate C_0-semigroups. This hones our attention to the class of bounded linear operators (cf. Thrm. 3.3.4). Of course, we must narrow our search considerably. We seek to define a sequence of bounded linear operators that converges to A in a sufficiently strong sense. To this end, we introduce the following sequence.

Definition 3.5.2. Let $A : \text{dom}(A) \subset \mathscr{X} \to \mathscr{X}$ be a linear operator such that
i.) A is closed and densely-defined, and
ii.) For every $\lambda > 0$, $\lambda \in \rho(A)$ and $\|R_\lambda(A)\|_{\mathbb{B}(\mathscr{X})} \leq \frac{1}{\lambda}$.
The collection of operators $A_\lambda : \text{dom}(A_\lambda) \subset \mathscr{X} \to \mathscr{X}$ defined by

$$A_\lambda x = \lambda A R_\lambda(A)x, \tag{3.92}$$

where $\lambda > 0$, is called the *Yosida approximation* of A.

Complete the following exercise as a warm-up before Prop. 3.5.3.

Exercise 3.5.2. Assume that $A : \text{dom}(A) \subset \mathbb{C}([0,\infty);\mathbb{R}) \to \mathbb{C}([0,\infty);\mathbb{R})$ is defined by (3.23).
i.) We know that A satisfies Def. 3.5.2(i). (Why?) Verify that A satisfies Def. 3.5.2(ii).
ii.) Compute $R_\lambda(A)$.
iii.) Verify that the following properties involving $A_\lambda x$ hold, $\forall \lambda > 0$.
    a.) $\text{dom}(A_\lambda) = \mathbb{C}([0,\infty);\mathbb{R})$,
    b.) $\lim\limits_{\lambda \to \infty} \|A_\lambda x - Ax\|_{\mathscr{X}} = 0, \forall x \in \text{dom}(A)$,
    c.) $A_\lambda x = \lambda^2 R_\lambda(A)x - \lambda x$.
iv.) Why must A_λ generate a U.C. semigroup $\{e^{A_\lambda t} : t \geq 0\}$ on $\mathbb{C}([0,\infty);\mathbb{R})$?
v.) For any $x \in \mathbb{C}([0,\infty);\mathbb{R})$, conjecture the value of $\lim\limits_{t\to\infty} e^{A_\lambda t}x$?

The following list of properties involving the Yosida approximation of A are needed in the proof of the very important *Hille-Yosida theorem*.

Proposition 3.5.3. (Properties of Resolvents and Yosida Approximations)
i.) *For every* $\lambda > 0, \text{dom}(A_\lambda) = \mathscr{X}$.
ii.) *For every* $\lambda, \mu > 0$,
    a.) $R_\lambda(A)R_\mu(A) = R_\mu(A)R_\lambda(A)$.
    b.) *For every* $x \in \text{dom}(A), AR_\lambda(A)x = R_\lambda(A)Ax$.
    c.) $A_\lambda A_\mu = A_\mu A_\lambda$.
iii.) *For every* $x \in \mathscr{X}, \lim\limits_{\lambda\to\infty} \lambda R_\lambda(A)x = x$.
iv.) *For every* $x \in \mathscr{X}, A_\lambda x = \lambda^2 R_\lambda(A)x - \lambda x$.
v.) *For every* $x \in \text{dom}(A), \lim\limits_{\lambda\to\infty} \|A_\lambda x - Ax\|_{\mathscr{X}} = 0$.
vi.) *For every* $\lambda > 0, A_\lambda$ *generates a U.C. contractive semigroup* $\{e^{A_\lambda t} : t \geq 0\}$ *on* $\mathscr{X}$. *As such,* $\|e^{A_\lambda t}\|_{\mathbb{B}(\mathscr{X})} \leq 1, \forall \lambda > 0$ *and* $t \geq 0$.

vii.) *For every* $x \in \mathscr{X}$, $t \geq 0$, *and* $\lambda, \mu > 0$, $\left\| e^{A_\lambda t} x - e^{A_\mu t} x \right\|_{\mathscr{X}} \leq t \left\| A_\lambda x - A_\mu x \right\|_{\mathscr{X}}$.

viii.) *For every* $t \geq 0$, $\exists T(t) \in \mathbb{B}(\mathscr{X})$ *such that* $\forall x \in \mathscr{X}$,

$$T(t)x = \lim_{\lambda \to \infty} e^{A_\lambda t} x \text{ uniformly on bounded intervals (in } t\text{) of } [0, \infty).$$

ix.) $\{T(t) : t \geq 0\}$ *is a contractive* C_0*-semigroup on* $\mathscr{X}$.

Proof. We assume that $\lambda, \mu > 0$ throughout the proof.

<u>Proof of (i):</u> The operator A_λ is defined $\forall x \in \mathscr{X}$ since $R_\lambda(A) \in \mathbb{B}(\mathscr{X})$, and Def. 3.5.1 ensures that $\text{rng}(R_\lambda(A)) \subset \text{dom}(A)$.

<u>Proof of (ii)(a):</u> Observe that

$$(\mu I - A) = (\lambda I - A) + (\mu - \lambda) I = [I + (\mu - \lambda) R_\lambda(A)] (\lambda I - A)$$

so that applying $R_\lambda(A)$ on the right-side yields

$$(\mu I - A) R_\lambda(A) = [I + (\mu - \lambda) R_\lambda(A)] \underbrace{(\lambda I - A) R_\lambda(A)}_{=I}.$$

Subsequently, applying $R_\mu(A)$ on the left-side yields

$$\underbrace{R_\mu(A) (\mu I - A)}_{=I} R_\lambda(A) = R_\mu(A) [I + (\mu - \lambda) R_\lambda(A)]. \tag{3.93}$$

Applying linearity in (3.93) results in

$$R_\lambda(A) = R_\mu(A) + R_\mu(A) (\mu - \lambda) R_\lambda(A) = R_\mu(A) + (\mu - \lambda) R_\mu(A) R_\lambda(A),$$

or equivalently,

$$R_\lambda(A) - R_\mu(A) = (\mu - \lambda) R_\mu(A) R_\lambda(A). \tag{3.94}$$

Interchanging the roles of λ and μ in (3.94) yields

$$R_\mu(A) - R_\lambda(A) = (\lambda - \mu) R_\lambda(A) R_\mu(A). \tag{3.95}$$

We then arrive at the conclusion by first multiplying both sides of (3.95) by -1 and then equating that result to (3.94).

<u>Proof of (ii)(b):</u> For any $x \in \text{dom}(A)$, using linearity in Def. 3.5.1 yields

$$AR_\lambda(A)x = \lambda R_\lambda(A)x - x = R_\lambda(A)Ax. \tag{3.96}$$

<u>Proof of (ii)(c):</u> This follows directly from (ii)(a) and (ii)(b). (Tell how.)

<u>Proof of (iii):</u> Let $x \in \text{dom}(A)$ and $\varepsilon > 0$. There exists $\delta_0 > 0$ such that $\forall \lambda > \delta_0$,

$$\frac{1}{\lambda} \|Ax\|_{\mathscr{X}} < \varepsilon. \tag{3.97}$$

For such λ, using (3.96) and (3.97) yields

$$\|\lambda R_\lambda(A)x - x\|_{\mathscr{X}} = \|R_\lambda(A)Ax\|_{\mathscr{X}} \leq \|R_\lambda(A)\|_{\mathbb{B}(\mathscr{X})} \|Ax\|_{\mathscr{X}} \leq \frac{1}{\lambda} \|Ax\|_{\mathscr{X}} < \varepsilon.$$

This shows that

$$\lim_{\lambda \to \infty} \lambda R_\lambda (A) x = x, \; \forall x \in \text{dom}(A). \tag{3.98}$$

The density of $\text{dom}(A)$ in $\mathscr{X}$, together with the estimate $\|\lambda R_\lambda(A)\|_{\mathbb{B}(\mathscr{X})} \leq 1$, implies that (3.98) holds on the entire space $\mathscr{X}$.

<u>Proof of (iv)</u>: Observe that

$$\lambda^2 R_\lambda (A) - \lambda I = \lambda^2 R_\lambda(A) - \lambda \left(\lambda I - A \right) R_\lambda(A) = \lambda A R_\lambda(A) = A_\lambda.$$

<u>Proof of (v)</u>: Let $x \in \text{dom}(A)$. Part (iii) implies that

$$\lim_{\lambda \to \infty} A_\lambda x = \lim_{\lambda \to \infty} \lambda A R_\lambda(A) x = \lim_{\lambda \to \infty} \left(\lambda R_\lambda(A) \right) (A x) = A x.$$

<u>Proof of (vi)</u>: In light of part (i), since $A_\lambda \in \mathbb{B}(\mathscr{X})$, Thrm. 3.3.4 implies that A_λ generates a U.C. semigroup $\left\{ e^{A_\lambda t} : t \geq 0 \right\}$ on $\mathscr{X}$. The fact that Def 3.3.7 is satisfied follows easily by using part (iv) and the remark preceding Exer. 3.3.9. Indeed, observe that

$$\left\| e^{A_\lambda t} \right\|_{\mathbb{B}(\mathscr{X})} \leq \left\| e^{t\lambda^2 R_\lambda (A) - t\lambda I} \right\|_{\mathbb{B}(\mathscr{X})} \leq \left\| e^{t\lambda^2 R_\lambda (A)} \right\|_{\mathbb{B}(\mathscr{X})} \left\| e^{-t\lambda I} \right\|_{\mathbb{B}(\mathscr{X})}$$

$$\leq e^{t\lambda^2 (\frac{1}{\lambda})} e^{-t\lambda} = 1. \tag{3.99}$$

<u>Proof of (vii)</u>: Let $x \in \mathscr{X}$ and $0 < s < t$. Since

$$\frac{d}{ds} \left(e^{(t-s)A_\mu} e^{sA_\lambda} x \right) = e^{(t-s)A_\mu} e^{sA_\lambda} \left(A_\lambda x - A_\mu x \right),$$

integrating both sides over $(0, t)$ yields

$$e^{tA_\lambda} x - e^{tA_\mu} x = \int_0^t \frac{d}{ds} \left(e^{(t-s)A_\mu} e^{sA_\lambda} x \right) ds = \int_0^t e^{(t-s)A_\mu} e^{sA_\lambda} \left(A_\lambda x - A_\mu x \right) ds.$$

Using (3.99) and standard integral properties subsequently yields

$$\left\| e^{tA_\lambda} x - e^{tA_\mu} x \right\|_{\mathscr{X}} \leq \int_0^t \underbrace{\left\| e^{(t-s)A_\mu} \right\|_{\mathbb{B}(\mathscr{X})} \left\| e^{sA_\lambda} \right\|_{\mathbb{B}(\mathscr{X})}}_{\leq 1} \left\| A_\lambda x - A_\mu x \right\|_{\mathscr{X}} ds$$

$$\leq t \left\| A_\lambda x - A_\mu x \right\|_{\mathscr{X}}. \tag{3.100}$$

<u>Proof of (viii)</u>: Part (v) implies that the right-side of (3.100) goes to zero uniformly on bounded intervals (in t) as $\lambda, \mu \to \infty$. As such, $\forall t \geq 0$, $\left\{ e^{A_\lambda t} | \lambda \geq 1 \right\}$ is a Cauchy sequence in $\mathbb{B}(\mathscr{X})$. As such, by completeness, there exists $T \in \mathbb{B}(\mathscr{X})$ such that

$$T(t) x = \lim_{\lambda \to \infty} e^{A_\lambda t} x, \; \forall x \in \mathscr{X}. \tag{3.101}$$

<u>Proof of (ix)</u>: Part (vi) implies that

$$\| T(t) \|_{\mathbb{B}(\mathscr{X})} \leq 1, \tag{3.102}$$

so that $\{T(t) : t \geq 0\}$ is contractive. Verification of Def. 3.3.1 is left as an exercise. We verify the strong continuity. Let $x_0 \in \mathscr{X}$ and $\varepsilon > 0$. Fix $y_0 \in \mathfrak{B}_{\mathscr{X}}\left(x_0; \frac{\varepsilon}{4}\right)$ and note that part (viii) ensures the existence of $M > 0$ such that $\forall \lambda > M$,

$$\left\| T(h)y_0 - e^{hA_\lambda} y_0 \right\|_{\mathscr{X}} < \frac{\varepsilon}{4}. \tag{3.103}$$

The uniform continuity (in h) of $\{e^{hA_\lambda}\}$ guarantees the existence of $\delta > 0$ such that $\forall 0 < h < \delta$,

$$\left\| e^{hA_\lambda} - I \right\|_{\mathbb{B}(\mathscr{X})} < \frac{\varepsilon}{4\left(1 + \|y_0\|_{\mathscr{X}}\right)}. \tag{3.104}$$

So, for such values of h,

$$\left\| e^{tA_\lambda} y_0 - y_0 \right\|_{\mathscr{X}} \leq \left\| e^{hA_\lambda} - I \right\|_{\mathbb{B}(\mathscr{X})} \|y_0\|_{\mathscr{X}} < \frac{\varepsilon \|y_0\|_{\mathscr{X}}}{4\left(1 + \|y_0\|_{\mathscr{X}}\right)} < \frac{\varepsilon}{4}. \tag{3.105}$$

Using the fact that $\|x_0 - y_0\|_{\mathscr{X}} < \frac{\varepsilon}{4}$, along with (3.103) and (3.105), shows that $\forall 0 < h < \delta$,

$$\begin{aligned}
\|T(h)x_0 - x_0\|_{\mathscr{X}} &\leq \|T(h)x_0 - T(h)y_0\|_{\mathscr{X}} + \left\| T(h)y_0 - e^{hA_\lambda} y_0 \right\|_{\mathscr{X}} \\
&\quad + \left\| e^{hA_\lambda} y_0 - y_0 \right\|_{\mathscr{X}} + \|y_0 - x_0\|_{\mathscr{X}} \\
&< \varepsilon,
\end{aligned}$$

as needed. Thus, $\{T(t) : t \geq 0\}$ is a contractive C_0−semigroup on $\mathscr{X}$. This completes the proof. $\qquad\square$

Exercise 3.5.3. Verify that the family of operators $\{T(t) : t \geq 0\}$ defined by (3.101) satisfies Def. 3.3.1.

Having completed the necessary preparatory work, we now state and prove the first main result, namely the Hille-Yosida theorem. We focus only the case of a contractive semigroup, for simplicity. The proof is based on those in **[107, 318]**.

Theorem 3.5.4. (Hille-Yosida theorem)
A linear operator $A : \text{dom}(A) \subset \mathscr{X} \to \mathscr{X}$ generates a contractive C_0−semigroup $\{e^{At} : t \geq 0\}$ on $\mathscr{X}$ if and only if
i.) *A is closed and densely-defined, and*
ii.) *For every $\lambda > 0$, $\lambda \in \rho(A)$ and $\|R_\lambda(A)\|_{\mathbb{B}(\mathscr{X})} \leq \frac{1}{\lambda}$.*

Proof. ($\Longrightarrow$) Assume that the linear operator $A : \text{dom}(A) \subset \mathscr{X} \to \mathscr{X}$ generates a contractive C_0−semigroup $\{e^{At} : t \geq 0\}$ on $\mathscr{X}$. Theorem 3.3.6 (ix), (x) imply that criterion (i) of the present theorem holds. (Why?) To verify criterion (ii), $\forall \lambda > 0$, define the operator $\overline{R}(\lambda) : \mathscr{X} \to \mathscr{X}$ by

$$\overline{R}(\lambda)x = \int_0^\infty e^{-\lambda t} e^{At} x \, dt, \quad \forall x \in \mathscr{X}. \tag{3.106}$$

Since $\left\| e^{-\lambda t} e^{At} x \right\|_{\mathscr{X}} \le e^{-\lambda t} \left\| x \right\|_{\mathscr{X}}$, $\forall x \in \mathscr{X}$, the integral on the right-side of (3.106) is convergent. So, $\overline{R}(\lambda) \in \mathbb{B}(\mathscr{X})$. Moreover, $\forall x \in \mathscr{X}$,

$$\left\| \overline{R}(\lambda) x \right\|_{\mathscr{X}} \le \int_0^\infty e^{-\lambda t} \left\| x \right\|_{\mathscr{X}} dt \le \frac{1}{\lambda} \left\| x \right\|_{\mathscr{X}}.$$

Consequently,

$$\left\| \overline{R}(\lambda) \right\|_{\mathbb{B}(\mathscr{X})} = \inf \left\{ m \,\middle|\, \left\| \overline{R}(\lambda) x \right\|_{\mathscr{X}} \le m \left\| x \right\|_{\mathscr{X}}, \forall x \in \mathscr{X} \right\} \le \frac{1}{\lambda}. \tag{3.107}$$

Next, we argue that $\overline{R}(\lambda) = R_\lambda(A)$ by showing that

$$I = (\lambda I - A) \overline{R}(\lambda) = \overline{R}(\lambda)(\lambda I - A). \tag{3.108}$$

To this end, for $h > 0$, $x \in \mathscr{X}$, and $\lambda > 0$, a change of variable and use of the semi-group property yields

$$\begin{aligned}
\frac{e^{Ah} - I}{h} \overline{R}(\lambda) x &= \frac{1}{h} \int_0^\infty e^{-\lambda t} e^{(t+h)A} x \, dt - \frac{1}{h} \int_0^\infty e^{-\lambda t} e^{tA} x \, dt \\
&= \frac{1}{h} \int_h^\infty e^{-\lambda(u-h)} e^{uA} x \, du - \frac{1}{h} \int_0^\infty e^{-\lambda t} e^{tA} x \, dt \tag{3.109} \\
&= \frac{e^{\lambda h}}{h} \left[\int_0^\infty e^{-\lambda u} e^{uA} x \, du - \int_0^h e^{-\lambda u} e^{uA} x \, du \right] - \frac{1}{h} \int_0^\infty e^{-\lambda t} e^{tA} x \, dt \\
&= \frac{e^{\lambda h} - 1}{h} \int_0^\infty e^{-\lambda t} e^{tA} x \, dt - \frac{e^{\lambda h}}{h} \int_0^h e^{-\lambda t} e^{tA} x \, dt.
\end{aligned}$$

Taking the limit as $h \to 0^+$ in (3.109) (using Exer. 3.3.13) shows that $\overline{R}(\lambda) x \in \mathrm{dom}(A)$ and

$$A\overline{R}(\lambda) x = \left(\lim_{h \to 0^+} \frac{e^{hA} - I}{h} \right) \overline{R}(\lambda) x = \lambda \overline{R}(\lambda) x - x. \tag{3.110}$$

Thus, $A\overline{R}(\lambda) = \lambda \overline{R}(\lambda) - I$ and hence, $(\lambda I - A) \overline{R}(\lambda) = I$, thereby establishing the first equality in (3.108). As for the second equality in (3.108), let $x \in \mathrm{dom}(A)$. Theorem 3.3.6 and an integration by parts in (3.106) (with Ax in place of x) yields

$$\begin{aligned}
\overline{R}(\lambda) Ax &= \int_0^\infty e^{-\lambda t} e^{tA} Ax \, dt = \int_0^\infty e^{-\lambda t} \frac{d}{dt} \left(e^{tA} x \right) dt \\
&= \underbrace{\lim_{b \to \infty} e^{-\lambda b} e^{bA} x}_{=0} - x + \lambda \overline{R}(\lambda) x.
\end{aligned}$$

Hence,

$$\overline{R}(\lambda) Ax = \lambda \overline{R}(\lambda) x - x, \quad \forall x \in \mathrm{dom}(A),$$

as needed. This completes the proof of the forward implication.

($\Longleftarrow$) Here is a summary of the main steps of the proof:

<u>Step 1</u>: Construct a sequence of operators $\{A_\lambda\}$ "close to" A.
<u>Step 2</u>: Use Step 1 to construct a family of semigroups $\{\{e^{tA_\lambda}\} : \lambda > 0\}$ "close to" $\{e^{tA} : t \geq 0\}$.
<u>Step 3</u>: For every $t \geq 0$, show that $\{e^{tA_\lambda}\}$ converges strongly (in $\mathscr{X}$) as $\lambda \to \infty$ to an operator $T(t)$. The family $\{T(t) : t \geq 0\}$ is itself a contractive C_0-semigroup on $\mathscr{X}$.
<u>Step 4</u>: Prove that the generator B of $\{T(t) : t \geq 0\}$ is actually the operator A.

Much of the technical work has been completed in stages in Prop. 3.5.3. It remains to verify Step 4. Assume that $B : \mathrm{dom}(B) \subset \mathscr{X} \to \mathscr{X}$ generates $\{T(t) : t \geq 0\}$. We must show that $\mathrm{dom}(B) = \mathrm{dom}(A)$ and that $Ax = Bx$, for all x in this common domain. To this end, begin with $x \in \mathrm{dom}(A)$. Let $h > 0$ and observe that for $0 < s < h$,

$$\left\| e^{sA_\lambda} A_\lambda x - T(s)Ax \right\|_{\mathscr{X}} = \left\| e^{sA_\lambda} A_\lambda x - e^{sA_\lambda} Ax + e^{sA_\lambda} Ax - T(s)Ax \right\|_{\mathscr{X}} \qquad (3.111)$$
$$\leq \left\| e^{sA_\lambda} \right\|_{\mathbb{B}(\mathscr{X})} \left\| A_\lambda x - Ax \right\|_{\mathscr{X}} + \left\| \left(e^{sA_\lambda} - T(s) \right) Ax \right\|_{\mathscr{X}} .$$

Proposition 3.5.3 (v), (viii) imply that the right-side of (3.111) goes to zero uniformly for $0 < s < h$ as $\lambda \to \infty$. Hence,

$$\lim_{\lambda \to \infty} e^{sA_\lambda} A_\lambda x = T(s)Ax. \qquad (3.112)$$

As such, taking the limit as $\lambda \to \infty$ in

$$e^{hA_\lambda} x - x = \int_0^h \frac{d}{ds} \left(e^{sA_\lambda} x \right) ds = \int_0^h e^{sA_\lambda} A_\lambda x \, ds$$

yields

$$T(h)x - x = \int_0^h T(s)Ax \, ds. \qquad (3.113)$$

Therefore, using Thrm. 3.3.6(v) and (3.113) yields

$$Ax = \lim_{h \to 0^+} \frac{1}{h} \int_0^h T(s)Ax \, ds = \lim_{h \to 0^+} \frac{T(h)x - x}{h} = Bx.$$

This shows that $x \in \mathrm{dom}(B)$. Hence, the operators A and B coincide on $\mathrm{dom}(A)$.

Next, let $x \in \mathrm{dom}(B)$. Since Prop. 3.5.3 (viii), (ix) imply that B generates a contractive C_0-semigroup on $\mathscr{X}$, the ($\Longrightarrow$) direction of the present theorem implies that

$$(I - B)^{-1} \text{ exists and maps } \mathscr{X} \text{ onto } \mathrm{dom}(B).$$

Since we also know that

$$(I - A)^{-1} \text{ exists and maps } \mathscr{X} \text{ onto } \mathrm{dom}(A),$$

and that (from the previous paragraph)

$$Ax = Bx, \ \forall x \in \mathrm{dom}(A),$$

we conclude that

$$(I-B)(\text{dom}(A)) = (I-A)(\text{dom}(A)) \underline{= \mathscr{X}} = (I-B)(\text{dom}(B)). \qquad (3.114)$$

The underlined portion of (3.114) shows that $(I-B)^{-1}(\mathscr{X}) = \text{dom}(A)$, as needed. This completes the proof of the theorem. $\qquad\qquad\qquad\qquad\qquad\qquad\qquad\square$

As an illustration of this theorem, consider the following example from **[318]**.

Example. Consider the operator $B_c : \text{dom}(B_c) = \text{dom}(A) \subset \mathbb{L}^2(0,a;\mathbb{R}) \to \mathbb{L}^2(0,a;\mathbb{R})$ defined by $B_c = c^2 A$, where $c > 0$ and A is defined by (3.44). The fact that B_c is a closed, densely-defined operator follows as in Prop. 3.1.17. Let $\lambda > 0$. In order to apply the Hille-Yosida theorem, we must argue that $R_\lambda(B_c)$ exists and that $\|R_\lambda(B_c)\|_{\mathbb{L}^2(0,a;\mathbb{R})} \le \frac{1}{\lambda}$. To this end, let $f \in \text{dom}(B_c)$. Taking the BCs into account, showing the existence of $R_\lambda(B_c)$ requires that we produce a unique solution $u \in \mathbb{L}^2(0,a;\mathbb{R})$ of the BVP

$$\begin{cases} (\lambda I - B_c)u(x) = f(x), \ 0 < x < a, \\ u'(0) = 0 = u(a), \end{cases} \qquad (3.115)$$

which is equivalent to

$$\begin{cases} \lambda u(x) - c^2 u''(x) = f(x), \ 0 < x < a, \\ u'(0) = 0 = u(a). \end{cases} \qquad (3.116)$$

(The existence and uniqueness of such a solution can be shown using the variation of parameters technique.) Multiplying both sides of (3.116) by $u(x)$ yields

$$\lambda u(x) \cdot u(x) - c^2 u''(x) \cdot u(x) = f(x) \cdot u(x), \ 0 < x < a. \qquad (3.117)$$

Now, integrate (3.117) over $(0,a)$. In doing so, an integration by parts yields

$$\int_0^a c^2 u''(x) \cdot u(x) dx = \underbrace{c^2 u(x) \cdot u'(x)\big|_0^a}_{=0 \text{ by BCs}} - c^2 \int_0^a (u'(x))^2 dx = -c^2 \|u'\|_{\mathbb{L}^2(0,a;\mathbb{R})}^2.$$

$$\qquad (3.118)$$

Substituting (3.118) into (3.117) then results in

$$\lambda \|u\|_{\mathbb{L}^2(0,a;\mathbb{R})}^2 + c^2 \|u'\|_{\mathbb{L}^2(0,a;\mathbb{R})}^2 = \langle f, u \rangle_{\mathbb{L}^2(0,a;\mathbb{R})}. \qquad (3.119)$$

Subsequently, using the Cauchy-Schwarz inequality (cf. Prop. 1.7.11(iv)) leads to the following string of implications which holds, $\forall f \in \text{dom}(B_c)$:

$$\lambda \|u\|_{\mathbb{L}^2(0,a;\mathbb{R})}^2 + c^2 \|u'\|_{\mathbb{L}^2(0,a;\mathbb{R})}^2 \le \|f\|_{\mathbb{L}^2(0,a;\mathbb{R})} \|u\|_{\mathbb{L}^2(0,a;\mathbb{R})} \implies$$

$$\|u\|_{\mathbb{L}^2(0,a;\mathbb{R})} \left(\lambda \|u\|_{\mathbb{L}^2(0,a;\mathbb{R})} - \|f\|_{\mathbb{L}^2(0,a;\mathbb{R})} \right) \le -c^2 \|u'\|_{\mathbb{L}^2(0,a;\mathbb{R})}^2 \le 0 \implies$$

$$\lambda \|u\|_{\mathbb{L}^2(0,a;\mathbb{R})} - \|f\|_{\mathbb{L}^2(0,a;\mathbb{R})} \le 0 \implies$$

$$\lambda \|R_\lambda(B_c)f\|_{\mathbb{L}^2(0,a;\mathbb{R})} \le \|f\|_{\mathbb{L}^2(0,a;\mathbb{R})}.$$

We conclude that $\|R_\lambda(B_c)\|_{\mathbb{B}(\mathbb{L}^2(0,a;\mathbb{R}))} \leq \frac{1}{\lambda}$. Hence, the Hille-Yosida theorem guarantees that B_c generates a contractive C_0−semigroup on $\mathbb{L}^2(0,a;\mathbb{R})$, a formula for which is provided in Exer. 3.3.4.

Exercise 3.5.4. Use the Hille-Yosida theorem to verify that the operator defined by (3.44) equipped instead with homogenous Neumann BCs (3.45) generates a contractive C_0−semigroup on $\mathbb{L}^2(0,a;\mathbb{R})$.

Exercise 3.5.5. Use the Hille-Yosida theorem to verify that the operator defined by (3.23) generates a C_0−semigroup on $\mathbb{C}([0,\infty);\mathbb{R})$.

Of course, not all C_0−semigroups on a given space $\mathscr{X}$ arising in practice are contractive. Unfortunately, Thrm. 3.5.4 does not apply to them. But, we still have the growth estimate $\|e^{At}\|_{\mathbb{B}(\mathscr{X})} \leq Me^{\omega t}$ (for some $M \geq 1$ and $\omega > 0$) that can be exploited to form an extension of the Hille-Yosida theorem that applies to *any* C_0−semi group. Such a more general characterization result, formulated by Feller, Miyadera, and Phillips, is similar in spirit to the Hille-Yosida theorem, but the lack of contractivity necessitates that hypothesis (ii) be replaced by a more technical counterpart. Consequently, the proof is somewhat more technical.

Theorem 3.5.5. (Feller-Miyadera-Phillips)
A linear operator $A : \mathrm{dom}(A) \subset \mathscr{X} \to \mathscr{X}$ generates a C_0−semigroup $\{e^{At} : t \geq 0\}$ on $\mathscr{X}$ (for which $\|e^{At}\|_{\mathbb{B}(\mathscr{X})} \leq Me^{\omega t}$, for some $M \geq 1$ and $\omega > 0$) if and only if
i.) *A is closed and densely-defined, and*
ii.) *For every $\lambda > \omega$, $\lambda \in \rho(A)$ and $\|R_\lambda^n(A)\|_{\mathbb{B}(\mathscr{X})} \leq \frac{M}{(\lambda-\omega)^n}$, $\forall n \in \mathbb{N}$.*

Remark. Proving that the operator A is densely-defined in $\mathscr{X}$ can be nontrivial and tedious. But, the requirement technically is not essential since $\mathrm{cl}_{\mathscr{X}}(\mathrm{dom}(A))$ when equipped with the same norm as $\mathscr{X}$, is a Banach space (since a closed subset of a Banach space is complete). Hence, we can assume instead that $A : \mathrm{dom}(A) \subset \mathscr{X} \to \mathscr{X}$ generates a C_0−semigroup on $\mathrm{cl}_{\mathscr{X}}(\mathrm{dom}(A))$ and remove the densely-defined portion of criterion (i).

The two aforementioned theorems characterize exactly when an unbounded linear operator A generates a C_0−semigroup $\{e^{At} : t \geq 0\}$. In the affirmative case, we can invoke Thrm. 3.4.2 to conclude that the classical solution of (3.87) is given by $u(t) = e^{At}u_0, t \geq 0$. Luckily, we were able to also derive an explicit formula for the semigroup in each of the models considered in Section 3.2. However, determining such a formula is often intractable. Sometimes, the complexity of the operator involved simply presents an insurmountable hurdle. That said, certain operators *are* nice enough as to generate a semigroup that can be described by a formula similar in nature to the one produced by the Putzer algorithm in the finite-dimensional case (cf. Prop. 2.2.3). One such class of operators defined on a separable Hilbert space is described below.

For the remainder of this subsection, $\mathscr{H}$ denotes a separable Hilbert space equipped with an inner product $\langle \cdot, \cdot \rangle_{\mathscr{H}}$ and orthonormal basis $\{\phi_n : n \in \mathbb{N}\}$. Suppose that a linear operator $A : \mathrm{dom}(A) \subset \mathscr{H} \to \mathscr{H}$ can be expressed in the form

$$Az = \sum_{n=1}^{\infty} \lambda_n \langle z, \phi_n \rangle_{\mathscr{H}} \phi_n \qquad (3.120)$$

$$\mathrm{dom}(A) = \left\{ z \in \mathscr{H} : \sum_{n=1}^{\infty} |\lambda_n \langle z, \phi_n \rangle_{\mathscr{H}}|^2 < \infty \right\},$$

where $\{\lambda_n : n \in \mathbb{N}\} \subset \mathbb{R}$. We have encountered a similar decomposition formula for $\mathbf{A} \in \mathbb{M}^N(\mathbb{R})$ whose eigenvectors span $\mathbb{R}^N$. And, in such case, the structure of $\mathbf{A}$ was strongly linked to the representation formula for the associated matrix exponential $\{e^{\mathbf{A}t} : t \geq 0\}$. In light of this connection, does the same sort of relationship hold in the present setting for operators expressed in the form (3.120)? The next three results, taken together, constitute an answer to this question.

Lemma 3.5.6. *For every $t \geq 0$, the operator $T(t) : \mathscr{H} \to \mathscr{H}$ defined by*

$$T(t)[z] = \sum_{n=1}^{\infty} e^{\lambda_n t} \langle z, \phi_n \rangle_{\mathscr{H}} \phi_n \qquad (3.121)$$

is a bounded linear operator on $\mathscr{H}$ if and only if $\sup_{n \in \mathbb{N}} \lambda_n < \infty$.

Exercise 3.5.6. Prove Lemma 3.5.6.

The proof of the following lemma involves the use of Thrm. 3.5.5 (and not Thrm. 3.5.4) since it is not necessarily the case that $(0, \infty) \subset \rho(A)$. (See **[103]** for details.)

Lemma 3.5.7. *The operator $A : \mathrm{dom}(A) \subset \mathscr{H} \to \mathscr{H}$ defined in (3.120) generates a C_0–semigroup on $\mathscr{H}$.*

These two lemmas are used to verify the following structure result. The proof is adapted from **[103]**.

Proposition 3.5.8. *If $\sup_{n \in \mathbb{N}} \lambda_n < \infty$, then the family of operators $\{T(t) : t \geq 0\}$ defined by (3.121) is a C_0–semigroup on $\mathscr{H}$ generated by the operator A defined by (3.120).*

Proof. We first show that $\{T(t) : t \geq 0\}$ defined by (3.121) generates a C_0–semigroup on $\mathscr{H}$. Let $t, s \geq 0$. The fact that $T(t) \in \mathbb{B}(\mathscr{H})$ follows from Lemma 3.5.6. Let $z \in \mathscr{H}$ and observe that

$$T(t+s)[z] = \sum_{n=1}^{\infty} e^{\lambda_n(t+s)} \langle z, \phi_n \rangle_{\mathscr{H}} \phi_n = \sum_{n=1}^{\infty} e^{\lambda_n t} e^{\lambda_n s} \langle z, \phi_n \rangle_{\mathscr{H}} \phi_n$$

$$= \sum_{n=1}^{\infty} e^{\lambda_n t} \left\langle \sum_{m=1}^{\infty} e^{\lambda_m s} \langle z, \phi_m \rangle_{\mathscr{H}} \phi_m, \phi_n \right\rangle_{\mathscr{H}} \phi_n.$$

Using the fact that

$$\langle \phi_n, \phi_m \rangle_{\mathscr{H}} = 0 \text{ if and only if } n \neq m$$

(cf. Def. 1.7.13), we continue this string of equalities to conclude that

$$T(t+s)[z] = \sum_{n=1}^{\infty} e^{\lambda_n t} \langle T(s)z, \phi_n \rangle_{\mathscr{H}} \phi_n = T(t)T(s)[z].$$

Further, since $\{\phi_n : n \in \mathbb{N}\}$ is an orthonormal basis of $\mathscr{H}$, we also have

$$T(0)[z] = \sum_{n=1}^{\infty} \langle z, \phi_n \rangle_{\mathscr{H}} \phi_n = z. \tag{3.122}$$

Thus, the conditions of Def. 3.3.1 are satisfied.

Next, we verify the strong continuity of $\{T(t) : t \geq 0\}$ on $\mathscr{H}$. Let $z \in \mathscr{H}$. We can restrict our attention to $0 \leq t \leq 1$ since we need only to consider sufficiently small values of t. (Why?) Using the second equality in (3.122) yields $\forall N \in \mathbb{N}$,

$$\|T(t)z - z\|_{\mathscr{H}}^2 = \left\| \sum_{n=1}^{\infty} e^{\lambda_n t} \langle z, \phi_n \rangle_{\mathscr{H}} \phi_n - z \right\|_{\mathscr{H}}^2$$

$$= \left\| \sum_{n=1}^{\infty} \left(e^{\lambda_n t} - 1 \right) \langle z, \phi_n \rangle_{\mathscr{H}} \phi_n \right\|_{\mathscr{H}}^2$$

$$\leq \sum_{n=1}^{\infty} \left| e^{\lambda_n t} - 1 \right|^2 |\langle z, \phi_n \rangle_{\mathscr{H}}|^2 \underbrace{\|\phi_n\|_{\mathscr{H}}^2}_{=1} \tag{3.123}$$

$$= \sum_{n=1}^{N} \left| e^{\lambda_n t} - 1 \right|^2 |\langle z, \phi_n \rangle_{\mathscr{H}}|^2 + \sum_{n=N+1}^{\infty} \left| e^{\lambda_n t} - 1 \right|^2 |\langle z, \phi_n \rangle_{\mathscr{H}}|^2$$

$$\leq \left(\sup_{\substack{1 \leq n \leq N \\ 0 \leq t \leq 1}} \left| e^{\lambda_n t} - 1 \right|^2 \right) \sum_{n=1}^{N} |\langle z, \phi_n \rangle_{\mathscr{H}}|^2$$

$$+ \left(\sup_{\substack{n \geq N+1 \\ 0 \leq t \leq 1}} \left| e^{\lambda_n t} - 1 \right|^2 \right) \sum_{n=N+1}^{\infty} |\langle z, \phi_n \rangle_{\mathscr{H}}|^2.$$

Let $\varepsilon > 0$ and

$$M = \sup \left\{ \left| e^{\lambda_n t} - 1 \right|^2 : n \in \mathbb{N} \wedge 0 \leq t \leq 1 \right\}.$$

There exists $N_0 \in \mathbb{N}$ such that

$$\sum_{n=N_0+1}^{\infty} |\langle z, \phi_n \rangle_{\mathscr{H}}|^2 < \frac{\varepsilon}{2M+1}. \tag{3.124}$$

(Why?) Also, $\exists 0 < \delta < 1$ such that $\forall 0 \leq t < \delta < 1$,

$$\sup_{1 \leq n \leq N_0} \left| e^{\lambda_n t} - 1 \right|^2 \leq \frac{\varepsilon}{2 \left(\|z\|^2_{\mathscr{H}} + 1 \right) \left(\sum_{n=1}^{N_0} |\langle z, \phi_n \rangle_{\mathscr{H}}|^2 + 1 \right)}. \tag{3.125}$$

(Why?) For such values of t, using (3.124) and (3.125) in (3.123) yields

$$\|T(t)z - z\|^2_{\mathscr{H}} < \frac{\varepsilon}{2 \left(\|z\|^2_{\mathscr{H}} + 1 \right) \left(\sum_{n=1}^{N_0} |\langle z, \phi_n \rangle_{\mathscr{H}}|^2 + 1 \right)} \sum_{n=1}^{N_0} |\langle z, \phi_n \rangle_{\mathscr{H}}|^2 + \frac{\varepsilon M}{2M + 1}$$

$$< \varepsilon,$$

thereby establishing the strong continuity.

Finally, note that Lemma 3.5.7 guarantees that A generates *some* C_0−semigroup $\{ e^{At} : t \geq 0 \}$ on $\mathscr{H}$. We now show that this semigroup coincides with (3.121). Indeed, using Thrm. 3.3.6 and the orthonormality of $\{ \phi_n : n \in \mathbb{N} \}$ leads to

$$\frac{d}{dt} \left(e^{At} \phi_m \right) = e^{At} A \phi_m = e^{At} \sum_{n=1}^{\infty} \lambda_n \langle \phi_m, \phi_n \rangle_{\mathscr{H}} \phi_n$$

$$= e^{At} \left(\lambda_m \phi_m \right) = \lambda_m \left(e^{At} \phi_m \right). \tag{3.126}$$

Solving the ODE defined by the underlined portion of (3.126) yields $e^{At} \phi_m = e^{\lambda_m t} \phi_m$ (cf. Section 1.9). Taking into account Lemma 3.5.6, we conclude that $\forall z \in \mathscr{H}$,

$$e^{At} z = \sum_{n=1}^{\infty} e^{\lambda_n t} \langle z, \phi_n \rangle_{\mathscr{H}} \phi_n = T(t)z,$$

as desired. This completes the proof. $\qquad\square$

Example. Recall that the semigroup associated with the one-dimensional heat conduction IBVP equipped with homogenous Dirichlet BCs is given by

$$e^{At}[z_0][x] = \sum_{m=1}^{\infty} e^{-\left(\frac{m\pi}{a} \right)^2 t} \left\langle z_0(\cdot), \sin \left(\frac{m\pi}{a} \cdot \right) \right\rangle_{\mathbb{L}^2(0,a;\mathbb{R})} \sin \left(\frac{m\pi}{a} x \right).$$

Since $\{1\} \cup \left\{ \sqrt{2} \sin \left(\frac{m\pi}{a} x \right) : m \in \mathbb{N} \right\}$ forms an orthonormal basis of $\mathbb{L}^2(0,a;\mathbb{R})$ and $\lambda_m = -\left(\frac{m\pi}{a} \right)^2$, $\forall m \in \mathbb{N}$, Prop. 3.5.8 guarantees that $\{ e^{At} : t \geq 0 \}$ generates a C_0−semigroup on $\mathbb{L}^2(0,a;\mathbb{R})$ with generator given by (3.120). This is precisely the decomposition of the operator A in (3.44), as needed.

In light of Prop. 3.5.8, we take an additional step back and ask which operators can be written in the form (3.120). Is there some readily verifiable condition that ensures an operator admits such a representation? Considerable attention is given to this question in Hilbert space theory. One well-known result involving compact operators (cf. Section 5.1) is as follows. (See **[118]** for a thorough discussion.)

Proposition 3.5.9. *Any linear operator $A : \mathrm{dom}(A) \subset \mathscr{H} \to \mathscr{H}$ which is negative and self-adjoint for which there exists λ_0 such that $R_{\lambda_0}(A)$ is a compact operator can be written in the form (3.120).*

3.5.2 A First Look at Dissipative Operators

Monotonicity is a useful tool in analysis. This notion can be extended in a natural way to operators in a more abstract setting, like a Hilbert space or Banach space. As it turns out, this notion is closely related to the range condition (ii) of the Hille-Yosida theorem and is often easier to verify in applications. Moreover, it is central to the development of the theory when A is a *nonlinear* unbounded operator. We begin with the following definitions connected to monotonicity formulated in a real Hilbert space.

Definition 3.5.10. An operator $B : \text{dom}(B) \subset \mathcal{H} \to \mathcal{H}$ is
i.) *accretive* if $\|Bx - By\|_{\mathcal{H}} \geq \|x - y\|_{\mathcal{H}}$, $\forall x, y \in \text{dom}(B)$.
ii.) *nonexpansive* if $\|Bx - By\|_{\mathcal{H}} \leq \|x - y\|_{\mathcal{H}}$, $\forall x, y \in \text{dom}(B)$.
iii.) *monotone* if $\langle Bx - By, x - y \rangle_{\mathcal{H}} \geq 0$, $\forall x, y \in \text{dom}(B)$.
iv.) *dissipative* if $\langle Bx, x \rangle_{\mathcal{H}} \leq 0$, $\forall x \in \text{dom}(B)$.
v.) *m-dissipative* if B is dissipative and $\forall \alpha > 0$, $\text{rng}(I - \alpha B) = \mathcal{H}$.

Remark. Definitions 3.5.10(i) and (ii) are defined in the same manner in any normed linear space. If $\mathcal{H}$ is a complex Hilbert space, then $\langle \cdot, \cdot \rangle_{\mathcal{H}}$ is replaced by $\text{Re} \langle \cdot, \cdot \rangle_{\mathcal{H}}$ in (iii) and (iv). Also, (iii) - (v) can be extended to a Banach space, but the inner product is replaced by the so-called *duality map* (cf. **[39, 40, 340]**).

Exercise 3.5.7. Explain why Def. 3.5.10(iii) is a direct generalization of the notion of an increasing real-valued function.

Many natural connections exist among the various notions in Def. 3.5.10, two of which are as follows:

Proposition 3.5.11. *Let $\mathcal{H}$ be a real Hilbert space.*
i.) *An operator $B : \text{dom}(B) \subset \mathcal{H} \to \mathcal{H}$ is monotone if and only if $I + \alpha B$ is accretive, $\forall \alpha > 0$.*
ii.) *An operator $B : \text{dom}(B) \subset \mathcal{H} \to \mathcal{H}$ is accretive if and only if $(I + \alpha B)^{-1}$ exists, $\forall \alpha > 0$, and is nonexpansive.*

Exercise 3.5.8. Prove Prop. 3.5.11.

Examples.
1.) The operator $\mathscr{A} : \mathbb{R}^N \to \mathbb{R}^N$ defined by

$$\mathscr{A}\mathbf{x} = \begin{bmatrix} \alpha_1 & 0 & \cdots & 0 \\ 0 & \alpha_2 & & 0 \\ \vdots & & \ddots & \vdots \\ 0 & 0 & \cdots & \alpha_n \end{bmatrix} \mathbf{x}$$

where $\alpha_i \geq 0$, $1 \leq i \leq N$, is a monotone operator on $\mathbb{R}^N$. (Why?) In general, if $\mathbf{B} \in \mathbb{M}^N(\mathbb{R})$ has eigenvalues which are all nonnegative, then the operator $\mathscr{C} : \mathbb{R}^N \to \mathbb{R}^N$ defined by $\mathscr{C}\mathbf{x} = \mathbf{B}\mathbf{x}$ is monotone. (Why?)

2.) If $\mathscr{D} : \mathrm{dom}(\mathscr{D}) \subset \mathscr{H} \to \mathscr{H}$ is nonexpansive, then $I - \mathscr{D}$ is monotone. (Why?)

3.) The operator $\mathscr{G} : \mathrm{dom}(\mathscr{G}) \subset \mathbb{L}^2(0, a; \mathbb{R}) \to \mathbb{L}^2(0, a; \mathbb{R})$ defined by

$$\mathscr{G}f = -\frac{d^2 f}{dx^2}$$

$$\mathrm{dom}(\mathscr{G}) = \left\{ f \in \mathbb{L}^2(0, a; \mathbb{R}) \,\middle|\, \frac{df}{dx}, \frac{d^2 f}{dx^2} \in \mathbb{L}^2(0, a; \mathbb{R}) \wedge f(0) = f(a) = 0 \right\}.$$

is monotone because

$$\langle f, \mathscr{G}f \rangle_{\mathbb{L}^2(0,a;\mathbb{R})} = -\int_0^a f(x) \cdot \frac{d^2 f}{dx^2}(x) dx = \int_0^a \left(\frac{df}{dx}(x) \right)^2 dx \geq 0.$$

More generally, the Laplacian $\triangle : \mathrm{dom}(\triangle) \subset \mathbb{L}^2\left(\prod_{i=1}^n (0, a_i); \mathbb{R}\right) \to \mathbb{L}^2\left(\prod_{i=1}^n (0, a_i); \mathbb{R}\right)$ defined by

$$\triangle f = -\left(\frac{\partial^2 f}{\partial x_1^2} + \ldots + \frac{\partial^2 f}{\partial x_n^2} \right)$$

$$\mathrm{dom}(\triangle) = \left\{ f \in \mathbb{L}^2\left(\prod_{i=1}^n (0, a_i); \mathbb{R}\right) \,\middle|\, \frac{\partial f}{\partial x_i}, \frac{\partial^2 f}{\partial x_i^2} \in \mathbb{L}^2\left(\prod_{i=1}^n (0, a_i); \mathbb{R}\right) \right.$$

$$\left. i = 1, \ldots, n \wedge f = 0 \text{ on } \partial \left(\prod_{i=1}^n (0, a_i) \right) \right\}$$

is monotone. (Prove this.)

4.) If $\mathbf{I}$ denotes the identity operator on $\mathbb{R}^N$, then $-\mathbf{I}$ is dissipative since $\forall \mathbf{x} \in \mathbb{R}^N$,

$$\langle \mathbf{x}, -\mathbf{I}\mathbf{x} \rangle_{\mathbb{R}^N} = \mathbf{x} \cdot (-\mathbf{x}) = -\|\mathbf{x}\|_{\mathbb{R}^N}^2 \leq 0.$$

5.) The operator $\mathscr{J} : \mathrm{dom}(\mathscr{J}) \subset \mathbb{L}^2(0, a; \mathbb{R}) \to \mathbb{L}^2(0, a; \mathbb{R})$ defined by

$$\mathscr{J}f = f',$$
$$\mathrm{dom}(\mathscr{J}) = \left\{ f \in \mathbb{L}^2(0, a; \mathbb{R}) \,\middle|\, f(0) = 0 \right\}$$

is dissipative because

$$\langle f, \mathscr{J}f \rangle_{\mathbb{L}^2(0,a;\mathbb{R})} = \int_0^a f(s) f'(s) ds = -\frac{1}{2}(f(0))^2 = 0.$$

The following technical characterization of dissipative operators is useful in the proof of the Lumer-Phillips theorem. The proof is adapted from **[318]**.

Proposition 3.5.12. *Let $\mathscr{H}$ be a real Hilbert space and $B : \mathrm{dom}(B) \subset \mathscr{H} \to \mathscr{H}$. Definition 3.5.10 (iv) is equivalent to*

$$\lambda \|x\|_{\mathscr{H}} \leq \|(\lambda I - B)x\|_{\mathscr{H}}, \ \forall \lambda > 0 \text{ and } x \in \mathrm{dom}(B). \tag{3.127}$$

Proof. Let $x \in \text{dom}(B)$ and $\lambda > 0$. Assume that (3.127) holds. Squaring both sides of (3.127) and using standard properties of inner products yields the following string of implications:

$$\lambda \|x\|_{\mathscr{H}} \le \|(\lambda I - B)x\|_{\mathscr{H}} \implies$$
$$\lambda^2 \|x\|_{\mathscr{H}}^2 \le \|(\lambda I - B)x\|_{\mathscr{H}}^2 \implies$$
$$\lambda^2 \langle x,x \rangle_{\mathscr{H}} \le \langle (\lambda I - B)x, (\lambda I - B)x \rangle_{\mathscr{H}} \implies \qquad (3.128)$$
$$0 \le \langle \lambda x, -Bx \rangle_{\mathscr{H}} + \langle -Bx, \lambda x \rangle_{\mathscr{H}} + \langle -Bx, -Bx \rangle_{\mathscr{H}} \implies$$
$$0 \le -\lambda \langle x, Bx \rangle_{\mathscr{H}} - \lambda \langle Bx, x \rangle_{\mathscr{H}} + \lambda^2 \|Bx\|_{\mathscr{H}}^2 \implies$$
$$0 \le -2\lambda \langle x, Bx \rangle_{\mathscr{H}} + \lambda^2 \|Bx\|_{\mathscr{H}}^2 \implies$$
$$\langle x, Bx \rangle_{\mathscr{H}} \le \frac{\lambda}{2} \|Bx\|_{\mathscr{H}}^2 .$$

Taking the limit as $\lambda \to 0^+$ in the last inequality in (3.128) proves that (3.127) holds. Moreover, all of the above implications reverse since $\lambda > 0$. This shows that (3.127) implies Def. 3.5.10 (iv), thereby completing the proof. $\qquad \square$

The following result is an equivalent formulation of the Hille-Yosida theorem. Its benefit lies in the fact that the range condition (ii) in the Hille-Yosida theorem is replaced by an easier-to-verify dissipativity one.

Theorem 3.5.13. (Lumer-Phillips theorem)
A linear operator $A : \text{dom}(A) \subset \mathscr{H} \to \mathscr{H}$ generates a contractive C_0-semigroup $\{e^{At} : t \ge 0\}$ on $\mathscr{H}$ if and only if
i.) *A is densely-defined, and*
ii.) *A is m-dissipative.*

Proof. ($\impliedby$) We verify the hypotheses of the Hille-Yosida theorem. Let $\lambda > 0$.
Step 1: Show that R_λ exists.
Let $y, z \in \text{dom}(A)$. Observe that

$$(\lambda I - A)y = (\lambda I - A)z \implies$$
$$(\lambda I - A)(y - z) = 0 \implies$$
$$\|(\lambda I - A)(y - z)\|_{\mathscr{H}} = 0 \implies$$
$$\|y - z\|_{\mathscr{H}} = 0 \implies$$
$$y = z,$$

where Prop. 3.5.11 was used in the second-to-last line. So, $(\lambda I - A)$ is 1-1.
Next, we verify that $(\lambda I - A)$ is onto. The m-dissipativity of A implies that $\text{rng}(I - \lambda A) = \mathscr{H}, \forall \lambda > 0$. We need to show that $\text{rng}(\lambda I - A) = \mathscr{H}, \forall \lambda > 0$. To this end, let $y \in \mathscr{H}$. Then, $\exists z \in \text{dom}(A)$ such that $(I - \lambda A)z = y$, or equivalently

$$(\lambda I - A)\left(\frac{z}{\lambda}\right) = y. \qquad (3.129)$$

Since Thrm. 3.3.6 (i) implies that $\frac{z}{\lambda} \in \mathrm{dom}(A)$, we conclude that $(\lambda I - A)$ is onto. This establishes Step 1.

<u>Step 2</u>: Show that $\|R_\lambda\|_{\mathbb{B}(\mathscr{H})} \leq \frac{1}{\lambda}$.
Let $y \in \mathscr{H}$. Note that

$$(I - \lambda A)z = y \Longleftrightarrow z = (I - \lambda A)^{-1} y = R_\lambda y \tag{3.130}$$

is equivalent to

$$(\lambda I - A)\left(\frac{z}{\lambda}\right) = y \Longleftrightarrow z = \lambda (\lambda I - A)^{-1} y = \lambda R_\lambda y. \tag{3.131}$$

Hence,

$$\|R_\lambda y\|_{\mathscr{H}} = \left\|\frac{z}{\lambda}\right\|_{\mathscr{H}} \leq \frac{1}{\lambda} \|(I - \lambda A)z\|_{\mathscr{H}} = \frac{1}{\lambda} \|y\|_{\mathscr{H}}, \forall y \in \mathscr{H}.$$

So, we conclude that $\|R_\lambda\|_{\mathbb{B}(\mathscr{H})} \leq \frac{1}{\lambda}$.

As a result of Steps 1 and 2, we can invoke the Hille-Yosida theorem to conclude that A generates a contractive C_0−semigroup $\{e^{At} : t \geq 0\}$ on $\mathscr{H}$.

($\Longrightarrow$) Assume that A generates a contractive C_0−semigroup $\{e^{At} : t \geq 0\}$ on $\mathscr{H}$. Invoking the Hille-Yosida theorem, we see that $\mathrm{rng}\,(\lambda I - A) = \mathscr{H}, \forall \lambda > 0$, so that by (3.129), the second portion of the definition of m-dissipativity holds. To see why A is dissipative, observe that $\forall x \in \mathrm{dom}(A)$ and $h > 0$,

$$\left\langle e^{hA}x, x \right\rangle_{\mathscr{H}} \leq \left\|e^{hA}x\right\|_{\mathscr{H}} \|x\|_{\mathscr{H}} \leq \|x\|_{\mathscr{H}}^2 \Longrightarrow$$

$$\left\langle e^{hA}x - x, x \right\rangle_{\mathscr{H}} = \left\langle e^{hA}x, x \right\rangle_{\mathscr{H}} - \|x\|_{\mathscr{H}}^2 \leq 0 \Longrightarrow$$

$$\frac{1}{h} \left\langle e^{hA}x - x, x \right\rangle_{\mathscr{H}} = \left\langle \frac{e^{hA}x - x}{h}, x \right\rangle_{\mathscr{H}} \leq 0 \Longrightarrow$$

$$\lim_{h \to 0^+} \left\langle \frac{e^{hA}x - x}{h}, x \right\rangle_{\mathscr{H}} = \left\langle \lim_{h \to 0^+} \frac{e^{hA}x - x}{h}, x \right\rangle_{\mathscr{H}} \leq 0 \Longrightarrow$$

$$\langle Ax, x \rangle_{\mathscr{H}} \leq 0.$$

This completes the proof. □

Exercise 3.5.9. Justify all implications in the proof of ($\Longrightarrow$) in Thrm. 3.5.13.

3.6   A Useful Perturbation Result

Suppose that a linear operator A generates a C_0−semigroup on $\mathscr{X}$ and a "sufficiently well-behaved" operator B is added to it. Intuitively, as long as $A + B$ is "relatively close" to A, it seems that $A + B$ ought to generate a C_0−semigroup on $\mathscr{X}$. Is this true? Some related questions that naturally arise are:

1.) If both A and B generate C_0−semigroups on $\mathscr{X}$, say $\{e^{At} : t \geq 0\}$ and $\{e^{Bt} : t \geq 0\}$, respectively, must $A + B$ also generate a C_0−semigroup on $\mathscr{X}$?

2.) Must B generate a C_0−semigroup on $\mathscr{X}$ in order for $A + B$ to generate one?

The answer to (1) in the $\mathbb{R}^N$−setting is yes, since every bounded linear operator on $\mathbb{R}^N$ is identified with a member of $\mathbb{M}^N(\mathbb{R})$, and we know the matrix exponential is defined for all members of $\mathbb{M}^N(\mathbb{R})$. The answer to (2) is no. The combination of these two observations leads to the following proposition.

Proposition 3.6.1. (A Perturbation Result)
If $A : \mathrm{dom}(A) \subset \mathscr{X} \to \mathscr{X}$ generates a C_0−semigroup $\{e^{At} : t \geq 0\}$ on $\mathscr{X}$ and $B \in \mathbb{B}(\mathscr{X})$, then $A + B$ generates a C_0−semigroup $\left\{e^{(A+B)t} : t \geq 0\right\}$ on $\mathscr{X}$ given by $e^{(A+B)t} = \sum_{n=0}^{\infty} u_n(t)$, where $\{u_n\}$ is defined recursively by

$$\begin{cases} u_n(t) = \int_0^t e^{A(t-s)} B u_{n-1}(s)ds, \\ u_0(t) = e^{At}. \end{cases}$$

The proof is not difficult and can be found in **[103, 142]**.

Exercise 3.6.1. Let $\mathbf{A} = \begin{bmatrix} \alpha & 0 \\ 0 & \beta \end{bmatrix}$, where $0 > \alpha > \beta$, and $\mathbf{B} = \begin{bmatrix} 2 & 0 \\ 0 & 3 \end{bmatrix}$.

i.) Determine an explicit formula for $e^{(\mathbf{A}+\mathbf{B})t}$ using Prop. 3.6.1.

ii.) What is a more efficient way to compute $e^{(\mathbf{A}+\mathbf{B})t}$ in this case? Verify that you get the same result as in (i).

iii.) Show that $\left\|e^{(\mathbf{A}+\mathbf{B})t}\right\|_{\mathbb{B}(\mathbb{R}^N)} \leq e^{(\alpha+3)t}$, $\forall t \geq 0$.

Exercise 3.6.2. Assuming the context of Prop. 3.6.1, show that if $\left\|e^{At}\right\|_{\mathbb{B}(\mathscr{X})} \leq Me^{\omega t}$, then $\left\|e^{(A+B)t}\right\|_{\mathbb{B}(\mathscr{X})} \leq Me^{\left(\omega + M\|B\|_{\mathbb{B}(\mathscr{X})}\right)t}$, $\forall t \geq 0$.

It follows from Prop. 3.6.1 together with Thrm. 3.4.2 that the IVP

$$\begin{cases} u'(t) = (A+B)u(t), \ t > 0, \\ u(0) = u_0, \end{cases} \tag{3.132}$$

has a unique classical solution given by $u(t) = e^{(A+B)t}u_0$, $\forall t \geq 0$.

Example. Consider the following extension of (3.25):

$$\begin{cases} \frac{\partial}{\partial t}z(x,t) = k\frac{\partial^2}{\partial x^2}z(x,t) + z(x,t), \ 0 < x < a, t > 0, \\ z(x,0) = z_0(x), \ 0 < x < a, \\ z(t,0) = z(t,a) = 0. \end{cases} \tag{3.133}$$

This IBVP can be written in the abstract form (3.132), where A and x_0 are defined in (3.44). Also, $B : \mathbb{L}^2(0,a;\mathbb{R}) \to \mathbb{L}^2(0,a;\mathbb{R})$ is the identity operator, which is clearly a member of $B \in \mathbb{B}(\mathbb{L}^2(0,a;\mathbb{R}))$. Hence, (3.133) has a unique classical solution.

Exercise 3.6.3. The following is a more general version of the Sobolev IBVP introduced in Model VI.1:

$$
\begin{cases}
\frac{\partial}{\partial t}\left(z(x,t) - \frac{\partial^2 z}{\partial x^2}(x,t)\right) = \frac{\partial^2 z}{\partial x^2}(x,t) + \int_0^\pi K(t,s)z(x,s)\,ds,\ 0 < x < \pi, t > 0, \\
z(x,0) = z_0(x),\ 0 < x < \pi, \\
z(0,t) = z(\pi,t) = 0,\ t > 0.
\end{cases}
$$

(3.134)

Must (3.134) have a unique classical solution on $(0,\pi)$? Explain.

3.7   Some Approximation Theory

The discussion in Section 3.5 resulted in various characterization theorems that described precisely which linear operators generate C_0-semigroups. But, unless we are dealing with a very nice subclass of these operators (like those of Prop.3.5.9), an explicit formula for the semigroup is often untenable. As such, an approximation method is helpful to gain insight into its structure.

If the hypotheses of the Hille-Yosida theorem are satisfied, then the semigroup $\left\{e^{At} : t \ge 0\right\}$ generated by A can be obtained through various limiting procedures, one of which closely resembles the familiar relation

$$
e^{at} = \lim_{n\to\infty}\left(1 - \frac{at}{n}\right)^{-n},\ \text{where } a,t \in \mathbb{R}.
$$

The strategy is to define a sequence of operators (each of which is known to generate an easy-to-compute semigroup) which approximates the as-of-yet-uncomputable semigroup of interest. We restrict our discussion to a small batch of such theorems, the first of which was actually established in the proof of Thrm. 3.5.4. We state it as the following corollary.

Corollary 3.7.1. *Suppose that a linear operator $A : \mathrm{dom}(A) \subset \mathscr{X} \to \mathscr{X}$ generates a contractive C_0-semigroup $\left\{e^{At} : t \ge 0\right\}$ on $\mathscr{X}$. Then,*

$$
e^{At}x = \lim_{\lambda\to\infty} e^{A_\lambda t}x,\ \forall x \in \mathscr{X},
$$

(3.135)

where $\{A_\lambda : \lambda \ge 0\}$ is the Yosida approximation of A defined in (3.92).

Since the Yosida operators $A_\lambda, \lambda \ge 0$, are bounded, the structure of the semigroup $\left\{e^{A_\lambda t}\right\}$ is completely known, thereby rendering the computation of the limit in (3.135) viable.

Next, since the phrase "*A generates a C_0-semigroup*" means that $\forall x \in \mathrm{dom}(A)$,

$$
Ax = \lim_{h\to 0^+} \frac{e^{Ah}x - x}{h},
$$

it follows directly that the sequence of operators $B_h : \mathcal{X} \to \mathcal{X}$ defined by

$$B_h x = \frac{e^{Ah}x - x}{h} \tag{3.136}$$

approximates Ax (strongly in $\mathcal{X}$) more accurately as $h \to 0^+$. Moreover, since $B_h \in \mathbb{B}(\mathcal{X})$, $\forall h > 0$, Thrm. 3.3.4 ensures that B_h generates a U.C. semigroup $\left\{ e^{B_h t} : t \geq 0 \right\}$. It seems natural that this sequence of semigroups would, in turn, serve as a good approximation of $\left\{ e^{At} : t \geq 0 \right\}$. Indeed it does, as the following result indicates.

Proposition 3.7.2. *Assume that* $\left\{ e^{At} : t \geq 0 \right\}$ *is a* C_0−*semigroup on* $\mathcal{X}$. *Then,* $\forall x \in \mathcal{X}$ *and* $t \geq 0$,

$$e^{At}x = \lim_{h \to 0^+} e^{B_h t} x \text{ uniformly on bounded intervals in } t.$$

Proof. We provide an outline of the proof. Let $a > 0$. We must show that

$$\lim_{h \to 0^+} \left(\sup_{0 \leq t \leq a} \left\| e^{At}x - e^{B_h t}x \right\|_{\mathcal{X}} = 0 \right), \ \forall x \in \mathcal{X}. \tag{3.137}$$

To this end, let $0 \leq t \leq a$ be fixed and observe that

$$e^{B_h t} = e^{-\frac{t}{h}} \sum_{n=0}^{\infty} \frac{t^n e^{A(nh)}}{h^n n!}. \tag{3.138}$$

Then, Thrm. 3.3.6 is used to determine real constants $M \geq 1$ and $\omega > 0$ such that

$$\left\| e^{B_h t} \right\|_{\mathbb{B}(\mathcal{X})} \leq M e^{t(e^\omega - 1)}. \tag{3.139}$$

Fix $x_0 \in \text{dom}(A)$ and let $\varepsilon > 0$. There exists $\delta > 0$ such that $\forall 0 < h < \delta$,

$$\left\| e^{At}x_0 - e^{B_h t}x_0 \right\|_{\mathcal{X}} < \varepsilon. \tag{3.140}$$

This holds true $\forall 0 \leq t \leq a$. Since $\text{dom}(A)$ is dense in $\mathcal{X}$, the theorem is proved. $\square$

Exercise 3.7.1. Carefully verify (3.138) - (3.140) in the proof of Prop. 3.7.2.

Exercise 3.7.2. Verify Prop. 3.7.2 when **A** is a diagonal $N \times N$ matrix.

Exercise 3.7.3. Verify Prop. 3.7.2 for the Sobolev IBVP explored in Exer. 3.3.5.

Exercise 3.7.4. Verify Prop. 3.7.2 for the advection IBVP explored in Exer. 3.3.6.

One notable theorem connecting the convergence of a sequence of C_0−semigroups to the convergence of the sequence of their respective generators is the *Trotter-Kato Approximation* theorem, stated below. (See **[267]** for a proof.)

Theorem 3.7.3. (Trotter-Kato Approximation theorem)
Assume that A and $\{A_n : n \in \mathbb{N}\}$ are linear operators on $\mathscr{X}$ for which $\exists M \geq 1$ and $\omega \in \mathbb{R}$ such that $\forall t \geq 0$,
i.) *$A : \operatorname{dom}(A) \subset \mathscr{X} \to \mathscr{X}$ generates a C_0-semigroup $\{e^{At} : t \geq 0\}$ on $\mathscr{X}$ such that $\left\| e^{At} \right\|_{\mathbb{B}(\mathscr{X})} \leq M e^{\omega t}$, and*
ii.) *For each $n \in \mathbb{N}$, $A_n : \operatorname{dom}(A_n) \subset \mathscr{X} \to \mathscr{X}$ generates a C_0-semigroup $\{e^{A_n t} : t \geq 0\}$ on $\mathscr{X}$ such that $\left\| e^{A_n t} \right\|_{\mathbb{B}(\mathscr{X})} \leq M e^{\omega t}$.*
Then, the following are equivalent:
a.) *For all $x \in \mathscr{X}$ and $\lambda > \omega$, $\lim_{n \to \infty} R_\lambda(A_n)x = R_\lambda(A)x$.*
b.) *For all $x \in \mathscr{X}$ and $t \geq 0$, $\lim_{n \to \infty} e^{A_n t} x = e^{At} x$ uniformly on bounded intervals in t.*

Exercise 3.7.5. Formulate and prove an analog of Prop. 2.4.1 for the abstract setting of this chapter. Then, illustrate your result by applying it to the models discussed in Section 3.2.

3.8   A Brief Glimpse at Long-Term Behavior

Understanding the long-term behavior (as $t \to \infty$) of the classical solution of (3.87) is important to attaining a more complete picture of the evolution of the solution of the IBVPs expressed in the form of (3.87). We investigated such behavior in the finite-dimensional case and discovered a very strong link between the long-term behavior of the solution and the eigenvalues of **A**. The existence of such a link is not particularly surprising, since the eigenvalues of **A** played a crucial role in the structure of the matrix exponential, and the classical solution of (3.1) was, in turn, expressed uniquely in terms of this matrix exponential. Specifically, the classical solution of (3.1) was shown to be exponentially stable if and only if the real part of all eigenvalues of **A** were negative.

Can we establish such a result for the abstract IVP (3.87) in a Banach space $\mathscr{X}$? In response to this question, there is a substantive theory devoted to describing the relationships among the so-called *spectral properties* of a generator A, the structure of its semigroup, and the stability of the semigroup (and, in turn, the classical solution of (3.87)). A complete discussion of this elegant theory would take us far afield of our present goals. A more complete discussion of the spectral properties of A and their link to stability can be found in a number of monographs (see **[93, 123, 142, 179, 267]**), and a readable initial account of *Lyapunov stability theory* applicable to the problems considered in the current text developed from a topological dynamics viewpoint can be found in **[320]**.

The following notions of stability are analogous to those in Def. 3.5.2.

Definition 3.8.1. A C_0-semigroup $\{e^{At} : t \geq 0\}$ on $\mathscr{X}$ is
i.) *uniformly stable* if $\lim_{t \to \infty} \left\| e^{At} \right\|_{\mathbb{B}(\mathscr{X})} = 0$;

ii.) *exponentially stable* if $\exists M \geq 1$ and $\alpha > 0$ such that $\left\| e^{At} \right\|_{\mathbb{B}(\mathscr{X})} \leq M e^{-\alpha t}$, $\forall t \geq 0$;

iii.) *strongly stable* if $\lim_{t \to \infty} \left\| e^{At} z \right\|_{\mathscr{X}} = 0$, $\forall z \in \mathscr{X}$.

Exercise 3.8.1. Is the classical solution of (3.56) exponentially stable? Explain.

Exercise 3.8.2. Is the classical solution of (3.47) strongly stable? Explain.

Exercise 3.8.3. Suppose that a linear operator $A : \text{dom}(A) \subset \mathscr{X} \to \mathscr{X}$ generates a contractive C_0−semigroup $\left\{ e^{At} : t \geq 0 \right\}$ on $\mathscr{X}$. Explain why that if this semigroup is exponentially stable, then the classical solution of (3.87) must tend to zero exponentially fast.

3.9   An Important Look Back

Before moving onto the next leg of our journey, we pause to take a critical look at the theoretical development thus far and examine the necessity of the underlying hypotheses.

Exercise 3.9.1. (A Look Back at Chapter 2 - Boundedness)
Revisit Chapter 2 and identify all occurrences in which the boundedness of the operator **A** played a crucial role in the argument. Compare the result to its analog in Chapter 3, and if a counterexample is not provided to show that the version of the result in Chapter 2 does not extend in the same manner to the setting of Chapter 3, try to provide one.

Exercise 3.9.2. (A Look Back at Chapter 3 - Linearity)
Revisit Chapter 3 and identify exactly where in the proofs the linearity of either the operator A or the semigroup it generated was used. (This exercise also serves as preparation for our study of nonlinear evolution equations later in Chapter 9, and will be recalled at that time to point out where the difficulties might lie in the extension of the theory to the nonlinear case.)

3.10   Looking Ahead

By way of preparation for Chapter 4, we consider some modified versions of IBVP (3.25), now with increased complexity.

Suppose an external space-dependent term is added to the right-side of the PDE in

(3.25) to account for an external heat source. This leads to the modified IBVP

$$\begin{cases} \frac{\partial}{\partial t} z(x,t) = \frac{\partial^2}{\partial x^2} z(x,t) + f(x), \ 0 < x < a, t > 0, \\ z(x,0) = 0, \ 0 < x < a, \\ z(0,t) = z(a,t) = 0, \ t > 0, \end{cases} \quad (3.141)$$

where the forcing term $f : [0,a] \to \mathbb{R}$ can be given by, for instance,

$$f(x) = \begin{cases} x, & 0 \le x \le \frac{a}{2}, \\ a - x, & \frac{a}{2} \le x \le a. \end{cases} \quad (3.142)$$

Exercise 3.10.1.
i.) Verify directly that a classical solution of (3.141) is given by

$$z(x,t) = \frac{4}{a} \sum_{n=1}^{\infty} \frac{(-1)^{n-1}}{(2n-1)^4} \left(1 - e^{-(2n-1)^2 t}\right) \sin\left((2n-1)x\right). \quad (3.143)$$

ii.) Does $\exists \lim_{t \to \infty} |z(x,t)|$? Explain.
iii.) Comment on the differences between (3.143) and (3.41) (when $f = 0$).

We would like to formulate such an IBVP abstractly in a Banach space and then express its classical solution using a suitable semigroup. To provide a slightly clearer view of this issue, recall that the unique classical solution of (3.25) coupled with the homogenous Neumann BCs is $u(t) = e^{At}[u_0], t \ge 0$, where the operators $e^{At} : \mathbb{L}^2(0,a;\mathbb{R}) \to \mathbb{L}^2(0,a;\mathbb{R})$ are defined by

$$e^{At}(g)[x] = \sum_{m=0}^{\infty} \left(\frac{2}{a} \int_0^a g(w) \cos\left(\frac{m\pi}{a}w\right) dw\right) e^{-\left(\frac{m\pi}{a}\right)^2 kt} \cos\left(\frac{m\pi}{a}x\right). \quad (3.144)$$

We now incorporate a forcing term $f : [0,a] \times [0,\infty) \to \mathbb{R}$ into the heat conduction model to obtain the following more general IBVP:

$$\begin{cases} \frac{\partial}{\partial t} z(x,t) = k \frac{\partial^2}{\partial x^2} z(x,t) + f(x,t), \ 0 < x < a, t > 0, \\ z(x,0) = z_0(x), \ 0 < x < a, \\ \frac{\partial z}{\partial x}(0,t) = \frac{\partial z}{\partial x}(a,t) = 0, \ t > 0. \end{cases} \quad (3.145)$$

Exercise 3.10.2. Try to rewrite (3.145) as an abstract evolution equation. What new difficulties do you encounter?

Assuming that f is sufficiently smooth, it can be expressed uniquely in the form

$$f(x,t) = \sum_{m=1}^{\infty} f_m(t) \sin\left(\frac{m\pi}{a}x\right), \ 0 < x < a, t > 0, \quad (3.146)$$

where

$$f_m(t) = \frac{2}{a} \int_0^a f(w,t) \sin\left(\frac{m\pi}{a}w\right) dw, t > 0. \quad (3.147)$$

(This follows from an extension of the Fourier series discussion in Section 1.7.2. See [121] for more details.) Assuming that the solution of (3.145) is of the form

$$z(x,t) = \sum_{m=1}^{\infty} z_m(t) \sin\left(\frac{m\pi}{a}x\right), \ 0 < x < a, \ t > 0, \tag{3.148}$$

complete the following exercise.

Exercise 3.10.3.
i.) Substitute (3.148) into (3.145) to show that $\forall m \in \mathbb{N}$, $z_m(t)$ satisfies the IVP

$$\begin{cases} z_m'(t) + \frac{m^2\pi^2}{a^2}z_m(t) = f_m(t), \ t > 0, \\ z(0) = z_m^0. \end{cases} \tag{3.149}$$

ii.) Use the variation of parameters technique to show that the solution of (3.149) is given by

$$z_m(t) = z_m^0 e^{-\left(\frac{m^2\pi^2}{a^2}\right)t} + \int_0^t e^{-\left(\frac{m^2\pi^2}{a^2}\right)(t-s)} f_m(s)ds, \ t > 0. \tag{3.150}$$

iii.) Use (3.144) and (3.147) in (3.150) to show that the solution of (3.149) can be simplified to

$$z(x,t) = e^{At}(z_0)[x] \tag{3.151}$$
$$+ \int_0^t \int_0^a \frac{2}{a} \sum_{m=1}^{\infty} e^{-\left(\frac{m^2\pi^2}{a^2}\right)(t-s)} \sin\left(\frac{m\pi}{a}w\right) \sin\left(\frac{m\pi}{a}x\right) f(w,s)dwds.$$

iv.) Finally, use (3.151) to further express the solution of (3.145) in the form

$$z(\cdot,t) = e^{At}[z_0][\cdot] + \int_0^t e^{A(t-s)}f(s,\cdot)ds, \ t > 0. \tag{3.152}$$

Equation (3.152) is reminiscent of the familiar variation of parameters formula (cf. (1.110)), and suggests that if we identified the terms correctly between (3.145) and the abstract evolution equation

$$\begin{cases} u'(t) = Au(t) + F(t), \ t > 0, \\ u(0) = u_0, \end{cases} \tag{3.153}$$

then a viable representation formula for the classical solution might be

$$u(t) = e^{At}[u_0] + \int_0^t e^{A(t-s)}F(s)ds, \ t > 0.$$

Does such a formula make sense? If so, what conditions can be imposed on F to ensure that u given by this formula actually satisfies (3.153)? Such questions are the focus of our investigation in the next chapter.

3.11   Guidance for Exercises

3.11.1   Level 1: A Nudge in a Right Direction

3.1.1. i.) Verify the conditions of Def. 1.7.3.

  ii.) Use Def. 3.1.1(ii) with standard norm properties.

3.1.2. For each $m \in \mathbb{N}$, find a ...

3.1.3. i.) Show $\exists m > 0$ such that

$$\sup_{\|x\|_{\mathbb{C}}=1} \left\| \int_a^b g(\cdot,s)x(\cdot)\,ds \right\|_{\mathbb{C}([a,b];\mathbb{R})} \leq m \, \|x\|_{\mathbb{C}([a,b];\mathbb{R})} \,.$$

  ii.) Use the $\mathbb{L}^2$−norm in the above inequality. Does Prop. 1.8.15(i) help?

3.1.4. In general, this operator is not linear (Why?), but it is when $g\left(x,y,z(y)\right) = \overline{g}(x,y)z(y)$. (Why?) Prove this operator is bounded in a manner similar to the one used in Exer. 3.1.3(i).

3.1.5. i.) Use $g_n(x) = x^n$, where $n \in \mathbb{N}$.

  ii.) Carefully use the definition of norm.

3.1.6. Apply the hint for Exer. 3.1.5(i), where $n \geq 2$. (How?)

3.1.7. i.) Note that $\|\mathscr{F}x_n - \mathscr{F}x\|_{\mathscr{Y}} = \|\mathscr{F}(x_n - x)\|_{\mathscr{Y}}$. (Now what?)

3.1.8. Linearity follows easily since $\mathscr{F}$ and $\mathscr{G}$ are both linear. Use the first of two versions of Def. 3.1.1(ii) with Exer. 3.1.1(ii).

3.1.9. Adapt the hint provided for Exer. 3.1.7 to this situation.

3.1.10. i.) ($\Longrightarrow$) $\mathscr{F}$ must be one-to-one, so $\mathscr{F}x = \mathscr{F}y$ implies what?
  ($\Longleftarrow$) $\mathscr{F}x = \mathscr{F}y \Longrightarrow \mathscr{F}(x-y) = 0$. (So what?)

  ii.) Let $\alpha, \beta \in \mathbb{R}$ and choose $x = \mathscr{F}z_1$ and $y = \mathscr{F}z_2$ in the expression $\mathscr{F}^{-1}(\alpha x + \beta y)$.

3.1.11. Prove that $\{S_n\}$ is a Cauchy sequence in $\mathscr{X}$.

3.1.12. For any $\alpha \neq 0$, find an operator $\mathscr{B}_\alpha$ such that $\mathscr{B}_\alpha(\alpha \mathscr{A}) = (\alpha \mathscr{A})\mathscr{B}_\alpha = I$.

3.1.13. Verify the properties in Def. 1.7.3 directly to prove that $\|\cdot\|_{\mathscr{X} \times \mathscr{Y}}$ is a norm. Completeness of $\mathscr{X} \times \mathscr{Y}$ follows from the completeness of $\mathscr{X}$ and $\mathscr{Y}$. (How?)

3.1.14. Adapt the argument used in the example directly following Def. 3.1.14.

3.1.15. Determine the graph of $\mathscr{B}^{-1}$.

3.1.16. If $\mathscr{X}_1 \subset \mathscr{X}_2 \subset \mathscr{X}$ and all are equipped with the same norm $\|\cdot\|_{\mathscr{X}}$, then $\mathrm{cl}_{\mathscr{X}}(\mathscr{X}_1) = \mathscr{X}$. (So what?)

3.1.17. Yes, because every $\mathbb{L}^2$−function has a unique Fourier representation. (So what?)

3.2.1. If $z = g(x,y)$, $x = x(t,s)$, and $y = y(t,s)$, then $\frac{\partial z}{\partial t} = \frac{\partial g}{\partial x}\frac{\partial x}{\partial t} + \frac{\partial g}{\partial y}\frac{\partial y}{\partial t}$.

3.2.2. Since (3.21), and hence (3.20), is not satisfied.

3.2.3. i.) Use the choices similar to those used in the example. Note that $\mathrm{dom}(A) \subset \mathscr{X}$. (Why?) For what other reason is this a good choice for $\mathrm{dom}(A)$?

  ii.) $z_0(x) = u(0)[x]$. Now determine the space $\mathscr{X}$.

3.2.4. i.) The process is practically the same, but be careful in steps (3.34) and (3.35).

  ii.) Integrate by parts to compute b_m.

iii.) The only change from the previous example occurs in how dom(A) is defined.
3.2.5. Substituting (3.48) into (3.47) leads to

$$k\left(X''(x)Y(y)T(t) + X(x)Y''(y)T(t)\right) = T'(t)X(x)Y(y).$$

Dividing all terms in this equality by $X(x)Y(y)T(t)$ yields

$$\underbrace{k\left(\frac{X''(x)}{X(x)} + \frac{Y''(y)}{Y(y)}\right)}_{\text{Function of only } x \text{ and } y} = \lambda = \underbrace{\frac{T'(t)}{T(t)}}_{\text{Function of } t} . \text{ (Now what?)}$$

3.2.6. Use $\mathscr{X} = \mathbb{L}^2\left((0,a) \times (0,b)\right)$ equipped with the usual inner product

$$\langle f,g \rangle_{\mathbb{L}^2} = \int_0^b \int_0^a f(x,y)g(x,y)dxdy.$$

Defining A in a natural way, how does this choice affect the definition of dom(A)?
3.2.7. i.) The main change is that the cosine terms remain instead of the sine terms. There might be one extra Fourier coefficient also, so be careful.

ii.) Everything remains the same except for one change in dom(A). What is it?
3.2.8. Let $z = z(x_1,x_2,\ldots,x_n,t)$. The IBVP becomes

$$\begin{cases} \frac{\partial z}{\partial t} = k\sum_{i=1}^n \frac{\partial^2 z}{\partial x_i^2}, \ (x_1,x_2,\ldots,x_n) \in [0,a_1] \times \ldots \times [0,a_n], t > 0, \\ \frac{\partial z}{\partial x_i}(x_1,\ldots,x_{i-1},0,x_{i+1},\ldots,x_n,t) = \frac{\partial z}{\partial x_i}(x_1,\ldots,x_{i-1},a_i,x_{i+1},\ldots,x_n,t) = 0, \end{cases}$$

(3.154)

where $i = 1,\ldots,n$, $t > 0$. In order to find the solution of (3.154), suitably modify (3.50) - (3.54). How many different separations of variable does this entail?
3.2.9. You would need to apply B^{-1} on both sides. But, what must be true in order to justify such action?
3.2.10. Use the fact that the series converges uniformly. (How?) The BC and IC are easily verified.
3.3.1. All instances involve the use of $\|A\|_{\mathbb{M}^N}$. (Why?)
3.3.2. The domain of the matrix operator $\mathbf{A}$ (that is, the set of vectors in $\mathbb{R}^N$ for which multiplication on the left by $\mathbf{A}$ yields a well-defined member of $\mathbb{R}^N$) is what set?
3.3.3. Use Prop. 1.7.11(iv) with $\langle f,f \rangle_{\mathbb{L}^2} = \|f\|_{\mathbb{L}^2}^2$. Now, determine $\overline{M}$.
3.3.4. i.) Follow the discussion in the example, but use (3.41) instead of (3.46).

ii.) This is the same as the example, but with more variables. Just keep the book-keeping straight with the double sum

iii.) See Exer. 3.2.8 and the hint provided.
3.3.5. Verify the steps in the example using (3.59) instead.
3.3.6. Linearity and Def. 3.3.1(i) both hold trivially. (Why?) As for Def. 3.3.1(ii), apply the composition correctly.
3.3.8. Apply the semigroup property to $e^{A(t+h)}$.
3.3.9. i.) For (3.65), what is the power series for $e^{t\|A\|_{\mathbb{B}(\mathscr{X})}}$?
ii.) Look at (3.66).

iii.) <u>First equality</u>: Start by using the semigroup property.

<u>Second equality</u>: Consider the cases when $0 < \delta < h$ and $0 < h < \delta$ separately and use interval additivity.

iv.) First, use the semigroup property. Then, factor p times using difference of squares. (Now what?)

3.3.10. How are strong continuity in $\mathscr{X}$ and $\mathbb{B}(\mathscr{X})$ related?

3.3.11. Refer to Exer. 3.3.6 for the portion involving Def. 3.1.1. Verifying the strong continuity is somewhat technical, but try it!

3.3.12. i.) Make certain that the inputs t used in e^{At} are positive. The argument is very similar to the case shown.

ii.) Use part (iv) of the theorem, then (3.76) and (3.77) together. Why is bringing the limit inside the integral justified?

iii.) Use parts (iii) and (xi) of the theorem.

3.3.13. Define the function $G : [0, \infty) \to \mathscr{X}$ by $G(t) = f(t)e^{At}x$ and compute $G'(0)$ using two different methods. (Now what?)

3.3.14. Use the definitions of semigroup and generator directly.

3.3.15. Use the so-called *Parseval equality*: $\sum_{m=0}^{\infty} b_m^2 = \|z_0\|_{\mathbb{L}^2(0,a)}^2$. (See **[121]** for a discussion and proof.)

3.3.16. Use Def. 3.3.2 directly. You must have $\alpha > 0$. (Why?)

3.4.1. The right-side of (3.87) must be defined. What is dom(A) if $A \in \mathbb{B}\left(\mathbb{R}^N\right)$?

3.4.2. Follow the proof of Thrm. 2.3.2. Are there any changes?

3.4.3. Applying (3.87) here implies that $\frac{\partial u}{\partial t} = \frac{\partial u}{\partial s}$ and $u(0, s) = u_0(s)$. (So what?)

3.4.4. Look at Thrm. 3.3.4.

3.4.5. For each $n \in \mathbb{N}$, compute $\sup \left\{ \|e^{At}u_0^n - e^{At}u_0\|_{\mathscr{X}} : 0 \leq t \leq T \right\}$.

3.4.6. What if $\left\{ e^{At} : t \geq 0 \right\}$ were contractive?

3.4.7. i.) Use the definition directly.

ii.) Argue inductively in n.

iii.) Verifying most parts is straightforward. At some point you will need to use the fact that A is closed.

3.4.8. i.) <u>part (ii)</u>: $\bigcap_{n=1}^{\infty} \text{dom}(A^n) \subset \text{dom}(A^{n_0})$, $\forall n_0 \in \mathbb{N}$.

<u>part (iii)</u>: Use the fact that A generates $\left\{ e^{At} : t \geq 0 \right\}$. When A is restricted to dom(A^n), use Exer. 3.4.7 to argue that $\left\{ e^{At} |_{\text{dom}(A^n)} : t \geq 0 \right\}$ retains its properties and maps into dom(A^n).

ii.) At the base step of the induction argument, consider the IVP:

$$\begin{cases} w'(t) & = Aw(t), \ t > 0, \\ w(0) & = -Au_0. \end{cases} \tag{3.155}$$

Why must there exist a classical solution w of (3.155)? Find an expression for it. (Now what?)

3.4.9. i.) Apply Prop. 3.4.7 directly.

ii.) There are several choices here. Look carefully at Prop. 3.4.7 and try to identify the minimal assumption that works.

3.5.1. i.) Find the eigenvalues of $\mathbf{A}$.

ii.) Use the definition of $R_\lambda(\mathbf{A}) = (I - \lambda\mathbf{A})^{-1}$ directly.

iii.) $\lambda R_\lambda(\mathbf{A})\mathbf{x} = \frac{\lambda}{(1-\lambda)(2-\lambda)} \begin{bmatrix} 2-\lambda & 0 \\ 0 & 1-\lambda \end{bmatrix} \mathbf{x}$. (Now what?)

3.5.2. i.) & ii.) Look ahead at Exer. 3.5.5 for a hint.

iii.)(d) $R_\lambda(A) \in \mathbb{B}(\mathscr{X})$. (So what?)

iii.)(e) Come on now... Guess!

3.5.3. Use (3.99) to verify Def. 3.3.1(i). Part (ii) is trivial and (iii) works because $\{e^{A_\lambda t} : t \ge 0\}$ is a semigroup, $\forall \lambda > 0$.

3.5.4. The main change occurs in the BC imposed. Does (3.118) still hold?

3.5.5. A is closed: Take $\{f_n\} \subset \operatorname{dom}(A)$ such that

$$\|f_n - f\|_{C([0,\infty);\mathbb{R})} \longrightarrow 0 \text{ and } \|f'_n - g\|_{C([0,\infty);\mathbb{R})} \longrightarrow 0.$$

Show that $f \in \operatorname{dom}(A)$ and $Af = g$.

$R_\lambda(A)$ exists: Let $\lambda \in \rho(A)$ and $g_0 \in C([0,\infty);\mathbb{R})$. Prove that $\exists f \in \operatorname{dom}(A)$ such that $(\lambda I - A)f = g_0$.

$R_\lambda(A) \in \mathbb{B}(C([0,\infty);\mathbb{R}))$: Estimate $\|R_\lambda(A)(g_0)\|_{C([0,\infty);\mathbb{R})}$ using $R_\lambda(A)(g_0)(x) = \int_0^\infty e^{\lambda(x-z)}g_0(z)dz$.

3.5.6. Use convergence of the series and the linearity-type property of the inner product to verify the linearity of $T(t)$. For boundedness, use Prop. 1.7.11(iv). (Now what?)

3.5.7. A real-valued function f is increasing on $\operatorname{dom}(f)$ iff

$$x \le y \implies f(x) \le f(y), \forall x,y \in \operatorname{dom}(f). \text{ (So what?)}$$

3.5.8. i.) $(\implies)$ $\|(x-y) + \alpha(Bx - By)\|_{\mathscr{X}}^2 = \langle \_\_, \_\_ \rangle_{\mathscr{X}} = \cdots$

$(\impliedby)$ If $a < b + \varepsilon$, $\forall \varepsilon > 0$, how must a and b compare?

ii.) $I + \alpha B$ is one-to-one because if $x \ne y$, then the left-side of the inequality in the definition is positive. (Now what?)

3.5.9. Use Prop. 1.7.11(iv), the contractivity of $\{e^{At} : t \ge 0\}$, and the continuity of the inner product. (How?)

3.6.1. ii.) $\mathbf{A} + \mathbf{B}$ is a diagonal matrix. (So what?)

iii.) Use an appropriate matrix norm in (ii).

3.6.3. Argue that the operator $B : L^2(0,\pi;\mathbb{R}) \to L^2(0,\pi;\mathbb{R})$ defined by $B[z](\cdot) = \int_0^\pi K(\cdot,s)z(x,s)\,ds$ belongs to $\mathbb{B}(L^2(0,\pi;\mathbb{R}))$. (So what?)

3.7.1. (3.138): Start by using elementary exponential properties.

(3.139): Use Thrm. 3.3.6(ii) and estimate (3.138).

(3.140): Use the fact that $\lim_{h\to 0^+} B_h x_0 = A x_0$. (How?)

3.7.2. Let A be given by (2.27). Then,

$$B_h[t] = \begin{bmatrix} \frac{(e^{a_{11}h})t-t}{h} & 0 & 0 & \cdots & 0 \\ 0 & \frac{(e^{a_{22}h})t-t}{h} & 0 & \cdots & 0 \\ \vdots & \vdots & \ddots & & \vdots \\ 0 & 0 & & \ddots & 0 \\ 0 & 0 & 0 & \cdots & \frac{(e^{a_{NN}h})t-t}{h} \end{bmatrix}$$

3.7.3. Using $S(t)$ from Exer. 3.3.5, show that $S(t)[f](x) = \lim\limits_{h \to 0^+} e^{B_h t}[f](x)$, where

$$B_h[g] = \frac{S(h)[g](x) - g(x)}{h}, \, g \in \mathbb{L}^2(0, \pi; \mathbb{R}).$$

3.7.4. Repeat Exer. 3.7.3 using $S(t)$ from Exer. 3.3.6. The regularity of f will enter into the computation.

3.7.5. <u>Statement</u>: Let $\{T_n(t)\}$ be a sequence of contractive C_0-semigroups on $\mathscr{X}$ with respective generators $\{A_n\} \subset \mathbb{B}(\mathscr{X})$ such that for some $\{T_0(t) | t \geq 0\} \subset \mathbb{B}(\mathscr{X})$,

$$A_n T_0(t) = T_0(t) A_n, \, \forall n \in \mathbb{N}, t > 0.$$

If $\exists \lim\limits_{n \to \infty} A_n x = A_0 x, \, \forall x \in \mathrm{dom}(A_0)$, then

$$\lim\limits_{n \to \infty} T_n(t) x = T_0(t) x \text{ uniformly on bounded intervals (in } t) \text{ of } [0, \infty).$$

3.8.1. Use the semigroup formula in the example directly following Def. 3.3.1.

3.8.2. Use the fact that the solution is given by $u(t) = e^{At}[u_0]$ to assist you in computing $\lim\limits_{t \to \infty} \|u(t)\|_{\mathscr{X}}$.

3.9.1. Consider any result when an estimate involving $\|\mathbf{A}\|_{\mathbb{M}^N}$ is used.

3.9.2. Establishing the elementary properties of linear semigroups involves linearity. Since the properties are used heftily in the development of the rest of the theory, this dependence on linearity filters down. Identify specific instances of this.

3.10.1. i.) Why can you bring the derivative inside the sum?

ii.) Yes, because of the negative exponential term.

3.10.2. Of course, the presence of the new term $f(x,t)$ is the only difference. Assuming that all other identifications remain the same as the homogenous case studied in Section 3.2, how would you handle this term?

3.10.3. The calculations are not difficult, but they are tedious. If you need substantive hints at this point, proceed to Section 4.2.

3.11.2 Level 2: An Additional Thrust in a Right Direction

3.1.1. For both parts, apply standard norm properties in $\mathscr{X}$ or $\mathscr{Y}$, whichever is appropriate, together with properties of sup and inf.

3.1.2. $\ldots x \in \mathscr{X}$ such that $\|\mathscr{F}(x)\|_{\mathscr{Y}} > m \|x\|_{\mathscr{X}}$.

3.1.3. i.) Show that $\|\mathscr{F}\|_{\mathbb{B}(\mathbb{C})} \leq M_g(b-a)$, where

$$M_g = \sup\{|g(t,s)| : (t,s) \in [a,b] \times [a,b]\}.$$

ii.) $\int_a^b \left| \int_a^b g(t,s)x(s)\,ds \right|^2 dt \leq \int_a^b \left(\int_a^b g^2(t,s)\,ds \right)^{1/2} \|x\|_{\mathbb{L}^2}\,dt.$ (So what?)

3.1.4. Show that $\|\mathscr{F}\|_{\mathbb{B}(\mathbb{C})} \leq m_g(b-a).$ (So what?)

3.1.5. i.) $g_n'(x) = nx^{n-1}$, where $n \in \mathbb{N}$, and $\|g_n'\|_{C([0,a];\mathbb{R})} = na^{n-1}.$ (So what?)

ii.) $\|\mathscr{F}(g)\|_{\mathbb{L}^2}^2 = \int_0^a |g'(x)|^2\,dx \leq a\|g\|_{C^1([0,a];\mathbb{R})}$

3.1.6. For any $n \geq 2$, compute $\|g''\|_{C([0,a];\mathbb{R})}$. What can you conclude?

3.1.7. Use Exer. 3.1.1(ii).

3.1.8. Show that $\|\mathscr{F}\mathscr{G}x\|_{\mathscr{X}} \leq \|\mathscr{F}\|_{\mathbb{B}(\mathscr{X})}\|\mathscr{G}\|_{\mathbb{B}(\mathscr{X})}\|x\|_{\mathscr{X}}, \forall x \in \mathscr{X}$.

3.1.9. $\|\mathscr{F}_n(x) - \mathscr{F}(x)\|_{\mathscr{Y}} = \|(\mathscr{F}_n - \mathscr{F})(x)\|_{\mathscr{Y}} \leq \|\mathscr{F}_n - \mathscr{F}\|_{\mathbb{B}(\mathscr{X})}\|x\|_{\mathscr{X}}, \forall x \in \mathscr{X}$.

3.1.10. i.) $(\Longrightarrow) x = 0$. (Why?) and $(\Longleftarrow)x - y = 0$. (Why?)

  ii.) Simplify to get $\alpha\mathscr{F}^{-1}x + \beta\mathscr{F}^{-1}y$.

3.1.11. Observe that $\forall p \in \mathbb{N}, \|S_{n+p} - S_n\|_{\mathscr{X}} \leq \sum_{k=n+1}^{n+p}\|x_k\|_{\mathscr{X}}$. (So what?)

3.1.12. Use $\mathscr{B}_\alpha = \frac{1}{\alpha}\mathscr{A}^{-1}$.

3.1.13. Start with a Cauchy sequence $\{(x_n, y_n)\}$ in $\mathscr{X} \times \mathscr{Y}$ and use the completeness of $\mathscr{X}$ and $\mathscr{Y}$ independently to produce the natural candidate $(x, y) \in \mathscr{X} \times \mathscr{Y}$ to which the sequence converges.

3.1.14. Adapt the argument used in the example directly following Def. 3.1.14.

3.1.15. graph $(\mathscr{B}^{-1}) = \{(\mathscr{B}x, x) \mid x \in \text{dom}(\mathscr{B})\}$ is closed. (Why?)

3.1.16. It must be the case that $\text{cl}_{\mathscr{Z}}(\mathscr{X}_2) = \mathscr{Z}$. (Why?)

3.1.17. You can show that the set used in Prop. 3.1.17 is dense in $\mathbb{L}^2(a, b; \mathbb{R})$.

3.2.1. To apply the chain rule, let $w(x, t) = x + vt$. Then, $c(x, t) = c_0(w(x, t))$. (Now what?)

3.2.3. i.) This choice imposes just enough regularity on the functions without presuming additional smoothness, and the BC are satisfied.

  ii.) $z_0 \in \mathbb{L}^2(0, a; \mathbb{R})$, at least. Using $\mathbb{L}^1(0, a; \mathbb{R})$ is insufficient, which can be seen by appealing to Prop. 1.8.15(i). (How?)

3.2.4. iii.) Incorporate the BCs into the definition of $\text{dom}(A)$. Is the resulting set contained within $\mathscr{X}$?

3.2.5. Now separate X from Y. This explains why we have both λ_x and λ_y. (Why?) The remaining calculations are routine.

3.2.6. Use $A[z] = \left(\frac{\partial^2}{\partial x^2} + \frac{\partial^2}{\partial y^2}\right)[z]$. Let $R = [0, a] \times [0, b]$. Taking into account the BCs, define

$$\text{dom}(A) = \left\{z \in \mathbb{H}^2((0, a) \times (0, b); \mathbb{R}) \mid z|_{\partial R} = 0\right\}.$$

3.2.7. i.) Be careful at steps (3.49) and (3.50); you will get cosines instead of sines. Also, since $\cos(0) = 1$, you get one additional term in the solution formula.

  ii.) $\text{dom}(A) = \left\{z \in \mathbb{H}^2((0, a) \times (0, b); \mathbb{R}) \mid \frac{\partial z}{\partial x}, \frac{\partial z}{\partial y}\big|_{\partial R} = 0\right\}$. (The condition on the partials amounts to saying that the outward normal to the boundary is zero.)

3.2.8. (3.50) becomes

$$\begin{cases} (X_i)_{m_i}(x_i) = \cos\left(\frac{m_i\pi_i}{a_i}x_i\right), 0 < x_i < a_i, m_i \in \mathbb{N}, i = 1, 2, \ldots \\ T_{m_1\ldots m_n}(t) = Ce^{-kt\sum_{i=1}^n\left(\frac{m_i\pi_i}{a_i}\right)^2}, t > 0. \end{cases}$$

(3.51) becomes

$$z_{m_1\ldots m_n}(x_1, \ldots, x_n, t) = b_{m_1\ldots m_n}T_{m_1\ldots m_n}(t)\prod_{i=1}^n(X_i)_{m_i}(x_i).$$

(3.52) becomes

$$z(x_1, \ldots, x_n, t) = \sum_{m_1=1}^\infty \cdots \sum_{m_n=1}^\infty z_{m_1\ldots m_n}(x_1, \ldots, x_n, t),$$

where

$$b_{m_1...m_n} = \frac{2^n}{a_1 \cdots a_n} \int_0^{a_n} \cdots \int_0^{a_1} z_0\left(w_1,\ldots,w_n,t\right) \prod_{i=1}^{n} \cos\left(\frac{m_i \pi_i}{a_i} w_i\right) dw_1 \cdots dw_n.$$

Finally, use $\mathscr{X} = \mathbb{L}^2\left((0,a_1) \times (0,a_n)\right)$ along with the natural modifications to A, dom(A), etc. in order to write the IBVP in the abstract form (3.11).

3.2.9. The domains of A and B must be "compatible." Additional restrictions must be imposed on the operator $B^{-1}A$ in order for the theory that we are about to develop to be applicable. (See Chapter 7.)

3.2.10. You can differentiate the series term-by-term. Now, the calculation is routine.

3.3.2. $\mathbb{R}^N$, which is finite dimensional.

3.3.3. Use $\overline{M} = r\left[1 + \frac{2M}{a} \sum_{m=1}^{\infty} e^{-\left(\frac{m\pi}{a}\right)^2 2kt}\right]$.

3.3.4. i.) You need to use a different orthonormal basis for $\mathbb{L}^2\left(0,a\right)$.

iii.) See Exer. 3.2.8 and make a similar generalization to the one used in going from (i) to (ii).

3.3.6. $S\left(t_1 + t_2\right)[f](x) = f\left(x + v\left(t_1 + t_2\right)\right) = f\left(\underbrace{(x + vt_1)}_{\text{Treat as a single input}} + vt_2\right)$

3.3.8. Note that $\left\|e^{At} - e^{A(t+h)}\right\|_{\mathbb{B}(\mathscr{X})} = \left\|e^{At}\left(I - e^{Ah}\right)\right\|_{\mathbb{B}(\mathscr{X})}$.

3.3.9. i.) For (3.66), expand the power series for $e^{t\|A\|_{\mathbb{B}(\mathscr{X})}}$ and compare term-by-term to the series $t\sum_{n=1}^{\infty} \frac{t^{n-1}\|A\|_{\mathbb{B}(\mathscr{X})}^n}{n!}$. (Now what?)

ii.) What must be true in order for the operator A^2 to be defined?

iii.) First equality: Since $T(h)T(t) = T(h+t)$ and let $u = h+t$.

Second equality: For $0 < \delta < h$,

$$\int_\delta^{\delta+h} T(u)du - \int_0^\delta T(u)du = \int_\delta^{\delta+h} T(u)du - \left[\int_\delta^h T(u)du + \int_0^\delta T(u)du\right]$$

iv.) Now, use (3.73) and the boundedness of the product.

3.3.11. See **[318]** for details.

3.3.12. ii.) The integrand is bounded. So, use Prop. 1.8.16, suitably modified.

iii.) Use Prop. 1.8.6(v).

3.3.13. Equate the two expressions for $G'(0)$.

3.3.15. For any $f \in \mathbb{L}^2(0,a)$, use standard inequalities with the first hint to show

$$\|S(t)[f]\|_{\mathbb{L}^2(0,a)} \leq e^{-\alpha t}\|f\|_{\mathbb{L}^2(0,a)} \leq 1, \forall t > 0.$$

3.3.16. The parameter of the semigroup must be positive.

3.4.1. dom$(A) = \mathbb{R}^N$ (So what?)

3.4.2. Nothing significant. Just be careful to justify all steps using the properties in Thrm. 3.3.6.

3.4.3. The only classical solution of (3.87) that can exist is of the form $u(t,s) =$

$u_0(t+s)$. This does it. (Why?)

3.4.4. $\text{dom}(A) = \mathscr{X}$ in such case.

3.4.5. Use Thrm. 3.3.6(ii) to conclude. (How?) .

3.4.7. ii.) Use (3.76) in the inductive step. (How?)

    iii.) Use the fact that A is closed in the verification of completeness.

3.4.8. i.) part (ii): $\mathscr{X} = \text{cl}_{\mathscr{X}}\left(\bigcap_{n=1}^{\infty}\text{dom}(A^n)\right) \subset \text{cl}_{\mathscr{X}}\left(\text{dom}(A^n)\right) \subset \mathscr{X}, \forall n \in \mathbb{N}$.

ii.) Note that $w(t) = -S(t)Au_0 = \frac{d}{dt}\left[u_0 + \int_0^t w(s)ds\right]$. Argue that w and Au both belong to $\bigcap_{k=0}^{n-1} C^{n-1-k}\left([0,\infty);\text{dom}(A^k)\right)$. (How do you conclude?)

3.4.9. ii.) How about $u_0 \in \text{dom}(A^3)$?

3.5.1. i.) For all $\lambda \neq 1,2$. (Why?)

    ii.) $(I - \lambda\mathbf{A})^{-1} = \frac{1}{(1-\lambda)(2-\lambda)}\begin{bmatrix} 2-\lambda & 0 \\ 0 & 1-\lambda \end{bmatrix}$.

    iii.) Compute the limit entrywise.

3.5.2. iii.)(d) Use part (c).

    iii.)(e) If we are lucky, it will be $S(t)x$, where $\{S(t) : t \geq 0\}$ is the semigroup generated by A.

3.5.3. Use the continuity of the norm to complete the verification of (i). The other two follow easily.

3.5.4. Yes, it still holds. In fact, it would hold even if we were to use mixed homogenous Dirichlet and Neumann BC. (Why?)

3.5.5. *A closed*: Conclude that $f' = g$. (How?) (So what?)

    $R_\lambda(A)$ *exists*: Use the variation of parameters method that $f(x) = \int_0^\infty e^{\lambda(x-z)}g_0(z)dz$ is the solution of $(\lambda I - A)f = g_0$ that we seek.

    $R_\lambda(A) \in \mathbb{B}\left(\mathbb{C}\left([0,\infty);\mathbb{R}\right)\right)$: Show that $\|R_\lambda(A)(g_0)\|_{\mathbb{C}([0,\infty);\mathbb{R})} \leq \frac{1}{\lambda}\|g_0\|_{\mathbb{C}([0,\infty);\mathbb{R})}$, $\forall g_0 \in \mathbb{C}\left([0,\infty);\mathbb{R}\right)$. What can you conclude about $\|R_\lambda(A)\|_{\mathbb{B}(\mathbb{C}([0,\infty);\mathbb{R}))}$?

3.5.6. Use (3.120) also to verify boundedness.

3.5.7. What is true about the sign of $(f(y) - f(x))(y-x)$?

3.5.8. i.) Continue using properties of inner product (cf. Section 1.7.2).

    ii.) Now apply the definition of nonexpansive with $x = (I + \alpha B)^{-1}z$ and $y = (I + \alpha B)^{-1}w$.

3.5.9. Use the fact that $\|x\|_{\mathscr{X}}^2 = \langle x,x \rangle_{\mathscr{X}}$ and the definition of the generator A.

3.6.1. ii.) Use Exer. 2.2.3.

3.6.2. See [**103**].

3.6.3. Mimic Exer. 3.1.4 and invoke Prop. 3.6.1 to conclude that (3.134) must have a unique classical solution.

3.7.1. (3.138): You will need to use the power series representation of the exponential as the penultimate step.

    (3.139): Use the power series representation for the exponential and the fact that $h \mapsto \frac{e^{\omega h}-1}{h}$ is increasing.

    (3.140): Estimate $\left\|e^{At}x_0 - e^{B_h t}x_0\right\|_{\mathscr{X}} = \left\|\int_0^t \frac{d}{ds}e^{B_h(t-s)}e^{As}x_0 ds\right\|$ using the previous hint and (3.139).

3.7.2. Calculate $e^{B_h t}$ and compute $\lim_{h \to 0^+}$ entrywise using l'Hopital's rule.

3.7.3. Compute the limit using linearity, accounting for the convergence of the series.

3.7.5. <u>Proof:</u> For every $x \in \mathrm{dom}(A_0)$, $0 \leq t \leq T$, and $n \in \mathbb{N}$, show that

$$T_n(t)x - T_0(t)x = \int_0^t T_n(t-s)T_0(s)(A_n x - A_0 x)\,ds.$$

Then, $\|T_n(t)x - T_0(t)x\|_{\mathscr{X}} \leq t\|A_n x - A_0 x\|_{\mathscr{X}}$. (Why?) Let $\varepsilon > 0$. There exists $N \in \mathbb{N}$ such that

$$n \geq N \Longrightarrow \|A_n x - A_0 x\|_{\mathscr{X}} < \frac{\varepsilon}{T}.$$

The conclusion now follows. (How?)

3.8.2. Note that $0 \leq \lim\limits_{t\to\infty} \|u(t)\|_{\mathscr{X}} \leq \lim\limits_{t\to\infty} Me^{-\alpha t} u_0 = 0$, which is exponentially fast.

3.9.2. For instance, pay attention to any calculation where, within the norm, we use $\|A_n x - Ax\|_{\mathscr{X}} = \|(A_n - A)x\|_{\mathscr{X}}$.

3.10.2. Define $F : [0,T] \to \mathscr{X}$ by $F[t](x) = f(x,t)$, $0 < x < a$, $t > 0$.

Chapter 4

Nonhomogenous Linear Evolution Equations

Overview

More often than not, significant external forces impact the evolution of the process. A mathematical model of this phenomenon must account for this. How does one incorporate such external forces into the IBVPs and subsequent abstract formulation? What effect does this have on the solution in the sense of existence, continuous dependence, long-term behavior, etc.? We focus on these questions in this chapter.

4.1 Finite-Dimensional Setting

4.1.1 Motivation by Models

Model II.2 Pharmacokinetics with time-varying drug dosage

Suppose that we now incorporate a term $D(t)$ into IVP (2.4) describing the time variability of the drug dosage from the GI tract viewpoint. Since the rate at which the quantity y changes is directly affected by variations in the dosage over time, it is reasonable to add this forcing term to the right-side of the differential equation describing $\frac{dy}{dt}$ (Why?) As such, the resulting IVP is

$$\begin{cases} \frac{dy}{dt} = -ay(t) + D(t), t > 0, \\ \frac{dz}{dt} = ay(t) - bz(t), t > 0, \\ y(0) = y_0, \ z(0) = 0. \end{cases} \quad (4.1)$$

Note that $a, b > 0$ and $D(\cdot)$ should decrease to zero as $t \to \infty$. Since y_0 is the initial full dosage, it should coincide with $D(0)$. System (4.1) in matrix form is

$$\begin{cases} \begin{bmatrix} y(t) \\ z(t) \end{bmatrix}' = \begin{bmatrix} -a & 0 \\ a & -b \end{bmatrix} \begin{bmatrix} y(t) \\ z(t) \end{bmatrix} + \begin{bmatrix} D(t) \\ 0 \end{bmatrix}, t > 0, \\ \begin{bmatrix} y(0) \\ z(0) \end{bmatrix} = \begin{bmatrix} y_0 \\ 0 \end{bmatrix}. \end{cases} \quad (4.2)$$

Of particular interest is the effect that the presence of $D(t)$ has on the behavior of the solution of (4.1), in comparison to the solution of (2.4). One important related question is, "When is the solution time-periodic?"

Exercise 4.1.1. Reformulate (4.2) in vector form.

Model III.2 Spring Mass System with External Force
   Suppose that an external force described by the function $f : [0, \infty) \to \mathbb{R}$ acts on a spring-mass system, resulting in the following IVP:

$$\begin{cases} x''(t) + \omega^2 x(t) = f(t),\ t > 0, \\ x(0) = x_0,\ x'(0) = x_1. \end{cases} \tag{4.3}$$

The IVP (4.3) can be written equivalently in matrix form as

$$\begin{cases} \begin{bmatrix} y(t) \\ z(t) \end{bmatrix}' = \begin{bmatrix} 0 & 1 \\ -\omega^2 & 0 \end{bmatrix} \begin{bmatrix} y(t) \\ z(t) \end{bmatrix} + \begin{bmatrix} 0 \\ f(t) \end{bmatrix},\ t > 0, \\ \begin{bmatrix} y(0) \\ z(0) \end{bmatrix} = \begin{bmatrix} x_0 \\ x_1 \end{bmatrix}. \end{cases}$$

The variation of parameters technique (cf. Section 1.9.3) can be used to show that the classical solution of (4.3) is

$$x(t) = x_0 \cos(\omega t) + \frac{x_1}{\omega} \sin(\omega t) + \int_0^t \frac{1}{\omega} f(s) \sin(\omega(t - s))\, ds,\ t \geq 0, \tag{4.4}$$

assuming that the function f is sufficiently smooth. What if the forcing term is discontinuous? Does this adversely affect the existence and regularity of this solution? If so, how? Also, when is the motion time-periodic? These questions are explored in some particular cases in the following exercise.

Exercise 4.1.2. Consider the forcing terms $f : [0, \infty) \to \mathbb{R}$ listed below and answer the questions that follow in each case:
   I) Simplify the expression for the solution (4.4).
   II) Is $x \in \mathbb{C}^1((0, \infty); \mathbb{R})$?
   III) Does x satisfy (4.3)?
   IV) Is x time-periodic?
   V) Determine if $\lim_{t \to \infty} x(t)$ exists; if it does, determine its value.
   i.) $f(t) = C$, where $C > 0$ is a constant.

   ii.) Let $0 < a < b$ and $C > 0$. Define $f(t) = \begin{cases} 0,\ 0 \leq t < a, \\ C,\ a \leq t \leq b, \\ 0,\ t > b. \end{cases}$

   iii.) $f(t) = \alpha \sin(\omega t)$, where $\alpha, \omega > 0$.
   iv.) $f(t) = \int_0^t k(t, s) g(s)\, ds$, where $g : [0, \infty) \to \mathbb{R}$ is continuous and $k : [0, \infty) \times [0, \infty) \to$

$\mathbb{R}$ is continuous and bounded.

The above IVPs are not drastically different from those discussed in Chapter 2. In fact, it is easy to see that they can be rewritten in the abstract vector form

$$\begin{cases} \mathbf{U}'(t) = \mathbf{A}\mathbf{U}(t) + \mathbf{f}(t), t > 0, \\ \mathbf{U}(0) = \mathbf{U}_0, \end{cases} \tag{4.5}$$

where $\mathbf{A} \in \mathbb{M}^N(\mathbb{R})$, $\mathbf{f} : [0, \infty) \to \mathbb{R}^N$, and $\mathbf{U}_0 \in \mathbb{R}^N$. As before, we shall formulate a theory for (4.5) and then recover the results for the particular IVPs as corollaries.

4.1.2 One-Dimensional Case

We begin our investigation of (4.5) in $\mathbb{R}$, so that A and U_0 are simply real numbers and $U(\cdot)$ and $f(\cdot)$ are real-valued functions. In such case, (4.5) consists of a single ODE coupled with an IC. We wish to derive a representation formula for a solution of (4.5). For the moment, we hold off on imposing specific hypotheses on $f(\cdot)$ and tacitly assume whatever is needed to ensure that all steps of the derivation are justified and all quantities involved are well-defined. As you work through the derivation, keep track of these conditions in preparation for formulating an existence result.

Starting with (4.5), we proceed as follows:

$$U'(s) - AU(s) = f(s)$$
$$e^{-As}\left(U'(s) - AU(s)\right) = e^{-As}f(s)$$
$$\frac{d}{ds}\left(e^{-As}U(s)\right) = e^{-As}f(s) \tag{4.6}$$
$$\int_0^t \frac{d}{ds}\left(e^{-As}U(s)\right)ds = \int_0^t e^{-As}f(s)ds$$
$$e^{-At}U(t) - e^{-A(0)}U(0) = \int_0^t e^{-As}f(s)ds$$
$$e^{-At}U(t) = U_0 + \int_0^t e^{-As}f(s)ds$$
$$U(t) = e^{At}U_0 + e^{At}\int_0^t e^{-As}f(s)ds$$
$$U(t) = e^{At}U_0 + \int_0^t e^{A(t-s)}f(s)ds \tag{4.7}$$

Formula (4.7) is called a *variation of parameters* formula for a solution of (4.5).

If we now begin with (4.7), the question is whether or not it must actually "satisfy" (4.5). This would require the following:

(a) $U(\cdot)$ is differentiable on $(0, \infty)$,

(b) $U(\cdot)$ is continuous on $[0, \infty)$, and

(c) Both the equation and IC in (4.5) are satisfied.

As an illustration of the potential difficulty that can arise, consider this example.

Example. Consider the IVP

$$\begin{cases} U'(t) = 2U(t) + f(t), t > 0, \\ U(0) = 2, \end{cases} \tag{4.8}$$

where

$$f(t) = \begin{cases} t, 0 \le t < 1, \\ 0, t \ge 1. \end{cases}$$

Does the presence of the discontinuity of $f(\cdot)$ at $t = 1$ prevent the solution candidate for (4.8), as given by (4.7), from being continuous? Worse yet, does it prevent differentiability of the solution? Although the integral operator has a smoothing effect (cf. Section 1.8.4), the extent of the irregularity on the integrand enters into the picture. The continuity of the solution $U(\cdot)$ of (4.8), as expressed by (4.7), is easily verified at all times $t \ne 1$. (Do so!) Focusing on the behavior at $t = 1$, we see that $U(\cdot)$ must satisfy the following two IVPs:

$$\begin{cases} U'(t) = 2U(t) + t, 0 \le t < 1, \\ U(0) = 2, \end{cases} \tag{4.9}$$

$$\begin{cases} U'(t) = 2U(t), t \ge 1, \\ U(1) = 2e^2 + e - 2. \end{cases} \tag{4.10}$$

Exercise 4.1.3.
i.) Verify that the solution $U(\cdot)$ of (4.8) is

$$U(t) = \begin{cases} 2e^{2t} + e^t - t - 1, 0 \le t < 1, \\ \left(2 + \frac{1}{e} - \frac{2}{e^2}\right) e^{2t}, t \ge 1. \end{cases}$$

ii.) Show that while $U(\cdot)$ continuous on $[0, \infty)$, $\frac{d^+}{dt} U(t)|_{t=1} \ne \frac{d^-}{dt} U(t)|_{t=1}$. As such, $U(\cdot)$ is not differentiable on $(0, \infty)$ and hence cannot be a solution of (4.8) in the classical sense (i.e., satisfying (a) - (c) above). However, if we are willing to weaken criterion (a) to allow differentiability *almost everywhere* on $(0, \infty)$ (cf. Section 1.8.4), then $U(\cdot)$ would be a solution in this modified classical sense.

Consequently, even a single point of irregularity in the forcing term $f(\cdot)$ can create complications. What conditions can be imposed on U_0 and/or $f(\cdot)$ to ensure that the function $U(\cdot)$ given by (4.7) is continuous on $[0, \infty)$? differentiable on $(0, \infty)$?

Since the mapping $t \mapsto e^{At}$ is differentiable, the regularity of $U(\cdot)$ given by (4.7) depends on the regularity of the mapping $t \mapsto \int_0^t e^{A(t-s)} f(s) ds$. Precisely, we have:

Proposition 4.1.1. *Define* $\Phi : [0, \infty) \to \mathbb{R}$ *by* $\Phi(t) = \int_0^t e^{A(t-s)} f(s) ds$.
i.) *If* $f(\cdot)$ *is integrable on* $(0, \infty)$, *then* $\Phi(\cdot)$ *is continuous on* $[0, \infty)$.
ii.) *If* $f(\cdot)$ *is continuous on* $(0, \infty)$, *then* $\Phi(\cdot)$ *is continuously differentiable on* $(0, \infty)$.

Proof. (i) Proposition 1.8.16 can be used to show that $\Phi(t)$ is well-defined, $\forall t > 0$. The continuity can be verified using Def. 1.8.4(i). Finally, an application of Prop. 1.8.13(vii) yields (ii). ☐

Exercise 4.1.4. Provide the details in the proof of Prop. 4.1.1.

Proposition 4.1.1 guarantees that if the forcing term $f(\cdot)$ is at least continuous, then (4.7) should satisfy (4.5) in the classical sense. However, if the forcing term is less regular, then the differentiability of $U(\cdot)$ in (4.7) is not guaranteed and this prevents us from showing that $U(\cdot)$ satisfies (4.5). The need for an alternative notion of solution is apparent since such functions arise naturally in practice.

Definition 4.1.2. A function $U:[0,T] \to \mathbb{R}$ is a
i.) *classical solution* of (4.5) if
    a.) $U(\cdot)$ is continuous on $[0,T]$,
    b.) $U(\cdot)$ is differentiable on $(0,T)$,
    c.) $U'(t) = AU(t) + f(t)$, $\forall t \in [0,T]$,
    d.) $U(0) = U_0$.
ii.) *mild solution* of (4.5) if
    a.) $U(\cdot)$ is continuous on $[0,T]$,
    b.) $U(\cdot)$ is given by (4.7), $\forall t \in [0,T]$.

Exercise 4.1.5. Describe the relationship between the two notions of a solution provided in Def. 4.1.2.

Proposition 4.1.3. *If a classical solution of (4.5) exists, then it is unique. In such case, it is given by (4.7).*

Proof. Suppose that $U(\cdot)$ and $V(\cdot)$ are both solutions of (4.5). Then, the function $Z:[0,T] \to \mathbb{R}$ defined by $Z(t) = U(t) - V(t)$ satisfies Def. 4.1.2(i)(a) and (b). (Why?). Moreover, $Z'(t) = AZ(t)$ and $Z(0) = 0$. Hence, $Z(t) = 0$, $\forall t \in [0,T]$. (Why?) This completes the proof. ☐

Exercise 4.1.6. If (4.5) is only guaranteed to have a mild solution, must it be unique?

An interesting observation is that the mild solution of (4.5) can be viewed as a uniform limit of the classical solutions to an aptly-chosen sequence of approximating IVPs. For definiteness, consider IVP (4.8) on the interval $[0,2]$. Since each expression in the definition of $f(\cdot)$ is continuous, it has a unique Fourier representation given by

$$f(t) = \sum_{n=0}^{\infty} b_n \sin\left(\frac{n\pi t}{2}\right), \quad 0 \le t \le 2, \tag{4.11}$$

where $b_n = \int_0^2 f(t) \sin\left(\frac{n\pi t}{2}\right) dt$. Using the partial sums associated with (4.11), we define the sequence of approximating IVPs

$$\begin{cases} U_N'(t) = 2U_N(t) + \sum_{n=0}^{N} b_n \sin\left(\frac{n\pi t}{2}\right), & 0 \le t \le 2, \\ U_N(0) = 2. \end{cases} \tag{4.12}$$

For any $N \in \mathbb{N}$, the IVP (4.12) has a unique classical solution U_N. (Why?) Also, since (4.11) means

$$f(t) = \lim_{N \to \infty} \sum_{n=0}^{N} b_n \sin\left(\frac{n\pi t}{2}\right), \text{ uniformly for } t \in [0, 2],$$

Prop. 1.8.16 implies that

$$\lim_{N \to \infty} \int_0^t e^{A(t-s)} \sum_{n=0}^{N} b_n \sin\left(\frac{n\pi s}{2}\right) ds = \int_0^t e^{A(t-s)} f(s) ds. \tag{4.13}$$

As such, it is not difficult to argue that

$$\lim_{N \to \infty} \|U_N - U\|_{C([0,2];\mathbb{R})} = 0. \tag{4.14}$$

Exercise 4.1.7. Supply the missing details in the above discussion.

When considering an IVP such as (4.5), we are tacitly assuming that the precise nature of the forcing term is known at every time. However, this is unrealistic, since due to the existence of imprecision in measurement, we can only say that for any time $t > 0$, the *actual value* of the forcing term, denoted $\bar{f}(t)$, lies somewhere in the interval $(f(t) - \varepsilon_1, f(t) + \varepsilon_1)$. Similarly, the actual value of the IC, denoted $\overline{U}_0$, is somewhere in the interval $(U_0 - \varepsilon_2, U_0 + \varepsilon_2)$. (Even worse, the error terms ε_i can be time-dependent.) But, what difference does this error make in the solution to (4.5)? Consider the IVP

$$\begin{cases} \overline{U}'(t) = A\overline{U}(t) + \bar{f}(t), t > 0, \\ \overline{U}(0) = \overline{U}_0. \end{cases} \tag{4.15}$$

We are hopeful that for any time interval $[0, T]$, the difference between the solutions U and $\overline{U}$ of (4.5) and (4.15), respectively, in the sense of the sup norm on $C([0, T]; \mathbb{R})$ can be controlled by ensuring that the error terms ε_i are sufficiently small. In such case, we would say that the solution *depends continuously* on the data. This is summarized in the following result:

Proposition 4.1.4. *Let $\varepsilon_1, \varepsilon_2, T > 0$. Assume that $f, \bar{f} : [0, T] \to \mathbb{R}$ are integrable and $U_0, \overline{U}_0 \in \mathbb{R}$. If*

$$\|f - \bar{f}\|_{C([0,T];\mathbb{R})} < \varepsilon_1 \text{ and } |U_0 - \overline{U}_0| < \varepsilon_2, \tag{4.16}$$

then

$$\|U - \overline{U}\|_{C([0,T];\mathbb{R})} < e^{AT}(\varepsilon_1 + T\varepsilon_2), \tag{4.17}$$

where the mild solutions U and $\overline{U}$ of (4.5) and (4.15), respectively.

Exercise 4.1.8.
i.) Prove Prop. 4.1.4.
ii.) If we assume in (4.16) that $\|f - \bar{f}\|_{\mathbb{L}^1(0,T;\mathbb{R})} < \varepsilon_1$, how would (4.17) change?

The long-term behavior of the solution U of (4.5), in the sense of $\lim_{t \to \infty} |U(t)|$, is also of interest. Consider the following exercise.

Exercise 4.1.9.
i.) What condition(s) can be imposed on f : $[0, \infty) \to \mathbb{R}$ to ensure that $\lim_{t \to \infty} |U(t)| < \infty$?
ii.) If A > 0, can $\lim_{t \to \infty} |U(t)| < \infty$?
iii.) If $\lim_{t \to \infty} f(t) = L$ and A < 0, what, if anything, can be said about $\lim_{t \to \infty} |U(t)|$?

4.1.3 Extension of Theory to $\mathbb{R}^N$

All of the IVPs mentioned in Section 4.1.1 can be reformulated as the abstract evolution equation (4.5), where $\mathbf{A} \in \mathbb{M}^N(\mathbb{R})$, $\mathbf{f} : [0, \infty) \to \mathbb{R}^N$, and $\mathbf{U}_0 \in \mathbb{R}^N$. Given that the calculus of vector-valued functions is performed componentwise (and hence resembles the real-valued case for each component), the properties of $\{e^{\mathbf{A}t} : t \geq 0\}$ established in Chapter 2 can be used to argue exactly as in Section 4.1.2 to conclude that a mild solution of (4.5) (in the more general setting of $\mathbb{R}^N$) exists and is given by (4.7).

Exercise 4.1.10. Carefully justify all steps leading to (4.7) for the $\mathbb{R}^N$ setting.

Exercise 4.1.11.
i.) What does it mean for $\mathbf{f} \in \mathbb{L}^1(0, T; \mathbb{R}^N)$?
ii.) Prove that if $\mathbf{f} \in \mathbb{L}^1(0, T; \mathbb{R}^N)$, then the function $\mathbf{U}$ given by (4.7) belongs to $\mathbb{C}([0, T]; \mathbb{R}^N)$.

The fact that $\mathbb{R}^N$ (for $N > 1$) and $\mathbb{R}$ are equipped with essentially the same norm topology suggests that Def. 4.1.2 should extend in an analogous manner without issue, as should the various results established in Section 4.1.2 with possibly some minor modifications. The following exercises explore various facets of this generalization.

Exercise 4.1.12. (Existence and Uniqueness) Prove that if $\mathbf{f} \in \mathbb{C}([0, T]; \mathbb{R}^N)$, then (4.5) has a unique classical solution given by (4.7).

Exercise 4.1.13. (Continuous Dependence) Formulate and prove results in the spirit of Prop. 4.1.4 and of Exer. 4.1.8 (ii).

Exercise 4.1.14. (Approximation) Let $\bar{\mathbf{f}} \in \mathbb{L}^1(0, T; \mathbb{R}^N)$, $\overline{\mathbf{U}_0} \in \mathbb{R}^N$, and $\mathbf{U}$ be the mild solution of (4.15). Use the fact that $\mathbb{C}([0, T]; \mathbb{R}^N)$ is dense in $\mathbb{L}^1(0, T; \mathbb{R}^N)$ to argue the existence of a sequence $\{\mathbf{U}_n\}$ of classical solutions (of a sequence of aptly-constructed IVPs) such that $\lim_{n \to \infty} \|\mathbf{U}_n - \mathbf{U}\|_{\mathbb{C}([0,T];\mathbb{R}^N)} = 0$.

Exercise 4.1.15. (Long-term Behavior - Part 1) Consider the following three Jordan forms in $\mathbb{M}^2(\mathbb{R})$:

$$\mathbf{A}_1 = \begin{bmatrix} \alpha & 0 \\ 0 & \beta \end{bmatrix}, \ \alpha \neq \beta \neq 0,$$

$$\mathbf{A}_2 = \begin{bmatrix} \alpha & 1 \\ 0 & \alpha \end{bmatrix}, \ \alpha \neq 0, \tag{4.18}$$

$$\mathbf{A}_3 = \begin{bmatrix} \alpha & \beta \\ \beta & \alpha \end{bmatrix}, \ \alpha \neq \beta \neq 0.$$

i.) Calculate $e^{\mathbf{A}_i t}$, for $i = 1, 2, 3$.
ii.) In each case, what must be true in order to ensure that

$$\lim_{n \to \infty} \left\| e^{\mathbf{A}_i t} \mathbf{y} \right\|_{\mathbb{R}^2} = 0, \ \forall \mathbf{y} \in \mathbb{R}^2?$$

iii.) In light of your responses in (ii) and assuming that $\mathbf{f}$ is bounded on $[0, \infty)$, determine a sufficient condition guaranteeing that $\int_0^t e^{\mathbf{A}_i(t-s)} \mathbf{f}(s) ds < \infty$, $\forall t > 0$.
iv.) Let $\mathbf{U}(\cdot)$ be a mild solution of (4.15) and assume that $\mathbf{f}$ is bounded on $[0, \infty)$. Formulate sufficient conditions to ensure that $\lim_{t \to \infty} \|\mathbf{U}(t)\|_{\mathbb{R}^2} < \infty$.

Exercise 4.1.16. (**Long-term Behavior - Part 2**)
i.) Assume that $\mathbf{A}_1$ in (4.18) is such that Exer. 4.1.15(ii) holds. Define $\mathbf{f} : [0, \infty) \to \mathbb{R}^2$ by $\mathbf{f}(t) = \begin{bmatrix} 1 + e^{-t} \\ te^{-2t} - 1 \end{bmatrix}$.
  a.) Determine $\mathbf{f}_0 \in \mathbb{R}^2$ for which $\lim_{t \to \infty} \|\mathbf{f}(t) - \mathbf{f}_0\|_{\mathbb{R}^2} = 0$.
  b.) Simplify (4.7) for the IVP (4.5) corresponding to $\mathbf{A}_1$ and $\mathbf{f}(\cdot)$ as above.
  c.) Compute $\lim_{t \to \infty} \mathbf{U}(t)$.
ii.) Consider $\mathbf{A}_3 = \begin{bmatrix} -2 & 3 \\ 3 & -2 \end{bmatrix}$ and $\mathbf{f}(\cdot)$ as in (i).
  a.) Simplify (4.7) for the IVP (4.5) corresponding to $\mathbf{A}_3$ and $\mathbf{f}(\cdot)$.
  b.) Compute $\lim_{t \to \infty} \mathbf{U}(t)$.
  c.) Describe the nature of $\lim_{t \to \infty} \mathbf{U}(t)$ for the more general matrix $\mathbf{A}_3$ in (4.18).
iii.) What role does the invertibility of $\mathbf{A}$ enter into this general discussion?

We now apply the above results to the models discussed in Section 4.1.1, beginning with spring-mass systems.

Exercise 4.1.17. Suppose that the forcing term $f : [a, b] \to \mathbb{R}$ has finitely many jump discontinuities. It can be shown that there exists a sequence $\{f_n\} \subset C^1((a, b); \mathbb{R})$ such that $f_n \to f$ uniformly on $[a, b]$. For each $n \in \mathbb{N}$, form a new IVP by equipping (4.3) with f_n instead of f; denote its mild solution by x_n. Must there exist $x \in C^2((a, b); \mathbb{R})$ for which $x_n \to x$ uniformly on $[a, b]$?

Exercise 4.1.18. Show directly using (4.7) that if $f \in C^1(\mathbb{R}; \mathbb{R})$ is periodic with period $\frac{2\pi}{\omega}$, then the mild solution x of (4.3) is also periodic with period $\frac{2\pi}{\omega}$.

Exercise 4.1.19. Let $\varepsilon > 0$. Assume that $f_\varepsilon \in C^1(\mathbb{R}; \mathbb{R})$ is periodic with period $\frac{2\pi}{\omega_\varepsilon}$ and that $\lim_{\varepsilon \to 0^+} \omega_\varepsilon = \omega$.

i.) For every $\varepsilon > 0$, denote the mild solution of the IVP obtained by equipping (4.3) with f_ε instead of f by x_ε. Is x_ε periodic? Explain.

ii.) Does $\exists x \in C(\mathbb{R}; \mathbb{R})$ that satisfies (4.3) for which $x_\varepsilon \to x$ uniformly on compact subsets of $\mathbb{R}$ as $\varepsilon \to 0^+$? Must $x \in C^1(\mathbb{R}; \mathbb{R})$? If so, x be time-periodic?

We continue our applied exploration in pharmacokinetics.

Execise 4.1.20. Suppose that the drug dosage term $D(t)$ in (4.1) is defined by

$$D(t) = \begin{cases} \hat{D}(t), & 0 \le t \le 12, \\ 0, & t > 12, \end{cases}$$

where $\hat{D}(t)$ decreases to zero as $t \to 12^-$.

i.) If $t \mapsto \hat{D}(t)$ is differentiable and decreases to zero exponentially as $t \to 12^-$, must (4.1) have a unique classical solution $\begin{bmatrix} y \\ z \end{bmatrix}$? If so, compute $\lim_{t \to \infty} \begin{bmatrix} y(t) \\ z(t) \end{bmatrix}$, if it exists.

ii.) Assuming the presence of error in measuring $\hat{D}(t)$ at any time t, interpret the estimate (4.17) in Prop. 4.1.4 for (4.1).

iii.) Suppose that $t \mapsto D(t)$ is time-periodic with period 12. Must the solution $t \mapsto \begin{bmatrix} y(t) \\ z(t) \end{bmatrix}$ also be periodic? Explain?

iv.) Not all pills are created equal and so, the initial dosage y_0 will likely change from one pill to another. Consider (4.1) for two different ICs x_0 and y_0, where $|x_0 - y_0| < \delta$. Establish an estimate on the difference of the corresponding solutions of (4.1). Interpret the result in the context of this model.

4.2   Infinite-Dimensional Setting

External forces naturally arise in IBVPs as well. The hope is that we can reformulate such IBVPs abstractly as the evolution equation (4.5) in a general Banach space $\mathscr{X}$, and then adapt the theory developed in Section 4.1 to handle this more general form without much difficulty.

4.2.1   Motivation by Models

Model V.2 Heat Conduction with External Source

Suppose that a heat source is positioned in proximity to one edge of a rectangular slab of material for which we are monitoring the temperature over time. Assuming that its intensity increases with time, a possible generalization of IBVP (3.47) describing this scenario is given by

$$
\begin{cases}
\frac{\partial}{\partial t}z(x,y,t) & = k\triangle z(x,y,t) + t^2 + x + 2y,\ 0 < x < a,\ 0 < y < b,\ t > 0, \\
z(x,y,0) & = \sin 2x + \cos 2y,\ 0 < x < a,\ 0 < y < b, \\
z(x,0,t) & = 0 = z(x,b,t),\ 0 < x < a,\ t > 0, \\
z(0,y,t) & = 0 = z(a,y,t),\ 0 < y < b,\ t > 0,
\end{cases}
\tag{4.19}
$$

where $z(x,y,t)$ represents the temperature at the point (x,y) on the plate at time t.

Model VI.2 Sobolev Equation with Forcing Term

External forces naturally arise when modeling fluid flow through fissured rocks. A simplified one-dimensional version of an IBVP arising in the modeling of such a scenario is given by

$$
\begin{cases}
\frac{\partial}{\partial t}\left(z(x,t) - \frac{\partial^2}{\partial x^2}z(x,t)\right) = \frac{\partial^2}{\partial x^2}z(x,t) + 1 + x,\ 0 < x < \pi,\ t > 0, \\
z(x,0) = 1 + x^3,\ 0 < x < \pi, \\
z(0,t) = z(\pi,t) = 0,\ t > 0.
\end{cases}
\tag{4.20}
$$

Exercise 4.2.1. Express both (4.19) and (4.20) as an abstract evolution equation of the form

$$
\begin{cases}
u'(t) = Au(t) + f(t),\ t > 0, \\
u(0) = u_0,
\end{cases}
\tag{4.21}
$$

in a Banach space $\mathscr{X}$, where $f : [0,\infty) \to \mathscr{X}$ and $u_0 \in \mathscr{X}$. What slightly modified version of (4.21) could be used as an abstract formulation of (4.20)?

To what extent can the theory from Section 4.1 be extended to handle (4.21)?

4.2.2   Theory in a General Banach Space $\mathscr{X}$

The fact that forming the theoretical framework for the homogenous version of (4.21) followed naturally from the development of Chapter 2 gives reason for optimism in our current situation. Assuming that A generates a C_0−semi-group on $\mathscr{X}$, we begin with the variation of parameters formula. Given that each term of this formula is now a member of a general Banach space $\mathscr{X}$, the integral is $\mathscr{X}$-valued (cf. Section 1.8.4). Upon tracing through the steps leading to (4.7) we find that we simply need to reinterpret the notion of "multiplying both sides of an equality by the term e^{-At}" as "applying the operator e^{-At} to both sides." All steps of the derivation then remain valid. As such, assuming that $f \in \mathbb{L}^1(0,T;\mathscr{X})$, Thrm. 3.3.6 implies that all calculations are justified <u>provided that</u> $u(t) \in \mathrm{dom}(A)$, $\forall t > 0$. In such case,

$$
u(t) = e^{At}u_0 + \int_0^t e^{A(t-s)} f(s)ds.
\tag{4.22}
$$

Exercise 4.2.2. Why is the condition "$u(t) \in \text{dom}(A), \forall t > 0$" needed?

We modify Def. 4.1.2 to account for this requirement.

Definition 4.2.1. A function $u : [0,T] \to \mathscr{X}$ is a
i.) *classical solution* of (4.21) on $[0,T]$ if
    a.) $u(t) \in \text{dom}(A), \forall t \in [0,T]$,
    b.) $u(\cdot)$ is continuous on $[0,T]$,
    c.) $u(\cdot)$ is differentiable on $(0,T)$,
    d.) (4.21) is satisfied.
ii.) *mild solution* of (4.21) on $[0,T]$ if
    a.) $u(\cdot)$ is continuous on $[0,T]$,
    b.) $u(\cdot)$ is given by (4.22), $\forall t \in [0,T]$.

Exercise 4.2.3. Must $u(t) \in \text{dom}(A), \forall t > 0$, in order for u to be a mild solution of (4.21) ? Explain.

The derivation and strategy used in the proof of Prop. 4.1.3 also apply to prove the following result. (Show this.)

Proposition 4.2.2. *If (4.21) has a classical solution, then it is unique and is given by (4.22).*

It is natural to ask whether or not every result developed in Section 4.1 can be extended to this more general setting in the study of (4.21). Optimistically, since the extension from $\mathbb{R}$ to $\mathbb{R}^N$ was effortless, perhaps going one step further from $\mathbb{R}^N$ to a general Banach space $\mathscr{X}$ will be similar. But, we must be careful, since the norm used for $\mathscr{X}$ can, in general, bear no resemblance to that of $\mathbb{R}^N$. Indeed, it was the similarity of the topological structures that facilitated the extension from $\mathbb{R}$ to $\mathbb{R}^N$. Indeed, consider the following scenario observed in **[267]**:

Exercise 4.2.4. Let $\mathscr{X}$ be a Banach space and assume that $A : \text{dom}(A) \subset \mathscr{X} \to \mathscr{X}$ generates a semigroup $\{e^{At} : t \geq 0\}$ on $\mathscr{X}$. Suppose that $\exists y_0 \in \mathscr{X}$ for which $\{e^{At} y_0 : t \geq 0\} \cap \text{dom}(A) = \emptyset$ and define $f_0 : [0,\infty) \to \mathscr{X}$ by $f_0(t) = e^{At} y_0$.
i.) Prove that f_0 is continuous.
ii.) Define $g : [0,\infty) \to \mathscr{X}$ by $g(t) = t f_0(t)$. Prove that g is not differentiable, $\forall t > 0$.
iii.) Prove that the IVP

$$\begin{cases} u'(t) & = Au(t) + f_0(t), \ t > 0, \\ u(0) & = 0, \end{cases} \tag{4.23}$$

does not have a classical solution.
iv.) Does (4.23) have a mild solution on $[0,\infty)$?
v.) Consider the operator A defined in (3.44). Does $\exists y_0 \in \mathbb{L}^2(0,a;\mathbb{R})$ for which the above argument applies? Explain.
vi.) Repeat part (v) for the operator A defined in (3.57) in $\mathbb{L}^2(0,\pi;\mathbb{R})$.

This presents a hurdle to overcome when trying to formulate the analog of Prop. 4.1.1 in the present setting. We must impose a higher degree of regularity on f to ensure the differentiability of (4.22). The main difference from the $\mathbb{R}^N$ setting is that for a general Banach space $\mathscr{X}$, it need not be the case that $\mathrm{dom}(A) = \mathscr{X}$, whereas equality <u>must</u> occur when $\mathscr{X} = \mathbb{R}^N$. This difference is precisely what made the above scenario possible.

This raises the question, "What is sufficient to guarantee the differentiability of (4.22)?" Certainly, we need $u_0 \in \mathrm{dom}(A)$ in order for $t \mapsto e^{At}u_0$ to be differentiable (Why?), but what about the second term in (4.22)? Since we ultimately need $u \in \mathbb{C}^1((0,T);\mathrm{dom}(A))$, a natural sufficient condition would be to require that the forcing term $f \in \mathbb{C}([0,T];\mathrm{dom}(A))$ be such that $\int_0^t e^{A(t-s)}f(s)ds \in \mathbb{C}^1((0,T);\mathrm{dom}(A))$. This suggests the following result, the proof of which has been adapted from **[267, 341]**.

Proposition 4.2.3. *If $u_0 \in \mathrm{dom}(A)$ and $f \in \mathbb{C}([0,T];\mathrm{dom}(A))$ is such that*

$$\int_0^t e^{A(t-s)}f(s)ds \in \mathbb{C}^1((0,T);\mathrm{dom}(A)),$$

then (4.21) has a unique classical solution given by (4.22).

Proof. Uniqueness was established earlier. We must argue that u given by (4.22) satisfies Def. 4.2.1.

The continuity of u is clear. (Why?) We begin by showing that $u(t) \in \mathrm{dom}(A), \forall t \in [0,T]$. This requires that we verify that $\exists \lim\limits_{h \to 0} \frac{e^{Ah}u(t)-u(t)}{h}$. Since Thrm. 3.3.6 implies that $e^{At}u_0 \in \mathrm{dom}(A), \forall t \in [0,T]$, and $\mathrm{dom}(A)$ is a linear subspace of $\mathscr{X}$, we need only to argue the existence of

$$\lim_{h \to 0} \frac{e^{Ah}-I}{h}\left(\int_0^t e^{A(t-s)}f(s)ds\right).$$

(Tell why.) Observe that

$$
\begin{aligned}
A\int_0^t e^{A(t-s)}f(s)ds &= \lim_{h \to 0}\frac{e^{Ah}-I}{h}\left(\int_0^t e^{A(t-s)}f(s)ds\right) \\
&= \lim_{h \to 0}\int_0^t \frac{e^{Ah}-I}{h}\left(e^{A(t-s)}f(s)\right)ds \\
&= \lim_{h \to 0}\int_0^t \frac{e^{Ah}e^{A(t-s)}f(s)-e^{A(t-s)}f(s)}{h}ds \qquad (4.24) \\
&= \lim_{h \to 0}\left[\frac{\int_0^t e^{A(t+h-s)}f(s)ds - \int_0^t e^{A(t-s)}f(s)ds}{h}- \right.\\
&\qquad\qquad \left. \frac{1}{h}\int_t^{t+h} e^{A(t+h-s)}f(s)ds\right] \\
&= \frac{d}{dt}\left(\int_0^t e^{A(t-s)}f(s)ds\right)-f(t).
\end{aligned}
$$

The right-side of the last equality in (4.24) is a well-defined member of dom(A) (by assumption) and $f(t) \in$ dom(A). Thus, $u(t) \in$ dom(A), $\forall t \in [0, T]$.

Finally, we show that u given by (4.22) satisfies (4.21). Observe that (4.24) implies

$$\frac{d}{dt}\left(\int_0^t e^{A(t-s)}f(s)ds\right) = A\int_0^t e^{A(t-s)}f(s)ds + f(t). \qquad (4.25)$$

Also,

$$\frac{d}{dt}\left(e^{At}u_0\right) = Ae^{At}u_0. \qquad (4.26)$$

Using the linearity of A, adding (4.25) and (4.26) yields

$$u'(t) = \frac{d}{dt}\left(e^{At}u_0 + \int_0^t e^{A(t-s)}f(s)ds\right) = A\left(e^{At}u_0 + \int_0^t e^{A(t-s)}f(s)ds\right) + f(t)$$
$$= Au(t) + f(t),$$

and $u(0) = u_0$. This completes the proof. $\qquad \square$

Exercise 4.2.5. Assume that if $f \in \mathbb{L}^1(0, T; \text{dom}(A)) \cap \mathbb{C}([0, T]; \mathscr{X})$. Prove that $t \mapsto \int_0^t e^{A(t-s)}f(s)ds$ belongs to $\mathbb{C}^1((0, T); \mathscr{X})$.

Exercise 4.2.6.
i.) Let $p > 1$ and assume that $f \in \mathbb{L}^p(0, T; \text{dom}(A))$. Must $t \mapsto \int_0^t e^{A(t-s)}f(s)ds$ belong to $\mathbb{C}^1((0, T); \mathscr{X})$?
ii.) If $f \in \mathbb{C}^1((0, T); \text{dom}(A))$, must $t \mapsto \int_0^t e^{A(t-s)}f(s)ds$ belong to $\mathbb{C}^1((0, T); \text{dom}(A))$?

There are other criteria that ensure (4.22) is a classical solution (see **[267, 318, 341]**). Motivated by Exercises 4.1.14 and 4.1.7, we can approximate a mild solution of (4.21) (in the $\mathbb{C}$-norm) by a sequence of classical solutions of aptly-chosen IVPs. Indeed, consider (4.21), where $u_0 \in \mathscr{X}$ and $f \in \mathbb{L}^1(0, T; \mathscr{X})$. There exist $\{u_0^n\} \subset$ dom(A) and $\{f_n\} \subset \mathbb{C}^1((0, T); \mathscr{X})$ (equipped with the $\mathbb{L}^1$-norm) such that

$$\lim_{n \to \infty} \|u_0^n - u_0\|_{\mathscr{X}} = 0, \qquad (4.27)$$

$$\lim_{n \to \infty} \|f_n - f\|_{\mathbb{L}^1(0, T; \mathscr{X})} = 0. \qquad (4.28)$$

For any $n \in \mathbb{N}$, the IVP

$$\begin{cases} u_n'(t) &= Au_n(t) + f_n(t), \ 0 < t < T, \\ u_n(0) &= u_0^n, \end{cases} \qquad (4.29)$$

has a unique classical solution $u_n : [0, T] \to \mathscr{X}$. It is not difficult to show that

$$\lim_{n \to \infty} \|u_n - u\|_{\mathbb{C}([0,T];\mathscr{X})} = 0. \qquad (4.30)$$

Exercise 4.2.7. Verify the details in the above discussion.

Exercise 4.2.8. Assume that $A : \text{dom}(A) \subset \mathscr{X} \to \mathscr{X}$ generates a C_0−semigroup $\{e^{At} : t \geq 0\}$ on $\mathscr{X}$. Let $\{B_n\} \subset \mathbb{B}(\mathscr{X})$, $\{f_n\} \subset \mathbb{C}^1\left((0,T);\mathscr{X}\right)$, and $\{u_0^n\} \subset \text{dom}(A)$ be such that

$$\lim_{n\to\infty} \|u_0^n - u_0\|_{\mathscr{X}} = 0, \quad \lim_{n\to\infty} \|f_n\|_{\mathbb{L}^1(0,T;\mathscr{X})} = 0, \quad \lim_{n\to\infty} \|B_n - B\|_{\mathbb{B}(\mathscr{X})} = 0,$$

where $u_0 \in \text{dom}(A)$ and $B \in \mathbb{B}(\mathscr{X})$. Consider the following IVPs:

$$\begin{cases} u_n'(t) &= (A + B_n)\, u_n(t) + f_n(t), \ 0 < t < T, \\ u_n(0) &= u_0^n, \end{cases} \tag{4.31}$$

$$\begin{cases} u'(t) &= (A + B)\, u(t), \ 0 < t < T, \\ u(0) &= u_0. \end{cases} \tag{4.32}$$

Does $\lim\limits_{n\to\infty} \|u_n - u\|_{\mathbb{C}([0,T];\mathscr{X})} = 0$?

Regarding the long-term behavior (as $t \to \infty$) of a mild solution, Exercises 4.1.9, 4.1.15, and 4.1.16 in the finite-dimensional setting suggested that the boundedness of **f** was sufficient to guarantee that $\int_0^\infty e^{\mathbf{A}(t-s)}\mathbf{f}(s)ds < \infty$, $\forall t > 0$, provided that $\lim\limits_{t\to\infty} \left\|e^{\mathbf{A}t}\right\|_{\mathbb{B}(\mathbb{R}^N)} = 0$. How does this translate into the present setting? Observe that if

$$\left\|e^{At}\right\|_{\mathbb{B}(\mathscr{X})} \leq M_A e^{\omega t}, \text{ where } \omega < 0, \tag{4.33}$$

and

$$\sup_{s\geq 0}\|f(s)\|_{\mathscr{X}} \leq M_f, \tag{4.34}$$

then $\forall t, s \geq 0$,

$$\left\|e^{A(t-s)}f(s)\right\|_{\mathscr{X}} \leq \left\|e^{A(t-s)}\right\|_{\mathbb{B}(\mathscr{X})} \|f(s)\|_{\mathscr{X}} \leq M_A M_f e^{\omega(t-s)}. \tag{4.35}$$

Exercise 4.2.9. Why does (4.35) ensure that $\int_0^\infty e^{A(t-s)}f(s)ds < \infty$, $\forall t > 0$?

If, in addition, f is continuous and

$$\lim_{t\to\infty} \|f(t) - L\|_{\mathscr{X}} = 0, \tag{4.36}$$

then (4.34) holds, and we have more information to aid us in calculating $\lim\limits_{t\to\infty} \|u(t)\|_{\mathscr{X}}$. Precisely, in the spirit of Exer. 4.1.9, we have the following result.

Proposition 4.2.4. *Let $u_0 \in \text{dom}(A)$, $f : [0,\infty) \to \mathscr{X}$, and u be the mild solution of (4.21).*
i.) *If (4.33) and (4.34) hold, then $\lim\limits_{t\to\infty} \|u(t)\|_{\mathscr{X}} = 0$.*
ii.) *If (4.33) and (4.36), then $\lim\limits_{t\to\infty} \left\|u(t) - A^{-1}L\right\|_{\mathscr{X}} = 0$.*

Exercise 4.2.10. Prove Prop. 4.2.4 (i). (A proof of (ii) can be found in **[267]**.)

Exercise 4.2.11. Consider the following modification of the advection equation (3.9):

$$\begin{cases} \frac{\partial}{\partial t}c(z,t) & = V\frac{\partial}{\partial z}c(z,t) + \arctan\left(1 + \sqrt{1+t^2}\right), \ z > 0, t > 0, \\ c(z,0) & = 1 + 2z, \ z > 0, \\ c(0,t) & = 0, \ t > 0. \end{cases} \tag{4.37}$$

i.) Reformulate (4.37) as the abstract evolution equation (4.21) in an appropriate space.
ii.) Argue that (4.37) has a unique mild solution. Find an explicit formula for it.
iii.) Calculate A^{-1}, if it exists.
iv.) Does the conclusion of Prop. 4.2.4(ii) hold?

4.3 Introducing Two New Models

Model VII.1 Classical Wave Equations

The evolution over time of the vertical displacement of a vibrating string of length L subject to small vibrations can be described by the so-called *wave equation*. Precisely, suppose that the deflection of the string at position x along the string at time t is given by $z(x,t)$. An argument based on elementary physical principles (see **[109, 273]**) yields the following IBVP:

$$\begin{cases} \frac{\partial^2}{\partial t^2}z(x,t) + c^2\frac{\partial^2}{\partial x^2}z(x,t) & = 0, \ 0 < x < L, t > 0, \\ z(x,0) = z_0(x), \ \frac{\partial z}{\partial t}(x,0) & = z_1(x), \ 0 < x < L, \\ z(0,t) = z(L,t) & = 0, t > 0. \end{cases} \tag{4.38}$$

Recall from our discussion of (2.6) that the IVP

$$\begin{cases} u''(t) + ku(t) = 0, \ t > 0, \\ u(0) = u_0, \ u'(0) = u_1, \end{cases} \tag{4.39}$$

in $\mathbb{R}^N$ (where $k > 0$) was converted into a system of first-order IVPs that could be viewed as a special case (2.10). The solution of (4.39) was identified with the first row of the solution of (2.10), namely

$$u(t) = \text{first row of } e^{\begin{bmatrix} 0 & 1 \\ -k & 0 \end{bmatrix} t}\begin{bmatrix} u_0 \\ u_1 \end{bmatrix} \tag{4.40}$$

$$= \frac{u_1}{\sqrt{k}}\sin\left(\sqrt{k}t\right) + u_0\cos\left(\sqrt{k}t\right).$$

Does the approach and ultimate form of the solution extend to handle (4.38)?

Our exploration follows in a manner similar to our discussion of the heat equation in Chapter 3. We first apply the separation of variables technique to argue that the solution $z(x,t)$ of (4.38) is given by

$$\sum_{n=1}^{\infty} 2 \left[\langle z_0(\cdot), e_n(\cdot) \rangle_{\mathbb{L}^2} \cos(\lambda_n ct) + \frac{1}{L\lambda_n} \langle z_1(\cdot), e_n(\cdot) \rangle_{\mathbb{L}^2} \sin(\lambda_n ct) \right] \cdot \sin(\lambda_n x),$$

$$(4.41)$$

where $e_n(\cdot) = \sin(\lambda_n \cdot)$ and $\lambda_n = \sqrt{\frac{n\pi}{L}}, \forall n \in \mathbb{N}$.

Exercise 4.3.1. Provide the details leading to (4.41).

Now, we argue that (4.38) can be written as the abstract evolution equation (3.87) by suitably choosing the state space $\mathscr{X}$ and the operator A. We do this by adapting the approach used in the finite-dimensional case. Applying the change of variable

$$v_1 = z, \qquad v_2 = \frac{\partial z}{\partial t}$$

$$\frac{\partial v_1}{\partial t} = v_2, \ \frac{\partial v_2}{\partial t} = -c^2 \frac{\partial^2 v_1}{\partial x^2} \qquad (4.42)$$

enables us to express (4.38) as the equivalent system

$$\begin{cases} \frac{\partial}{\partial t} \begin{bmatrix} v_1 \\ v_2 \end{bmatrix}(x,t) = \begin{bmatrix} 0 & I \\ -c^2 \frac{\partial^2}{\partial x^2} & 0 \end{bmatrix} \begin{bmatrix} v_1 \\ v_2 \end{bmatrix}(x,t), \ 0 < x < L, t > 0, \\ \begin{bmatrix} v_1 \\ v_2 \end{bmatrix}(x,0) = \begin{bmatrix} z_0 \\ z_1 \end{bmatrix}(x,0), \ 0 < x < L, \\ v_1(0,t) = v_1(L,t) = 0, \ t > 0. \end{cases} \qquad (4.43)$$

We can deduce from this form possible choices for the unknown u and operator A, but the main difference this time is that the state space must be a product space $K_1 \times K_2$ since the unknown is a vector consisting of two components. Looking back at Model V.1, the choice of $\mathbb{L}^2(0,L;\mathbb{R})$ for the state space worked well from both the mathematical modeling and physical viewpoints when handling the term $-c^2 \frac{\partial^2}{\partial x^2}$. As such, this space, or some closed subspace thereof, might be a reasonable choice for K_2. But what about K_1?

At first glance, some natural choices for K_1 include $\mathbb{L}^2(0,L;\mathbb{R})$ (to mirror the first component) and $\mathbb{W}^{2,1}(0,L;\mathbb{R})$ (since we need at least one time derivative). However, it turns out that neither one of these is suitable for our present purposes (see **[320]** for an argument). (Now what?)

At this point, we must make use of other information inherent to the phenomenon being described; this time, we appeal to underlying physical principles and proceed as in **[320]**. Indeed, the total mechanical energy at time $t > 0$ given by

$$\frac{1}{2L} \int_0^L \left[\left| \frac{\partial z}{\partial x}(x,t) \right|^2 + |z(x,t)|^2 \right] dx = \frac{1}{2} \left[\left\| \frac{\partial z}{\partial x}(\cdot,t) \right\|_{\mathbb{L}^2}^2 + \|z(\cdot,t)\|_{\mathbb{L}^2}^2 \right] \qquad (4.44)$$

is bounded above by the initial energy of the system (and is therefore finite). (So what?) Well, as it turns out, the space $\mathbb{H}_0^1(0,L;\mathbb{R})$ equipped with the norm

$$\|z\|_{\mathbb{H}_0^1}^2 \equiv \left\| \frac{\partial z}{\partial x} \right\|_{\mathbb{L}^2}^2 + \|z\|_{\mathbb{L}^2}^2 \tag{4.45}$$

is a Hilbert space. (Show this.) Consequently, the following function space meaningful:

$$\mathcal{H} = \mathbb{H}_0^1(0,L;\mathbb{R}) \times \mathbb{L}^2(0,L;\mathbb{R}) \tag{4.46}$$

$$\left\langle \begin{bmatrix} v_1 \\ v_2 \end{bmatrix}, \begin{bmatrix} v_1^\star \\ v_2^\star \end{bmatrix} \right\rangle_{\mathcal{H}} \equiv \int_0^L \left[\frac{\partial v_1}{\partial x} \frac{\partial v_1^\star}{\partial x} + v_2 v_2^\star \right] dx.$$

Now, define the operator $A : \mathrm{dom}(A) \subset \mathcal{H} \to \mathcal{H}$ by

$$A \begin{bmatrix} v_1 \\ v_2 \end{bmatrix} = \begin{bmatrix} 0 & I \\ -c^2 \frac{\partial^2}{\partial x^2} & 0 \end{bmatrix} \begin{bmatrix} v_1 \\ v_2 \end{bmatrix} = \begin{bmatrix} v_1 \\ -c^2 \frac{\partial^2 v_2}{\partial x^2} \end{bmatrix} \tag{4.47}$$

$$\mathrm{dom}(A) = \left(\mathbb{H}^2(0,L;\mathbb{R}) \cap \mathbb{H}_0^1(0,L;\mathbb{R}) \right) \times \mathbb{H}_0^1(0,L;\mathbb{R}).$$

Remember, we must also incorporate the BCs into the definition of $\mathrm{dom}(A)$ in a manner for which $(A, \mathrm{dom}(A))$ satisfies the hypotheses of the Lumer-Phillips theorem.

Proposition 4.3.1. *The space $\mathcal{H}$ given by (4.46) and $A : \mathrm{dom}(A) \subset \mathcal{H} \to \mathcal{H}$ defined in (4.47) satisfy the following:*
i.) *$\mathcal{H}$ is a Hilbert space.*
ii.) *A is a densely-defined, closed, m-accretive on $\mathcal{H}$.*
iii.) $\mathrm{rng}\,(I+A) = \mathcal{H}$.

Proof. Part (i) and the fact that $\mathrm{dom}(A)$ is dense in $\mathcal{H}$ follow from standard arguments (see **[125, 320, 341]**), and part (iii) is argued as in the example directly following Def. 3.3.1. (Try it!) We show that A is *m*-accretive on $\mathcal{H}$. Using the BCs, an integration by parts yields

$$\left\langle A \begin{bmatrix} v_1 \\ v_2 \end{bmatrix}, \begin{bmatrix} v_1 \\ v_2 \end{bmatrix} \right\rangle_{\mathcal{H}} = \left\langle \begin{bmatrix} v_1 \\ -c^2 \frac{\partial^2 v_2}{\partial x^2} \end{bmatrix}, \begin{bmatrix} v_1 \\ v_2 \end{bmatrix} \right\rangle_{\mathcal{H}}$$

$$= \int_0^L \frac{\partial v_1}{\partial x} \cdot \frac{\partial v_1}{\partial x} dx - c^2 \int_0^L v_2 \cdot \frac{\partial^2 v_2}{\partial x^2} dx \tag{4.48}$$

$$= \int_0^L \left(\frac{\partial v_1}{\partial x} \right)^2 dx + c^2 \int_0^L \left(\frac{\partial v_2}{\partial x} \right)^2 dx$$

$$\geq 0.$$

Thus, (ii)(b) holds. Proving that A is a closed operator on $\mathcal{H}$ requires that we show that for any sequence $\{y_k\} \subset \mathrm{dom}(A)$, if $y_k \to y$ and $Ay_k \to w$ in the sense of the $\mathbb{L}^2$−norm, then $Ay = w$. (Try this as an exericse.) $\square$

Exercise 4.3.2. Complete the proof of Prop. 4.3.1.

In light of Prop. 4.3.1 and the Hille-Yosida theorem, we conclude that A generates a C_0−semigroup $\{e^{At} : t \geq 0\}$ on $\mathscr{H}$. (Actually, A generates a *group* on $\mathscr{H}$ in the sense that e^{At} is also defined $\forall t < 0$.) Further, if (4.43) is viewed as the abstract evolution equation (3.87) using the above identifications, then Thrm. 3.4.2 ensures that $\forall u_0 = \begin{bmatrix} z_0 \\ z_1 \end{bmatrix} \in \mathscr{H}$, (4.43) has a unique mild solution given by

$$\begin{bmatrix} z(t) \\ \frac{\partial z}{\partial t}(t) \end{bmatrix} = e^{At} \begin{bmatrix} z_0 \\ z_1 \end{bmatrix}, \tag{4.49}$$

which is also a classical solution whenever $\begin{bmatrix} z_0 \\ z_1 \end{bmatrix} \in \text{dom}(A)$. As in Model III.1, the second row of (4.49) is redundant.

Exercise 4.3.3. Precisely define what is meant by a classical solution of (4.38).

Is a representation formula for the semigroup tenable in this scenario? Uniqueness ensures that the first row of (4.49) must coincide with (4.41). Working backwards from (4.41) and using its time derivative as the second row yields

$$e^{At} \begin{bmatrix} z_0(x,0) \\ z_1(x,0) \end{bmatrix} = \tag{4.50}$$

$$\begin{bmatrix} \sum_{n=1}^{\infty} 2 \left[\langle z_0(\cdot), e_n(\cdot) \rangle_{\mathbb{L}^2} \cos(\lambda_n c t) + \frac{1}{L\lambda_n} \langle z_1(\cdot), e_n(\cdot) \rangle_{\mathbb{L}^2} \sin(\lambda_n c t) \right] \cdot \sin(\lambda_n x) \\ \sum_{n=1}^{\infty} 2\lambda_n c \left[-\langle z_0(\cdot), e_n(\cdot) \rangle_{\mathbb{L}^2} \sin(\lambda_n c t) + \frac{1}{L} \langle z_1(\cdot), e_n(\cdot) \rangle_{\mathbb{L}^2} \cos(\lambda_n c t) \right] \cdot \sin(\lambda_n x) \end{bmatrix}.$$

Note that $\forall t \geq 0$, e^{At} must be a 2×2 matrix of operators which acts on $\mathscr{H}$. (Why?) As such, we can loosely obtain a representation of e^{At} by appealing to (4.40). Indeed, we write

$$e^{At} = e^{\begin{bmatrix} 0 & I \\ -c^2 \frac{\partial^2}{\partial x^2} & 0 \end{bmatrix} t}$$

$$= \cos\left(t \left(-c^2 \frac{\partial^2}{\partial x^2} \right)^{1/2} \right) \begin{bmatrix} I & 0 \\ 0 & I \end{bmatrix} +$$

$$\left(-c^2 \frac{\partial^2}{\partial x^2} \right)^{-1/2} \sin\left(t \left(-c^2 \frac{\partial^2}{\partial x^2} \right)^{1/2} \right) \begin{bmatrix} 0 & I \\ -c^2 \frac{\partial^2}{\partial x^2} & 0 \end{bmatrix}$$

$$= \begin{bmatrix} \cos\left(t A^{1/2} \right) & A^{-1/2} \sin\left(t A^{1/2} \right) \\ A \left(A^{-1/2} \sin\left(t A^{1/2} \right) \right) & \cos\left(t A^{1/2} \right) \end{bmatrix} \tag{4.51}$$

$$= \begin{bmatrix} C(t) & S(t) \\ AS(t) & C(t) \end{bmatrix}.$$

Of course, this is merely a symbolic identification at this stage, since we have not precisely defined what is meant by the sine and cosine of an operator. Indeed, what does the term $\cos\left(tA^{1/2}\right)$ mean when A is an operator? Arguably, earlier success in defining the matrix exponential suggests that the use of a power series representation might be useful. All computations in (4.51) can be made precise, thereby providing an alternative abstract formulation of the solution to our IBVP for the wave equation. The families of operators $\{C(t) : t \in \mathbb{R}\}$ and $\{S(t) : t \in \mathbb{R}\}$ are appropriately called *cosine* and *sine families*, respectively. Using (4.51) in (4.49) shows that the unique mild solution of (4.43) can be expressed equivalently in $\mathscr{H}$ as

$$u(t) = C(t)z_0 + S(t)z_1. \tag{4.52}$$

A complete discussion of cosine and sine families and their utility in studying abstract second-order evolution equations is provided in **[128, 160, 240, 277, 310, 311]**.

Exercise 4.3.4. Work through the above discussion (up through (4.50)), but now using the Neumann BCs $\frac{\partial z}{\partial x}(0,t) = \frac{\partial z}{\partial x}(L,t) = 0$. Interpret this physically.

Generalizing the above discussion from a one-dimensional spatial domain to a bounded domain $\Omega \subset \mathbb{R}^N$ with smooth boundary $\partial\Omega$ is not difficult. Indeed, the resulting IBVP (4.38) is given by

$$\begin{cases} \frac{\partial^2}{\partial t^2}z(x,t) + c^2\triangle z(x,t) = 0, \ x \in \Omega, \ t > 0, \\ z(x,0) = z_0(x), \ \frac{\partial z}{\partial t}(x,0) = z_1(x), \ x \in \Omega, \\ z(x,t) = 0, \ x \in \partial\Omega, \ t > 0. \end{cases} \tag{4.53}$$

Transforming (4.53) into a system comparable to (4.43) amounts to using the more general matrix operator $\begin{bmatrix} 0 & I \\ -c^2\triangle & 0 \end{bmatrix}$ and subsequently replacing every occurrence of the interval $(0,L)$ by Ω. The resulting function spaces are Hilbert spaces. Of course, the detail-checking becomes more involved; specifically, showing that the new operator A generates a C_0-semigroup on $\mathbb{H}_0^1(\Omega;\mathbb{R}) \times \mathbb{L}^2(\Omega;\mathbb{R})$ relies partly on the Lax-Milgram Theorem and the theory of elliptic PDEs (see **[242, 329]**). But, the process closely resembles the one used in the one-dimensional setting. We summarize these observations below; see **[140, 141, 142]** for details.

Proposition 4.3.2. *Let* $\Omega \subset \mathbb{R}^N$ *be a bounded domain with smooth boundary* $\partial\Omega$ *and* $\mathscr{H} = \mathbb{H}_0^1(\Omega;\mathbb{R}) \times \mathbb{L}^2(\Omega;\mathbb{R})$. *The operator* $A : \text{dom}(A) \subset \mathscr{H} \to \mathscr{H}$ *defined by*

$$A = \begin{bmatrix} 0 & I \\ -c^2\triangle & 0 \end{bmatrix} \tag{4.54}$$

$$\text{dom}(A) = \left(\mathbb{H}^2(\Omega;\mathbb{R}) \cap \mathbb{H}_0^1(\Omega;\mathbb{R})\right) \times \mathbb{H}_0^1(\Omega;\mathbb{R})$$

generates a C_0-*semigroup on* $\mathscr{H}$.

Next, incorporating *viscous damping* into the model (as we did for the spring-mass system model) leads to the following variant of (4.38):

$$\begin{cases} \frac{\partial^2}{\partial t^2} z(x,t) + \alpha \frac{\partial}{\partial t} z(x,t) + c^2 \frac{\partial^2}{\partial x^2} z(x,t) = 0, \ 0 < x < L, t > 0, \\ z(x,0) = z_0(x), \ \frac{\partial z}{\partial t}(x,0) = z_1(x), \ 0 < x < L, \\ z(0,t) = z(L,t) = 0, t > 0, \end{cases} \quad (4.55)$$

where $\alpha > 0$ is the damping coefficient.

Exercise 4.3.5. Assume that z_0 and z_1 are sufficiently smooth.
i.) Find the solution of (4.55) using the separation of variables technique.
ii.) What can be said about the regularity of the solution if $z_0, z_1 \in \text{dom}(A^2)$?
iii.) Describe the limiting behavior of the solution of (4.55).
iv.) Rewrite (4.55) as a system in $\mathscr{H}$ (defined in (4.46)) and show that the new operator A is *m*-accretive on $\mathscr{H}$.

Incorporating a forcing term into (4.55) further yields the IBVP

$$\begin{cases} \frac{\partial^2}{\partial t^2} z(x,t) + \alpha \frac{\partial}{\partial t} z(x,t) + c^2 \frac{\partial^2}{\partial x^2} z(x,t) = F(x,t), \ 0 < x < L, t > 0, \\ z(x,0) = z_0(x), \ \frac{\partial z}{\partial t}(x,0) = z_1(x), \ 0 < x < L, \\ z(0,t) = z(L,t) = 0, t > 0, \end{cases} \quad (4.56)$$

where $F : [0,L] \times [0,\infty) \to \mathbb{R}$ is a given mapping. Mimicking (4.42), the equation portion of (4.56) can be written as

$$\frac{\partial}{\partial t} \begin{bmatrix} z \\ \frac{\partial z}{\partial t} \end{bmatrix} + \begin{bmatrix} 0 & I \\ c^2 \frac{\partial^2}{\partial x^2} & \alpha I \end{bmatrix} \begin{bmatrix} z \\ \frac{\partial z}{\partial t} \end{bmatrix} = \begin{bmatrix} 0 \\ F(x,t) \end{bmatrix}. \quad (4.57)$$

(Why?) Using the space $\mathscr{H}$ from (4.46), defining $(A,\text{dom}(A))$ as in Exer. 4.3.5, and identifying the mapping $f : [0,\infty) \to \mathscr{H}$ as

$$f(t)(\cdot) = \begin{bmatrix} 0 \\ F(\cdot,t) \end{bmatrix}. \quad (4.58)$$

enables us to express (4.56) in the form (4.21) in $\mathscr{H}$.

Exercise 4.3.6. Impose conditions on $F : [0,L] \times [0,\infty) \to \mathbb{R}$ to ensure that the mapping f given by (4.58) is well-defined. What additional regularity assumptions can be imposed to ensure that Prop. 4.2.3 can be applied to (4.58)?

Assuming the conditions imposed in Exer. 4.3.6, we can invoke Prop. 4.2.3 to conclude that (4.56) has a unique classical solution.

Exercise 4.3.7. Form an analog of IBVP (4.56) in N spatial dimensions in the spirit of (4.53). Indicate the changes that arise in the subsequent discussion when reformulating the IBVP as an abstract evolution equation.

Refer to [53, 55] for additional detail for second-order Cauchy problems.

Model VIII.1 Advection 2 - Gas Flow in a Large Container

A linearized system governing the flow of gas in a large container (as discussed in [150]) is given by

$$
\begin{cases}
\frac{\partial v}{\partial t}(x,t) + c^2 \frac{\partial p}{\partial x}(x,t) = 2t \sin^3 \left(x^2 + 1\right), \ 0 < x < \infty, \ t > 0, \\
\frac{\partial p}{\partial t}(x,t) + c^2 \frac{\partial v}{\partial x}(x,t) = -t \cos \left(x^2 + 1\right), \ 0 < x < \infty, \ t > 0, \\
p(x,0) = h_1(x), \ v(x,0) = h_2(x), \ 0 < x < \infty, \\
p(0,t) = v(0,t) = 0, \ t > 0,
\end{cases}
\tag{4.59}
$$

where v is the velocity of the gas and p is the variation in density. This IBVP can be expressed equivalently as

$$
\begin{cases}
\frac{\partial}{\partial t} \begin{bmatrix} v \\ p \end{bmatrix}(x,t) = \begin{bmatrix} 0 & -c^2 \frac{\partial}{\partial x} \\ -\frac{\partial}{\partial x} & 0 \end{bmatrix} \begin{bmatrix} v \\ p \end{bmatrix}(x,t) + \begin{bmatrix} 2t \sin^3 \left(x^2 + 1\right) \\ -t \cos \left(x^2 + 1\right) \end{bmatrix}, \ x > 0, t > 0, \\
\begin{bmatrix} v \\ p \end{bmatrix}(x,0) = \begin{bmatrix} h_1(x) \\ h_2(x) \end{bmatrix}, \ x > 0
\end{cases}
\tag{4.60}
$$

The structure of this IBVP resembles a cross between the wave equation (due to the 2×2 matrix of operators involved) and a diffusion equation (due to the presence of the first-order time derivative). A combination of the approaches used when reformulating these IBVPs abstractly can be implemented here. To view (4.60) as the abstract evolution equation (4.21), consider the Hilbert space

$$
\mathscr{H} = \mathbb{L}^2 (0, \infty; \mathbb{R}) \times \mathbb{L}^2 (0, \infty; \mathbb{R})
\tag{4.61}
$$

equipped with the inner product

$$
\left\langle \begin{bmatrix} v_1 \\ v_2 \end{bmatrix}, \begin{bmatrix} v_1^\star \\ v_2^\star \end{bmatrix} \right\rangle_{\mathscr{H}} \equiv \int_0^\infty [v_1 v_1^\star + v_2 v_2^\star] \, dx.
\tag{4.62}
$$

Define the operator $A : \operatorname{dom}(A) \subset \mathscr{H} \to \mathscr{H}$ by

$$
A \begin{bmatrix} v \\ p \end{bmatrix} = \begin{bmatrix} 0 & -c^2 \frac{\partial}{\partial x} \\ -\frac{\partial}{\partial x} & 0 \end{bmatrix} \begin{bmatrix} v \\ p \end{bmatrix}
$$
$$
\operatorname{dom}(A) = \mathbb{H}^1 (0, \infty; \mathbb{R}) \times \mathbb{H}^1 (0, \infty; \mathbb{R}),
\tag{4.63}
$$

the mapping $f : [0, \infty) \to \mathscr{H}$ by

$$
f(t) = \begin{bmatrix} 2t \sin^3 \left(\cdot^2 + 1\right) \\ -t \cos \left(\cdot^2 + 1\right) \end{bmatrix},
\tag{4.64}
$$

and the IC by

$$u_0 = \begin{bmatrix} h_1(\cdot) \\ h_2(\cdot) \end{bmatrix}. \tag{4.65}$$

Exercise 4.3.8.
i.) Consider (4.59) in the absence of the forcing term.
  a.) Verify directly that the solution of this homogenous IBVP is given by

$$\begin{cases} v(x,t) = \frac{c}{2}h_2(x-ct) + \frac{1}{2}h_1(x-ct) - \frac{c}{2}h_2(x+ct) + \frac{1}{2}h_1(x+ct) \\ p(x,t) = \frac{1}{2}h_2(x-ct) + \frac{1}{2c}h_1(x-ct) - \frac{1}{2c}h_2(x+ct) + \frac{1}{2}h_2(x+ct) \end{cases} \tag{4.66}$$

  b.) Show that the operator A defined by (4.63) is accretive on $\mathrm{dom}(A)$. (In fact, A is m-accretive on $\mathrm{dom}(A)$ and hence generates a C_0–semigroup on $\mathscr{H}$.)
ii.) Does Prop. 4.2.3 apply to (4.60) if h_1 and h_2 are sufficiently smooth? Explain.

4.4   Looking Ahead

External forces acting on a system are often state-dependent. For instance, if the forcing term represents temperature regulation of a material, then it necessarily takes into account the temperature of the material at various times t and makes appropriate adjustments. This is easily illustrated by the following adaption of the forced heat equation (4.19) discussed in Model V.2:

$$\begin{cases} \frac{\partial}{\partial t}z(x,y,t) & = k\triangle z(x,y,t) + \alpha e^{-\frac{\beta}{z(t,x,y)}}, 0 < x < a, 0 < y < b, t > 0, \\ z(x,y,0) & = \sin 2x + \cos 2y, \ 0 < x < a, 0 < y < b, \\ z(x,0,t) & = 0 = z(x,b,t), \ 0 < x < a, t > 0, \\ z(0,y,t) & = 0 = z(a,y,t), \ 0 < y < b, t > 0, \end{cases} \tag{4.67}$$

where $z(x,y,t)$ represents the temperature at the point (x,y) on the plate at time $t > 0$.

Exercise 4.4.1.
i.) Reformulate (4.67) as an abstract evolution equation. Indicate any new complications or changes that arise.
ii.) Conjecture a representation formula for a mild solution of the evolution equation formulated in (i).

Consider the abstract evolution equation formulated in Exer. 4.4.1 (i) and look back at the results developed in this chapter. What new obstacles are present that might complicate the extension of the theory needed to handle this evolution equation?

4.5   Guidance for Exercises

4.5.1   Level 1: A Nudge in a Right Direction

4.1.1. Mimic the transition from (2.5) to (2.10), for instance, but with an added term $\mathbf{f}(t) = \begin{bmatrix} D(t) \\ 0 \end{bmatrix}$.

4.1.2. (I) is straightforward for all choices of f.

(II) This is determined by the regularity of f. Think about elementary regularity properties of the integral, like the FTC. (See Section 1.8.4.)

(III) Since (4.4) is an integrated form of (4.3) in $\mathbb{R}^N$, the answer to (III) depends solely on the result of (II). (Why?)

(IV) Does $\exists p > 0$ such that $x(t+p) = x(t), \forall t > 0$?

(V) How does the presence of the expression $x_0 \cos(\omega t) + \frac{x_1}{\omega} \sin(\omega t)$ affect this?

4.1.3. i.) Solve (4.9) and (4.10) separately using variation of parameters. You need only check the continuity of $U(\cdot)$ at $t = 1$. (Why?)

ii.) $\frac{d^+}{dt} U(t)|_{t=1} = 4e^2 + e - 1$ while $\frac{d^-}{dt} U(t)|_{t=1} = 4e^2 + 2e - 4$.

4.1.4. i.) Use Prop. 2.2.5(iii) to form the dominator needed when applying Prop. 1.8.16. Use Def. 1.8.4(i) to verify the continuity of Φ.

ii.) Since $\Phi(t) = e^{At} \int_0^t e^{-As} f(s) ds$ (Why?), apply the FTC to the integral portion to conclude. (Tell how.)

4.1.5. Does one imply the other?

4.1.6. Suppose there are two such solutions, x and y. Subtract them. (Now what?)

4.1.7. Use the variation of parameters formulae for the mild solutions of (4.12) and (4.8) together with the definition of the sup norm for $\mathbb{C}([0,2];\mathbb{R})$ to establish (4.14).

4.1.8. i.) Subtract the variation of parameter formulae for U and $\overline{U}$, take the sup norm of the difference, and then apply the triangle inequality. (Now what?)

ii.) Compare the $\mathbb{L}^1$-norm to the $\mathbb{C}$-norm. The presence of the integral operator in the definition of the $\mathbb{L}^1$-norm might suggest where the difference occurs.

4.1.9. i.) You must ensure the convergence of both $\int_0^\infty e^{A(t-s)} f(s) ds$ and $e^{At} U_0$.

ii.) Refer to (i).

iii.) Try computing $\lim_{t \to \infty} \left| \int_0^t e^{A(t-s)} f(s) ds \right|$. (Now what?)

4.1.10. The calculus extends to $\mathbb{R}^N$ componentwise. Justify each step carefully.

4.1.11. i.) Suitably modify the $\mathbb{L}^1(0,T;\mathbb{R})$ norm.

ii.) What did we do when $f \in \mathbb{L}^1(0,T;\mathbb{R})$?

4.1.12. The key is to extend Prop. 4.1.1 to the $\mathbb{R}^N$ setting. Once done, mimic the proof of Prop. 4.1.3.

4.1.13. Change $\mathbb{R}$ to $\mathbb{R}^N$ in the statements and implement a suitable change of norm in the proofs. How does (4.17) change since $\mathbf{A}$ is now a matrix?

4.1.14. Let $\{\mathbf{g}_n\} \subset \mathbb{C}([0,T];\mathbb{R}^N)$ be such that $\|\mathbf{g}_n - \mathbf{f}\|_{\mathbb{L}^1(0,T;\mathbb{R}^N)} \longrightarrow 0$ as $n \to \infty$. For every $n \in \mathbb{N}$, consider (4.15) with $\mathbf{f}$ and U replaced by $\mathbf{g}_n$ and U_n, respectively. Why must each of these IVPs have a classical solution? (Now what?)

4.1.15. i.) This is routine. Refer to the exercises and examples in Section 2.2.

ii.) You need an estimate for $\left\|e^{\mathbf{A}_i t}\mathbf{y}\right\|_{\mathbb{R}^2}$ using the forms in (i). In this case, the estimate $\left\|e^{\mathbf{A}_i t}\mathbf{y}\right\|_{\mathbb{R}^2} \leq e^{t\|\mathbf{A}_i\|_{\mathbb{M}^2}}\|\mathbf{y}\|_{\mathbb{R}^2}$ does not help here. (Why?)

iii.) Since this integral is computed componentwise, we must ensure that each component of $e^{\mathbf{A}_i(t-s)}\mathbf{f}(s)$ is controlled appropriately. (How?)

iv.) Think about the eigenvalues of $\mathbf{A}$ and use (iii). (How?)

4.1.16. i.) (a) Use the fact that such limits are computed componentwise.

(b) Let $\mathbf{U}_0 = \begin{bmatrix} \mathbf{x}_0 \\ \mathbf{y}_0 \end{bmatrix}$. Substitute in the various terms and simplify componentwise using $e^{\mathbf{A}_1 t} = \begin{bmatrix} e^{\alpha t} & 0 \\ 0 & e^{\beta t} \end{bmatrix}$.

(c) Compute this componentwise using the representation formula developed in (b). The limit might not exist for all choices of α and β.

ii.) (a) See Section 2.2 for $e^{\mathbf{A}_3 t}$.

(b) Compute the limit componentwise.

iii.) Draw the conclusion using (i)(c) and (ii)(b).

4.1.17. Verify that $\{x_n\}$ converges in $\mathbb{C}\left([a,b]\,;\mathbb{R}\right)$. Must the limit candidate be unique? (Now what?)

4.1.18. Compute the variation of parameters formula at $t + \frac{2\pi}{\omega}$. What complications arise?

4.1.19. i.) Use Exer. 4.1.18.

ii.) Determine the natural limit candidate. Then, ponder whether or not the partial sums of a Fourier series must converge pointwise to a periodic function.

4.1.20. i.) Determine a formula for the mild solution and invoke Prop. 4.1.1. Regarding the limit as $t \to \infty$, can you argue as in Exer. 4.1.16(ii)(b)?

ii.) This quantifies variability in measurement. (How?)

iii.) It depends on the structure of $e^{\mathbf{A}t}$.

iv.) Estimate in a manner similar to Exer. 4.1.13.

4.2.1. For (4.19), use Exer. 3.2.6; for (4.20), use (3.57). Make certain to apply both of these with the appropriate choices of f. The other identifications are easily determined.

4.2.2. A particular property in Thrm. 3.3.6 is being used. (Which one?)

4.2.3. Think about Exer. 4.2.2 in the context of Def. 4.2.1.

4.2.4. i.) Use Thrm. 3.3.6(iii).

ii.) Use Thrm. 3.3.6(vi).

iii.) If u is a classical solution, it must also be a mild solution. What is the formula for a mild solution of (4.23)?

iv.) See (iii) and check Def. 4.2.1(ii).

v.) & vi.) Determine dom(A) in both cases and use the representation formulae developed earlier.

4.2.5. Compute the difference quotient and use properties of the integral. Along the way, you will need to compute $\lim_{h\to 0^+} \frac{e^{Ah}-I}{h} f(s)$ and $\lim_{h\to 0^+} \frac{1}{h} \int_t^{t+h} e^{A(t+h-s)} f(s)ds$.

4.2.6. i.) How smooth must a member of $\mathbb{L}^p\left(0,T;\mathscr{X}\right)$ be?

ii.) Verify that $\frac{d}{dt}\left(\int_0^t e^{A(t-s)} f(s)ds\right) = \frac{d}{dt}\left(\int_0^t e^{As} f(t-s)ds\right)$. Compute this.

4.2.7. Subtract the variation of parameters formulae for mild solutions x_n and x, and estimate the $\mathbb{C}$-norm of the difference.

4.2.8. First, verify that each of (4.31) and (4.32) has a unique mild solution. Then, subtract their variation of parameters formulae. The integral term is the most difficult one with which to work.

4.2.9. Check $\int_0^\infty M_A M_f e^{\omega(t-s)} ds$ for a fixed value of t.

4.2.10. Use the variation of parameters for a mild solution of (4.21), take the $\mathscr{X}$-norm, and perform appropriate calculations to conclude.

4.2.11. i.) Use the same identifications and space as for (3.9), and define $f : [0, T] \to \mathscr{X}$ by $f(t) = \arctan\left(1 + \sqrt{1 + t^2}\right)$. (Now what?)

ii.) We need only to check the regularity of f. (Why?)

iii.) Consider the FTC. (So what?)

iv.) Check (4.33) and (4.36).

4.3.1. Assume that $u(x,t) = X(x)T(t)$ and proceed as in Section 3.1 to see that for $\lambda > 0$,

$$X(x) = c_1 \cos\left(\sqrt{\lambda}x\right) + c_2 \sin\left(\sqrt{\lambda}x\right),$$
$$T(t) = b_1 \cos\left(\sqrt{\lambda}ct\right) + b_2 \sin\left(\sqrt{\lambda}ct\right).$$

(Now what?)

4.3.2. Use the fact $\|z_1 - z_2\|_{\mathscr{H}}^2 = \langle z_1 - z_2, z_1 - z_2 \rangle_{\mathscr{H}}$.

4.3.3. We need $u \in \mathbb{C}\left([0,T];\mathbb{H}^2 \cap \mathbb{H}_0^1\right) \cap \mathbb{C}^1\left((0,T);\mathbb{H}_0^1\right)$. (What else?)

4.3.4. This is mostly routine, but be careful to make appropriate changes to the formula (4.50) to account for the new ICs.

4.3.5. i.) The ODE for $T(t)$ will be different, but the general approach is the same. Compare this to Exer. 3.2.4.

ii.) Use Prop. 3.4.7. (How?)

iii.) We need a growth estimate on the semigroup?

iv.) Use (4.47).

4.3.6. We need for the mapping $t \mapsto F(\cdot,t)$ to belong to $\mathbb{C}^1\left((0,T);\mathscr{X}\right)$. Must we assume that $F(\cdot,t) \in \mathbb{L}^2(0,L)$, $\forall t > 0$, in order for this to be the case? (Explain.)

4.3.7. This is similar in nature to Exer. 3.2.8. Examine that exercise and determine what must be adapted for (4.56).

4.3.8. i.)(a) Apply the multivariable chain rule.

(b) Observe that $\left\langle A \begin{bmatrix} v \\ p \end{bmatrix}, \begin{bmatrix} v \\ p \end{bmatrix} \right\rangle_{\mathscr{H}} = \int_0^\infty \left(c^2 \frac{\partial^2 p}{\partial x^2} \frac{\partial v}{\partial x} + \frac{\partial p}{\partial x} \frac{\partial^2 v}{\partial x^2} \right) dx$. (Now what?)

ii.) It can be shown that A generates a C_0-semigroup $\{e^{At} : t \geq 0\}$ on $\mathscr{H}$ given by

$$e^{At} \begin{pmatrix} v \\ \rho \end{pmatrix} (x) = \begin{pmatrix} v(x,t) \\ \rho(x,t) \end{pmatrix}$$

(So what?)

4.4.1. i.) This is very similar to (4.21), but now incorporate $f(t,z)[x,y] = \alpha e^{-\frac{\beta}{z(t,x,y)}}$ abstractly.

ii.) Does a formula similar to (4.7) work?

4.5.2   Level 2: An Additional Thrust in a Right Direction

4.1.1. You should arrive at (4.5) with the natural identifications.
4.1.2. (II) If g is continuous, then $t \mapsto \int_0^t g(s)ds$ is differentiable. What if g is discontinuous?

   (IV) (i) and (iii) are periodic. Find their periods.

4.1.3. ii.) Note that $U(t)$ is differentiable everywhere except $t = 1$.
4.1.4. i.) For every $t > 0$, $\left| e^{A(t-s)} f(s) \right| \le e^{|A|(t-s)} |f(s)|$ and the right-side belongs to
$\mathbb{L}^1 (0, \infty; \mathbb{R})$, $\forall t > 0$. Also, it follows from Exer. 2.4.1(ii) that $\Phi(t) = e^{At} \int_0^t e^{-As} f(s)ds$.
(Why?) In order to argue continuity, let $t_n \to t$ and show that $\lim\limits_{n \to \infty} \int_0^{t_n} e^{-As} f(s)ds = \int_0^t e^{-As} f(s)ds$. Then, the continuity of Φ will follow from Prop. 2.2.7(i) and the fact
that the product of continuous functions is continuous. (Tell how.)

   ii.) The product of differentiable functions is differentiable. (So what?)
4.1.5. Prove that $(i) \Longrightarrow (ii)$.
4.1.6. Such a solution is unique, since the portion involving f drops out when subtracting. Thus, $z(t) = x(t) - y(t) = 0$. (So what?)
4.1.7. Subtract the formulae for the mild solutions of (4.12) and (4.8) and estimate the two resulting differences separately.
4.1.8. i.) Use the estimate for $\left| e^{At} \right|$ with appropriate properties of the integral.

   ii.) Does $e^{At} (\varepsilon_1 + \varepsilon_2)$ work?
4.1.9. i.) Must $A \le 0$ in order for this to occur?

   iii.) Show that $\lim\limits_{t \to \infty} |U(t)| = -A^{-1}L$.

4.1.11. i.) $\|f\|_{\mathbb{L}^1(0,T;\mathbb{R}^N)} = \int_0^T \|f(s)\|_{\mathbb{R}^N} ds$.

   ii.) Apply the FTC as extended to $\mathbb{R}^N$.
4.1.12. Make a suitable change of norm in the details of Exer. 4.1.4 in order to verify the new version of Prop. 4.1.1. Why does the proof of Prop. 4.1.3 then work with essentially no change? (Explain.)
4.1.13. The estimate $\left\| \int_0^t e^{A(t-s)} (f(s) - \bar{f}(s)) \, ds \right\|_{\mathbb{R}^N} \le e^{T \|A\|_{MN}} \int_0^t \|f(s) - \bar{f}(s)\|_{\mathbb{R}^N} ds$
suggests how to alter (4.17), as well as Exer. 4.1.8(ii).
4.1.14. Subtract the variation of parameters formulae for the mild solutions and estimate as in Exer. 4.1.13.
4.1.15. ii.) We need to ensure that all components of $e^{A_i t}$ are zero or negative exponentials to ensure the limit exists. (Why?) How does this relate to the eigenvalues of A_i? (Note that the suggested estimate in the first hint fails because it does not allow us to take advantage of the negative exponentials that appear to be critical here.)

   iii.) Is it sufficient to assume that all eigenvalues of A have negative real parts? (Explain.)

4.1.16. i.) (a) $f_0 = \begin{bmatrix} 1 \\ -1 \end{bmatrix}$

(b) Note that $e^{\mathbf{A}_1 t}\mathbf{U}_0 = \begin{bmatrix} e^{\alpha t}x_0 \\ e^{\beta t}y_0 \end{bmatrix}$ and

$$\int_0^t e^{\mathbf{A}_1(t-s)}\mathbf{f}(s)ds = \begin{bmatrix} \int_0^t e^{\alpha(t-s)}\left(1+e^{-s}\right)ds \\ \int_0^t e^{\beta(t-s)}\left(se^{-2s}-1\right)ds \end{bmatrix}$$

Now, compute.
  (c) Prove that the limit is $-\mathbf{A}_1^{-1}\mathbf{f}_0$.
ii.) (b) Prove that the limit is $-\mathbf{A}_3^{-1}\mathbf{f}_0$.
  (c) This has the same form as in (b).
iii.) Yes.
4.1.17. Determine the regularity of x_n and try to verify that $\{x_n\}$ converges in that space. Does an extension of Prop. 4.1.1 enter into the discussion?
4.1.18. You need to use a change of variable to handle the integral term.
4.1.19. ii.) The limit function need not be periodic (and usually is not periodic).
4.1.20. i.) Check to see if the limit is $-\mathbf{A}^{-1}\mathbf{0}$.
  iii.) Must $e^{\mathbf{A}(t+12)} = e^{\mathbf{A}t}$, $\forall t > 0$. If so, then what?
  iv.) This describes how far apart the two measurements of drug concentration corresponding to x_0 and y_0 can be.
4.2.1. For (4.19), let $\mathcal{X} = \mathbb{L}^2(0,a;\mathbb{R}) \times \mathbb{L}^2(0,b;\mathbb{R})$ and identify $f : [0,\infty) \to \mathcal{X}$ by $f(t)(x,y) = t^2 + x + 2y$. For a fixed $t^\star \in [0,\infty)$, observe that $\int_0^a \int_0^b f(t^\star)(x,y)dydx < \infty$. (Why?) For (4.20), let $\mathcal{X} = \mathbb{L}^2(0,\pi;\mathbb{R})$ and use the natural choice for f as for (4.19). Finally, check in both cases that $u_0 \in \mathcal{X}$.
4.2.2. Use (3.76).
4.2.3. No. This condition is only needed when differentiating the variation of parameters formula. (Why?)
4.2.4. ii.) The fact that $y_0 \notin \mathrm{dom}(A)$ is critical. (Tell why.)
  iii.) A mild solution of (4.23) is given by $u(t) = \int_0^t e^{A(t-s)}e^{As}xds = g(t)x$. (So what?)
  iv.) Yes.
4.2.5. Use Prop. 1.8.16 and Thrm. 3.3.6 appropriately.
4.2.6. i.) There exists $f \in \mathbb{L}^p(0,T;\mathcal{X}) \setminus \mathbb{C}([0,T];\mathcal{X})$. (So what?)
  ii.) Continuing the equality, we have $e^{At}f(0) + \int_0^t e^{A(t-s)}f'(s)ds$. (So what?)
4.2.8. Does the convergence theorem from Section 3.7 help?
4.2.9. This integral converges.
4.2.10. Use the fact that $\forall \omega < 0$, $\lim\limits_{t\to\infty} e^{\omega t} = 0$.
4.2.11. ii.) Use the variation of parameters formula with the representation formula for e^{At}.
  iii.) It should be an integral operator of the form $\int_0^x g(z)dz$. (Why?)
4.3.1. Show that

$$u(x,t) = \sum_{n=0}^{\infty}\left[a_n \cos\left(\sqrt{\frac{n\pi}{L}}ct\right) + b_n \sin\left(\sqrt{\frac{n\pi}{L}}ct\right)\right]\sin\left(\sqrt{\frac{n\pi}{L}}x\right)$$

where a_n and b_n are Fourier coefficients. Then, apply the $\mathbb{L}^2$−norm.

4.3.2. Let $y_k = \begin{bmatrix} y_k^1 \\ y_k^2 \end{bmatrix}$ and $y = \begin{bmatrix} y^1 \\ y^2 \end{bmatrix}$, and carefully use (4.46) and (4.47) with the continuity properties of the inner product.

4.3.3. We also need $\frac{\partial u}{\partial t} \in C\left([0,T];\mathbb{H}_0^1\right) \cap C^1\left((0,T);\mathbb{L}^2\right)$.

4.3.5. i.) The solution is

$$u(x,t) = e^{-\frac{\alpha t}{2}} \sum_{n=0}^{\infty} \left[a_n \cos\left(\mu_n t\right) + b_n \sin\left(\mu_n t\right)\right] \sin\left(\lambda_n x\right),$$

where $a_n = \frac{2}{L}\int_0^L z_0(x)\sin\left(\lambda_n x\right)dx$, $b_n = \frac{2}{L}\int_0^L z_1(x)\sin\left(\lambda_n x\right)dx + \frac{\alpha}{2}a_n$, $\lambda_n = \frac{n\pi}{L}$ and $\mu_n = 4c^2\lambda_n^2 - \alpha^2 > 0$.

4.3.7. The operator A and space $\mathscr{X}$ are the most notable changes, but the form of the abstract evolution equation is the same.

4.3.8. i.)(b) Use the BC and an equivalent form of the expression $\frac{\partial}{\partial x}\left[\frac{\partial \rho}{\partial x}\frac{\partial v}{\partial x}\right]$. How can you use this?

ii.) Yes.

4.4.1. i.) Now, $f(t)$ in (4.21) will be replaced by $f(t,u(t))$.

Chapter 5

Semi-Linear Evolution Equations

Overview

We develop an extension of the theory of Chapter 4 that enables us to formally study mathematical models whose forcing terms are state-dependent. The general nature of such perturbations creates various complications, even when trying to establish the existence of a mild solution. The focus of the current chapter is to make precise how these complications can be overcome.

5.1   Motivation by Models

We briefly consider variants of several models explored in earlier chapters and introduce two new applications. Our investigation of them will progress gradually as we move through the chapter.

5.1.1   Some Models Revisited

Model II.3 Pharmacokinetics with concentration-dependent dosage

The rate of absorption of a drug into the bloodstream is affected by the dosage $D(t)$ at any time $t > 0$. In turn, the dosage might depend on the concentration of the drug in the GI tract and/or bloodstream, which in turn varies with time. Abstractly, this leads to the following *semi-linear* version of (4.1) in Model II.2:

$$\begin{cases} \frac{dy}{dt} = -ay(t) + D(t, y(t), z(t)), t > 0, \\ \frac{dz}{dt} = ay(t) - bz(t) + \overline{D}(t, y(t), z(t)), t > 0, \\ y(0) = y_0, \ z(0) = 0, \end{cases} \tag{5.1}$$

where the term $\overline{D}$ has been added in the second equation as a result of the state-dependence within the term D in the first equation. Though somewhat more general, the basic form of (5.1) is not too dissimilar from (4.1).

Exercise 5.1.1. Rewrite (5.1) in vector form.

Model VII.2 Semi-linear Wave Equations

Dispersion can be incorporated into the classical wave equation by altering the form of the forcing term. Precisely, consider the following extension of IBVP (4.56), where $\alpha, \beta > 0$:

$$\begin{cases} \frac{\partial^2}{\partial t^2} z(x,t) + \alpha \frac{\partial}{\partial t} z(x,t) + c^2 \frac{\partial^2}{\partial x^2} z(x,t) = z(x,t),\ 0 < x < L, t > 0, \\ z(x,0) = z_0(x),\ \frac{\partial z}{\partial t}(x,0) = z_1(x),\ 0 < x < L, \\ \frac{\partial z}{\partial x}(0,t) = \frac{\partial z}{\partial x}(L,t) = 0, t > 0. \end{cases} \tag{5.2}$$

Exercise 5.1.2. Reformulate (5.2) as an abstract evolution equation in the space given by (4.46). How does the resulting form compare to (4.57)?

Similarly, a system of weakly coupled damped wave equations with nonlinear dispersion can be described by

$$\begin{cases} \frac{\partial^2}{\partial t^2} z(x,t) + \alpha \frac{\partial}{\partial t} z(x,t) + c^2 \frac{\partial^2}{\partial x^2} z(x,t) = f_1(z) + g_1(w),\ 0 < x < L, t > 0, \\ \frac{\partial^2}{\partial t^2} w(x,t) + \beta \frac{\partial}{\partial t} w(x,t) + c^2 \frac{\partial^2}{\partial x^2} w(x,t) = f_2(z) + g_2(w),\ 0 < x < L, t > 0, \\ z(x,0) = z_0(x),\ \frac{\partial z}{\partial t}(x,0) = z_1(x),\ 0 < x < L, \\ w(x,0) = w_0(x),\ \frac{\partial w}{\partial t}(x,0) = w_1(x),\ 0 < x < L, \\ \frac{\partial z}{\partial x}(0,t) = \frac{\partial z}{\partial x}(L,t) = \frac{\partial w}{\partial x}(0,t) = \frac{\partial w}{\partial x}(L,t) = 0, t > 0. \end{cases} \tag{5.3}$$

Exercise 5.1.3. Reformulate (5.3) as an abstract evolution equation in the space $\mathscr{H} = \left(\mathbb{H}_0^1(0,L;\mathbb{R}) \times \mathbb{L}^2(0,L;\mathbb{R}) \right)^2$. What hurdles do you encounter?

Model V.3 Heat Conduction with State-Dependent Heat Source

The heat production source can change with time and the temperature of the material being heated. For instance, it is reasonable for certain chemical reactions to exhibit an exponentially decaying heat source. The following IBVP for heat conduction through a metal sheet takes this into account and constitutes a natural generalization of (4.19):

$$\begin{cases} \frac{\partial}{\partial t} z(x,y,t) = k \triangle z(x,y,t) + \alpha e^{-\frac{\beta}{z(t,x,y)}},\ 0 < x < a, 0 < y < b, t > 0, \\ z(x,y,0) = z_0(x,y),\ 0 < x < a, 0 < y < b, \\ \frac{\partial z}{\partial y}(x,0,t) = 0 = \frac{\partial z}{\partial y}(x,b,t),\ 0 < x < a, t > 0, \\ \frac{\partial z}{\partial x}(0,y,t) = 0 = \frac{\partial z}{\partial x}(a,y,t),\ 0 < y < b, t > 0. \end{cases} \tag{5.4}$$

Exercise 5.1.4. Reformulate (5.4) as an abstract evolution equation.

5.1.2 Introducing Two New Models

Model IX.1 Neural Networks

Hopfield initiated the study of neural networks in 1982. Applications range from the modeling of physiological functions to using artificial neural networks to perform parallel computations. (See [133, 173, 203, 274, 314, 332]). A heuristic development of a basic neural network is as follows.

Suppose that we begin with M neurons interconnected via a network of synapses. Every neuron receives an input signal from the other $M - 1$ neurons, each with a varying degree of strength. The neuron acts on the total signal and produces an output which is subsequently broken down and emitted as inputs into the other $M - 1$ neurons comprising the network. To form the mathematical model, label the neurons as $1 \leq i \leq M$, and for each i and time $t \geq 0$, let

$x_i(t) =$	voltage of input from i at time t
$\omega_{ij}(t) =$	strength of signal that j contributes to i at time t
$\sum_{j=1}^{M} \omega_{ij}(t)x_j(t) =$	total input signal from the networked neurons
$g\left(\sum_{j=1}^{M} \omega_{ij}(t)x_j(t)\right) =$	output signal at time t from i due to internal activity

Momentarily, we assume for simplicity that the output signal can be expressed as

$$g\left(\sum_{j=1}^{M} \omega_{ij}(t)x_j(t)\right) = \sum_{j=1}^{M} \omega_{ij}(t)g_j(x_j(t))$$

so that it is clear how the neuron acts on each input signal individually. (We note that this is somewhat unrealistic because it assumes that we can distinguish among the M inputs.) These quantities evolve over time. As such, the rates at which the voltages of these M neurons change is governed by a system of ODEs formulated under the assumption that the rate of voltage change for each neuron is proportional to its present voltage and its output signal. Precisely, we have the system

$$\begin{cases} \frac{d}{dt}x_1(t) = a_1 x_1(t) + \sum_{j=1}^{M} \omega_{1j}(t)g_j(x_j(t)), \\ \vdots \\ \frac{d}{dt}x_M(t) = a_M x_M(t) + \sum_{j=1}^{M} \omega_{Mj}(t)g_j(x_j(t)), \\ x_i(0) = x_{i,0},\, i = 1,\ldots,M. \end{cases} \tag{5.5}$$

Exercise 5.1.5. Reformulate (5.5) as an abstract evolution equation.

Model X.1 Spatial Pattern Formation

Diffusion, without the intervening effects of kinetics, disperses a pattern. Chemicals react and diffuse in different ways, thereby resulting in a distribution of varying concentrations which can be viewed as a distinct spatial pattern. Morphogenesis is the development of pattern and form in a living thing, and arises naturally in ecology by way of describing migratory patterns, the formation of animal coatings (e.g., dispersion and pattern of spots on a leopard), butterfly wing patterns, etc. There are differing viewpoints as to how, biologically, such patterns are formed. In some manner, though, the reaction-diffusion equations enter into the mathematical modeling of this phenomenon. In 1952, Turing asserted that diffusion need not lead to a uniformly distributed concentration, but rather certain perturbations could redirect its action to form patterns. (See [**120, 244, 252, 284, 335**]). We discuss a two-dimensional version of his model below.

Let Ω be a bounded domain in $\mathbb{R}^3$ with smooth boundary $\partial\Omega$ and suppose that there are N chemicals interacting within this domain. For each $\mathbf{x} = (x,y,z) \in \Omega$, $t \geq 0$, and $1 \leq i \leq N$, let

$C_i = C_i(\mathbf{x},t) =$	Concentration of chemical i at $(\mathbf{x},t)$
$\alpha_i =$	Diffusion coefficient for chemical i
$f_i(t,C_1,\ldots,C_N) =$	Reaction among N chemicals affecting how C_i changes

We describe the dynamics of these N chemicals within Ω as the following system of diffusion equations:

$$\begin{cases} \frac{\partial C_1}{\partial t}(\mathbf{x},t) & = \alpha_1 \triangle C_1 + f_1(t,C_1,\ldots,C_N), \mathbf{x} \in \Omega, t > 0, \\ \quad \vdots \\ \frac{\partial C_N}{\partial t}(\mathbf{x},t) & = \alpha_N \triangle C_N + f_N(t,C_1,\ldots,C_N), \mathbf{x} \in \Omega, t > 0, \\ C_i(\mathbf{x},0) & = C_i^*(\mathbf{x}), \mathbf{x} \in \Omega, 1 \leq i \leq N, \\ C_i(\mathbf{x},t) & = 0, \mathbf{x} \in \partial\Omega, t > 0, 1 \leq i \leq N. \end{cases} \tag{5.6}$$

We can express the equation portion of (5.6) in matrix form as

$$\frac{\partial}{\partial t} \underbrace{\begin{bmatrix} C_1 \\ \vdots \\ C_N \end{bmatrix}}_{=\mathbf{C}} = \underbrace{\begin{bmatrix} \alpha_1\triangle & 0 & \cdots & 0 \\ 0 & \ddots & & \vdots \\ \vdots & & \ddots & 0 \\ 0 & \cdots & 0 & \alpha_N\triangle \end{bmatrix}}_{=A} \underbrace{\begin{bmatrix} C_1 \\ \vdots \\ C_N \end{bmatrix}}_{=\mathbf{C}} + \underbrace{\begin{bmatrix} f_1(t,C_1,\ldots,C_N) \\ \vdots \\ f_N(t,C_1,\ldots,C_N) \end{bmatrix}}_{=\mathbf{f}(t,\mathbf{C})\text{ Kinetics}} \tag{5.7}$$

or more succinctly as

$$\frac{\partial \mathbf{C}}{\partial t}(\mathbf{x},t) = A\mathbf{C}(\mathbf{x},t) + \mathbf{f}(t,\mathbf{C}(\mathbf{x},t)). \tag{5.8}$$

For simplicity, we consider the case in which there are only two chemicals (i.e., $N = 2$ above) interacting in Ω.

Exercise 5.1.6. On what Hilbert space would it be natural to reformulate (5.7) abstractly when $N = 2$?

Different kinetic terms f_i have been derived theoretically and experimentally by researchers in the field. One classical model of activator-inhibitor type [**120**] describes the kinetics by

$$\begin{cases} f_1(C_1,C_2) & = \beta_1 - \beta_2 C_1 + \beta_3 C_1^2 C_2, \\ f_2(C_1,C_2) & = \beta_4 - \beta_3 C_1^2 C_2, \end{cases} \tag{5.9}$$

where β_i $(i = 1, 2, 3, 4)$ are the rate constants. Several examples illustrating how imposing different conditions on β_i guarantees diffusive instability leading to pattern formation are discussed in the references cited within **[120, 252]**.

Common Theme: All of the IBVPs considered in this section can be reformulated as the abstract evolution equation

$$\begin{cases} u'(t) = Au(t) + f(t, u(t)), \ t > 0, \\ u(0) = u_0, \end{cases} \tag{5.10}$$

in an appropriate Banach space $\mathscr{X}$. Mere symbolic identification suggests that a variation of parameters formula for a mild solution of (5.10) might be given by

$$u(t) = e^{At}u_0 + \int_0^t e^{A(t-s)} f(s, u(s)) ds, \ t > 0. \tag{5.11}$$

This is intuitive, but try tracing through the derivation of (4.22). The dependence of the forcing term on the state $u(s)$ creates a self-referential situation in (5.11) that was not present before. Somehow, we need a technique that allows us to temporarily suspend this interdependence in order to make use of the theory in Chapter 4. Understanding this technique requires additional tools from functional analysis, discussed in the next section.

5.2   More Tools from Functional Analysis

We need a handful of new tools to help navigate the next leg of our journey. We present the highlights of the theory and refer you to **[146, 185, 202, 340]** for a more detailed discussion. Throughout this chapter, $(\mathscr{X}, \|\cdot\|_{\mathscr{X}})$ is a real Banach space, unless otherwise specified.

5.2.1   Fixed-Point Theory

One of our central strategies involves the use of so-called *fixed-point theory*. This broad approach is based on a very straightforward strategy whose utility is especially evident in the study of existence theory. We present several useful results.

5.2.1.1   The Contraction Mapping Principle

The most common result is based on the notion of a contraction, defined below.

Definition 5.2.1. A mapping $\Phi : \mathscr{X} \to \mathscr{X}$ is a *contraction* if $\exists 0 < \alpha < 1$ such that

$$\|\Phi x - \Phi y\|_{\mathscr{X}} \le \alpha \|x - y\|_{\mathscr{X}}, \ \forall x, y \in \mathscr{X}. \tag{5.12}$$

Note that a contraction is automatically uniformly continuous. (Why?)

Theorem 5.2.2. Contraction Mapping Principle
If $\Phi : \mathscr{X} \to \mathscr{X}$ is a contraction, then there exists a unique $z^\star \in \mathscr{X}$ such that $\Phi(z^\star) = z^\star$. (We call $z^\star$ a fixed-point of Φ.)

Proof. Let $z_0 \in \mathscr{X}$ and define the sequence $\{z_n : n \in \mathbb{N}\} \subset \mathscr{X}$ by

$$z_n = \Phi(z_{n-1}), \ n \in \mathbb{N}. \tag{5.13}$$

If $z_1 = z_0$, then we are done. (Why?) If not, then prove inductively that the following two statements hold, $\forall m, k \geq 0$.

$$\|z_{m+1} - z_m\|_{\mathscr{X}} \leq \alpha^m \|z_1 - z_0\|_{\mathscr{X}}, \tag{5.14}$$

$$\|z_{m+k} - z_m\|_{\mathscr{X}} \leq \sum_{j=0}^{k-1} \alpha^{n+j} \|z_1 - z_0\|_{\mathscr{X}} \leq \frac{\alpha^m}{1-\alpha} \|z_1 - z_0\|_{\mathscr{X}}. \tag{5.15}$$

Let $\varepsilon > 0$. There exists $M \in \mathbb{N}$ such that

$$m \geq M \implies \alpha^m < \frac{(1-\alpha)\varepsilon}{\|z_1 - z_0\|_{\mathscr{X}}}. \tag{5.16}$$

As such, (5.15) and (5.16) together imply that

$$m \geq M \implies \|z_{m+k} - z_m\|_{\mathscr{X}} < \varepsilon, \ \forall k \geq 0. \tag{5.17}$$

So, $\{z_n : n \in \mathbb{N}\}$ is a Cauchy sequence in $\mathscr{X}$ and hence convergent to some $z^\star \in \mathscr{X}$. The fact that $z^\star$ is a fixed point of Φ follows from the fact that

$$\lim_{n \to \infty} \Phi(z_{n-1}) = \lim_{n \to \infty} z_n = z^\star.$$

As for uniqueness, suppose that both $x^\star$ and $y^\star$ are fixed-points of Φ and argue that $\|\Phi x - \Phi y\|_{\mathscr{X}} < \|x - y\|_{\mathscr{X}}$. (So what?) ☐

Corollary 5.2.3. *If $\Phi : \mathscr{X} \to \mathscr{X}$ is a mapping for which $\exists 0 < \alpha < 1$ such that for some $n_0 \in \mathbb{N}$,*

$$\|\Phi^{n_0} x - \Phi^{n_0} y\|_{\mathscr{X}} \leq \alpha \|x - y\|_{\mathscr{X}}, \ \forall x, y \in \mathscr{X}, \tag{5.18}$$

then Φ has a unique fixed-point in $\mathscr{X}$.

Exercise 5.2.1. Prove Cor. 5.2.3.

Remark. The above two results are particularly useful when establishing the existence and uniqueness of mild solutions of many classes of evolution equations. Often, it is beneficial to apply them on a closed subspace of $\mathscr{X}$, specifically a closed ball $\mathscr{B}_{\mathscr{X}}(x_0; \varepsilon)$, rather than on the entire space $\mathscr{X}$. The theorems are still applicable since a closed metric subspace of a Banach space is complete.

Exercise 5.2.2. Was completeness of $\mathscr{X}$ an essential ingredient of Thrm. 5.2.2 and Cor. 5.2.3? Explain.

5.2.1.2 Schauder's Fixed Point Theorem

Not every continuous operator $\Phi : \mathscr{X} \to \mathscr{X}$ is a contraction. As such, Thrm. 5.2.2 is not always applicable. But, is it necessary for Φ to be a contraction in order for it to have a fixed-point?

Even in the one-dimensional case there is no shortage of continuous functions $f : \mathbb{R} \to \mathbb{R}$ that do not have fixed-points. For instance, take $f(x) = x - 1$. We also know from Exer. 1.8.6(i) that a continuous function $f : [a,b] \to [a,b]$ *must* possess a fixed-point, while continuity is not strong enough to ensure a function $g : (a,b) \to \mathbb{R}$ has a fixed-point. (Why?) This collection of examples suggests that the structure of the domain and range plays an important role. The following theorem due to Brouwer [146, 202] provides a complete answer for the $\mathbb{R}^N$-setting.

Theorem 5.2.4. Brouwer's Fixed-Point Theorem
Let $\mathscr{D} \subset \mathbb{R}^N$ be a closed, convex, and bounded set. If $\Psi : \mathscr{D} \to \mathscr{D}$ is continuous, then Ψ has at least one fixed point in $\mathscr{D}$.

This theorem does not hold true in the infinite-dimensional setting because while a closed ball $\mathfrak{B}_{\mathbb{R}^N}(x_0; R)$ in $\mathbb{R}^N$ must be compact, it need not be in a general Banach space. This fact significantly enters into the proof of Thrm.5.2.4, as well as in the construction of a counterexample that illustrates the falsity of the theorem in an infinite-dimensional space. (For the latter, see [146].) In light of this failure, we must impose more stringent conditions on the nature of the set $\mathscr{D}$ in order to obtain the desired extension of Thrm. 5.2.4. This is where the notion of a *compact set* in a Banach space comes into play. Some of the following notions were introduced in Section 1.4.4 and are recalled here for convenience.

Definition 5.2.5. Let $(\mathscr{X}, \|\cdot\|_{\mathscr{X}})$ be a normed linear space. A set $\mathscr{K} \subset \mathscr{X}$ is called
i.) *convex* if $\forall 0 \leq t \leq 1$ and $x, y \in \mathscr{K}$, $tx + (1-t)y \in \mathscr{K}$;
ii.) *bounded* if $\exists M > 0$ such that $\|x\|_{\mathscr{X}} \leq M$, $\forall x \in \mathscr{K}$;
iii.) *compact* if every sequence $\{x_n\} \subset \mathscr{K}$ contains a subsequence $\{x_{n_k}\}$ which converges to a member of $\mathscr{K}$;
iv.) *precompact* if $\mathrm{cl}_{\mathscr{X}}(\mathscr{K})$ is a compact subset of $\mathscr{X}$.

A compact set $\mathscr{K} \subset \mathscr{X}$ must be closed and bounded (Why?), but not conversely, as shown in the following exercise.

Exercise 5.2.3. Let $\mathscr{K} = \left\{ \cos\left(\frac{2n\pi x}{b-a}\right) : n \in \mathbb{N} \wedge a \leq x \leq b \right\}$. Show that while $\mathscr{K}$ is a closed and bounded subset of $\mathbb{L}^2(a,b;\mathbb{R})$, it is not compact in $\mathbb{L}^2(a,b;\mathbb{R})$.

We use the notions introduced in Def. 5.2.5 to formulate the following extension of Thrm. 5.2.4:

Theorem 5.2.6. (Schauder's Fixed-Point Theorem) *Let $(\mathscr{X}, \|\cdot\|_{\mathscr{X}})$ be a normed linear space and $\mathscr{K}$ a closed, convex subset of $\mathscr{X}$. If $\Psi : \mathscr{K} \to \mathscr{K}$ is a continuous mapping for which $\Psi(\mathscr{K})$ is precompact in $\mathscr{K}$, then Ψ has at least one fixed-point in $\mathscr{K}$.*

Remark. One consequence of weakening the hypotheses on Ψ to mere continuity is that the uniqueness of the fixed-point is no longer guaranteed, even if $\mathscr{X} = \mathbb{R}$. For instance, the function $f : [0, 1] \to [0, 1]$ defined by $f(x) = x$ has infinitely many fixed-points.

The set $\mathscr{K}$ is often taken to be a closed ball in $\mathscr{X}$, possibly equipped with an extra condition resembling a characteristic that you wish for a fixed-point to possess. Since a closed ball is automatically a closed, convex set, and verifying the continuity of Ψ is typically standard, we are left with the more difficult task of verifying the precompactness of $\Psi(\mathscr{K})$. This depends on the underlying topology of $\mathscr{X}$ since the closure is taken in $\mathscr{X}$. The so-called Arzela-Ascoli theorem offers a method of attack. To state it, we begin with:

Definition 5.2.7. A set $\mathscr{Z} \subset \mathbb{C}([a,b]; \mathscr{Y})$ is *equicontinuous* at $t_0 \in [a,b]$ if $\forall \varepsilon > 0$, $\exists \delta > 0$ (depending on ε and t_0) such that

$$s \in [a,b] \text{ with } |s - t_0| < \delta \implies \|z(t_0) - z(s)\|_{\mathscr{Y}} < \varepsilon, \ \forall z \in \mathscr{Z}.$$

Theorem 5.2.8. Arzela-Ascoli in $\mathbb{C}([a,b]; \mathscr{Y})$
A set $\mathscr{Z} \subset \mathbb{C}([a,b]; \mathscr{Y})$ is precompact if and only if the following hold, $\forall t \in [a,b]$:
i.) $\{z(t) \,|\, z \in \mathscr{Z}\}$ *is precompact in $\mathscr{Y}$.*
ii.) $\{z(t) \,|\, z \in \mathscr{Z}\}$ *is equicontinuous.*

(See **[318]** for a proof.) This result reduces our task to verifying precompactness of a somewhat more manageable set. Even so, verifying condition (i) by appealing directly to the definition (involving closures) can be tedious. Often, the following condition and resulting proposition are more easily verified.

Definition 5.2.9. A set $\mathscr{K} \subset \mathscr{X}$ is *totally bounded* if $\forall \varepsilon > 0$, there exists a finite set $\{x_1, \dots, x_{p_\varepsilon}\} \subset \mathscr{X}$ such that $\mathscr{K} \subset \bigcup_{i=1}^{p_\varepsilon} B_{\mathscr{X}}(x_i; \varepsilon)$.

Proposition 5.2.10. *Let $\mathscr{X}$ be a Banach space. A set $\mathscr{K} \subset \mathscr{X}$ is precompact if and only if $\mathscr{K}$ is totally bounded.*

We leave the proof as an exercise and remark that Prop. 5.2.10 need not hold when $\mathscr{X}$ is merely a normed linear space. The following strategy is often useful.

Strategy $\mathscr{P}$: Occasionally, the situation arises in which we can identify a set $\mathscr{Y}$ which (i) is known to be precompact in $\mathscr{X}$ (and hence, totally bounded), and (ii) is "close" to the set $\mathscr{K}$ which we seek to show is precompact in $\mathscr{X}$. By "close," we mean that

$$\forall \varepsilon > 0 \text{ and } x \in \mathscr{K}, \ \exists y_x \in \mathscr{K} \text{ such that } \|x - y_x\|_{\mathscr{X}} < \tfrac{\varepsilon}{2}.$$

Since the total boundedness of $\mathscr{Y}$ guarantees the existence of $\{\overline{y_1}, \dots, \overline{y_{p_\varepsilon}}\} \subset \mathscr{X}$ for which $\mathscr{Y} \subset \bigcup_{i=1}^{p_\varepsilon} B_{\mathscr{X}}(\overline{y_i}; \tfrac{\varepsilon}{2})$, using the triangle inequality implies that $\mathscr{K} \subset$

$\bigcup_{i=1}^{p_\varepsilon} B_{\mathscr{X}} (\overline{y_i}; \varepsilon)$ (Why?), so that $\mathscr{K}$ is totally bounded and hence precompact in $\mathscr{X}$.

Looking ahead, we shall typically define the operator Φ to which we intend to apply a particular fixed-point theorem in a manner that guarantees that the fixed-point coincides with a mild solution of the IVP under investigation. Since a mild solution must be continuous by definition, the mapping Φ is typically defined on the space $\mathbb{C}([a,b]; \mathscr{X})$. In such case, the above discussion is useful. However, in some settings (cf. Chapters 9 and 10), the complexity of the operators arising when expressing the IBVP abstractly requires that we seek a solution in a sense weaker than our current notion of a mild solution. This is manifested in different ways depending on the approach used, one of which is to show that an aptly-chosen operator $\Phi : \mathbb{L}^1(a,b; \mathscr{X}) \to \mathbb{L}^1(a,b; \mathscr{X})$ has a fixed-point. Since $\mathbb{L}^1(a,b; \mathscr{X})$ is a larger space than $\mathbb{C}([a,b]; \mathscr{X})$ and is equipped with a different norm, Thrm. 5.2.8 and the concomitant discussion no longer apply. We need a Thrm. 5.2.8-type result for $\mathbb{L}^1(a,b; \mathscr{X})$. Such results have been established and are proven in a similar manner (see [318]) .

5.2.1.3   Compact Operators and Schaefer's Fixed-Point Theorem

We discuss one more fixed-point theorem that requires the following strengthening of the notion of a bounded operator.

Definition 5.2.11. Let $\mathscr{X}$ and $\mathscr{Y}$ be Banach spaces. An operator $B : \mathscr{X} \to \mathscr{Y}$ is *compact* if for every bounded set $\mathscr{D} \subset \mathscr{X}$, the image $B(\mathscr{D})$ is precompact in $\mathscr{Y}$.

The following properties of compact operators are useful.

Proposition 5.2.12. *Let $\mathscr{X}$ and $\mathscr{Y}$ be Banach spaces.*
i.) *Every compact operator $B : \mathrm{dom}(B) \subset \mathscr{X} \to \mathscr{Y}$ is bounded, but not conversely.*
ii.) *Let $B \in \mathbb{B}(\mathscr{X}, \mathscr{Y})$. If either $\mathrm{dom}(B)$ or $\mathrm{rng}(B)$ is finite dimensional, then B is compact.*
iii.) *Let $B_1 \in \mathbb{B}(\mathscr{X}, \mathscr{Y})$ and $B_2 \in \mathbb{B}(\mathscr{Y}, \mathscr{Z})$. If at least one of B_1 or B_2 is compact, then $B_2 B_1$ is compact.*

Exercise 5.2.4. Prove Prop. 5.2.12. Is completeness of either $\mathscr{X}$ or $\mathscr{Y}$ necessary?

Examples.
1. Let $\mathbf{A} \in \mathbb{M}^N(\mathbb{R})$. It follows immediately from Prop. 5.2.12 that $\forall t \geq 0$, the matrix exponential $e^{\mathbf{A}t} : \mathbb{R}^N \to \mathbb{R}^N$ is compact. (Why?)
2. Let $g_0 \in \mathbb{C}([a,b]; \mathbb{R})$. The operator $\Psi : \mathbb{C}([a,b]; \mathbb{R}) \to \mathbb{C}([a,b]; \mathbb{R})$ defined by

$$\Psi[f](x) = \int_a^x f(u) g_0(u) du, \ a \leq x \leq b,$$

is compact.

We shall encounter variants of this type of operator later in Section 5.7 and will provide a detailed verification of compactness at that time. The following result follows the approach used in [146].

Proposition 5.2.13. *Let* $\Psi \in \mathbb{B}(\mathscr{X}, \mathscr{Y})$. *If* $\{\Psi_n\}$ *is a family of compact operators such that* $\lim\limits_{n \to \infty} \|\Psi_n - \Psi\|_{\mathbb{B}} = 0$, *then* Ψ *is compact.*

Proof. We must show that for any bounded set $\mathscr{D} \subset \mathscr{X}$, $\{\Psi x \,|\, x \in \mathscr{D}\}$ is precompact in $\mathscr{Y}$. This amounts to showing that every sequence $\{x_n\} \subset \mathscr{D}$ contains a subsequence $\{x_{n_k}\}$ for which $\{\Psi(x_{n_k})\}$ converges in $\mathscr{Y}$.

We employ a diagonalization argument similar to the one used to prove the Arzela-Ascoli theorem (see [146]). Let $\{x_m\} \subset \mathscr{D}$ be such that $\sup \{\|x_m\|_{\mathscr{X}} \,|\, m \in \mathbb{N}\} = M < \infty$. Since Ψ_1 is compact, there exists a subsequence of $\{x_m\}$, denoted by $\{y_k^1\}$, for which $\{\Psi_1(y_k^1)\}$ is convergent in $\mathscr{Y}$. Similarly, the compactness of Ψ_2 guarantees the existence of a subsequence of $\{y_k^1\}$, denoted by $\{y_k^2\}$ for which $\{\Psi_2(y_k^2)\}$ is convergent in $\mathscr{Y}$. Continuing in this manner, $\forall n \in \mathbb{N}$, the compactness of Ψ_n enables us to extract a subsequence of $\{y_k^{n-1}\}$, denoted by $\{y_k^n\}$ for which $\{\Psi_n(y_k^n)\}$ is convergent in $\mathscr{Y}$. Now, consider the diagonal subsequence $\{y_n^n\}$ and note that $\sup \{\|y_n^n\|_{\mathscr{X}} \,|\, n \in \mathbb{N}\} \leq M$.

Let $\varepsilon > 0$. There exists $N_1 \in \mathbb{N}$ such that

$$n \geq N_1 \implies \|\Psi - \Psi_n\|_{\mathbb{B}} < \frac{\varepsilon}{3M+1}. \tag{5.19}$$

Since Ψ_{N_1} is compact, $\exists N_2 \in \mathbb{N}$ such that

$$m, n \geq N_2 \implies \|\Psi_{N_1}(y_n^n) - \Psi_{N_1}(y_m^m)\|_{\mathscr{Y}} < \frac{\varepsilon}{3}. \tag{5.20}$$

(Why?) Hence, $\forall m, n \geq \max \{N_1, N_2\}$,

$$\begin{aligned}
\|\Psi(y_n^n) - \Psi(y_m^m)\|_{\mathscr{Y}} &\leq \|\Psi(y_n^n) - \Psi_{N_1}(y_n^n)\|_{\mathscr{Y}} + \|\Psi_{N_1}(y_n^n) - \Psi_{N_1}(y_m^m)\|_{\mathscr{Y}} \\
&\quad + \|\Psi_{N_1}(y_m^m) - \Psi(y_m^m)\|_{\mathscr{Y}} \\
&< \varepsilon.
\end{aligned}$$

(Why?) Thus, $\{\Psi(y_n^n)\}$ is a Cauchy sequence in $\mathscr{Y}$ and hence convergent. This completes the proof. $\qquad\square$

Example. Consider the semigroup of operators $\{e^{At} \,|\, t \geq 0\}$ arising in Exer. 3.2.8 for the N-dimensional homogenous heat equation with Neumann BCs. Precisely, $\forall t \geq 0$, $e^{At} : \mathbb{L}^2\left(\prod_{i=1}^N [0, a_i]; \mathbb{R}\right) \to \mathbb{L}^2\left(\prod_{i=1}^N [0, a_i]; \mathbb{R}\right)$ is given by

$$e^{At}[f](x_1, \ldots, x_n) = \sum_{k_1=1}^{\infty} \cdots \sum_{k_N=1}^{\infty} b_{k_1} \cdots b_{k_N} e^{-kt} \sum_{i=1}^{N} \left(\frac{k_i \pi}{a_i}\right)^2 \prod_{i=1}^{N} \cos\left(\frac{k_i \pi}{a_i} x_i\right) \tag{5.21}$$

where

$$b_{k_1} \cdots b_{k_N} = \frac{2^N}{a_1 \cdots a_N} \int_0^{a_1} \cdots \int_0^{a_N} f(u_1, \ldots, u_n) \prod_{i=1}^{N} \cos\left(\frac{k_i \pi}{a_i} u_i\right) du_1 \cdots du_N. \tag{5.22}$$

We claim that $\forall t_0 > 0$, e^{At_0} is a compact operator. We proceed by induction on N. Let $N = 1$ and note that (5.21) and (5.22) simplify to

$$e^{At_0}[f](x_1) = \sum_{k_1=1}^{\infty} b_{k_1} e^{-\left(\frac{k_1\pi}{a_1}\right)^2 kt_0} \cos\left(\frac{k_1\pi}{a_1}x_1\right)$$

$$= \lim_{M \to \infty} \sum_{k_1=1}^{M} b_{k_1} e^{-\left(\frac{k_1\pi}{a_1}\right)^2 kt_0} \cos\left(\frac{k_1\pi}{a_1}x_1\right) \quad (5.23)$$

$$= \lim_{M \to \infty} e_M^{At_0}[f](x_1),$$

where $b_{k_1} = \left(\frac{2}{a_1} \int_0^{a_1} f(u_1) \cos\left(\frac{k_1\pi}{a_1}u_1\right) du_1\right)$.

The linearity of e^{At_0} is easily verified. (Show this.) To argue boundedness, we first note that applying Holder's inequality yields

$$|b_{k_1}| \le \frac{2}{a_1} \|f\|_{\mathbb{L}^2(0,a_1;\mathbb{R})} \left[\int_0^{a_1} \cos^2\left(\frac{k_1\pi}{a_1}u_1\right) du_1\right]^{1/2} = \sqrt{\frac{2}{a_1}} \|f\|_{\mathbb{L}^2(0,a_1;\mathbb{R})}.$$

Using this fact together with another application of Holder's inequality yields

$$\left\|e_M^{At_0}[f]\right\|_{\mathbb{L}^2(0,a_1;\mathbb{R})}^2 \le \|f\|_{\mathbb{L}^2(0,a_1;\mathbb{R})}^2 \int_0^{a_1} \left[\sum_{k_1=1}^{M} \frac{2}{a_1} e^{-\left(\frac{k_1\pi}{a_1}\right)^2 kt_0} \cos\left(\frac{k_1\pi}{a_1}x_1\right)\right]^2 dx_1$$

$$\le C\|f\|_{\mathbb{L}^2(0,a_1;\mathbb{R})}^2, \quad (5.24)$$

for some $C > 0$ (independent of f and x_1). Thus, $e_M^{At_0} \in \mathbb{B}\left(\mathbb{L}^2(0,a_1;\mathbb{R})\right)$. Consequently, $\forall f \in \mathbb{L}^2(0,a_1;\mathbb{R})$, rng $\left(e_M^{At_0}[f]\right)$ is finite-dimensional. So, Prop. 5.2.12(ii) implies that $e_M^{At_0}$ is compact. In light of this fact and (5.23), Prop. 5.2.13 implies that e^{At_0} is compact. This proves the assertion for $N = 1$. Now, complete the inductive step as an exercise. □

Exercise 5.2.5. Verify (5.24) and complete the inductive step above.

The proof of the following fixed-point theorem, commonly referred to as both *Schaefer's fixed-point theorem* and the *Leray-Schauder alternative* was established in **[285]**.

Theorem 5.2.14. Schaefer's Fixed-Point Theorem
Let $\Psi : \mathscr{X} \to \mathscr{X}$ be a continuous, compact operator and let

$$\xi(\Psi) = \{x \in \mathscr{X} \mid \exists \lambda \ge 1 \text{ such that } \lambda x = \Psi x\}. \quad (5.25)$$

If $\xi(\Psi)$ is bounded, then Ψ has at least one fixed-point.

We shall encounter one other fixed-point theorem, namely *Krasnoselskii's fixed-point theorem*, in Chapter 7 to handle a very particular situation. We postpone a discussion of this theorem until that time.

5.2.1.4   The Fixed-Point Approach

A standard technique used to establish existence theorems of the type in which we are interested is commonly known as the *fixed-point approach*. Heuristically speaking, the heart of this approach lies in defining a mapping $\Phi : \mathscr{Y} \to \mathscr{Y}$ using the foresight that a fixed-point of Φ is a solution (of some type) of an underlying IVP of interest. The choice of the space $\mathscr{Y}$ is essential to obtaining the level of regularity that we wish the solution to possess.

Once we define Φ, we need to verify that

i.) Φ is indeed a well-defined mapping; and

ii.) the conditions of the fixed-point theorem to be applied are satisfied.

If we are successful, the conclusion of the argument will be the existence of a fixed-point which coincides with (or is at least related to in a significant way) a suitable solution of our IVP. Depending on the fixed-point theorem used, multiple fixed-points could exist, thereby producing more than one solution of the IVP.

5.2.2   A Handful of Integral Inequalities

Establishing *a priori* estimates is a crucial step in the proofs of most existence results. Such estimates often take the form of an upper bound of the state process in some function space. We begin with the following classical inequality from which many others are derived, proved as in **[97]**.

Theorem 5.2.15. Gronwall's Lemma
Let $t_0 \in (-\infty, T)$, $x \in \mathbb{C}\left([t_0, T]; \mathbb{R}\right)$, $K \in \mathbb{C}\left([t_0, T]; [0, \infty)\right)$, and M be a real constant. If

$$x(t) \leq M + \int_{t_0}^{t} K(s)x(s)ds, \ t_0 \leq t \leq T, \tag{5.26}$$

then

$$x(t) \leq M e^{\int_{t_0}^{t} K(s)ds}, \ t_0 \leq t \leq T. \tag{5.27}$$

Proof. Define $y : [t_0, T] \to \mathbb{R}$ by

$$y(t) = M + \int_{t_0}^{t} K(s)x(s)ds.$$

Observe that $y \in \mathbb{C}^1\left([t_0, T]; \mathbb{R}\right)$ and satisfies the IVP

$$\begin{cases} y'(t) & = K(t)x(t), t_0 \leq t \leq T, \\ y(t_0) & = M. \end{cases} \tag{5.28}$$

(Tell why.) By assumption, $x(t) \leq y(t)$, $\forall t_0 \leq t \leq T$. Multiplying both sides of this inequality by $K(t)$ and then substituting into (5.28) yields

$$\begin{cases} y'(t) \leq K(t)y(t), t_0 \leq t \leq T, \\ y(t_0) = M. \end{cases} \tag{5.29}$$

As such,

$$y'(t) - K(t)y(t) \leq 0, t_0 \leq t \leq T,$$

so that multiplying both sides by $e^{-\int_{t_0}^{t} K(s)ds}$ yields

$$\frac{d}{dt} \left[e^{-\int_{t_0}^{t} K(s)ds} y(t) \right] \leq 0, t_0 \leq t \leq T. \tag{5.30}$$

(Why?) Consequently,

$$e^{-\int_{t_0}^{t} K(s)ds} y(t) \leq y(t_0) = M, t_0 \leq t \leq T,$$

and so

$$y(t) \leq M e^{\int_{t_0}^{t} K(s)ds}, t_0 \leq t \leq T. \tag{5.31}$$

Since $x(t) \leq y(t)$, $\forall t_0 \leq t \leq T$, the conclusion follows from (5.31). $\square$

This form of Gronwall's lemma applies only for $t \geq t_0$, but we might need an estimate that holds for $t < t_0$. It is not difficult to establish such an estimate, as suggested by the following corollary.

Corollary 5.2.16. *Let* $t_0 \in \mathbb{R}, x \in \mathbb{C}((-\infty, t_0]; \mathbb{R}), K \in \mathbb{C}((-\infty, t_0]; [0, \infty))$, *and M be a real constant. If*

$$x(t) \leq M + \int_{t}^{t_0} K(s)x(s)ds, -\infty < t \leq t_0, \tag{5.32}$$

then

$$x(t) \leq M e^{\int_{t}^{t_0} K(s)ds}, -\infty < t \leq t_0. \tag{5.33}$$

Proof. Let $t \in (-\infty, t_0]$ and substitute $t = t_0 - s$ into (5.32) to obtain

$$x(t_0 - s) \leq M + \int_{t_0-s}^{t_0} K(\tau)x(\tau)d\tau, s > 0. \tag{5.34}$$

Implementing the change of variable $\tau = t_0 - \xi$ in (5.34) yields

$$x(t_0 - s) \leq M + \int_{0}^{s} K(t_0 - \xi)x(t_0 - \xi)d\xi, s > 0. \tag{5.35}$$

(Tell why.) Now, let $\overline{x}(s) = x(t_0 - s)$ and $\overline{K}(s) = K(t_0 - s)$ in (5.35) to obtain

$$\overline{x}(s) \leq M + \int_{0}^{s} \overline{K}(\xi)\overline{x}(\xi)d\xi, s > 0.$$

Applying Thrm. 5.2.15 then yields the estimate

$$\overline{x}(s) \leq M e^{\int_{0}^{s} \overline{K}(\xi)d\xi}, s > 0. \tag{5.36}$$

Going back to the original variable then results in (5.33). $\square$

While these integral inequalities are veritable workhorses in practice, it might not be possible (or feasible) to verify the hypotheses of the lemma. One natural question is what to do if the constant M were allowed to vary with t. In such case, the resulting estimate would depend on the regularity of M. One such result is as follows.

Proposition 5.2.17. *Let $t_0 \in (-\infty, T)$, $x \in C([t_0, T]; \mathbb{R})$, $K \in C([t_0, T]; [0, \infty))$, and $M : [t_0, T] \to [0, \infty)$. Assume that*

$$x(t) \le M(t) + \int_{t_0}^{t} K(s)x(s)ds, \ t_0 \le t \le T. \tag{5.37}$$

i.) *If $M \in C([t_0, T]; [0, \infty))$, then*

$$x(t) \le M(t) + \int_{t_0}^{t} M(s)K(s)e^{\int_s^t K(\tau)d\tau}ds, \ t_0 \le t \le T. \tag{5.38}$$

ii.) *If $M \in C^1((t_0, T); [0, \infty))$, then*

$$x(t) \le e^{\int_{t_0}^t K(s)ds}\left[M(t_0) + \int_{t_0}^{t} M'(s)e^{-\int_{t_0}^s K(\tau)d\tau}ds\right], \ t_0 \le t \le T. \tag{5.39}$$

Exercise 5.2.6. Prove Prop. 5.2.17.

A rich source of such inequalities is the text [261]. The following inequality, stated without proof, is useful when imposing more general growth conditions on the forcing term.

Proposition 5.2.18. *Let $z, a, b, m \in C([t_0, T]; [0, \infty))$ and $p \ge 1$. If*

$$z(t) \le a(t) + b(t)\left[\int_0^t m(s)z^p(s)ds\right]^{1/p}, \ 0 \le t \le T, \tag{5.40}$$

then

$$z(t) \le a(t) + b(t)\left(\frac{\left[\int_0^t m(s)\varsigma(s)a^p(s)ds\right]^{1/p}}{\left[1 - (1 - \varsigma(t))^{1/p}\right]}\right), \ 0 \le t \le T, \tag{5.41}$$

where $\varsigma(t) = e^{-\int_0^t m(s)b^p(s)ds}$.

One more useful integral inequality, also proved in [261], is as follows.

Proposition 5.2.19. *Let $w, \psi_1, \psi_2,$ and $\psi_3 \in C([0, \infty); [0, \infty))$ and $w_0 \ge 0$. If, $\forall t > 0$,*

$$w(t) \le w_0 + \int_0^t \psi_1(s)w(s)ds + \int_0^t \psi_1(s)\left(\int_0^s \psi_2(\tau)w(\tau)d\tau\right)ds$$
$$+ \int_0^t \psi_1(s)\left(\int_0^s \psi_2(\tau)\left(\int_0^\tau \psi_3(\theta)w(\theta)d\theta\right)d\tau\right)ds, \tag{5.42}$$

then $\forall t > 0$,

$$w(t) \leq w_0 \left[1 + \int_0^t \left\{ \psi_1(s) e^{\int_0^s \psi_1(\tau)d\tau} \left(1 + \right. \right. \right.$$
$$\left. \left. \left. \int_0^s \psi_2(\tau) e^{\int_0^\tau [\psi_2(\theta) + \psi_3(\theta)]d\theta} d\tau \right) \right\} ds \right]. \tag{5.43}$$

5.2.3 Frechet Differentiability

Various notions of "the derivative" can be formulated depending on the application being investigated and the underlying topology used to conduct the investigation. The variant that we now introduce is the so-called *Frechet derivative,* a natural extension of the usual notion of differentiability of functions $\mathbf{f} : \mathbb{R}^N \to \mathbb{R}^N$. Precisely, we have

Definition 5.2.20. Let $\mathscr{X}$ and $\mathscr{Y}$ be Banach spaces. The mapping $F : \mathscr{X} \to \mathscr{Y}$ is *(Frechet) differentiable* at $x_0 \in \mathscr{X}$ if there exists a linear operator $F'(x_0) : \mathscr{X} \to \mathscr{Y}$ such that

$$\underbrace{\overbrace{F(x_0 + \triangle x)}^{\text{in } \mathscr{X}} - F(x_0)}_{\text{in } \mathscr{Y}} = \underbrace{F'(x_0)}_{\text{mapping}} \overbrace{(\triangle x)}^{\text{applied to} \triangle x} + \underbrace{\varepsilon(x_0, \triangle x)}_{\text{Error term in } \mathscr{Y}}$$

where $\displaystyle \lim_{\|\triangle x\|_{\mathscr{X}} \to 0} \frac{\|\varepsilon(x_0, \triangle x)\|_{\mathscr{Y}}}{\|\triangle x\|_{\mathscr{X}}} = 0$.

Remarks.
1. A useful estimate: Let $\eta > 0$. There exists $\delta > 0$ such that

$$0 < \|\triangle x\|_{\mathscr{X}} < \delta \implies \|\varepsilon(x_0, \triangle x)\|_{\mathscr{Y}} < \eta \|\triangle x\|_{\mathscr{X}}$$

(Why?) and so,

$$\left\| (F(x_0 + \triangle x) - F(x_0)) - F'(x_0)(\triangle x) \right\|_{\mathscr{Y}} < \eta \|\triangle x\|_{\mathscr{X}}. \tag{5.44}$$

2. We shall use the notation $\frac{\partial f}{\partial x}(x_0)$ interchangeably with $F'(x_0)$. When it is not confusing to do so, we omit the prefix "Frechet" when referring to this type of derivative.

Proposition 5.2.21. Properties of $F'(x_0)$
Let $\mathscr{X}, \mathscr{Y}, \mathscr{Z}$ be Banach spaces and $F : \mathscr{X} \to \mathscr{Y}$ and $G : \mathscr{Y} \to \mathscr{Z}$ given mappings.
i.) *If $F'(x_0)$ exists, then it is unique.*
ii.) (**Chain Rule**) *If G is differentiable at x_0 and F is differentiable at $G(x_0)$, then $F \circ G$ is differentiable at x_0 and*

$$(F \circ G)'(x_0) = F'(G(x_0)) G'(x_0).$$

iii.) *If F is strongly continuous, then $F'(x_0) \in \mathbb{B}(\mathscr{X}, \mathscr{Y})$.*

Proof. We prove (iii) and refer you to **[205]** for the proofs of the others. We must argue that

$$\sup_{\triangle x \in \mathscr{X}} \frac{\|F'(x_0)(\triangle x)\|_{\mathscr{Y}}}{\|\triangle x\|_{\mathscr{X}}} < \infty. \tag{5.45}$$

Let $\varepsilon > 0$. There exists $\delta_1 > 0$ for which (5.43) holds for $\eta = \frac{\varepsilon}{2}$. Also, since F is continuous at x_0, $\exists 0 < \delta_2 < 1$ such that

$$\|\triangle x\|_{\mathscr{X}} < \delta_2 \implies \|F(x_0 + \triangle x) - F(x_0)\|_{\mathscr{Y}} < \frac{\varepsilon}{2}. \tag{5.46}$$

Let $\delta = \min\{\delta_1, \delta_2\}$. For $\triangle x$ such that $0 < \|\triangle x\|_{\mathscr{X}} < \delta$,

$$\left\|(F(x_0 + \triangle x) - F(x_0)) - F'(x_0)(\triangle x)\right\|_{\mathscr{Y}} < \frac{\varepsilon}{2} \|\triangle x\|_{\mathscr{X}} < \frac{\varepsilon}{2}. \tag{5.47}$$

As such,

$$\|\triangle x\|_{\mathscr{X}} < \delta \implies \left\|F'(x_0)(\triangle x)\right\|_{\mathscr{Y}} < \varepsilon. \tag{5.48}$$

(Why?) So, $\forall x^\star \neq 0$ in $\mathscr{X}$,

$$\frac{\delta}{2\|x^\star\|_{\mathscr{X}}} \left\|F'(x_0)(x^\star)\right\|_{\mathscr{Y}} \leq \left\|F'(x_0)\left(\frac{x^\star}{\|x^\star\|_{\mathscr{X}}} \cdot \frac{\delta}{2}\right)\right\|_{\mathscr{Y}} < \varepsilon. \tag{5.49}$$

Hence, $\forall x^\star \in \mathscr{X}$,

$$\left\|F'(x_0)(x^\star)\right\|_{\mathscr{Y}} < \frac{2\varepsilon}{\delta} \|x^\star\|_{\mathscr{X}},$$

from which we conclude that $F'(x_0) \in \mathbb{B}(\mathscr{X}, \mathscr{Y})$. (Why?) $\square$

Exercise 5.2.7. Justify (5.49).

5.3   Some Essential Preliminary Considerations

We shall take a slight departure from our usual tack in that we will not develop the entire theory first for the finite-dimensional case and then for the case of a general Banach space. Rather, we shall explore certain special cases as the need arises to spark our intuition as to what a concept or result "ought to be." Of course, upon completion of the development of our theory, we will effortlessly recover the results for finite-dimensional ODEs as a special case, at times under weaker hypotheses. In the latter case, we will explore various improvements of the theory by critically analyzing the proof in order to identify where, and how, the hypotheses can be weakened. Consider the abstract IVP (5.10).

Exercise 5.3.1. Before moving on, try to formulate natural definitions of a mild and classical solution of (5.10).

Suppose that $u : [0,\infty) \to \mathscr{X}$ satisfies (5.10) in a Banach space $\mathscr{X}$, assuming whatever level of regularity seems necessary to render (5.10) meaningful. Proceeding as in our development of (4.7), we can derive the variation of parameters formula

$$u(t) = e^{At}u_0 + \int_0^t e^{A(t-s)} f(s, u(s)) ds. \qquad (5.50)$$

(Do so!) The new struggle we face is the self-referential nature of (5.50). In essence, this is simply another equation to solve, albeit one of a different type. We have simply managed to replace the solvability of (5.10) by the solvability of (5.50), which is hopefully easier.

When does (5.50) have a solution? In Chapter 4, we merely needed the right-side of the variation of parameters formula to be continuous and to belong to a suitable Banach space. But, (5.50) is more complicated, since the forcing term f now changes according to a second variable. Solvability then naturally boils down to the behavior of the mapping $t \mapsto \int_0^t e^{A(t-s)} f(s, u(s)) ds$.

Exercise 5.3.2. Conjecture a sufficient condition on f to ensure that (5.50) is a classical solution of (5.10).

The dependence of the forcing term f on the state process u opens the door to possibilities that did not arise in Chapter 4. Indeed, we encounter new issues, some of which are explored in the following exercise.

Exercise 5.3.3. For simplicity, consider (5.10) with $A = 0$ and $\mathscr{X} = \mathbb{R}$.
i.) Show that the following IVP has more than one solution:

$$\begin{cases} x'(t) = 3[x(t)]^{5/8}, t > 0, \\ x(0) = 0. \end{cases} \qquad (5.51)$$

(In fact, this IVP has infinitely many solutions!)
ii.) Consider the IVP

$$\begin{cases} x'(t) = (1 + 2x(t))^4, t > 0, \\ x(0) = x_0. \end{cases} \qquad (5.52)$$

a.) Show that a solution of (5.52) is given by

$$x(t) = \frac{1}{2}\left[-1 + \left((2x_0 + 1)^{-3} - 6t\right)^{-1/3}\right].$$

b.) Let $T = \frac{1}{6}(2x_0 + 1)^{-3}$. Assuming that $x_0 > -\frac{1}{2}$, show that $\lim_{t \to T^-} |x(t)| = \infty$. As such, (5.52) does not have a continuous solution on $[0,\infty)$.

Note that the right-side $f(t,x(t))$ of both IVPs in Exer. 5.3.3 is a continuous function of both variables. This was sufficient to guarantee the existence and uniqueness of a mild solution of (4.21) on $[0,\infty)$ when $f(t,x(t)) = f(t)$. However, this is false for (5.51), and in a big way for (5.52). As such, we are faced with several questions right off the bat:

    i.) When does (5.50) have a unique solution on a given interval?
    ii.) Under what conditions is a mild solution also a classical solution?
    iii.) What is the largest interval on which a mild solution exists?

We shall explore these questions and along the way develop various strategies of attack that will be used throughout the remainder of the text.

5.4   Growth Conditions

Exercise 5.3.3 revealed that mere continuity of f in both variables is insufficient to guarantee uniqueness of a mild solution of (5.10) on $[0,\infty)$, even when $\mathscr{X} = \mathbb{R}$. Even worse, such continuity does not guarantee the existence of a mild solution when $\mathscr{X}$ is general Hilbert space. (See **[267]** for an example.) As such, it is sensible to ask what conditions could be coupled with continuity to ensure the existence (and possibly uniqueness) of a mild solution of (5.10) on at least some interval $[0,T_0)$. There is a plentiful supply of such conditions that can be imposed on f which further control its "growth." We introduce several common ones in this section and investigate how they are interrelated.

At the restrictive end of the spectrum, we can require f to have a smooth derivative on its domain, say $f \in \mathbb{C}^1 (\text{dom}(f); \mathscr{X})$. In such case, it would follow that for a given continuous function $u \in \mathbb{C}([0,T_0); \mathscr{X})$,

$$\int_0^t e^{A(t-s)} f(s,u(s))ds \in \mathbb{C}^1 ([0,T_0); \mathscr{X}). \tag{5.53}$$

Hence, the function $u(\cdot)$ given by (5.50) would be differentiable, and so has a chance at being a classical solution of (5.10) (assuming that u_0 belongs to a suitable space).

Exercise 5.4.1. Verify (5.53).

While imposing this level of regularity leads to a desirable conclusion, it prevents us from considering many forcing terms that arise naturally in practice (e.g., $f(u) = |u|$). A close investigation of (5.51) reveals that the curve corresponding to the forcing term was sufficiently steep in a vicinity of (0,0) as to enable us to construct a sequence of chord lines, all passing through (0,0), whose slopes became infinitely large.

Exercise 5.4.2. Show that the sequence of chord line slopes for $f(x) = 3x^{5/8}$ connecting $(0,0)$ to $\left(\frac{1}{n}, f\left(\frac{1}{n}\right)\right)$ approaches infinity as $n \to \infty$.

As such, close by to the initial starting point, the behavior of f changes very quickly, and this in turn affects the behavior of x' in a short interval of time. Moreover, this worsens the closer you get to the origin. Thus, if we were to try to generate the solution path on a given time interval $[0, T]$ numerically, refining the partition of $[0, T]$ (in order to increase the number of time points used to construct the approximate solution) would subsequently result in a sequence of paths which does not approach a single recognizable continuous curve.

The presence of the cusp in the graph is the troublemaker! Certainly, continuously differentiable functions cannot exhibit such behavior, but can we somehow control the chord line slopes without demanding that f be so nice? The search for such control over chord line slopes prompts us to make the following definition.

Definition 5.4.1. A function $f : \mathscr{X} \to \mathscr{Y}$ is *globally Lipschitz on* $\mathscr{D} \subset \mathscr{X}$ if $\exists M_f > 0$ such that

$$\|f(x) - f(y)\|_{\mathscr{Y}} \le M_f \|x - y\|_{\mathscr{X}}, \quad \forall x, y \in \mathscr{D}. \tag{5.54}$$

(M_f is called a *Lipschitz constant* for f.)

This definition is easily adapted to functions of more than one independent variable, but we must carefully indicate to which of the independent variables we intend the condition to apply. Functions of the form $g : [0, T] \times \mathscr{X} \to \mathscr{Y}$ commonly arise in practice. We introduce the following modification of Def. 5.4.1 as it applies to such functions.

Definition 5.4.2. A function $g : [0, T] \times \mathscr{X} \to \mathscr{Y}$ is *globally Lipschitz on* $\mathscr{D} \subset \mathscr{X}$ *(uniformly in t)* if $\exists M_g > 0$ (independent of t) such that

$$\|g(t, x) - g(t, y)\|_{\mathscr{Y}} \le M_g \|x - y\|_{\mathscr{X}}, \quad \forall t \in [0, T] \text{ and } x, y \in \mathscr{D}. \tag{5.55}$$

Exercise 5.4.3. Interpret (5.55) geometrically. For simplicity, assume that $\mathscr{X} = \mathscr{Y} = \mathbb{R}$. How does this interpretation change if M_g depends on t.

The space $\mathscr{X}$ could be a product space $\mathscr{X}_1 \times \cdots \times \mathscr{X}_n$. Assuming that it is equipped with the usual norm (cf. (3.5)), (5.55) becomes

$$\|g(t, x_1, \ldots, x_n) - g(t, y_1, \ldots, y_n)\|_{\mathscr{Y}} \le M_g \sum_{i=1}^{n} \|x_i - y_i\|_{\mathscr{X}}, \tag{5.56}$$

$\forall t \in [0, T]$ and $(x_1, \ldots, x_n), (y_1, \ldots, y_n) \in \mathscr{D} \subset \mathscr{X}_1 \times \cdots \times \mathscr{X}_n$.

Exercise 5.4.4. Let $f : \mathscr{X} \to \mathscr{Y}$ be given.
i.) If f is Frechet differentiable on $\mathscr{X}$, must f be globally Lipschitz on $\mathscr{X}$? Why or why not?

ii.) Show that the converse of (i) is false.

Exercise 5.4.5. Let $f \in \mathbf{L}^1 (a,b;\mathbb{R})$ and define $g : [a,b] \to \mathbb{R}$ by $g(x) = \int_a^x f(z)dz$.
i.) Is g globally Lipschitz on (a,b)? If not, try to identify the least amount of additional regularity that could be imposed on f to ensure that g is globally Lipschitz.
ii.) Does your response in (i) change if $\mathbb{R}$ is replaced by a Banach space $\mathscr{X}$?

Exercise 5.4.6. Suppose $f : [0,T] \times \mathbb{R} \to \mathbb{R}$ is continuous and $v \in \mathbb{C}^1 ((0,\infty);(0,\infty))$.
Define $h : [0,T] \times (0,\infty) \to \mathbb{R}$ by $h(t,x) = \int_0^{v(x)} f(t,z)dz$.
i.) Is h globally Lipschitz on $(0,\infty)$ (uniformly in t)? If not, try to identify the least amount of additional regularity that could be imposed on f and v to ensure that h is globally Lipschitz.
ii.) Let $k \in \mathbb{C}^1 (\mathbb{R};\mathbb{R})$ and define $h_k : [0,\infty) \times \mathbb{R} \to \mathbb{R}$ by $h_k(t,x) = \int_0^{v(x)} f(t,k(z))dz$. Is h_k globally Lipschitz on $[0,\infty)$ (uniformly in t)?

Exercise 5.4.7. Let $f : [0,T] \times \mathbb{R} \times \mathbb{R} \to \mathbb{R}$ and $g : \mathbb{R} \to \mathbb{R}$ be given mappings. Define $j : [0,T] \times \mathbb{R} \to \mathbb{R}$ by $j(t,x) = f\left(t,x,\int_0^x g(z)dz\right)$. Provide sufficient conditions on f and g that ensure that j is globally Lipschitz on $\mathbb{R}$ (uniformly in t).

Exercise 5.4.8. Must a finite linear combination of functions $f_i : \mathscr{D} \subset \mathscr{X} \to \mathscr{Y}$, $1 \leq i \leq n$, which are globally Lipschitz on $\mathscr{D}$ also be globally Lipschitz on $\mathscr{D}$?

Of course, imposing such a Lipschitz condition on a function f over an *entire space* like $[0,T] \times \mathbb{R}$ is still restrictive since the same Lipschitz constant is used throughout the space, which essentially demands that f grow no faster than a linear function on this space. This eliminates many functions from consideration, including relatively tame examples such as $f(t,x) = e^x$ or $f(t,x) = tx^2$. (Why?) Perhaps we can weaken the condition slightly so that rather than on the whole space, we can demand that the function be Lipschitz on any closed ball contained within in, with the caveat that the Lipschitz constant can change from ball to ball. This suggests the following localized version of Def. 5.4.1.

Definition 5.4.3. A function $f : \mathscr{X} \to \mathscr{Y}$ is *locally Lipschitz on $\mathscr{X}$* if $\forall x_0 \in \mathscr{X}$ and $\varepsilon > 0$, $\exists$ a constant $M_{(x_0,\varepsilon)} > 0$ (depending on x_0 and ε) such that

$$\|f(x) - f(y)\|_{\mathscr{Y}} \leq M_{(x_0,\varepsilon)} \|x - y\|_{\mathscr{X}}, \ \forall x,y \in \mathscr{B}_{\mathscr{X}} (x_0;\varepsilon). \qquad (5.57)$$

Exercise 5.4.9. Formulate local versions of Def. 5.4.2 and (5.56) in the spirit of Def. 5.4.3.

Exercise 5.4.10. Must all continuous real-valued functions $f : \mathbb{R} \to \mathbb{R}$ be locally Lipschitz on $\mathbb{R}$? Explain.

The inequalities (5.55) and (5.57) used in Def. 5.4.1 and Def. 5.4.3, respectively, can be generalized in various ways, two of which are:

$$\|f(t,x) - f(t,y)\|_{\mathscr{Y}} \le k(t) \|x - y\|_{\mathscr{X}}, \; \forall t \in [0,T] \text{ and } x, y \in \mathscr{X}, \quad (5.58)$$

$$\|f(t,x) - f(t,y)\|_{\mathscr{Y}} \le k(t) \|x - y\|_{\mathscr{X}}^{p}, \; \forall t \in [0,T] \text{ and } x, y \in \mathscr{X}, \quad (5.59)$$

where $p > 1$ and k typically belongs to either $\mathbb{L}^1(0,T;\mathbb{R})$ or $\mathbb{C}([0,T];(0,\infty))$.

Exercise 5.4.11. If $f : [0,T] \times \mathscr{X} \to \mathscr{Y}$ satisfies Def. 5.4.2, must it satisfy either (5.58) or (5.59) for some $k \in \mathbb{C}([0,T];\mathbb{R})$? What if $k \in \mathbb{L}^1(0,T;\mathbb{R})$? How about the converse implications?

Exercise 5.4.12. If $f : \mathscr{X} \to \mathscr{Y}$ satisfies Def. 5.4.3, must it satisfy either (5.58) or (5.59) if $k \in \mathbb{C}([0,T];\mathbb{R})$? What if $k \in \mathbb{L}^1(0,T;\mathbb{R})$? How about the converse implications?

We have merely scraped the tip of the iceberg. More peculiar-looking growth conditions are also used in practice (see **[27, 28, 50, 97, 161, 257, 258]**). The end goal in all cases is to control the behavior of the forcing term as minimally as possible while maximizing the level of regularity exhibited by the mild solution of the IVP. We could continue exploring successively weaker versions of these growth conditions, but we point out that the further we stray away from the assumption that $f \in \mathbb{C}^1$, the less likely it becomes that we will be able to establish the existence of a unique mild solution, not to mention a classical solution. For this reason, the tendency is to focus more broadly on the entire term $\int_0^t e^{A(t-s)} f(s,u(s)) ds$. After all, we can exploit the "smoothing property" of the integral, as well as the presence of the members of the semigroup in the integral. Regarding the latter, a seesaw interaction between the properties of $\{e^{At} : t \ge 0\}$ and those of $f(t,u(t))$ comes into play.

Up to this point, we have assumed nothing beyond strong continuity of the semigroup. But, we can impose additional restrictions on $\{e^{At} : t \ge 0\}$ (and hence on the operators A to which the theorem will be applicable) to help counterbalance a less well-behaved forcing term f. It is important to make certain that the restrictions are realistic in the sense that the theoretical results will be applicable to concrete IB-VPs. We explore the growth conditions used in one such approach below, and will investigate more thoroughly in Section 5.7.

We now introduce some weaker growth conditions that can be imposed on a continuous forcing term when the semigroup possesses stronger properties. The following assumptions are amongst the more commonly-used conditions:

  i.) f is a compact mapping (cf. Def. 5.2.11), or
  ii.) f is monotone (cf. Def. 3.5.10), or
  iii.) f satisfies a sublinear growth condition of the following type:

Definition 5.4.4. A function $f : [0,T] \times \mathscr{X} \to \mathscr{Y}$ has *sublinear growth (uniformly in t)* if $\exists M_1, M_2 > 0$ such that

$$\|f(t,x)\|_{\mathscr{Y}} \le M_1 \|x\|_{\mathscr{X}} + M_2, \; \forall t \in [0,T] \text{ and } x \in \mathscr{X}. \quad (5.60)$$

More generally, (5.60) can be replaced by one of the following:

$$\|f(t,x)\|_{\mathscr{Y}} \le M_1(t)\|x\|_{\mathscr{X}} + M_2(t), \; \forall t \in [0,T] \text{ and } x \in \mathscr{X}, \tag{5.61}$$

$$\|f(t,x)\|_{\mathscr{Y}} \le M_1(t)\|x\|_{\mathscr{X}}^p + M_2(t), \; \forall t \in [0,T] \text{ and } x \in \mathscr{X}, \tag{5.62}$$

where $p > 1$ and $M_1(\cdot), M_2(\cdot)$ are typically assumed to belong to either $\mathbb{L}^q(0,T;\mathbb{R})$ (where $\frac{1}{p} + \frac{1}{q} = 1$) or $\mathbb{C}([0,T];\mathbb{R})$.

Exercise 5.4.13. Define $\mathbf{f} : \mathbb{R}^N \to \mathbb{R}^N$ by $\mathbf{f(x)} = \mathbf{Ax} + \mathbf{B}$, where $\mathbf{A} \in \mathbb{M}^N(\mathbb{R})$ and $\mathbf{B} \in \mathbb{R}^N$.
i.) Is $\mathbf{f}$ locally Lipschitz?
ii.) Is $\mathbf{f}$ a compact mapping?

Exercise 5.4.14. If $f : \mathscr{X} \to \mathscr{Y}$ is globally Lipschitz, must it be compact?

Exercise 5.4.15. Suppose that $f : [0,T] \times \mathscr{X} \to \mathscr{Y}$ has sublinear growth in the sense of one of (5.60), (5.61), or (5.62), where $M_1, M_2 \in \mathbb{C}([0,T];\mathbb{R})$. Let $\mathscr{D} \subset \mathscr{X}$ be a bounded set.
i.) Show that the image $f([0,T] \times \mathscr{D})$ is a bounded subset of $\mathscr{Y}$.
ii.) Must the image $f([0,T] \times \mathscr{D})$ also be precompact in $\mathscr{Y}$? Explain.

The growth conditions discussed in this section, when coupled with the correct technique, can be used to formulate a rich existence theory for (5.10). Unless otherwise specified, we impose the following standing assumption:

$(\mathbf{H_A})$ $A : \text{dom}(A) \subset \mathscr{X} \to \mathscr{X}$ generates a C_0-semigroup $\{e^{At} : t \ge 0\}$ on $\mathscr{X}$ for which $\exists M^\star > 0$ and $\alpha \in \mathbb{R}$ such that

$$\left\|e^{tA}\right\|_{\mathbb{B}(\mathscr{X})} \le M^\star e^{\alpha t}, \; \forall t \ge 0. \tag{5.63}$$

In particular, $\forall T > 0$, the Principle of Uniform Boundedness ensures that

$$\overline{M_A} = \sup_{0 \le t \le T} \left\|e^{At}\right\|_{\mathbb{B}(\mathscr{X})} < \infty. \tag{5.64}$$

5.5 Theory for Lipschitz-Type Forcing Terms

We begin our discussion with the nicest case. Standard references for the material developed in this section include **[86, 97, 237, 267, 278, 318, 341]**.

5.5.1 Existence and Uniqueness Results

We begin by precisely defining the notions of a mild and classical solution of (5.10). We do so on a subinterval $[0, T_0] \subset [0, T]$ because, in general, global existence is not guaranteed *a priori* (cf. Exer. 5.3.3).

Definition 5.5.1. A function $u : [0, T_0] \to \mathscr{X}$ is a
i.) *classical solution* of (5.10) on $[0, T_0]$ if u satisfies Def. 4.2.1(i) with T replaced by T_0 and $f(t)$ replaced by $f(t, u(t))$;
ii.) *mild solution* of (5.10) on $[0, T_0]$ if u satisfies Def. 4.2.1(ii) with with T replaced by T_0 and $f(t)$ replaced by $f(t, u(t))$.

Prior to developing the theory, we introduce two strategies of attack.

Approach 1: A Typical Contraction Argument

With the Fixed-Point Approach in mind, we use (5.11) to define the solution map $\Phi : \mathbb{C}([0, T_0]; \mathscr{X}) \to \mathbb{C}([0, T_0]; \mathscr{X})$ by

$$(\Phi u)(t) = e^{At} u_0 + \int_0^t e^{A(t-s)} f(s, u(s)) ds. \tag{5.65}$$

Exercise 5.5.1. Assume that f is continuous in the first variable and globally Lipschitz in the second (uniformly in t).
i.) Why is (5.65) a well-defined mapping?
ii.) Φ is not automatically a contraction, in general. What condition(s) could be imposed to render it one?
iii.) Alternatively, we can possibly avoid imposing restrictions of the type suggested in (ii) by considering iterates of Φ. Specifically,
a.) Determine a positive constant ξ_2 such that

$$\left\| \Phi^2 x - \Phi^2 y \right\|_{\mathbb{C}} \leq \xi_2 \left\| x - y \right\|_{\mathbb{C}}, \ \forall x, y \in \mathbb{C}([0, T_0]; \mathscr{X}).$$

b.) Similarly, determine a positive constant ξ_3 such that

$$\left\| \Phi^3 x - \Phi^3 y \right\|_{\mathbb{C}} \leq \xi_3 \left\| x - y \right\|_{\mathbb{C}}, \ \forall x, y \in \mathbb{C}([0, T_0]; \mathscr{X}).$$

c.) For every $n \in \mathbb{N}$, determine a positive constant ξ_n such that

$$\left\| \Phi^n x - \Phi^n y \right\|_{\mathbb{C}} \leq \xi_n \left\| x - y \right\|_{\mathbb{C}}, \ \forall x, y \in \mathbb{C}([0, T_0]; \mathscr{X}). \tag{5.66}$$

d.) Compute $\lim_{n \to \infty} \xi_n$. Why is this useful?

Upon completion of Exer. 5.5.1, we can conclude from Cor. 5.2.3 that (5.10) has a unique mild solution on $[0, T_0)$. (Why?) In fact, since there is no need to restrict the value of T_0 in such an argument, the resulting solution would be globally defined on $[0, T]$.

Since this technique merely establishes that the fixed-point u of Φ (which is the mild solution of (5.10)) is in $\mathbb{C}([0, T_0]; \mathscr{X})$, arguing that u is, in fact, a classical solution of (5.10) would require us to further argue that u is continuously differentiable and satisfies (5.10).

Exercise 5.5.2.
i.) Assuming that f is sufficiently regular, show that

$$u'(t) = e^{At} \left(f((0, u_0) - Au_0 \right) + \int_0^t e^{A(t-s)} \frac{\partial f}{\partial u}(s, u(s)) u'(s) ds \tag{5.67}$$

by direct differentiation in the variation of parameters formula (5.11).

ii.) If u were differentiable, how would the formula from (i) compare to what you obtain using Leibniz's rule?

Approach 2: A Typical Convergence Argument

Broadly speaking, the underlying idea of a convergence argument is to define a sequence of functions whose limit (taken in an appropriate sense and space) is a solution of the IVP under investigation. We have encountered a version of this method in the proofs of various approximation theorems (e.g., exercises 4.2.7 and 4.2.8). We shall illustrate such a convergence approach, which can be adapted to the present setting, by considering the following homogenous IVP in $\mathbb{R}^N$:

$$\begin{cases} \mathbf{U}'(t) = \mathbf{A}\mathbf{U}(t), 0 \le t \le T, \\ \mathbf{U}(0) = \mathbf{U}_0, \end{cases} \tag{5.68}$$

where $\mathbf{U}_0 \in \mathbb{R}^N$ and $\mathbf{A} \in \mathbb{M}^N(\mathbb{R})$. The desired solution $\mathbf{U}(t) = e^{\mathbf{A}t}\mathbf{U}_0$ is constructed using the iteration scheme described below.

We begin with the integrated form of (5.68), namely

$$\mathbf{U}(t) = \mathbf{U}_0 + \int_0^t \mathbf{A}\mathbf{U}(s)ds, 0 \le t \le T. \tag{5.69}$$

We must overcome the self-referential nature of (5.69). One way to do so is to replace $\mathbf{U}(s)$ on the right-side of (5.69) by an approximation of $\mathbf{U}(s)$. Presently, the only knowledge about $\mathbf{U}$ that we have is its value at $t = 0$, namely $\mathbf{U}_0$. So, naturally we use $\mathbf{U}(s) = \mathbf{U}_0$ as an initial approximation. Making this substitution yields the following crude approximation of $\mathbf{U}(t)$:

$$\mathbf{U}(t) \approx \mathbf{U}_0 + \int_0^t \mathbf{A}\mathbf{U}_0 ds = \mathbf{U}_0 + \mathbf{A}\mathbf{U}_0 t, 0 \le t \le T. \tag{5.70}$$

Let $\mathbf{U}_1(t) = \mathbf{U}_0 + \mathbf{A}\mathbf{U}_0 t$. Now, in order to improve the approximation, we can replace $\mathbf{U}(s)$ on the right-side of (5.69) by $\mathbf{U}_1(s)$ to obtain

$$\begin{aligned} \mathbf{U}(t) &\approx \mathbf{U}_0 + \int_0^t \mathbf{A}\mathbf{U}_1(s)ds \\ &= \mathbf{U}_0 + \int_0^t \mathbf{A}\left[\mathbf{U}_0 + \int_0^s \mathbf{A}\mathbf{U}_0 d\tau\right]ds \\ &= \mathbf{U}_0 + \int_0^t \mathbf{A}\mathbf{U}_0 ds + \int_0^t \mathbf{A}\left(\int_0^s \mathbf{A}\mathbf{U}_0 d\tau\right)ds \\ &= \mathbf{U}_0 + \mathbf{A}\mathbf{U}_0 t + \mathbf{A}^2 \mathbf{U}_0 \frac{t^2}{2} \\ &= \left(\mathbf{A}^0 t^0 + \mathbf{A}t + \mathbf{A}^2 \frac{t^2}{2}\right)\mathbf{U}_0. \end{aligned} \tag{5.71}$$

Exercise 5.5.3. Justify all steps in (5.71).

The above sequence of "successively better" approximations can be formally described by the recursive sequence

$$\mathbf{U}_m(t) = \mathbf{U}_0 + \int_0^t \mathbf{A}\mathbf{U}_{m-1}(s)ds, \ m \in \mathbb{N}. \tag{5.72}$$

Proceeding as in (5.71) leads to the following explicit formula for $\mathbf{U}_m(t)$:

$$\mathbf{U}_m(t) = \sum_{k=0}^{m} \left(\frac{\mathbf{A}^k t^k}{k!} \right) \mathbf{U}_0. \tag{5.73}$$

Moreover,

$$\lim_{m \to \infty} \left(\sup_{0 \le t \le T} \left\| \mathbf{U}_m(t) - e^{\mathbf{A}t} \mathbf{U}_0 \right\|_{\mathbb{R}^N} = 0 \right). \tag{5.74}$$

Exercise 5.5.4.
i.) Justify (5.73) and (5.74).
ii.) For a fixed $m_0 \in \mathbb{N}$, what does the quantity $\sup_{0 \le t \le T} \left\| \mathbf{U}_{m_0}(t) - e^{\mathbf{A}t} \mathbf{U}_0 \right\|_{\mathbb{R}^N}$ tell you?

Of course, this is old news since we already proved that $\mathbf{U}(t) = e^{\mathbf{A}t} \mathbf{U}_0$ is the unique mild solution of (5.68) in Chapter 3. But, it does suggest that studying the convergence of the sequence

$$u_n(t) = e^{At}u_0 + \int_0^t e^{A(t-s)} f(s, u_{n-1}(s))ds, \ 0 \le t \le T, \tag{5.75}$$

is a viable approach to establishing the existence of a mild solution of (5.10)

Exercise 5.5.5. Why must the limit u of (5.75) belong to $\mathbb{C}([0,T]; \mathcal{X})$?

Our first main existence result for (5.10) is as follows. For illustrative purposes, we provide proofs using both approaches.

Theorem 5.5.2. *Assume* $(\mathbf{H_A})$ *and consider the hypothesis*
(H5.1) $f : [0,T] \times \mathcal{X} \to \mathcal{X}$ *is continuously Frechet differentiable on the product space* $[0,T] \times \mathcal{X}$.
i.) *If* $u_0 \in \mathcal{X}$ *and f is continuous in the first variable and globally Lipschitz in the second variable (uniformly in t), then (5.10) has a unique mild solution on* $[0,T]$ *given by (5.11).*
ii.) *If* $u_0 \in \text{dom}(A)$ *and* **(H5.1)** *holds, then the mild solution in (i) is a classical solution of (5.10).*

Proof. **Proof of (i) (using Approach 1):**
Consider the solution map Φ defined in (5.65). By assumption, f satisfies (5.55) with Lipschitz constant M_f. We shall show that Φ has a unique fixed-point using Cor. 5.2.3.

Let $u \in C([0,T];\mathscr{X})$. The well-definedness of Φ follows from

$$\sup_{0 \leq t \leq T} \|(\Phi u)(t)\|_{\mathscr{X}} \leq \sup_{0 \leq t \leq T} \left[\|e^{At} u_0\|_{\mathscr{X}} + \left\| \int_0^t e^{A(t-s)} f(s,u(s)) ds \right\|_{\mathscr{X}} \right]$$

$$\leq \overline{M_A} \left[\|u_0\|_{\mathscr{X}} + \int_0^T \|f(s,u(s))\|_{\mathscr{X}} ds \right]$$

$$\leq \overline{M_A} \left[\|u_0\|_{\mathscr{X}} + \int_0^T \|f(s,u(s)) - f(s,u_0) + f(s,u_0)\|_{\mathscr{X}} ds \right]$$

$$\leq \overline{M_A} \left[\|u_0\|_{\mathscr{X}} + \int_0^T \left[M_f \|u(s) - u_0\|_{\mathscr{X}} + \|f(s,u_0)\|_{\mathscr{X}} \right] ds \right]$$

$$< \infty.$$

(Why?) Next, let $x,y \in C([0,T];\mathscr{X})$. Observe that $\forall 0 \leq t \leq T$,

$$\|(\Phi x)(t) - (\Phi y)(t)\|_{\mathscr{X}} \leq \left\| \int_0^t e^{A(t-s)} [f(s,x(s)) - f(s,y(s))] ds \right\|_{\mathscr{X}}$$

$$\leq \overline{M_A} M_f \int_0^t \|x(s) - y(s)\|_{\mathscr{X}} ds$$

$$\leq \overline{M_A} M_f \int_0^t \sup_{0 \leq s \leq T} \|x(s) - y(s)\|_{\mathscr{X}} ds \qquad (5.76)$$

$$\leq \left(\overline{M_A} M_f \right) t \|x - y\|_C.$$

Applying (5.76) with x and y replaced by $\Phi^{n-1} x$ and $\Phi^{n-1} y$, respectively, and arguing inductively yields the following estimate that holds $\forall 0 \leq t \leq T$ and $n \in \mathbb{N}$:

$$\|(\Phi^n x)(t) - (\Phi^n y)(t)\|_{\mathscr{X}} \leq \frac{\left(\overline{M_A} M_f t \right)^n}{n!} \|x - y\|_C.$$

Taking the supremum over $0 \leq t \leq T$ subsequently yields

$$\|\Phi^n x - \Phi^n y\|_C \leq \frac{\left(\overline{M_A} M_f T \right)^n}{n!} \|x - y\|_C.$$

Since $\lim_{n \to \infty} \frac{\left(\overline{M_A} M_f T \right)^n}{n!} = 0$ (Why?), $\exists n_0 \in \mathbb{N}$ such that $\frac{\left(\overline{M_A} M_f T \right)^{n_0}}{n_0!} < 1$. Hence, Φ^{n_0} is a contraction on $C([0,T];\mathscr{X})$ and so, Cor. 5.2.3 guarantees that Φ has a unique fixed point u which is the mild solution of (5.10) on $[0,T]$ that we seek. $\square$

Proof of (ii): This proof is modified from [267, 341].

Assume that $u_0 \in \text{dom}(A)$ and let u be the mild solution of (5.10) guaranteed to exist by (i). We argue that u given by (5.11) is differentiable, satisfies (5.10), and is such that $u(t) \in \text{dom}(A)$, $\forall 0 \leq t \leq T$.

Consider the auxiliary IVP

$$\begin{cases} z'(t) + Az(t) = \left[\frac{\partial}{\partial u} f(\cdot, u(\cdot)) \right](t) \cdot z(t), \ 0 < t < T, \\ z(0) = f(0,u_0) - Au_0. \end{cases} \qquad (5.77)$$

Since $\frac{\partial f}{\partial u}(\cdot, u(\cdot))$ is continuous (Why?) and $u \in \mathbb{C}([0,T]; \mathscr{X})$ (from part (i) of the present theorem), the mapping $z \mapsto \left[\frac{\partial f}{\partial u}(\cdot, u(\cdot))\right] z$ must be globally Lipschitz on $\mathbb{C}([0,T]; \mathscr{X})$. (Show this by appealing to (5.43).) Hence, (5.77) has a unique mild solution $z \in \mathbb{C}([0,T]; \mathscr{X})$ given by

$$z(t) = e^{At}(f(0, u_0) - Au_0) + \int_0^t e^{A(t-s)} \left[\frac{\partial}{\partial u} f(\cdot, u(\cdot))\right](s) \cdot z(s) ds, \ 0 \le t \le T. \tag{5.78}$$

We argue that $z(t) = u'(t)$, $\forall 0 \le t \le T$. In view of Exer. 5.5.2, we argue that

$$\lim_{h \to 0} \left\| \frac{u(t+h) - u(t)}{h} - z(t) \right\|_{\mathscr{X}} = 0, \ \forall 0 \le t \le T. \tag{5.79}$$

Let $\varepsilon > 0$ and $t \in [0,T]$. Using (5.11) and (5.78) in (5.79) yields (after a strategic grouping of terms)

$$\left\| \frac{u(t+h) - u(t)}{h} - z(t) \right\|_{\mathscr{X}} = \|J_1(h) + J_2(h) + J_3(h)\|_{\mathscr{X}}, \tag{5.80}$$

where

$$J_1(h) = \left(\frac{e^{A(t+h)} - e^{At}}{h} \right) u_0 - e^{At} A u_0,$$

$$J_2(h) = \frac{1}{h} \int_0^h e^{A(t+h-s)} f(s, u(s)) ds - e^{At} f(0, u_0),$$

$$J_3(h) = \frac{1}{h} \int_h^{t+h} e^{A(t+h-s)} f(s, u(s)) ds - \frac{1}{h} \int_0^t e^{A(t-s)} f(s, u(s)) ds$$
$$- \int_0^t e^{A(t-s)} \left[\frac{\partial}{\partial u} f(\cdot, u(\cdot))\right](s) \cdot z(s) ds.$$

Our main strategy is to apply Gronwall's lemma to an inequality of the form

$$\|K(t)\|_{\mathscr{X}} \le M(\varepsilon) + C \int_0^t \|K(s)\|_{\mathscr{X}} ds, \tag{5.81}$$

where $M(\varepsilon)$ depends only on ε, C is a positive constant, and $K(t) = \frac{u(t+h)-u(t)}{h} - z(t)$. Applying this lemma requires us to establish several estimates involving $\|J_i(h)\|_{\mathscr{X}}$, for $i = 1, 2, 3$.

First, for use in a subsequent estimate, we note that since $u, \frac{\partial f}{\partial u} \in \mathbb{C}([0,T]; \mathscr{X})$, $\exists \overline{M_f} > 0$ such that

$$\sup_{0 \le s \le T} \left\| \frac{\partial f}{\partial u}(s, u(s)) \right\|_{\mathscr{X}} \le \overline{M_f}. \tag{5.82}$$

Also, $\exists \delta_1, \delta_2 > 0$ for which

$$0 < |h| < \delta_1 \implies \|J_1(h)\|_{\mathscr{X}} < \frac{\varepsilon}{3 e^{\overline{M_f} M_A T}}, \tag{5.83}$$

$$0 < |h| < \delta_2 \implies \|J_2(h)\|_{\mathscr{X}} < \frac{\varepsilon}{3 e^{\overline{M_f} M_A T}}. \tag{5.84}$$

Estimating $\|J_3(h)\|_{\mathscr{X}}$ is more delicate. We first modify the present form of $J_3(h)$ by:

   (a) adding and subtracting the term $\int_0^t e^{A(t-s)} \left[\frac{\partial f}{\partial u}(\cdot, u(\cdot))\right](s) \left[\frac{u(s+h)-u(s)}{h}\right] ds$,

   (b) performing the change of variable $w = s - h$ in the first of the three integrals in the expression for $J_3(h)$ to obtain

$$\frac{1}{h}\int_h^{t+h} e^{A(t+h-s)} f(s, u(s)) ds = \frac{1}{h}\int_0^t e^{A(t-w)} f(w+h, u(w+h)) dw.$$

Implementing these modifications yields the following equivalent form of $\|J_3(h)\|_{\mathscr{X}}$:

$$J_3(h) = \frac{1}{h}\int_0^t e^{A(t-w)} f(w+h, u(w+h)) dw - \frac{1}{h}\int_0^t e^{A(t-s)} f(s, u(s)) ds$$

$$- \int_0^t e^{A(t-s)} \left[\frac{\partial f}{\partial u}(\cdot, u(\cdot))\right](s) \left[\frac{u(s+h)-u(s)}{h}\right] ds \qquad (5.85)$$

$$+ \int_0^t e^{A(t-s)} \left[\frac{\partial f}{\partial u}(\cdot, u(\cdot))\right](s) \left[\frac{u(s+h)-u(s)}{h} - z(s)\right] ds.$$

Since w is a dummy variable, we can relabel it as s and combine the first three integrals on the right-side of (5.85). Doing so, and then applying the triangle inequality, yields

$$\|J_3(h)\|_{\mathscr{X}} \le \|J_4(h)\|_{\mathscr{X}} + \|J_5(h)\|_{\mathscr{X}}, \qquad (5.86)$$

where

$$J_4(h) = \int_0^t e^{A(t-s)} \frac{1}{h} \Big\{ f(s+h, u(s+h)) - f(s, u(s))$$

$$- \left[\frac{\partial f}{\partial u}(\cdot, u(\cdot))\right](s) \left[u(s+h)-u(s)\right] \Big\} ds$$

$$J_5(h) = \int_0^t e^{A(t-s)} \left[\frac{\partial f}{\partial u}(\cdot, u(\cdot))\right](s) \left[\frac{u(s+h)-u(s)}{h} - z(s)\right] ds.$$

Upon making hefty use of the triangle inequality, together with the regularity of u and $\frac{\partial f}{\partial u}$ and (5.44), we can show that $\exists \delta_3 > 0$ such that

$$0 < |h| < \delta_3 \implies \|J_4(h)\|_{\mathscr{X}} < \frac{\varepsilon}{3e^{M_f M_A T}}. \qquad (5.87)$$

(See Exer. 5.5.6.) Further,

$$\|J_5(h)\|_{\mathscr{X}} \le \overline{M_f M_A} \int_0^t \left\| \frac{u(s+h)-u(s)}{h} - z(s) \right\|_{\mathscr{X}} ds. \qquad (5.88)$$

Let $\delta = \min\{\delta_1, \delta_2, \delta_3\} > 0$. Then, $\forall 0 < |h| < \delta$, using (5.83) - (5.87) in (5.80) yields

$$\left\| \frac{u(t+h)-u(t)}{h} - z(t) \right\|_{\mathscr{X}} \le \frac{\varepsilon}{e^{M_f M_A T}} \qquad (5.89)$$

$$+ M_f M_A \int_0^t \left\| \frac{u(s+h)-u(s)}{h} - z(s) \right\|_{\mathscr{X}} ds.$$

Hence, applying Gronwall's lemma in (5.89) results in

$$\left\| \frac{u(t+h)-u(t)}{h} - z(t) \right\|_{\mathscr{X}} \leq \frac{\varepsilon}{e^{\overline{M_f M_A} T}} e^{\int_0^t \overline{M_f M_A} ds} < \varepsilon, \forall 0 \leq t \leq T. \qquad (5.90)$$

Consequently, we conclude that u is continuously differentiable. Thus, the IVP

$$\begin{cases} y'(t) + Ay(t) = f(t, u(t)), \ 0 < t < T, \\ y(0) = u_0, \end{cases} \qquad (5.91)$$

has a <u>unique</u> classical solution (by Prop. 4.2.3) given by

$$y(t) = e^{At} u_0 + \int_0^t e^{A(t-s)} f(s, u(s)) ds. \qquad (5.92)$$

Since $u(\cdot)$ is unique and is expressed by the right-sides of both (5.11) and (5.92), these two expressions must be equal, thereby showing that $u = y$. As such, we conclude that u is the unique classical solution of (5.10). This completes the proof. $\square$

Exercise 5.5.6.
i.) Show that if $u_0 \in \text{dom}(A)$ and f is globally Lipschitz (uniformly in t), then the mild solution u of (5.10) is also globally Lipschitz on $[0, T]$.
ii.) Use (i) to argue that (5.87) holds.

Next, for illustrative purposes, we provide an alternative argument of Thrm. 5.5.2(i) using Approach 2.

Alternative Proof of Thrm. 5.5.2(i):
Consider the following sequence:

$$u_n(t) = e^{At} u_0 + \int_0^t e^{A(t-s)} f(s, u_{n-1}(s)) ds, n \in \mathbb{N},$$

$$u_0(t) = e^{At} u_0. \qquad (5.93)$$

We need to prove the existence of $u \in \mathbb{C}([0, T]; \mathscr{X})$ such that
  (a) $\lim_{n \to \infty} \|u_n - u\|_{\mathbb{C}([0,T];\mathscr{X})} = 0$, and
  (b) u is a mild solution of (5.10) on $[0, T]$.

<u>Claim 1</u>: $\{u_n | n \in \mathbb{N}\}$ is a uniformly bounded subset of $\mathbb{C}([0, T]; \mathscr{X})$.
  Let $0 \leq t \leq T$ be fixed. For any $n \in \mathbb{N}$,

$$\|u_n(t)\|_{\mathscr{X}} = \left\| e^{At} u_0 \right\|_{\mathscr{X}} + \int_0^t \left\| e^{A(t-s)} [f(s, u_{n-1}(s)) - f(s, 0) + f(s, 0)] \right\|_{\mathscr{X}} ds$$

$$\leq \overline{M_A} \left[\|u_0\|_{\mathscr{X}} + \int_0^t [\|f(s, u_{n-1}(s)) - f(s, 0)\|_{\mathscr{X}} + \|f(s, 0)\|_{\mathscr{X}}] ds \right]$$

$$\leq \overline{M_A} \left[\|u_0\|_{\mathscr{X}} + \|f(\cdot, 0)\|_{\mathbb{L}^1(0,T;\mathscr{X})} \right] \qquad (5.94)$$

$$+ \overline{M_A} M_f \int_0^t \|u_{n-1}(s)\|_{\mathscr{X}} ds.$$

Let $C_1 = \overline{M_A} \left[\|u_0\|_{\mathscr{X}} + \|f(\cdot,0)\|_{\mathbb{L}^1(0,T;\mathscr{X})} \right]$. For any $K \in \mathbb{N}$, taking supremums on both sides of (5.94) yields

$$\sup_{1 \le n \le K} \|u_n(t)\|_{\mathscr{X}} \le C_1 + \overline{M_A} M_f \int_0^t \sup_{1 \le n \le K} \|u_{n-1}(s)\|_{\mathscr{X}} \, ds$$

$$\le C_1 + \overline{M_A} M_f \int_0^t \left[\|u_0\|_{\mathscr{X}} + \sup_{1 \le n \le K} \|u_n(s)\|_{\mathscr{X}} \right] ds \qquad (5.95)$$

$$\le \left[C_1 + \overline{M_A} M_f \|u_0\|_{\mathscr{X}} T \right] + \overline{M_A} M_f \int_0^t \sup_{1 \le n \le K} \|u_n(s)\|_{\mathscr{X}} \, ds.$$

Applying Gronwall's lemma to (5.95) yields

$$\sup_{1 \le n \le K} \|u_n(t)\|_{\mathscr{X}} \le \left[C_1 + \overline{M_A} M_f \|u_0\|_{\mathscr{X}} T \right] e^{\left(\overline{M_A} M_f \right) t}$$

so that

$$\sup_{1 \le n \le K} \|u_n\|_{\mathbb{C}([0,T];\mathscr{X})} \le \left[C_1 + \overline{M_A} M_f \|u_0\|_{\mathscr{X}} T \right] e^{\left(\overline{M_A} M_f \right) T}. \qquad (5.96)$$

Since the right-side of (5.96) is independent of K, we conclude that $\{u_n | n \in \mathbb{N}\}$ is a uniformly bounded subset of $\mathbb{C}([0,T];\mathscr{X})$.$\diamondsuit$

<u>Claim 2</u>: There exists $u \in \mathbb{C}([0,T];\mathscr{X})$ such that $\lim_{n \to \infty} \|u_n - u\|_{\mathbb{C}} = 0$.

For every $m \in \mathbb{N}$ and $0 \le t \le T$, observe that

$$u_m(t) = u_0(t) + \sum_{k=0}^{m-1} [u_{k+1}(t) - u_k(t)]. \qquad (5.97)$$

Propositions 1.7.7 and 1.7.8 imply that it suffices to show that $\sum_{k=0}^{\infty} \|u_{k+1}(t) - u_k(t)\|_{\mathscr{X}}$ is uniformly convergent on $[0,T]$. (Why?) Arguing iteratively, we see that $\forall k \in \mathbb{N}$,

$$\|u_{k+1}(t) - u_k(t)\|_{\mathscr{X}} = \left\| \int_0^t e^{A(t-s)} [f(s,u_k(s)) - f(s,u_{k-1}(s))] \, ds \right\|_{\mathscr{X}}$$

$$\le \overline{M_A} M_f \int_0^t \|u_k(s_1) - u_{k-1}(s_1)\|_{\mathscr{X}} \, ds_1$$

$$\le \left(\overline{M_A} M_f \right)^2 \int_0^t \int_0^{s_1} \|u_{k-1}(s_2) - u_{k-2}(s_2)\|_{\mathscr{X}} \, ds_2 ds_1$$

$$\vdots \qquad (5.98)$$

$$\le \left(\overline{M_A} M_f \right)^k \int_0^t \cdots \int_0^{s_{k-1}} \|u_1(s_k) - u_0(s_k)\|_{\mathscr{X}} \, ds_k \ldots ds_1$$

$$\le \left(\overline{M_A} M_f \right)^k \int_0^t \cdots \int_0^{s_{k-1}} \left\| \int_0^t e^{A(s_k-\tau)} f(\tau, e^{A\tau} u_0) d\tau \right\|_{\mathscr{X}} ds_k \ldots ds_1$$

$$\le M_f^k \left(\overline{M_A} \right)^{k+1} \overline{M_f} \frac{t^{k+1}}{(k+1)!}.$$

Therefore, $\forall 0 \le t \le T$,

$$\sum_{k=0}^{\infty} \|u_{k+1}(t) - u_k(t)\|_{\mathscr{X}} \le \left(\frac{\overline{M_A M_f}}{M_f} \right) \sum_{k=0}^{\infty} \frac{(tM_f)^{k+1}}{(k+1)!} \le \left(\frac{\overline{M_A M_f}}{M_f} \right) e^{M_f T} \quad (5.99)$$

We conclude from (5.99) that

$$\lim_{k \to \infty} \|u_{k+1}(t) - u_k(t)\|_{\mathscr{X}} = 0 \text{ uniformly } \forall 0 \le t \le T. \quad (5.100)$$

Hence, by Prop. 1.7.8, we conclude that

$$u_m \longrightarrow u \text{ uniformly on } [0, T] \text{ as } m \to \infty. \quad (5.101)$$

(Tell how carefully.) $\Diamond$

Claim 3: Show that u is a mild solution of (5.10).
Let $\varepsilon > 0$. There exists $M \in \mathbb{N}$ such that

$$m \ge M \implies \|u_m - u\|_{\mathbb{C}} < \frac{\varepsilon}{\overline{M_A M_f} T + 1}. \quad (5.102)$$

For such m, it follows that

$$\left\| \int_0^t e^{A(t-s)} [f(s, u_{m-1}(s)) - f(s, u(s))] ds \right\|_{\mathscr{X}} \le \overline{M_A M_f} \int_0^t \|u_{m-1}(s) - u(s)\|_{\mathscr{X}} ds$$
$$\le \overline{M_A M_f} T \|u_{m-1} - u\|_{\mathbb{C}}$$
$$< \varepsilon.$$

Thus, $\forall t \in [0, T]$,

$$\lim_{m \to \infty} \int_0^t e^{A(t-s)} f(s, u_{m-1}(s)) ds = \int_0^t e^{A(t-s)} f(s, u(s)) ds. \quad (5.103)$$

Using (5.103), we now take a limit as $n \to \infty$ in (5.93) to conclude that u satisfies (5.11) and hence is a mild solution of (5.10). $\Diamond$

Claim 4: Verify that the mild solution of (5.10) is unique.
Let v_1 and v_2 be mild solutions of (5.10). Then, they both satisfy (5.11). Subtracting these two expressions yields, $\forall 0 \le t \le T$,

$$\|v_1(t) - v_2(t)\|_{\mathscr{X}} \le \overline{M_A M_f} \int_0^t \|v_1(s) - v_2(s)\|_{\mathscr{X}} ds. \quad (5.104)$$

An application of Gronwall's lemma in (5.104) leads us to conclude that $v_1(t) = v_2(t)$, $\forall 0 \le t \le T$. (Why?) Thus, uniqueness follows. $\Diamond$

This completes the proof. $\qquad\qquad\qquad\qquad\qquad\qquad\qquad\qquad\qquad\qquad\Box$

Exercise 5.5.7. If the assumption "f is continuously Frechet differentiable" is replaced by "f is globally Lipschitz," must both parts of Thrm. 5.5.2 still hold?

Exercise 5.5.8. Assume $(\mathbf{H_A})$ and that $\exists K \in \mathbb{C}([0,T];(0,\infty))$ for which $f : [0,T] \times \mathscr{X} \to \mathscr{X}$ satisfies (5.58).
i.) Prove that (5.10) has a unique mild solution on $[0,T]$.
ii.) Does the conclusion of (i) remain valid if we assume that $K \in \mathbb{L}^1(0,T;(0,\infty))$?

Exercise 5.5.9. Define $f : [0,T] \times \mathbb{L}^2(\mathbb{R};\mathbb{R}) \to \mathbb{L}^2(\mathbb{R};\mathbb{R})$ by $f(t,x) = g(t)h(x)$, where $g : [0,T] \to (0,\infty)$ is an impulse function defined by

$$g(t) = \begin{cases} 0, \ 0 \le t < a, \\ U, \ a \le t \le b, \\ 0, \ b < t < T, \end{cases}$$

where $0 \le a - 1 < a < b < b+1 \le T$ and $h \in \mathbb{C}^1([0,T];\mathbb{L}^2(\mathbb{R};\mathbb{R}))$ is such that $h(a) = h(b) = 0$. If $(\mathbf{H_A})$ holds, must (5.10) have a unique mild solution on $[0,T]$? Must such a mild solution be a classical solution if $u_0 \in \text{dom}(A)$?

Exercise 5.5.10. Assume $(\mathbf{H_A})$ and that $\exists M \in \mathbb{C}([0,T];(0,\infty))$ for which $f : [0,T] \times \mathscr{X} \to \mathscr{X}$ satisfies

$$\int_0^t \|f(s,x) - f(s,y)\|_{\mathscr{X}} \, ds \le M(t) \|x-y\|_{\mathscr{X}}, \ \forall 0 \le t \le T, x,y \in \mathscr{X}. \quad (5.105)$$

Must (5.10) have a unique mild solution on $[0,T]$? Must it be a classical solution if $u_0 \in \text{dom}(A)$?

Exercise 5.5.11. Assume $(\mathbf{H_A})$, $f : [0,T] \times \mathscr{X} \to \mathscr{X}$ is globally Lipschitz (uniformly in t), and $a \in \mathbb{C}([0,T];[0,\frac{1}{2}])$. Consider the IVP

$$\begin{cases} u'(t) = Au(t) + f\left(t, u(t), \int_0^t a(s)u(s)ds\right), 0 < t < T, \\ u(0) = u_0. \end{cases} \quad (5.106)$$

i.) Determine an expression for the solution map $\Phi : \mathbb{C}([0,T];\mathscr{X}) \to \mathbb{C}([0,T];\mathscr{X})$ for (5.106) in the spirit of (5.65).
ii.) Show that there exist positive constants C_1 and C_2 such that $\forall x,y \in \mathscr{X}$,

$$\|(\Phi x)(t) - (\Phi y)(t)\|_{\mathscr{X}} \le C_1 \int_0^t \|x(s) - y(s)\|_{\mathscr{X}} \, ds + C_2 \int_0^t \int_0^s \|x(\tau) - y(\tau)\|_{\mathscr{X}} \, d\tau ds.$$

iii.) Can Prop. 5.2.19 be used to argue that $\exists N \in \mathbb{N}$ such that Φ^N is a strict contraction? Clearly indicate any additional restrictions that can be imposed on the data $(u_0, M_f, \text{etc.})$ in order to make this possible.

Next, we weaken the growth condition imposed on f to a <u>local</u> Lipschitz restriction (cf. Def. 5.4.3). Such functions can grow faster than linear ones. This presents a hurdle when defining the solution map since the Lipschitz constant can change every time we consider a new bounded subset of $[0,T] \times \mathscr{X}$. In fact, the collection of all such constants for f need not be a bounded subset of $(0,\infty)$. (Can you think of an example that illustrates this?) Even worse, although there is a fixed Lipschitz constant that works throughout a given set $[a, a+\delta) \times \mathfrak{B}_{\mathscr{X}}(x_0;\varepsilon) \subset [0,T] \times \mathscr{X}$, there can exist $u \in \mathbb{C}([0,T];\mathscr{X})$ such that $u(s) \notin \mathfrak{B}_{\mathscr{X}}(x_0;\varepsilon)$, for some $s \in [0,T)$. In such case, we can no longer control the term $\int_0^t e^{A(t-s)} f(s,u(s))ds$! (Why?) So, what do we do?

To remedy the situation, we fix $R > 0$ *a priori.* Then, $\forall x_0 \in \mathscr{X}$, there exists a constant $M_f(x_0) > 0$ for which

$$\|f(t,x) - f(t,y)\|_{\mathscr{X}} \leq M_f(x_0) \|x-y\|_{\mathscr{X}}, \ \forall x,y \in \mathfrak{B}_{\mathscr{X}}(x_0;R), t \in [0,T]. \quad (5.107)$$

The key is to form a very specific space $\mathscr{Y}$ of continuous functions for which we can show, using (5.107), that the solution map Φ in (5.65) satisfies $\Phi(\mathscr{Y}) \subset \mathscr{Y}$. Since all functions of interest are equal to u_0 when $t = 0$ (because of the IC in (5.10)), we use

$$\mathscr{Y} = \mathbb{C}([0,t_0];\mathfrak{B}_{\mathscr{X}}(u_0;R)), \quad (5.108)$$

which is a Banach space when equipped with the usual sup norm

$$\|y\|_{\mathscr{Y}} = \sup_{0 \leq t \leq t_0} \|y(t)\|_{\mathscr{X}}. \quad (5.109)$$

If we use the Contraction Mapping Theorem, the strategy is to choose $t_0 > 0$ sufficiently small such that

$$\sup_{0 \leq t \leq t_0} \|(\Phi u)(t) - u_0\|_{\mathscr{X}} \leq R \quad (5.110)$$

(so that $\Phi(\mathscr{Y}) \subset \mathscr{Y}$), and then further restrict t_0 to ensure that there exists a constant $0 < C(t_0) < 1$ such that

$$\|(\Phi x) - (\Phi y)\|_{\mathscr{Y}} \leq C(t_0) \|x-y\|_{\mathscr{Y}}, \ \forall x,y \in \mathscr{Y}. \quad (5.111)$$

Since $R > 0$ was fixed, using a smaller value of t_0 does not adversely affect any step of the proof. (Explain why.) The resulting fixed-point u is a mild solution of (5.10) on $[0,t_0)$; if $t_0 < T$, then u is called a *local mild solution* .

The above strategy is used to prove the following proposition (as in [**142**])..

Proposition 5.5.3. *Assume* (**H$_A$**) *and*
(**H5.2**) $f : [0,T] \times \mathscr{X} \to \mathscr{X}$ *is continuous in the first variable and locally Lipschitz in the second variable (uniformly in t).*
Then, there exists a unique local mild solution u of (5.10).

Proof. Let $R > 0$ and $t_0 \in (0,T]$ and define $\Phi : \mathscr{Y} \to \mathbb{C}([0,T];\mathscr{X})$ by (5.65). There exist functions $\alpha_i : [0,T] \to \mathbb{R}^+$ $(i = 1,2,3)$ for which

$$\lim_{t_0 \to 0^+} \alpha_i(t_0) = 0, \quad (5.112)$$

$$x \in \mathscr{Y} \implies \|\Phi(x) - u_0\|_{\mathscr{Y}} \leq \alpha_1(t_0) + \alpha_2(t_0)R \leq R, \quad (5.113)$$

$$x,y \in \mathscr{Y} \implies \|(\Phi x) - (\Phi y)\|_{\mathscr{Y}} \leq \alpha_3(t_0) \|x-y\|_{\mathscr{Y}}. \quad (5.114)$$

Choose $t_0 \in (0, T]$ such that (5.113) holds and $\alpha_3(t_0) < 1$. The result then follows.
(Tell how.) $\qquad\qquad\qquad\qquad\qquad\qquad\qquad\qquad\qquad\qquad\qquad\qquad$ □

Exercise 5.5.12. Provide the details in the proof of Prop. 5.5.3.

Exercise 5.5.13. Provide an alternative proof of Prop. 5.5.3 following a suitably modified version of Approach 2.

Exercise 5.5.14. Consider the IVP

$$\begin{cases} u'(t) = Au(t) + g(t)h(u(t)), \ 0 < t < T, \\ u(0) = u_0, \end{cases} \tag{5.115}$$

in $\mathscr{X}$, where $g \in \mathbb{C}([0,T];\mathscr{X})$ and $h : \mathscr{X} \to \mathscr{X}$ is locally Lipschitz.
i.) Argue that (5.115) has a unique local mild solution.
ii.) If we assume, instead, that $g \in \mathbb{L}^1(0,T;\mathscr{X})$, can the proof used in (i) be adapted to reach the same conclusion? Explain.

5.5.2 Continuous Dependence

Consider (5.10) together with the IVP

$$\begin{cases} v'(t) = Av(t) + \widehat{f}(t,v(t)), \ 0 < t < T, \\ v(0) = v_0, \end{cases} \tag{5.116}$$

where $\widehat{f} : [0,T] \times \mathscr{X} \to \mathscr{X}$ and $v_0 \in \mathscr{X}$. The following continuous dependence result follows easily from an application of Gronwall's lemma.

Proposition 5.5.4. *Assume* $(\mathbf{H_A})$ *and*
(H5.3) f *and* $\widehat{f}$ *are globally Lipschitz on* $\mathscr{X}$ *(uniformly in* t*);*
(H5.4) *There exists* $\varepsilon_1 > 0$ *such that* $\displaystyle\sup_{0 \leq t \leq T} \left\| f(t,x) - \widehat{f}(t,x) \right\|_{\mathscr{X}} < \varepsilon_1, \ \forall x \in \mathscr{X};$
(H5.5) *There exists* $\varepsilon_2 > 0$ *such that* $\|u_0 - v_0\|_{\mathscr{X}} < \varepsilon_2$.
Then,

$$\|u(t) - v(t)\|_{\mathscr{X}} < \overline{M_A}(\varepsilon_2 + \varepsilon_1 T) e^{\overline{M_f} t}, \ \forall 0 \leq t \leq T. \tag{5.117}$$

Exercise 5.5.15.
i.) Prove Prop. 5.5.4.
ii.) Identify exactly where the global Lipschitz hypothesis on f and $\widehat{f}$ was used.

Exercise 5.5.16. Replace **(H5.3)** by
(H5.3b) There exist $K_f, K_{\widehat{f}} \in \mathbb{C}([0,T];(0,\infty))$ such that f and $\widehat{f}$ satisfy (5.58).
i.) Assume that $(\mathbf{H_A})$ and **(H5.4)** - **(H5.6)** hold. Formulate and prove a continuous dependence estimate for $\|u(t) - v(t)\|_{\mathscr{X}}$.
ii.) Repeat (i), but now assuming that $K_f, K_{\widehat{f}} \in \mathbb{L}^1(0,T;(0,\infty))$.
iii.) Repeat (i) and (ii), but now assuming in **(H5.6)** that (5.58) is replaced by (5.59),

for some $p > 1$.

A more general version of Prop. 5.5.4 can be formulated if we also replace the operator A by an operator $\widehat{A} : \text{dom}(\widehat{A}) \subset \mathscr{X} \to \mathscr{X}$. Specifically, consider the IVP

$$\begin{cases} w'(t) = \widehat{A}w(t) + \widehat{f}(t, w(t)), \, 0 < t < T, \\ w(0) = w_0, \end{cases} \tag{5.118}$$

where $\widehat{A}$ satisfies $(\mathbf{H}_{\widehat{A}})$ and denote the associated semigroup by $\left\{ e^{\widehat{A}t} : t \geq 0 \right\}$. Considering the difference between the variation of parameters formulae for (5.10) and (5.118), we see that we will need to control the term $\left\| e^{At} - e^{\widehat{A}t} \right\|_{\mathbb{B}(\mathscr{X})}$. (Why?) Assuming this is possible, a continuous dependence estimate in the spirit of Prop. 5.5.4 and Exer. 5.5.16 can be established in a similar manner. Indeed, we have

Proposition 5.5.5. *Assume* $(\mathbf{H_A})$, $(\mathbf{H}_{\widehat{A}})$, $(\mathbf{H5.3})$ - $(\mathbf{H5.5})$ *(with w_0 in place of v_0), and*

$(\mathbf{H5.6})$ *There exists $\varepsilon_3 > 0$ such that* $\sup \left\{ \left\| e^{At} - e^{\widehat{A}t} \right\|_{\mathbb{B}(\mathscr{X})} : 0 \leq t \leq T \right\} < \varepsilon_3$.

Then,

$$\|u(t) - w(t)\|_{\mathscr{X}} \leq \left(\overline{M_A} \left(\varepsilon_2 + \varepsilon_1 T \right) + \varepsilon_3 \left(M^\star + \|w_0\|_{\mathscr{X}} \right) \right) e^{\left(\overline{M_A} M_f \right) t}, \, \forall 0 \leq t \leq T, \tag{5.119}$$

where $M^\star = \sup\limits_{0 \leq t \leq T} \int_0^t \|f(s, w(s))\|_{\mathscr{X}} \, ds$.

Proof. Theorem 5.5.2 guarantees the existence and uniqueness of mild solutions u, w of (5.10) and (5.118), respectively. Subtracting the variation of parameters formulae for these solutions and then estimating as usual yields

$$
\begin{aligned}
\|u(t) - w(t)\|_{\mathscr{X}} &\leq \left\| e^{At} u_0 - e^{\widehat{A}t} w_0 \right\|_{\mathscr{X}} \\
&+ \int_0^t \left\| e^{A(t-s)} f(s, u(s)) - e^{\widehat{A}(t-s)} f(s, w(s)) \right\|_{\mathscr{X}} ds \\
&\leq \left\| e^{At} \right\|_{\mathbb{B}(\mathscr{X})} \|u_0 - w_0\|_{\mathscr{X}} + \left\| e^{At} - e^{\widehat{A}t} \right\|_{\mathbb{B}(\mathscr{X})} \|w_0\|_{\mathscr{X}} \\
&+ \int_0^t \left\| e^{A(t-s)} \right\|_{\mathbb{B}(\mathscr{X})} \|f(s, u(s)) - f(s, w(s))\|_{\mathscr{X}} ds \\
&+ \int_0^t \left\| e^{A(t-s)} - e^{\widehat{A}(t-s)} \right\|_{\mathbb{B}(\mathscr{X})} \|f(s, w(s))\|_{\mathscr{X}} ds \\
&+ \int_0^t \left\| e^{\widehat{A}(t-s)} \right\|_{\mathbb{B}(\mathscr{X})} \left\| f(s, w(s)) - \widehat{f}(s, w(s)) \right\|_{\mathscr{X}} ds \\
&\leq \left(\overline{M_A} \varepsilon_2 + \varepsilon_3 \|w_0\|_{\mathscr{X}} \right) + \int_0^t \overline{M_A} M_f \|u(s) - w(s)\|_{\mathscr{X}} ds \\
&+ \varepsilon_3 \int_0^t \|f(s, w(s))\|_{\mathscr{X}} ds + \overline{M_A} \varepsilon_1 T \\
&\leq \left(\overline{M_A} \left(\varepsilon_2 + \varepsilon_1 T \right) + \varepsilon_3 \left(M^{\star} + \|w_0\|_{\mathscr{X}} \right) \right) \int_0^t \overline{M_A} M_f \|u(s) - w(s)\|_{\mathscr{X}} ds.
\end{aligned}
$$

Applying Gronwall's inequality now yields the estimate (5.119). (How?) □

5.5.3 Extendability of Local Solutions

When the existence of only a local solution on $[0, T_0) \subset [0, T]$ is established, naturally the value of T_0 is directly related to the ball $\mathfrak{B}$ used to construct the space $\mathscr{Y}$. Using the same IC, had we chosen a different ball *a priori*, it is conceivable that the value of T_0 would also have been different, and possibly larger. As such, we should not abandon hope for a global solution simply because the *method* did not yield one immediately. The central questions we need to address are:

(a) Can a mild solution always be extended to a larger time interval?
(b) When can the solution be extended to the entire interval $[0, T]$?

By way of motivation, consider the following exercise.

Exercise 5.5.17. The following questions arise when considering the extendability of local solutions to larger intervals.
i.) Identify a condition to impose on $f : [0, T] \to \mathscr{X}$ that ensures the existence of the left-sided limit

$$
\lim_{t \to T_0^-} \left[e^{At} u_0 + \int_0^t e^{A(t-s)} f(s) ds \right]. \tag{5.120}
$$

How can this condition be modified if f is replaced by $g : [0, T] \times \mathscr{X} \to \mathscr{X}$? (Note that if the expression in (5.120) is continuous, then the limit must exist.) (Why?)

ii.) If f is bounded on $[0, T_0]$, show that the limit (5.120) exists. Formulate a similar result with f replaced by g.

iii.) Suppose that the limit (5.120) (where f is replaced by g) exists and has value L. Consider the IVP

$$\begin{cases} v'(t) = Av(t) + g(t, v(t)), \ T_0 < t < T, \\ v(T_0) = L. \end{cases} \tag{5.121}$$

Assuming that g is locally Lipschitz, must (5.121) possess a mild solution v? How can this be used to extend the mild solution u of (5.10) known to exist on $[0, T_0]$?

We now formally address the issues presented in Exer. 5.5.17. As mentioned above, the *method* used to establish the existence of a local mild solution can actually obscure the true interval on which the mild solution ultimately exists. But how can we tell? Intuitively, if the method yields existence on the interval $[0, T_0)$, but the solution map Φ (cf. (5.65)) is bounded at $t = T_0$, we could use the stopping point as the IC in the related IVP

$$\begin{cases} \bar{u}'(t) = A\bar{u}(t) + f(t, \bar{u}(t)), \ T_0 < t < T, \\ \bar{u}(T_0) = \lim\limits_{t \to T_0^-} (\Phi u)(t). \end{cases} \tag{5.122}$$

We could then apply the same method again to obtain a mild solution of (5.122) on $[T_0, T_0 + \varepsilon)$ and "paste" the two solutions together to construct the piecewise-defined function $U : [0, T_0 + \varepsilon) \to \mathscr{X}$ given by

$$U(t) = \begin{cases} u(t), & 0 \le t < T_0, \\ \bar{u}(t), & T_0 \le t < T_0 + \varepsilon. \end{cases}$$

As long as U is continuous, it will be a mild solution of (5.10) on $[0, T_0 + \varepsilon)$. (Why?)

We could conceivably repeat this procedure, each time extending the solution defined on $[0, t_{n-1})$ to a slightly larger interval $[0, t_n)$. But, does this process always yield a *global* mild solution of (5.10) on $[0, T]$? Consider Exer. 5.3.3(ii). If we sought a global mild solution of (5.52) on the interval $\left[0, \frac{1}{3}(2x_0 + 1)^{-3}\right]$, then in light of Exer. 5.3.3(ii)(b), such an attempt would at best yield the interval $\left[0, \frac{1}{6}(2x_0 + 1)^{-3}\right]$. In this case, the inability to obtain a global mild solution was due to the nature of the solution itself (i.e., the presence of a vertical asymptote) rather than a shortcoming of the method.

In light of the above discussion, we adjust our goal to finding the *maximal* time interval $[0, T_{max})$ to which a mild solution can be extended. It would be nice to have readily verifiable conditions, involving only the data, that tell us when a mild solution is globally defined. One such result stemming from the exploration in Exer. 5.5.17 is as follows:

Theorem 5.5.6. (Extendability of a Mild Solution)
Assume ($\mathbf{H_A}$) *and* (**H5.2**). *Then, the unique mild solution* $u : [0, T_{max}) \to \mathscr{X}$ *is such that either*

(i) $T_{\max} = \infty$, *or*

(ii) $T_{\max} < \infty$ *and* $\lim\limits_{t \to T_{\max}^-} \|u(t)\|_{\mathscr{X}} = \infty$.

This might seem as though we are stating the obvious, but proving this theorem involves some work (see **[142]**). We use Thrm. 5.5.6 to establish some particular situations in which global existence is ensured.

Proposition 5.5.7. *Assume* $(\mathbf{H_A})$ *and* $(\mathbf{H5.2})$. *Let* $R > 0$ *and assume, in addition, that for some* $0 < R^\star \leq R$,

$$\|u(t) - u_0\|_{\mathscr{X}} \leq R^\star < R, \ \forall 0 \leq t \leq T, \tag{5.123}$$

Then, (5.10) has a unique global mild solution on $[0,T]$.

Proof. Let $t_0 > 0$ be fixed. There exists $M_f(t_0) > 0$ such that

$$\|f(t,x) - f(t,y)\|_{\mathscr{X}} \leq M_f(t_0) \|x - y\|_{\mathscr{X}}, \ \forall x,y \in \mathfrak{B}_{\mathscr{X}}(u_0;R), \ 0 \leq t \leq t_0. \tag{5.124}$$

Choose $0 < t_0^\star \leq t_0$ such that

$$1 > \sup_{0 \leq t \leq t_0^\star} \left[M^\star e^{\alpha t_0} \frac{R^\star}{R} + \frac{\|e^{At} u_0 - u_0\|_{\mathscr{X}}}{R} \right. \tag{5.125}$$

$$\left. + M^\star e^{\alpha t_0} \left(1 - \frac{e^{-\alpha t_0}}{\alpha}\right) \left(M_f(t_0) + \frac{1}{R} \cdot \sup_{0 \leq s \leq t_0^\star} \|f(s,u_0)\|_{\mathscr{X}}\right)\right].$$

We argue as in Prop. 5.5.3 to establish the existence of a unique local mild solution u of (5.10) on $[0,t_0^\star)$. Since $\{f(t,u(t)) : 0 \leq t \leq t_0^\star\}$ is a bounded subset of $\mathscr{X}$, we know from Exer. 5.5.17(ii) that the following left-sided limit exists:

$$\lim_{t \to (t_0^\star)^-} u(t) = u^\star. \tag{5.126}$$

Hence, the mild solution u can be extended continuously to a function $\bar{u} : [0,t_0^\star] \to \mathscr{X}$ defined by

$$\bar{u}(t) = \begin{cases} u(t), & 0 \leq t < t_0^\star, \\ u^\star, & t = t_0^\star. \end{cases} \tag{5.127}$$

Moreover, $\bar{u}(t_0^\star) \in \mathfrak{B}_{\mathscr{X}}(u_0;R)$. (Why?)

Consider the IVP

$$\begin{cases} y'(t) = Ay(t) + f(t,y(t)), \ 0 < t < t_0^\star, \\ y(0) = u^\star. \end{cases} \tag{5.128}$$

Let $\mathscr{Y}$ be as in (5.107) and define the map $\Phi^\star : \mathscr{Y} \to \mathbb{C}\left([0,t_0^\star]; \mathscr{X}\right)$ by

$$\Phi^\star(y)(t) = e^{At} u^\star + \int_0^t e^{A(t-s)} f(s,y(s))ds. \tag{5.129}$$

Then, for any $y \in \mathscr{Y}$,

$$\|\Phi^\star(y)(t) - u_0\|_{\mathscr{X}} \le R, \; \forall 0 \le t \le t_0^\star, \tag{5.130}$$

(so that $\Phi^\star(\mathscr{Y}) \subset \mathscr{Y}$) and $\forall y, z \in \mathscr{Y}$,

$$\|\Phi^\star(y) - \Phi^\star(z)\|_{\mathscr{Y}} \le \|y - z\|_{\mathscr{Y}}. \tag{5.131}$$

(Tell why.) Hence, (5.128) has a unique mild solution y on $[0, t_0^\star)$. (Why?) As such, the function

$$u^\star(t) = \begin{cases} u(t), & 0 \le t < t_0^\star, \\ y(t), & t_0^\star \le t < 2t_0^\star, \end{cases} \tag{5.132}$$

is a mild solution of (5.10) on $[0, 2t_0^\star)$. (Check this!)

We can repeat the above process indefinitely, and since it always results in an extension of the existence interval by the same length, the original mild solution u of (5.10) is extended to $[0, T]$ after finitely many steps. This completes the proof. $\square$

Proposition 5.5.8. *Assume* (**H$_A$**), (**H5.2**), *and that there exist positive constants c_1, c_2 such that*

$$\|f(t, x)\|_{\mathscr{X}} \le c_1 \|x\|_{\mathscr{X}} + c_2, \; \forall t > 0, x \in \mathscr{X}. \tag{5.133}$$

Then, $\forall T > 0$, (5.10) has a unique mild solution on $[0, T]$.

Proof. By Thrm. 5.5.6, (5.10) has a unique mild solution u on $[0, T_{\max})$ expressed by (5.11). If $T_{\max} < \infty$, then

$$u \text{ is bounded on } [0, T_{\max}). \tag{5.134}$$

(Why?) Hence, $\lim_{t \to T_{\max}^-} u(t)$ exists and is finite, which contradicts the maximality of the interval. Hence, $T_{\max} = \infty$. $\square$

Exercise 5.5.18. Verify (5.134).

Exercise 5.5.19.
i.) Does the conclusion of Prop. 5.5.8 hold if $c_1, c_2 \in \mathbb{C}([0, \infty); (0, \infty))$?
ii.) Repeat (i), but now assuming that $c_1, c_2 \in \mathbb{L}^1([0, \infty); (0, \infty))$?

5.5.4 Long-Term Behavior

We now briefly investigate (5.10) under the assumption that a unique mild solution u exists on $[0, \infty)$. (In the context of Section 5.5.3, this means $T_{\max} = \infty$.) Typical questions of interest include:

(i) Does $\exists \lim_{t \to \infty} \|u(t)\|_{\mathscr{X}} < \infty$?
(ii) Does $\exists u^\star \in \mathbb{C}([0, \infty); \mathscr{X})$ such that $\lim_{t \to \infty} \|u(t) - u^\star(t)\|_{\mathscr{X}} = 0$?

(iii) Is the solution time-periodic?

Consider the variation of parameters formula (5.11). Various convergence issues regarding the integral term in this formula arise when studying these questions. The integrand must converge to 0 (in the $\mathscr{X}$-norm) sufficiently fast in order to guarantee the resulting improper integral converges. As part of our analysis, we must also make certain that the first term in the formula behaves properly. Indeed, even if we take $f = 0$, then while the integral term clearly converges, the mapping $t \mapsto e^{At}u_0$ can exhibit a variety of behaviors, some of which are far from being convergent. As such, we must further control the semigroup regardless of what we impose on the forcing term. This is explored in the following exercises.

Exercise 5.5.20. Consider (5.10). Assume $(\mathbf{H_A})$ and that the semigroup is contractive with $\omega < 0$ (cf. Def. 3.3.7).
i.) Show that if f is globally Lipschitz (uniformly for $t \in [0,\infty)$), then $\lim\limits_{t\to\infty} \|u(t)\|_{\mathscr{X}}$ exists and equals 0.
ii.) Assuming that (5.10) has a mild solution u on $[0,\infty)$, show that if f satisfies the sublinear growth condition (5.133), then it is still the case that $\lim\limits_{t\to\infty} \|u(t)\|_{\mathscr{X}} = 0$.
iii.) Does the conclusion of (ii) still hold if c_1 in (5.133) is assumed to belong to $\mathbb{L}^1(0,\infty;(0,\infty))$?

Exercise 5.5.21. Repeat Exer. 5.5.20 assuming the "p^{th}-power versions" of the growth conditions (like (5.59)). Do the same conclusions hold?

Exercise 5.5.22. Consider (5.10). Assume $(\mathbf{H_A})$ and that the semigroup is contractive with $\omega < 0$. For each of the following specific forms of $f(t,x)$, formulate sufficient conditions ensuring that $\lim\limits_{t\to\infty} \|u(t)\|_{\mathscr{X}} = 0$. Prove your assertions.
i.) $f(t,x) = g(t) + h(x)$
ii.) $f(t,x) = j\left(t,x,\int_0^x k(z)dz\right)$

Exercise 5.5.23. Assuming that (5.10) has a unique mild solution u on $[0,\infty)$ such that $\lim\limits_{t\to\infty} \|u(t)\|_{\mathscr{X}} = 0$, must $u \in \mathbb{L}^1(0,\infty;\mathscr{X})$? If not, impose additional restrictions ensuring that this is the case.

5.5.5   Models Revisited

We now apply the results established in Section 5.5 to draw some conclusions about each of the models discussed in Section 5.1, including some new variants.

Model II.3 Semi-linear Pharmacokinetics

Consider IVP (5.1) and complete the following exercises.

Exercise 5.5.24.

i.) Assume that D and $\overline{D}$ are globally Lipschitz. Argue that (5.1) has a unique classical solution $U = \begin{bmatrix} x \\ y \end{bmatrix}$.

ii.) Suppose D depends on an additional real parameter μ, which could be viewed as an environmental quantity that impacts concentration level. Formally, we replace $D(t,x(t),y(t))$ in (5.1) by $\widehat{D}: [0,\infty) \times \mathbb{R} \times \mathbb{R} \times \mathbb{R} \to \mathbb{R}$ given by $\widehat{D}(t,x(t),y(t),\mu)$. Assume that $\forall \mu_0$, there exists a positive constant $M(\mu_0)$ such that

$$\left| \widehat{D}(t,x,y,\mu_0) - \widehat{D}(t,\widehat{x},\widehat{y},\mu_0) \right| \leq M(\mu_0) \left[|x - \widehat{x}| + |y - \widehat{y}| \right], \tag{5.135}$$

$\forall x, \widehat{x}, y, \widehat{y} \in \mathbb{R}$ and $t > 0$.
a.) For every $\mu_0 \in \mathbb{R}$, show that (5.1) has a unique mild solution U_{μ_0} on $[0,\infty)$.
b.) Compute $\lim_{t \to \infty} U_{\mu_0}(t)$, assuming that $a < b$ in (5.1).
c.) Assume that $\mu \mapsto M(\mu)$ is a continuous mapping such that $\lim_{\mu \to 0} M(\mu) = 0$. Establish a continuous dependence estimate for (5.1) with $\widehat{D}$ in place of D.
d.) Show that $\mu \mapsto U_\mu(\cdot)$ is continuous. Explain the meaning of the result.

Exercise 5.5.25. For every $n \in \mathbb{N}$, consider the IVP

$$\begin{cases} y_n'(t) = \left(\alpha + \frac{1}{n}\right) y_n(t) + \sin(2t) f\left(y_n(t)\right) g\left(z_n(t)\right), 0 < t < T, \\ z_n'(t) = \left(\beta - \frac{1}{n}\right) z_n(t) + \sin(2t) \widehat{f}\left(y_n(t)\right) \widehat{g}\left(z_n(t)\right), 0 < t < T, \\ y_n(0) = y_0 + \frac{1}{n^2}, z_n(0) = z_0 + \frac{1}{n^2}. \end{cases} \tag{5.136}$$

Assume that $f, \widehat{f}, g, \widehat{g} \in \mathbb{C}^1(\mathbb{R}; \mathbb{R})$ and $\alpha, \beta < 0$.
i.) Prove that $\forall T > 0$, (5.136) has a unique mild solution on $[0,T]$. Determine an explicit formula for it.
ii.) Is the mild solution in (i) also a classical solution?
iii.) Let $\varepsilon > 0$. Determine a value of $N \in \mathbb{N}$ for which

$$n, m \geq N \implies \left\| \begin{bmatrix} y_n \\ z_n \end{bmatrix} - \begin{bmatrix} y_m \\ z_m \end{bmatrix} \right\|_{\mathbb{C}\left([0,T];\mathbb{R}^2\right)} < \varepsilon. \tag{5.137}$$

What does this tell you?
iv.) Does $\exists \begin{bmatrix} y^\star \\ z^\star \end{bmatrix} \in \mathbb{C}\left([0,T]; \mathbb{R}^2\right)$ such that

$$\lim_{n \to \infty} \left\| \begin{bmatrix} y_n \\ z_n \end{bmatrix} - \begin{bmatrix} y^\star \\ z^\star \end{bmatrix} \right\|_{\mathbb{C}\left([0,T];\mathbb{R}^2\right)} = 0? \tag{5.138}$$

To which related IVP is $\begin{bmatrix} y^\star \\ z^\star \end{bmatrix}$ a mild solution?

Model III.3 Semi-linear Spring-Mass Systems

We can account for damping and friction in the model by incorporating the term $\beta x'(t)$ into the left-side of the ODE (4.3). Specifically, consider the IVP

$$\begin{cases} x''(t) + \beta x'(t) + \omega^2 x(t) = 0, \ t > 0, \\ x(0) = x_0, \ x'(0) = x_1. \end{cases} \tag{5.139}$$

(Technically, $\beta = \frac{\bar{\beta}}{m}$, where $\bar{\beta}$ is the damping constant.)

Exercise 5.5.26.
i.) Reformulate (5.139) abstractly as (2.10).

ii.) Show that the eigenvalues of the matrix $\mathbf{A}$ arising in (i) are $\lambda = \frac{-\beta \pm \sqrt{\beta^2 - 4\omega^2}}{2}$. The nature of the eigenvalues depends on the sign of $\beta^2 - 4\omega^2$; there are three distinct cases to consider. (In physics, these are typically referred to as the under-damped, critically-damped, and over-damped cases.) Describe the long-term behavior of $x(t)$ in each case.

Next, equip (5.139) with an external force driven by two time-periodic functions to obtain the following generalization of (5.139):

$$\begin{cases} x''(t) + \beta x'(t) + \omega^2 x(t) = \alpha_1 \sin(\omega t) + \alpha_2 \sin(\Omega t), \ t > 0, \\ x(0) = x_0, \ x'(0) = x_1, \end{cases} \tag{5.140}$$

where $\omega, \Omega > 0$.

Exercise 5.5.27.
i.) Prove that (5.140) has a unique classical solution on $[0, T]$, $\forall T > 0$. Determine an explicit formula for this solution.

ii.) Is the solution obtained in (i) time-periodic?

Exercise 5.5.28.
i.) Replace the right-side of (5.140) by the more general function $g(x) + h(t)$. What condition(s) can be imposed on g and h to ensure the existence of a mild solution on $[0, T]$, $\forall T > 0$? Under such conditions, show that both (5.139) and (5.140) can be recovered as special cases.

ii.) Replace the right-side of (5.140) by $g_1(x) + g_1(x') + h(t)$ and repeat (i).

The motion of an oscillator in a double-well potential, described by $W(x) = -\frac{1}{2}\omega^2 x^2 + \frac{1}{4}\sigma^2 x^4$, can be modeled by the IVP

$$\begin{cases} x''(t) + \beta x'(t) = W(x), \ t > 0, \\ x(0) = x_0, \ x'(0) = x_1. \end{cases} \tag{5.141}$$

Exercise 5.5.29.
Must (5.141) have a unique global mild solution on $[0, T]$, $\forall T > 0$? If so, must it be a classical solution?

Model VII.4 Semi-linear Wave Equations

We have accounted for dispersion in a model of elementary waves and have studied a system of such coupled waves. The following is an extension of those IBVPs.

Exercise 5.5.30. Consider the following generalization of (5.3), where $z = z(x,t)$ and $w = w(x,t)$:

$$
\begin{cases}
\frac{\partial^2 z}{\partial t^2} + \alpha \frac{\partial z}{\partial t} + c^2 \frac{\partial^2 z}{\partial x^2} = f(t,x,z,w,z_t,w_t), \ 0 < x < L, t > 0, \\
\frac{\partial^2 w}{\partial t^2} + \beta \frac{\partial w}{\partial t} + c^2 \frac{\partial^2 w}{\partial x^2} = g(t,x,z,w,z_t,w_t), \ 0 < x < L, t > 0, \\
z(x,0) = z_0(x), \ \frac{\partial z}{\partial t}(x,0) = z_1(x), \ 0 < x < L, \\
w(x,0) = w_0(x), \ \frac{\partial w}{\partial t}(x,0) = w_1(x), \ 0 < x < L, \\
\frac{\partial z}{\partial x}(0,t) = \frac{\partial z}{\partial x}(L,t) = \frac{\partial w}{\partial x}(0,t) = \frac{\partial w}{\partial x}(L,t) = 0, t > 0.
\end{cases}
\tag{5.142}
$$

i.) Reformulate (5.142) abstractly as (5.10) in an appropriate space.
ii.) Formulate and prove an existence result for (5.142) in the spirit of Thrm. 5.5.2(i).
iii.) Formulate an IBVP in which the one-dimensional domain is replaced by a two-dimensional rectangular domain $\mathscr{R}$ in (5.142). Repeat (i) and (ii).
iv.) Formulate an IBVP similar to (5.142) for a system of N such coupled waves on a two-dimensional rectangular domain $\mathscr{R}$ described by $z_i = z_i(x,y,t)$, $i = 1,\ldots,N$. Repeat (i) and (ii).

Model V.4 Diffusion Revisited

The semi-linear modifications of the type suggested below can be applied to all diffusion models discussed thus far.

Exercise 5.5.31. Consider IBVP (5.4).
i.) Prove that (5.4) has a unique global mild solution on $[0,T]$, $\forall T > 0$.
ii.) Specify a regularity condition that could be imposed on z_0 ensuring that the mild solution obtained in (i) is a classical solution.

Next, consider the IBVP

$$
\begin{cases}
\frac{\partial z}{\partial t}(x,t) + \alpha \frac{\partial^2 z}{\partial x^2}(x,t) + \beta z(x,t) = f(t,z(x,t)), \ 0 < x < 2\pi, 0 < t < T, \\
\frac{\partial z}{\partial x}(0,t) = \frac{\partial z}{\partial x}(2\pi,t) = 0, \ 0 < t < T, \\
z(x,0) = \cos(2x), \ 0 < x < 2\pi,
\end{cases}
\tag{5.143}
$$

where $f : [0,T] \times \mathbb{R} \to \mathbb{R}$ is defined by

$$
f(t,w) = \begin{cases}
|\cos(t)| \sin(w), \ w > 0, \\
0, \ w \leq 0.
\end{cases}
\tag{5.144}
$$

Exercise 5.5.32.
i.) Prove that (5.143) has a unique mild solution on $[0,T]$. Must it be a classical

solution?

ii.) More generally, suppose that $f : [0, \infty) \times \mathbb{R} \to \mathbb{R}$ is defined by

$$f(t, w) = \begin{cases} g(t)h(w), & w > 0 \wedge t \geq 0, \\ 0, & w \leq 0 \wedge t \geq 0. \end{cases} \tag{5.145}$$

where $g : [0, \infty) \to \mathbb{R}$ is continuous and bounded, and $h : \mathbb{R} \to \mathbb{R}$ is globally Lipschitz. Must (5.143) have a global mild solution on $[0, \infty)$?

iii.) Let Ω denote the open unit disc in $\mathbb{R}^2$ and consider the IBVP

$$\begin{cases} \frac{\partial z}{\partial t} + \alpha \left(\frac{\partial^2 z}{\partial x^2} + \frac{\partial^2 z}{\partial y^2} \right) + \beta z = f(t, x, y, z), \ (x, y) \in \Omega, \ 0 < t < T, \\ \frac{\partial z}{\partial \mathbf{n}}(x, y, t) = 0, \ (x, y) \in \Omega, \ 0 < t < T, \\ z(x, y, 0) = z_0(x, y), \ (x, y) \in \partial\Omega = \{(x, y) : x^2 + y^2 = 1\}, \end{cases} \tag{5.146}$$

where $z = z(x, y, t)$. Here, $\frac{\partial}{\partial \mathbf{n}}$ is the outward unit normal to $\partial\Omega$ and $f : [0, T] \times \Omega \times \mathbb{R} \to \mathbb{R}$ is given by

$$f(t, x, y, z) = \begin{cases} \left| \cos(t) \cos^3(x) \cos^5(y) \right| \sin(z), & 0 \leq t \leq T, \ (x, y) \in \Omega, \ z > 0, \\ 0, & 0 \leq t \leq T, \ (x, y) \in \Omega, \ z \leq 0. \end{cases}$$

Show that (5.146) has a unique mild solution on $[0, T]$.

iv.) Formulate a continuous dependence result for (5.146) in the spirit of Prop. 5.5.4.

Model IX.1 Neural Networks continued

We now formally study IVP (5.5).

Exercise 5.5.33. Assume that $\forall 1 \leq j \leq M$, $g_j : [0, \infty) \to \mathbb{R}$ is locally Lipschitz and that $\forall 1 \leq i, j \leq M$, $\omega_{ij}(t)$ is continuous and bounded. (In fact, $\mathrm{rng}(\omega_{ij}) \subset [-1, 1]$ since ω_{ij} represents a proportion in the positive and negative directions.)

i.) Show that (5.5) has a unique local mild solution on $[0, T_0)$, for some $T_0 > 0$.

ii.) Discuss the extendability of mild solutions of (5.5).

Exercise 5.5.34. For every $\varepsilon > 0$, consider the IVP (5.5) in which ω_{ij} is replaced by an approximation $(\omega_{ij})_\varepsilon$. Assume that $\forall 1 \leq j \leq M$, $g_j : [0, \infty) \to \mathbb{R}$ is globally Lipschitz. Denote the corresponding unique mild solution of (5.5) on $[0, T]$ by $\mathbf{U}_\varepsilon$.

i.) If $\forall 1 \leq i, j \leq M$, $\lim_{\varepsilon \to 0^+} (\omega_{ij})_\varepsilon = \omega_{ij}$, show that $\forall T > 0$, $\exists \mathbf{U} \in \mathbb{C}\left([0, T]; \mathbb{R}^M\right)$ such that $\lim_{\varepsilon \to 0^+} \|\mathbf{U}_\varepsilon - \mathbf{U}\|_{\mathbb{C}} = 0$.

ii.) Establish a continuous dependence estimate for $\|\mathbf{U}_\varepsilon - \mathbf{U}\|_{\mathbb{C}}$ in terms of ε.

The forcing term g in some applications (e.g., computer science) is not necessarily Lipschitz continuous. In fact, it is common for the forcing term to model an on/off switch, expressed by a function $g : \mathbb{R} \to \{0, 1\}$ defined by

$$g \left(\sum_{j=1}^{M} \omega_{ij}(t) x_j(t) \right) = \begin{cases} 1, & \text{whenever } \sum_{j=1}^{M} \omega_{ij}(t) x_j(t) > 0, \\ 0, & \text{whenever } \sum_{j=1}^{M} \omega_{ij}(t) x_j(t) \leq 0. \end{cases} \tag{5.147}$$

Exercise 5.5.35. How would you approach (5.5) equipped with such a forcing term?

Model X.1 Spatial Pattern Formation revisited

Consider the IBVP (5.6) for $N = 2$ in which f_1 and f_2 are given by (5.9). This can be reformulated abstractly as (5.7) on the space $\mathscr{H} = \mathbb{L}^2(\Omega) \times \mathbb{L}^2(\Omega)$. (Why?)

Let $\mathbf{C} = (C_1, C_2)$. Motivated by (5.9), define the map $F : \mathscr{H} \to \mathscr{H}$ by

$$F(\mathbf{C})(\mathbf{x},t) = \begin{bmatrix} \beta_1 - \beta_2 C_1(\mathbf{x},t) + \beta_3 C_1^2(\mathbf{x},t) C_2(\mathbf{x},t) \\ \beta_4 - \beta_3 C_1^2(\mathbf{x},t) C_2(\mathbf{x},t) \end{bmatrix}. \tag{5.148}$$

For brevity, we shall write $F(\mathbf{C})$ in place of $F(\mathbf{C})(\mathbf{x},t)$.

Exercise 5.5.36.
i.) For each $R > 0$, show that

$$\|F(\mathbf{w}) - F(\mathbf{z})\|_{\mathscr{H}} \leq 2R^2 \|\mathbf{w} - \mathbf{z}\|_{\mathscr{H}}, \quad \forall \mathbf{w}, \mathbf{z} \in \mathfrak{B}_{\mathscr{H}}(0;R).$$

ii.) Deduce that (5.6) has a unique local mild solution and determine an explicit variation of parameters formula for it. To what space must the solution belong?

Without significantly increasing the complexity of the IBVP, we can replace f_1 in (5.9) by the following more general form of an activator inhibitor:

$$f_1(C_1, C_2) = \beta_1 - \beta_2 C_1 + \mu \frac{|g_1(C_1, C_2)|}{|1 + g_2(C_1, C_2)|}, \tag{5.149}$$

where $g_1 : \mathbb{R} \times \mathbb{R} \to \mathbb{R}$ is globally Lipschitz, $g_2 : \mathbb{R} \times \mathbb{R} \to (0, \infty)$ is continuous, and $\mu \in \mathbb{R}$. Such functions are used to describe the kinetics in various biological models. For instance, in the so-called *Thomas model*, the functions

$$\begin{cases} g_1(C_1, C_2) &= v_1 C_1 C_2 \\ g_2(C_1, C_2) &= v_2 C_1 + v_3 C_1^2, \end{cases} \tag{5.150}$$

are used, where $v_i > 0 \, (i = 1, 2, 3)$, while different functions appear in the *Gray-Scott model* (see **[120]**).

Exercise 5.5.37. Consider the IBVP (5.6) for $N = 2$ in which the forcing term F is now taken to be

$$F(\mathbf{C})(\mathbf{x},t) = \begin{bmatrix} \beta_1 - \beta_2 C_1(\mathbf{x},t) + \mu \frac{|g_1(C_1(\mathbf{x},t),C_2(\mathbf{x},t))|}{|1 + g_2(C_1(\mathbf{x},t),C_2(\mathbf{x},t))|} \\ \beta_4 - \beta_3 C_2(\mathbf{x},t) \end{bmatrix}. \tag{5.151}$$

Show that (5.6) has a unique local mild solution.

The dynamics of (5.6) depending on the rate constants β_i are interesting to study. As alluded to in **[120]**, certain conditions on β_i lead to the generation of a discernible pattern, while others produce apparently random behavior.

5.6   Theory for Non-Lipschitz-Type Forcing Terms

The results of the previous section were formulated under the assumption that the forcing term satisfied a Lipschitz-type growth condition on the space in which the mild solution of (5.10) was sought. This assumption was central to the development of the theory. But, forcing terms are not always that nicely behaved. For instance, consider a function $f : [0,T] \times (0,\infty) \to (0,\infty)$ satisfying

$$|f(t,x) - f(t,y)| \leq t\,|x-y|^4, \ \forall x,y \in (0,\infty), t \in [0,T], \qquad (5.152)$$

$$|f(t,x)| \leq p(t)\,|x|\ln\left(\frac{1}{|x|}\right), \ \forall x \in (0,\infty), t \in [0,T], \qquad (5.153)$$

where $p \in \mathbb{L}^1(0,\infty;(0,\infty))$.

Exercise 5.6.1. Show that f is not locally Lipschitz on its domain.

In light of Exer. 5.6.1, the theory developed in Section 5.5 does not apply to (5.10) when equipped with such a forcing term. In order to handle situations of this type, the following notion of convexity enters in.

Definition 5.6.1. Let $\mathscr{D}$ be an open subset of $\mathbb{R}$. A function $g : \mathscr{D} \subset \mathbb{R} \to \mathbb{R}$ is
i.) *convex* if

$$g\,(\alpha x + (1-\alpha)y) \leq \alpha g(x) + (1-\alpha)g(y), \ \forall 0 \leq \alpha \leq 1, x,y \in \mathscr{D}. \qquad (5.154)$$

ii.) *concave* if $-g$ is convex.

Exercise 5.6.2. It can be shown that a convex function $f : \mathscr{D} \subset \mathbb{R} \to \mathbb{R}$ is locally Lipschitz, but not conversely (see **[281]**). Show that the following functions are convex.
i.) $f_1 : \mathbb{R} \to (0,\infty)$ defined by $f_1(x) = |x|^p$, where $p \geq 1$.
ii.) $f_2 : (0,\infty) \to (0,\infty)$ defined by $f_2(x) = x\log(x)$.

Exercise 5.6.3. It can be shown that if $f : \mathbb{R} \to \mathbb{R}$ is convex and $g : \mathbb{R} \to \mathbb{R}$ is continuous, then $f \circ g$ is convex. Use this fact to show that the function $h : \mathbb{R} \to (0,\infty)$ defined by $h(x) = \left|\sin(2x) - x\cos\left(x^2\right)\right|^{5/3}$ is convex.

The remainder of this section is devoted to studying (5.10) equipped with a forcing term $f : [0,T] \times \mathscr{X} \to \mathscr{X}$ that satisfies the following hypotheses:

(H5.7) There exists $K_1 : [0,T] \times [0,\infty) \to [0,\infty)$ such that
    i.) $K_1(\cdot,x)$ is integrable, $\forall x \in [0,\infty)$,
    ii.) $K_1(t,\cdots)$ is continuous, nondecreasing, and concave, $\forall t \in [0,T]$,
    iii.) $\|f(t,x)\|_{\mathscr{X}} \leq K_1(t,\|x\|_{\mathscr{X}}), \forall t \in [0,T], x \in \mathscr{X}$.
(H5.8) There exists $K_2 : [0,T] \times [0,\infty) \to [0,\infty)$ such that

i.) $K_2(\cdot,x)$ is integrable, $\forall x \in [0,\infty)$,
ii.) $K_2(t,\cdots)$ is continuous, nondecreasing, and $K_2(t,0) = 0$, $\forall t \in [0,T]$,
iii.) $\|f(t,x) - f(t,y)\|_{\mathscr{X}} \le K_2(t,\|x-y\|_{\mathscr{X}})$, $\forall t \in [0,T]$, $x,y \in \mathscr{X}$.
(H5.9) Any function $w : [0,T] \to [0,\infty)$ which is continuous, $w(0) = 0$, and satisfies

$$w(t) \le \overline{M_A} \int_0^t K_2(s,w(s))\,ds, \quad \forall 0 \le t \le T^\star \le T, \tag{5.155}$$

must be identically 0 on $[0,T^\star]$.

Remarks. Some standard examples of functions K_1 satisfying **(H5.7)** are
1. $K_1(t,z) = az$. (In this case, a is actually a Lipschitz constant.)
2. $K_1(t,z) = p(t)\phi(z)$, where $p \in \mathbb{L}^1(0,\infty;(0,\infty))$ and $\phi : (0,\infty) \to (0,\infty)$ is a continuous, nondecreasing, concave function such that $\phi(0^+) = 0$ and $\int_0^\varepsilon \frac{1}{\phi(r)}\,dr = \infty$. Some typical choices for ϕ are

$$\phi(z) = z\ln\left(\frac{1}{z}\right) \tag{5.156}$$

$$\phi(z) = z\ln\left(\frac{1}{z}\right)\ln\left(\ln\left(\frac{1}{z}\right)\right). \tag{5.157}$$

(See **[122]**.)

Our approach begins with the sequence of successive approximations (5.93), but this time showing that $\{u_n\}$ converges to a unique function u which is a mild solution of (5.10) is more involved. This is equivalent to showing that $\{u_n\}$ is a Cauchy sequence. Taking a step back, if we could argue that

$$\|u_n - u_m\|_{\mathbb{C}([0,\tilde{T}];\mathscr{X})} \le C\int_0^t K^\star\left(\|u_n(s) - u_m(s)\|_{\mathscr{X}}\right)ds, \tag{5.158}$$

$\forall 0 \le t \le \tilde{T} \le T$, for some $K^\star$ for which the right-side of (5.158) goes to zero as $n,m \to \infty$, then we would conclude that $\{u_n\}$ is a Cauchy sequence in $\mathbb{C}\left([0,\tilde{T}];\mathscr{X}\right)$. This constitutes the foundation of our strategy. This technique has been used frequently in the study of ODEs (see **[96]**) and has recently been applied in the study of evolution equations (see **[38, 122, 229, 302, 333]**).

We shall make use of the following lemma established in **[96]**.

Lemma 5.6.2. *For every* $\beta_1, \beta_2 > 0$, $\exists 0 < T_1 \le T$ *such that the equation*

$$z(t) = \beta_1 + \beta_2 \int_0^t K_1(s,z(s))ds \tag{5.159}$$

has a continuous local solution $z : [0,T_1) \to [0,\infty)$.

Our main result is

Theorem 5.6.3. *If* $(\mathbf{H_A})$, *(H5.7) - (H5.9) hold, then* $\exists 0 < T^\star \leq T$ *such that (5.10) has a unique local mild solution* $u : [0, T^\star) \to \mathscr{X}$.

Proof. We use the successive approximations defined in (5.93), but under different assumptions. For any $\beta_1 > \overline{M_A} \|u_0\|_{\mathscr{X}}$, Lemma 5.6.2 guarantees the existence of $0 < T_1 \leq T$ for which the equation

$$z(t) = \beta_1 + \overline{M_A} \int_0^t K_1(s, z(s)) ds \tag{5.160}$$

has a unique solution $z : [0, T_1) \to [0, \infty)$.

We divide the proof into several claims.

<u>Claim 1</u>: For each $n \in \mathbb{N}$, $\|u_n(t)\|_{\mathscr{X}} \leq z(t)$, $\forall 0 \leq t < T_1 \leq T$.

Proof: By induction on n. For $n = 1$, observe that $\forall 0 \leq t < T_1$,

$$
\begin{aligned}
\|u_1(t)\|_{\mathscr{X}} &\leq \left\| e^{At} u_0 \right\|_{\mathscr{X}} + \int_0^t \left\| e^{A(t-s)} \right\|_{\mathbb{B}(\mathscr{X})} \|f(s, u_0)\|_{\mathscr{X}} ds \\
&\leq \overline{M_A} \|u_0\|_{\mathscr{X}} + \overline{M_A} \int_0^t K_1(s, \|u_0\|_{\mathscr{X}}) ds \\
&\leq \overline{M_A} \|u_0\|_{\mathscr{X}} + \overline{M_A} \int_0^t K_1(s, \overline{M_A} \|u_0\|_{\mathscr{X}}) ds \tag{5.161} \\
&\leq \overline{M_A} \|u_0\|_{\mathscr{X}} + \overline{M_A} \int_0^t K_1(s, z(s)) ds \\
&\leq z(t).
\end{aligned}
$$

Next, assume that $\|u_n(t)\|_{\mathscr{X}} \leq z(t)$, $\forall 0 \leq t < T_1$, and observe that

$$
\begin{aligned}
\|u_{n+1}(t)\|_{\mathscr{X}} &\leq \overline{M_A} \|u_0\|_{\mathscr{X}} + \int_0^t \left\| e^{A(t-s)} \right\|_{\mathbb{B}(\mathscr{X})} \|f(s, u_n(s))\|_{\mathscr{X}} ds \\
&\leq \overline{M_A} \|u_0\|_{\mathscr{X}} + \overline{M_A} \int_0^t K_1(s, z(s)) ds \text{ (by } (\mathbf{H5.7})(\mathbf{ii}), (\mathbf{iii})) \\
&\leq z(t).
\end{aligned}
$$

This proves the claim. $\Diamond$

<u>Claim 2</u>: For every $\delta_0 > 0$, $\exists 0 < T_2 \leq T_1$ such that $\forall n \in \mathbb{N}$,

$$\left\| u_n(t) - e^{At} u_0 \right\|_{\mathscr{X}} \leq \delta_0, \ \forall 0 \leq t < T_2 \leq T_1 \leq T. \tag{5.162}$$

Proof: Let $\delta_0 > 0$ be fixed. By induction on n. For $n = 1$, observe that

$$
\begin{aligned}
\left\| u_1(t) - e^{At} u_0 \right\|_{\mathscr{X}} &\leq \int_0^t \left\| e^{A(t-s)} \right\|_{\mathbb{B}(\mathscr{X})} \|f(s, u_0)\|_{\mathscr{X}} ds \\
&\leq \overline{M_A} \int_0^t K_1\left(s, \left\| e^{At} \right\|_{\mathbb{B}(\mathscr{X})} \|u_0\|_{\mathscr{X}}\right) ds \tag{5.163} \\
&\leq \overline{M_A} \int_0^t K_1(s, z(s)) ds.
\end{aligned}
$$

The continuity of z and the integrability of K_1 (and hence the absolute continuity property of the integral; cf. Section 1.8.4) guarantees the existence of $0 < T_2 \leq T_1$ such that

$$\overline{M_A} \int_0^t K_1(s, z(s)) \, ds \leq \delta_0, \; \forall 0 \leq t < T_2 \leq T_1 \leq T. \tag{5.164}$$

(Why?) As such, (5.162) holds for $n = 1$.

Next, assume that (5.164) holds for n and observe that $\forall 0 \leq t < T_2 \leq T_1 \leq T$,

$$\begin{aligned}
\left\| u_{n+1}(t) - e^{At} u_0 \right\|_{\mathscr{X}} &\leq \int_0^t \left\| e^{A(t-s)} \right\|_{\mathbb{B}(\mathscr{X})} \left\| f(s, u_n(s)) \right\|_{\mathscr{X}} ds \\
&\leq \overline{M_A} \int_0^t K_1(s, z(s)) \, ds \; \text{(by Claim 1)} \\
&\leq \delta_0 \; \text{(by (5.164))}.
\end{aligned}$$

This proves the Claim. $\Diamond$

Claim 3: For every $n, m \in \mathbb{N}$,

$$\left\| u_{n+m}(t) - u_n(t) \right\|_{\mathscr{X}} \leq \overline{M_A} \int_0^t K_2(s, 2\delta_0) \, ds, \; \forall 0 \leq t < T_2. \tag{5.165}$$

Proof: Let $n, m \in \mathbb{N}$. The monotonicity of K_2 implies that $\forall 0 \leq t < T_2 \leq T_1 \leq T$,

$$\begin{aligned}
\left\| u_{n+m}(t) - u_n(t) \right\|_{\mathscr{X}} &\leq \overline{M_A} \int_0^t \left\| f(s, u_{n+m-1}(s)) - f(s, u_{n-1}(s)) \right\|_{\mathscr{X}} ds \\
&\leq \overline{M_A} \int_0^t K_2(s, \left\| u_{n+m-1}(s) - u_{n-1}(s) \right\|_{\mathscr{X}}) \, ds \\
&\leq \overline{M_A} \int_0^t K_2\left(s, \left\| u_{n+m-1}(s) - e^{As} u_0 \right\|_{\mathscr{X}} \right. \\
&\quad + \left. \left\| e^{As} u_0 - u_{n-1}(s) \right\|_{\mathscr{X}} \right) ds \\
&\leq \overline{M_A} \int_0^t K_2(s, 2\delta_0) \, ds.
\end{aligned} \tag{5.166}$$

This proves the Claim. $\Diamond$

Now, define $\gamma_n : [0, T_2] \to (0, \infty)$ and $\theta_{m,n} : [0, T_2] \to (0, \infty)$ by

$$\gamma_1(t) = \overline{M_A} \int_0^t K_2(s, 2\delta_0) \, ds,$$

$$\gamma_n(t) = \overline{M_A} \int_0^t K_2(s, \gamma_{n-1}(s)) \, ds, \; n \geq 2, \tag{5.167}$$

$$\theta_{m,n}(t) = \left\| u_{n+m}(t) - u_n(t) \right\|_{\mathscr{X}}, \; n, m \in \mathbb{N}. \tag{5.168}$$

The integrability of K_2 (by **(H5.8)(ii)**) (and hence the absolute continuity of the integral) guarantees the existence of $0 < T_3 \leq T_2$ such that

$$\gamma_1(t) \leq 2\delta_0, \; \forall 0 \leq t < T_3. \tag{5.169}$$

Claim 4: For every $n \geq 2$,

$$\gamma_n(t) \leq \gamma_{n-1}(t) \leq \ldots \leq \gamma_1(t), \; \forall 0 \leq t < T_3. \tag{5.170}$$

Proof: By induction on n. For $n = 2$, we use (5.167) and (5.169) to conclude that $\forall 0 \leq t < T_3$,

$$\gamma_2(t) = \overline{M_A} \int_0^t K_2(s, \gamma_1(s)) \, ds \leq \overline{M_A} \int_0^t K_2(s, 2\delta_0) \, ds = \gamma_1(t).$$

Next, assuming that (5.170) holds for n, it follows immediately that $\forall 0 \leq t < T_3$,

$$\gamma_{n+1}(t) = \overline{M_A} \int_0^t K_2(s, \gamma_n(s)) \, ds \leq \overline{M_A} \int_0^t K_2(s, \gamma_{n-1}(s)) \, ds = \gamma_n(t).$$

This proves the claim. $\Diamond$

Claim 5: For every $n, m \in \mathbb{N}$,

$$\theta_{m,n}(t) \leq \gamma_n(t), \; \forall 0 \leq t < T_3. \tag{5.171}$$

Proof: By induction on n. For $n = 1$, note that $\forall m \in \mathbb{N}$ and $0 \leq t < T_3$, Claim 3 implies that

$$\theta_{m,1}(t) = \|u_{m+1}(t) - u_1(t)\|_{\mathscr{X}} \leq \overline{M_A} \int_0^t K_2(s, 2\delta_0) \, ds = \gamma_1(t).$$

Next, assume that (5.171) holds for a fixed n, uniformly $\forall m \in \mathbb{N}$. Observe that $\forall m \in \mathbb{N}$ and $0 \leq t < T_3$,

$$\begin{aligned}
\theta_{m,(n+1)}(t) &= \|u_{n+1+m}(t) - u_{n+1}(t)\|_{\mathscr{X}} \\
&\leq \overline{M_A} \int_0^t \|f(s, u_{n+m}(s)) - f(s, u_n(s))\|_{\mathscr{X}} \, ds \\
&\leq \overline{M_A} \int_0^t K_2(s, \|u_{n+m}(s) - u_n(s)\|_{\mathscr{X}}) \, ds \\
&= \overline{M_A} \int_0^t K_2(s, \theta_{m,n}(s)) \, ds \\
&\leq \overline{M_A} \int_0^t K_2(s, \gamma_n(s)) \, ds \\
&= \gamma_{n+1}(t).
\end{aligned}$$

This proves the claim. $\Diamond$

Observe that Claim 4 implies that

$$\begin{cases} \text{For every } t \in [0, T_3), \; \{\gamma_n(t)\} \text{ is decreasing in } n. \\ \text{For every } n_0 \in \mathbb{N}, \; \{\gamma_{n_0}(t)\} \text{ is increasing in } t. \end{cases} \tag{5.172}$$

<u>Claim 6</u>: There exists $u \in \mathbb{C}\left([0, T_3); \mathscr{X}\right)$ such that $\lim_{n \to \infty} \|u_n - u\|_{\mathbb{C}([0,T_3); \mathscr{X})} = 0$.

Proof: Define $\gamma: [0, T_3) \to \mathbb{R}$ by

$$\gamma(t) = \lim_{n \to \infty} \gamma_n(t) = \inf_{n \in \mathbb{N}} \gamma_n(t). \tag{5.173}$$

The well-definedness of $\gamma(\cdot)$ follows from (5.172). It is easy to see that γ is nonnegative and $\gamma(0) = 0$. It turns out that γ is also continuous. Indeed, observe that since $K_2\left(s, \gamma_n(s)\right) \leq K_2\left(s, \gamma_1(s)\right), \forall n \in \mathbb{N}$ and $\forall s \in [0, T_3)$, we can use LDC to obtain the following string of equalities:

$$
\begin{aligned}
\gamma(t) &= \lim_{n \to \infty} \gamma_{n+1}(t) \\
&= \lim_{n \to \infty} \overline{M_A} \int_0^t K_2\left(s, \gamma_n(s)\right) ds \\
&= \overline{M_A} \int_0^t \lim_{n \to \infty} K_2\left(s, \gamma_n(s)\right) ds \qquad (5.174) \\
&= \overline{M_A} \int_0^t K_2\left(s, \lim_{n \to \infty} \gamma_n(s)\right) ds \\
&= \overline{M_A} \int_0^t K_2\left(s, \gamma(s)\right) ds.
\end{aligned}
$$

Note that the continuity (in the second variable) of K_2 was used in going from line three to four in (5.174). Since the right-side of (5.174) is a continuous function of t, we conclude that γ is continuous. Moreover, γ satisfies

$$\gamma(t) \leq \overline{M_A} \int_0^t K_2\left(s, \gamma(s)\right) ds, \ \forall 0 \leq t < T_3.$$

Hence, **(H5.9)** implies that

$$\gamma(t) = 0, \ \forall 0 \leq t < T_3. \tag{5.175}$$

Further, using Claim 5 and (5.172) yields

$$\|u_{n+m} - u_n\|_{\mathbb{C}([0,T_3); \mathscr{X})} = \sup_{0 \leq t < T_3} \theta_{m,n}(t) \leq \sup_{0 \leq t < T_3} \gamma_n(t) \leq \gamma_n(T_3). \tag{5.176}$$

The right-side of (5.176) goes to zero as $n \to \infty$ by (5.173) and (5.175). Consequently, $\{u_n\}$ is a Cauchy sequence in $\mathbb{C}\left([0, T_3); \mathscr{X}\right)$, and hence convergent to some $u \in \mathbb{C}\left([0, T_3); \mathscr{X}\right)$. This proves the claim. $\Diamond$

<u>Claim 7</u>: The function u from Claim 6 is a mild solution of (5.10) on $[0, T_3)$.

Proof: Let

$$y(t) = e^{At} u_0 + \int_0^t e^{A(t-s)} f(s, u(s)) ds.$$

We must argue that $\lim_{n \to \infty} \|u_n - y\|_{\mathbb{C}([0,T_3);\mathscr{X})} = 0$. To this end, observe that

$$\|u_n(t) - y(t)\|_{\mathscr{X}} = \left\| \int_0^t e^{A(t-s)} \left[f(s, u_{n-1}(s)) - f(s, u(s)) \right] ds \right\|_{\mathscr{X}}$$

$$\leq \overline{M_A} \int_0^t K_2 \left(s, \|u_{n-1}(s) - u(s)\|_{\mathscr{X}} \right) ds$$

$$\leq \overline{M_A} \int_0^t K_2 \left(s, \|u_{n-1} - u\|_{\mathbb{C}([0,T_3);\mathscr{X})} \right) ds$$

and so,

$$\|u_n - y\|_{\mathbb{C}([0,T_3);\mathscr{X})} \leq \overline{M_A} \int_0^T K_2 \left(s, \|u_{n-1} - u\|_{\mathbb{C}([0,T_3);\mathscr{X})} \right) ds.$$

Taking the limit as $n \to \infty$ now yields (due to Claim 6, the continuity of K_2, and LDC)

$$0 \leq \lim_{n \to \infty} \|u_n - y\|_{\mathbb{C}([0,T_3);\mathscr{X})} \leq \overline{M_A} \int_0^T K_2(s,0)\, ds = 0,$$

as needed. This proves the claim. $\Diamond$

Claim 8: The mild solution of (5.10) on $[0, T_3)$ from Claim 7 is unique.
Proof: Let $x, y \in \mathbb{C}([0, T_3); \mathscr{X})$ be two mild solutions of (5.10). Subtracting their representation formulae and using **(H5.8)(iii)** yields

$$\|x(t) - y(t)\|_{\mathscr{X}} \leq \overline{M_A} \int_0^t K_2(s, \|x(s) - y(s)\|_{\mathscr{X}})\, ds, \ \forall 0 \leq t < T_3.$$

Thus, by **(H5.9)**, we conclude that $\|x(t) - y(t)\|_{\mathscr{X}} = 0$, $\forall t \in [0, T_3)$. This proves the claim. $\Diamond$

The proof of the theorem is complete. $\qquad\qquad\qquad\qquad\qquad\qquad\qquad$ $\square$

Exercise 5.6.4. What additional assumption can be imposed to ensure the solution can be extended to the closed interval $[0, T_3]$? Once this is done, clearly explain the process used to extend the solution to the entire interval $[0, T]$.

Remarks.
1. While we have consistently defined the notion of a mild solution on a closed interval, the definition remains the same if the endpoint(s) are excluded (as in Claim 7 in the above proof). That said, we do not continue this line of reasoning to define a mild solution on a union of disjoint intervals.
2. If $\left\{ e^{At} : t \geq 0 \right\}$ is contractive, then $\sup \left\{ \|e^{At}\|_{\mathbb{B}(\mathscr{X})} \leq 1 : 0 \leq t \leq T \right\} \leq 1$. In such case, the calculations in the proof of Thrm. 5.6.3 simplify slightly. (Tell how.)
3. More general versions of the models considered in Section 5.5.5 can be easily

formed under conditions **(H5.7)** - **(H5.9)**. The results are of interest since uniqueness is still guaranteed, but it is more difficult to work with the hypotheses. Indeed, consider the following open exercises and pay particular attention to the difficulties you encounter when attempting them.

Open Exercise 5.6.5. Consider the IVP

$$\begin{cases} u'(t) & = Au(t) + f(t, u(t)) + g(t, u(t)), \ 0 < t < T, \\ u(0) & = u_0, \end{cases} \tag{5.177}$$

where $f : [0, T] \times \mathscr{X} \to \mathscr{X}$ and $g : [0, T] \times \mathscr{X} \to \mathscr{X}$ each satisfy **(H5.7)** and **(H5.8)** separately (with possibly different K_1, K_2) and **(H5.9)** is suitably modified. Formulate and prove an analog of Thrm. 5.6.3 for (5.177).

Open Exercise 5.6.6. Assume that $(\mathbf{H_A})$ and **(H5.7)** - **(H5.9)** hold. Establish a continuous dependence result in the spirit of Prop. 5.5.4.

Open Exercise 5.6.7. For every $n \in \mathbb{N}$, consider the IVP

$$\begin{cases} u'_n(t) = Au_n(t) + f_n(t, u_n(t)), \ 0 < t < T, \\ u_n(0) = u_0^n. \end{cases} \tag{5.178}$$

Assume that $f_n : [0, T] \times \mathscr{X} \to \mathscr{X}$ satisfies **(H5.7)** - **(H5.9)** hold with K_1 and K_2 replaced by K_1^n and K_2^n, respectively, and that a global unique mild solution u_n of (5.178) exists. If

$$K_1^n(t, x) = p_n(t) x \ln\left(\frac{1}{x}\right) \ln\left(\ln\left(\frac{1}{x}\right)\right),$$

where $\{p_n\} \subset \mathbb{L}^1(0, \infty; \mathbb{R})$ is such that $\lim_{n \to \infty} \|p_n - p\|_{\mathbb{L}^1(0, \infty; \mathbb{R})} = 0$, prove that $\exists u \in \mathbb{C}([0, T]; \mathscr{X})$ such that $\lim_{n \to \infty} \|u_n - u\|_{\mathbb{C}([0, T]; \mathscr{X})} = 0$.

5.7 Theory under Compactness Assumptions

We now investigate (5.10) under weaker conditions on f that no longer ensure the uniqueness of a mild solution. Naively, there is a total amount of regularity needed to ensure the existence of a mild solution of (5.10), and it can be distributed between A and f. Indeed, in Sections 5.5 and 5.6 we assumed that A merely generated a C_0−semigroup on $\mathscr{X}$ at the expense of imposing strong growth restrictions on f. However, the forcing term arising in an application might be only continuous or even less regular. We naturally ask whether or not (5.10) must have a mild solution without imposing additional restrictions on A. The answer is NO, because the Heine-Borel theorem (cf. Thrm. 1.4.14) does not hold in a general Banach space. This leads to the following proposition.

Proposition 5.7.1. *For a general Banach space $\mathscr{X}$, if $f : [0,T] \times \mathscr{X} \to \mathscr{X}$ is merely continuous, then assumption* $(\mathbf{H_A})$ *does not guarantee the existence of a mild solution of (5.10).*

An example illustrating Prop. 5.7.1 can be found in **[267]**. Note, however, that the continuity of f is sufficient if $\mathscr{X} = \mathbb{R}^N$ because we do not encounter the problem mentioned above. The existence result in this case is called *Peano's theorem* and follows as a special case of the results developed later in this section.

Our only recourse is to impose additional restrictions on the operator A. But how? Consider the solution map Φ defined in (5.65) under the assumption that f is continuous. Certainly the Contraction Mapping Theorem will not be useful since we can no longer control the integral term in the same manner. (Why?) However, perhaps we could use a different fixed-point theorem, such as Thrm. 5.2.6 or Thrm. 5.2.14. We must identify an additional natural restriction that could be imposed on $\{e^{At} : 0 \leq t \leq T\}$ in order to facilitate the use of one of these theorems. One such possibility is to consider semigroups of the following type.

Definition 5.7.2. A C_0–semigroup $\{e^{At} : 0 \leq t \leq T\}$ is *compact* if $e^{At} : \mathscr{X} \to \mathscr{X}$ is a compact operator, $\forall 0 < t \leq T$.

The exclusion of $t = 0$ in Def. 5.7.2 is important because the identity operator $I : \mathscr{X} \to \mathscr{X}$ is compact iff $\mathscr{X}$ is finite-dimensional (see **[237]**), and we are trying to develop a theory for a general (possibly infinite-dimensional) Banach space $\mathscr{X}$.

The notion of a compact semigroup is not artificial and arises in practical applications. Indeed, we have the following:

Some Common Examples:
1. If $\mathbf{A} \in \mathbb{M}^N(\mathbb{R})$, then it follows immediately from Prop. 5.2.12 that $\{e^{At} : 0 \leq t \leq T\}$ is compact on $\mathbb{R}^N$.
2. Let Ω be a bounded domain in $\mathbb{R}^N$ with smooth boundary. The semigroup generated by the Laplacian operator $\triangle$ on $\mathbb{L}^2(\Omega)$ is compact (see **[118]**).
3. The Sobolev operator introduced in (3.56) - (3.57) generates a compact semigroup on $\mathbb{L}^2(0,\pi;\mathbb{R})$. (Why?)

We shall make use of the following hypothesis throughout this section.
$(\mathbf{H_A^{\star}})$ A generates a compact semigroup $\{e^{At} : 0 \leq t \leq T\}$ on $\mathscr{X}$.

The following is one of two main results presented in this section.

Theorem 5.7.3. *Assume* $(\mathbf{H_A^{\star}})$ *and*
(H5.10) $f : [0,T] \times \mathscr{X} \to \mathscr{X}$ *is a continuous map for which* $\exists k_1, k_2 \in \mathbb{C}([0,T];\mathbb{R})$ *such that*

$$\|f(t,x)\|_{\mathscr{X}} \leq k_1(t)\|x\|_{\mathscr{X}} + k_2(t), \ \forall t \in [0,T], x \in \mathscr{X}. \tag{5.179}$$

Then, (5.10) has at least one global mild solution on $[0,T]$.

Proof. Consider the solution map $\Phi : \mathbb{C}([0,T];\mathcal{X}) \to \mathbb{C}([0,T];\mathcal{X})$ defined by (5.65). For illustration purposes, we prove the theorem using two different approaches.

Approach 1: Using Schaefer's Fixed-Point Theorem

Step 1: Show that Φ is a compact operator.

Let $K > 0$ and define the ball $\mathfrak{B} = \{v \in \mathbb{C}([0,T];\mathcal{X}) : \|v\|_{\mathbb{C}} \leq K\}$. We shall show that $\Phi(\mathfrak{B})$ is precompact in $\mathbb{C}([0,T];\mathcal{X})$ using Thrm. 5.2.8.

<u>Claim 1</u>: For every $0 \leq t_0 \leq T$, $\{(\Phi v)(t_0) : v \in \mathfrak{B}\}$ is precompact in $\mathcal{X}$.
Proof: This is clear for $t_0 = 0$. (Why?) Let $0 < t_0 \leq T$ be fixed and let $0 < \varepsilon < t_0$. Define $\Phi_\varepsilon : \mathbb{C}([0,T];\mathcal{X}) \to \mathbb{C}([0,T];\mathcal{X})$ by

$$(\Phi_\varepsilon v)(t_0) = e^{At_0}u_0 + e^{A\varepsilon} \int_0^{t_0-\varepsilon} e^{A(t_0-\varepsilon-s)} f(s,v(s))ds. \qquad (5.180)$$

Since $\left\{ \int_0^{t_0-\varepsilon} e^{A(t_0-\varepsilon-s)} f(s,v(s))ds : v \in \mathfrak{B} \right\}$ is bounded (Why?), we know that $K_\varepsilon = \{(\Phi_\varepsilon v)(t_0) : v \in \mathfrak{B}\}$ is precompact in $\mathcal{X}$, and hence totally bounded in $\mathcal{X}$.

Next, observe that $\forall v \in \mathfrak{B}$,

$$\|(\Phi v)(t_0) - (\Phi_\varepsilon v)(t_0)\|_{\mathcal{X}} \leq \left\| \int_{t_0-\varepsilon}^{t_0} e^{A(t_0-s)} f(s,v(s))ds \right\|_{\mathcal{X}}$$

$$\leq M^* e^{\alpha t_0} \int_{t_0-\varepsilon}^{t_0} \|f(s,v(s))\|_{\mathcal{X}} \, ds$$

$$\leq M^* e^{\alpha t_0} \int_{t_0-\varepsilon}^{t_0} [k_1(s)\|v(s)\|_{\mathcal{X}} + k_2(s)] \, ds$$

$$\leq \xi \cdot \varepsilon,$$

where $\xi = M^* e^{\alpha t_0} [K\|k_1\|_{\mathbb{C}} + \|k_2\|_{\mathbb{C}}]$. Thus, by Strategy $\mathscr{P}$, $\{(\Phi v)(t_0) : v \in \mathfrak{B}\}$ is totally bounded in $\mathcal{X}$, and hence precompact in $\mathcal{X}$. This proves the claim. ◊

<u>Claim 2</u>: $\{\Phi(v) : v \in \mathfrak{B}\}$ is equicontinuous.
Proof: Let $\varepsilon > 0$. There exist $\delta_1, \delta_2 > 0$ such that

$$|s-t| < \delta_1 \implies \|e^{At} - e^{As}\|_{\mathbb{B}(\mathcal{X})} < \frac{\varepsilon}{3(\|u_0\|_{\mathcal{X}} + 1)}, \qquad (5.181)$$

$$|s-t| < \delta_2 \implies \|e^{A(t-\tau)} - e^{A(s-\tau)}\|_{\mathbb{B}(\mathcal{X})} < \frac{\varepsilon}{3T[K\|k_1\|_{\mathbb{C}} + \|k_2\|_{\mathbb{C}} + 1]} \qquad (5.182)$$

Let

$$\delta = \min\left\{ \delta_1, \delta_2, \frac{\varepsilon}{3\overline{M_A}[K\|k_1\|_{\mathbb{C}} + \|k_2\|_{\mathbb{C}} + 1]} \right\} > 0. \qquad (5.183)$$

Observe that $\forall 0 \le s < t \le T$ for which $|s-t| < \delta$, using (5.181) - (5.183) yields

$$
\begin{aligned}
\|(\Phi v)(t) - (\Phi v)(s)\|_{\mathscr{X}} &\le \left\|\left(e^{At} - e^{As}\right)u_0\right\|_{\mathscr{X}} \\
&\quad + \left\|\int_0^t e^{A(t-\tau)} f(\tau, v(\tau)) d\tau - \int_0^s e^{A(s-\tau)} f(\tau, v(\tau)) d\tau\right\|_{\mathscr{X}} \\
&\le \left\|\left(e^{At} - e^{As}\right)u_0\right\|_{\mathscr{X}} \\
&\quad + \int_0^s \left\|e^{A(t-\tau)} - e^{A(s-\tau)}\right\|_{\mathbb{B}(\mathscr{X})} \|f(\tau, v(\tau))\|_{\mathscr{X}} d\tau \\
&\quad + \int_s^t \left\|e^{A(t-\tau)}\right\|_{\mathbb{B}(\mathscr{X})} \|f(\tau, v(\tau))\|_{\mathscr{X}} d\tau \qquad (5.184) \\
&\le \left\|\left(e^{At} - e^{As}\right)u_0\right\|_{\mathscr{X}} \\
&\quad + \int_0^s \left\|e^{A(t-\tau)} - e^{A(s-\tau)}\right\|_{\mathbb{B}(\mathscr{X})} [K\|k_1\|_{\mathbb{C}} + \|k_2\|_{\mathbb{C}}] d\tau \\
&\quad + \overline{M_A} [K\|k_1\|_{\mathbb{C}} + \|k_2\|_{\mathbb{C}}] |t-s| \\
&< \varepsilon,
\end{aligned}
$$

$\forall v \in \mathfrak{B}$. This proves the claim. $\diamond$

So, by Thrm. 5.2.8, we conclude that $\Phi(\mathfrak{B})$ is precompact in $\mathbb{C}([0,T];\mathscr{X})$. Thus, Φ is a compact map. This concludes Step 1.

Step 2: Show that Φ is continuous.
Proof: Let $\{v_n\} \subset \mathbb{C}([0,T];\mathscr{X})$ be such that $\lim_{n\to\infty} \|v_n - v\|_{\mathbb{C}} = 0$. We must show that $\lim_{n\to\infty} \|\Phi(v_n) - \Phi(v)\|_{\mathbb{C}} = 0$. To this end, note that $\forall 0 \le s \le T$,

$$
\|f(s, v_n(s)) - f(s, v(s))\|_{\mathscr{X}} \le 2\|k_2\|_{\mathbb{C}} + \|k_1\|_{\mathbb{C}} [\|v_n(s)\|_{\mathscr{X}} + \|v(s)\|_{\mathscr{X}}]. \quad (5.185)
$$

Since $\lim_{n\to\infty} \|v_n - v\|_{\mathbb{C}} = 0$, $\exists \overline{K} > 0$ such that

$$
\|v_n\|_{\mathbb{C}} \le \overline{K}, \forall n \in \mathbb{N}, \text{ and } \|v\|_{\mathbb{C}} \le \overline{K}. \quad (5.186)
$$

Moreover, the continuity of f guarantees that $\forall 0 \le s \le T$,

$$
\lim_{n\to\infty} \|f(s, v_n(s)) - f(s, v(s))\|_{\mathscr{X}} = 0. \quad (5.187)
$$

As such, $\{\|f(s, v_n(s)) - f(s, v(s))\|_{\mathscr{X}} : n \in \mathbb{N}, s \in [0,T]\}$ is uniformly bounded, and so by LDC, $\exists N \in \mathbb{N}$ such that

$$
n \ge N \implies \int_0^T \|f(s, v_n(s)) - f(s, v(s))\|_{\mathscr{X}} ds < \frac{\varepsilon}{\overline{M_A}}. \quad (5.188)
$$

Hence, $\forall 0 \le t \le T$ and $n \ge N$,

$$
\|\Phi(v_n)(t) - \Phi(v)(t)\|_{\mathscr{X}} \le \overline{M_A} \int_0^T \|f(s, v_n(s)) - f(s, v(s))\|_{\mathscr{X}} ds \quad (5.189)
$$
$$
< \varepsilon.
$$

Hence, taking the supremum over $[0, T]$ in (5.189) yields

$$n \geq N \Longrightarrow \|\Phi(v_n) - \Phi(v)\|_{\mathbb{C}} \leq \varepsilon,$$

as desired. This completes Step 2.

Step 3: The set $\xi(\Phi) = \{v \in \mathbb{C}([0,T]; \mathscr{X}) : \exists \lambda \geq 1 \text{ such that } \lambda v = \Phi v\}$ is bounded. *Proof*: Let $v \in \xi(\Phi)$ be fixed. We must produce a positive constant C^* (independent of v and λ) such that $\|v\|_{\mathbb{C}} \leq C^*$. To this end, observe that

$$
\begin{aligned}
\|v(t)\|_{\mathscr{X}} &\leq \lambda \|v(t)\|_{\mathscr{X}} \\
&\leq \left\|e^{At} u_0\right\|_{\mathscr{X}} + \left\|\int_0^t e^{A(t-s)} f(s, v(s)) ds\right\|_{\mathscr{X}} \\
&\leq \overline{M_A} \|u_0\|_{\mathscr{X}} + \overline{M_A} \int_0^t [k_1(s) \|v(s)\|_{\mathscr{X}} + k_2(s)] ds \\
&\leq \overline{M_A} (\|u_0\|_{\mathscr{X}} + T \|k_2\|_{\mathbb{C}}) + \int_0^t (\overline{M_A} k_1(s)) \|v(s)\|_{\mathscr{X}} ds.
\end{aligned}
$$

Let $C_1 = \overline{M_A}(\|u_0\|_{\mathscr{X}} + T \|k_2\|_{\mathbb{C}})$. Applying Gronwall's lemma yields

$$\|v(t)\|_{\mathscr{X}} \leq C_1 e^{\int_0^t \overline{M_A} k_1(s) ds} \leq C_1 e^{T \|k_1\|_{\mathbb{C}} \overline{M_A}} = C^*, \tag{5.190}$$

$\forall 0 \leq t \leq T$. So, taking the supremum over $[0, T]$ in (5.190) enables us to conclude that $\|v\|_{\mathbb{C}} \leq C^*, \forall v \in \xi(\Phi)$. This completes Step 3.

By virtue of Schaefer's Fixed-Point Theorem, we conclude that Φ has at least one fixed-point, which coincides with a mild solution of (5.10). This completes the proof of Thrm. 5.7.3. □

Approach 2: Using Schauder's Fixed-Point Theorem

We provide an outline of the main steps used in this approach and leave the details for you to complete. We first argue the existence of a local mild solution of (5.10). Let $R > 0$. There exist $0 < T_1, T_2, T_3 \leq T$ such that

$$\left\|e^{At} u_0 - u_0\right\|_{\mathscr{X}} < \frac{R}{3}, \forall 0 \leq t \leq T_1, \tag{5.191}$$

$$\int_0^t M^* e^{\alpha s} k_1(s) ds \leq \frac{R}{3}, \forall 0 \leq t \leq T_2, \tag{5.192}$$

$$\int_0^t M^* e^{\alpha s} [k_1(s) \|u_0\|_{\mathscr{X}} + k_2(s)] ds \leq \frac{R}{3}, \forall 0 \leq t \leq T_3. \tag{5.193}$$

Let $T^* = \min\{T_1, T_2, T_3\}$ and consider the set

$$\mathscr{D} = \{y \in \mathbb{C}([0, T^*); \mathscr{X}) \mid y(0) = u_0 \wedge \|y(t) - u_0\|_{\mathscr{X}} \leq R, \forall 0 \leq t \leq T^*\}. \tag{5.194}$$

Observe that $\mathscr{D}$ is a nonempty, closed, convex subset of $\mathbb{C}([0, T^*); \mathscr{X})$ such that $\Phi(\mathscr{D}) \subset \mathscr{D}$. (Why?) The continuity of Φ and precompactness of $\Phi(\mathscr{D})$ are argued as in Approach 1. (Do so!) Hence, we conclude from Schauder's fixed-point theorem that (5.10) has at least one mild solution on $[0, T^\star)$.

Next, we must argue that such a solution can be extended to $[0, T]$. To do so, begin by proving the existence of $\lim\limits_{t \to (T^\star)^-} \|u(t)\|_{\mathscr{X}}$. Then, using $t = T^\star$ as the new starting time, argue as before that the solution can be extended to the interval $[T^\star, 2T^\star]$, and then to $[2T^\star, 3T^\star]$, and so on to the entire interval $[0, T]$ in finitely many steps. This completes the proof. $\quad\square$

The sublinear growth condition used in **(H5.10)** can be generalized in various ways, each of which leads to a slightly different existence result. For instance, consider the following proposition.

Proposition 5.7.4. *Assume* $(\mathbf{H_A^\star})$ *and*
(H5.11) $f : [0, T] \times \mathscr{X} \to \mathscr{X}$ *is a continuous map such that*
i.) *For every* $k \in \mathbb{N}$, $\exists g_k \in \mathbb{C}([0,T];(0,\infty))$ *such that*

$$\sup_{\|x\|_{\mathscr{X}} \le k} \|f(t,x)\|_{\mathscr{X}} \le g_k(t), \tag{5.195}$$

ii.) $\lim\limits_{k \to \infty} \frac{1}{k} \int_0^T g_k(s)ds = \alpha < \infty.$
If $\overline{M_A}\,\alpha < 1$, *then (5.10) has at least one mild solution on* $[0,T]$.

Proof. We outline the main steps of the proof.

Consider the solution map Φ defined in (5.65). The continuity and compactness of Φ are established as in the proof of Thrm. 5.7.3. For every $n \in \mathbb{N}$ and consider the ball $\mathfrak{B}_n = \{x \in \mathbb{C}([0,T];\mathscr{X}) : \|x\|_{\mathbb{C}} \le n\}$. Suppose, by way of contradiction, that there does not exist $n_0 \in \mathbb{N}$ such that $\Phi(\mathfrak{B}_{n_0}) \subset \mathfrak{B}_{n_0}$. Then, $\forall k \in \mathbb{N}$, $\exists z_k \in \mathfrak{B}_k$ such that $\|\Phi(z_k)\|_{\mathbb{C}} > k$. As such,

$$1 < \frac{1}{k}\|\Phi(z_k)\|_{\mathbb{C}}, \ \forall k \in \mathbb{N}, \tag{5.196}$$

so that

$$1 \le \lim_{k \to \infty} \frac{1}{k}\|\Phi(z_k)\|_{\mathbb{C}}. \tag{5.197}$$

For every $0 \le t \le T$,

$$\|\Phi(z_k)(t)\|_{\mathscr{X}} \le \overline{M_A}\|u_0\|_{\mathscr{X}} + \overline{M_A}\int_0^T \|f(s,z_k(s))\|_{\mathscr{X}}\,ds. \tag{5.198}$$

The fact that $z_k \in \mathfrak{B}_k$ implies that $\|z_k(s)\|_{\mathscr{X}} \le k$, $\forall 0 \le s \le T$. Hence, **(H5.10)(i)** guarantees the existence of $g_k \in \mathbb{C}([0,T];\mathscr{X})$ such that

$$\|f(s,z_k(s))\|_{\mathscr{X}} \le g_k(s), \ \forall 0 \le s \le T. \tag{5.199}$$

Using (5.199) in (5.198) yields

$$\|\Phi(z_k)\|_{\mathbb{C}} \le \overline{M_A}\left(\|u_0\|_{\mathscr{X}} + \int_0^T g_k(s)ds\right)$$

and subsequently

$$\lim_{k\to\infty} \frac{1}{k}\|\Phi(z_k)\|_{\mathbb{C}} \leq \lim_{k\to\infty}\left[\frac{1}{k}\overline{M_A}\|u_0\|_{\mathscr{X}} + \overline{M_A}\frac{1}{k}\int_0^T g_k(s)ds\right] = \overline{M_A}\alpha < 1.$$

But, this contradicts (5.197). As such, there must exist $n_0 \in \mathbb{N}$ such that $\Phi\left(\mathscr{B}_{n_0}\right) \subset \mathscr{B}_{n_0}$. The result then follows from an application of Schauder's fixed-point theorem. □

Remarks.
1. The assumption "$\overline{M_A}\alpha < 1$" simplifies to "$\alpha < 1$" if $\{e^{At} : t \geq 0\}$ is contractive.
2. An alternate proof using a so-called Peano approximation scheme can be found in **[267, 318]**.

Exercise 5.7.1. If we assume instead that $g_k \in \mathbb{L}^1(0,T;(0,\infty))$ in **(H5.11)**, does the conclusion of Prop. 5.7.4 still hold? Explain carefully.

Exercise 5.7.2.
i.) Let $f(t,u) = g(t)h(u)$ in (5.10). Impose sufficient conditions on g and h in the spirit of Prop. 5.7.4 that ensure that (5.10) has at least one mild solution on $[0,T]$.
ii.) If we take $h(u) = \cos\left(\sqrt{|u|}\right)$ in (i), is the existence of at least one global mild solution of (5.10) guaranteed? Is such a solution continuable to $[0,\infty)$?

Exercise 5.7.3. Let $h \in \mathbb{C}([0,T];\mathscr{X})$ and $\alpha > 0$. Define $f : [0,T] \times \mathscr{X} \to \mathscr{X}$ by

$$f(t,u) = \frac{h(u)}{1 + \alpha\|h(u)\|_{\mathscr{X}}}. \tag{5.200}$$

i.) Let $\mathscr{D}$ be a bounded subset of $\mathscr{X}$. Show that $f([0,T] \times \mathscr{D})$ is a bounded subset of $\mathscr{X}$.
ii.) Is (i) sufficient to guarantee the existence of at least one global mild solution of (5.10) under assumption $(\mathbf{H}_{\mathbf{A}}^{\star})$? Explain.

Exercise 5.7.4. Assume that $A : \text{dom}(A) \subset \mathscr{X} \to \mathscr{X}$ and $B : \text{dom}(B) \subset \mathscr{Y} \to \mathscr{Y}$ satisfy $(\mathbf{H}_{\mathbf{A}}^{\star})$ on $\mathscr{X}$ and $\mathscr{Y}$, respectively. Consider the system

$$\begin{cases} u'(t) &= Au(t) + f(t,u(t),v(t)), \ 0 < t < T, \\ v'(t) &= Bv(t) + g(t,u(t),v(t)), \ 0 < t < T, \\ u(0) &= u_0, \ v(0) = v_0, \end{cases} \tag{5.201}$$

in $\mathscr{X} \times \mathscr{Y}$. Here, $u : [0,T] \to \mathscr{X}$ and $v : [0,T] \to \mathscr{Y}$.
i.) If $f : [0,T] \times \mathscr{X} \times \mathscr{Y} \to \mathscr{X}$ and $g : [0,T] \times \mathscr{X} \times \mathscr{Y} \to \mathscr{Y}$ are continuous, compact mappings, must (5.201) have a global mild solution $\begin{bmatrix} u \\ v \end{bmatrix} \in \mathbb{C}([0,T];\mathscr{X} \times \mathscr{Y})$?
ii.) Replace the operator B by the operator $B + D$, where $D \in \mathbb{B}(\mathscr{Y})$, and repeat (i).

iii.) Apply the results of parts (i) and (ii) to the spatial pattern formation IBVP (5.6).

Exercise 5.7.5. Explain how Prop. 5.7.4 can be applied to the two-dimensional wave equation described by IBVP (5.146). Carefully specify conditions on the forcing term f.

5.8   A Summarizing Look Back

The following table outlines the main existence theory developed in this chapter.

Regularity on f $\mathbb{C}^1$		$\left\{ e^{At} : t \geq 0 \right\}$		Type and Nature of Solution
$\mathbb{C}^1$		C_0		Global classical
global Lipschitz		C_0		Global mild
generalized-Lipschitz		C_0		Local mild
local Lipschitz	+	C_0	$\implies$	Local mild
continuity + sublinear		compact		Global mild
continuity		compact		Global mild

Many other results can be added to this table if we are willing to further weaken the notion of a solution or impose different conditions (such as differentiability or analyticity) on the semigroup. (See **[86, 142, 246, 267, 318]** for some examples.)

5.9   Looking Ahead

The form of the forcing term arising in the mathematical modeling of applications can be too complicated to conveniently treat as a special case of the semi-linear term $f(t,u)$. For instance, consider the following two IBVPs arising in the modeling of pharmacokinetics and wave phenomena in which the forcing term describes a build-up over time of toxin in the GI tract (in the case of pharmacokinetics) and an external force (in the case of the wave equation):

Pharmacokinetics:

$$
\begin{cases}
\frac{dy}{dt} = -ay(t) + \int_0^t k(s)D(s,y(s),z(s))ds, \, t > 0, \\
\frac{dz}{dt} = ay(t) - bz(t) + \overline{D}(t,y(t),z(t)), \, t > 0, \\
y(0) = y_0, \, z(0) = 0,
\end{cases}
\tag{5.202}
$$

where $k \in \mathbb{L}^1(0,T;(0,\infty))$ is a positive, bounded function and $D, \overline{D}$ are continuous in their first variables and globally Lipschitz in their second and third variables (uniformly in t).

Coupled Wave Equations:
Let $z = z(x,t)$, and $w = w(x,t)$. Consider the IBVP

$$
\begin{cases}
\frac{\partial^2 z}{\partial t^2} + \alpha \frac{\partial z}{\partial t} + c^2 \frac{\partial^2 z}{\partial x^2} = \int_0^t a_1(t-s) f_1(s,x,z,w) ds, \, 0 \le x \le L, t > 0, \\
\frac{\partial^2 w}{\partial t^2} + \beta \frac{\partial w}{\partial t} + c^2 \frac{\partial^2 w}{\partial x^2} = \int_0^t a_2(t-s) f_2(s,x,z,w) ds, \, 0 \le x \le L, t > 0, \\
z(x,0) = z_0(x), \, \frac{\partial z}{\partial t}(x,0) = z_1(x), \, 0 \le x \le L, \\
w(x,0) = w_0(x), \, \frac{\partial w}{\partial t}(x,0) = w_1(x), \, 0 \le x \le L, \\
\frac{\partial z}{\partial x}(0,t) = \frac{\partial z}{\partial x}(L,t) = \frac{\partial w}{\partial x}(0,t) = \frac{\partial w}{\partial x}(L,t) = 0, t > 0.
\end{cases}
\tag{5.203}
$$

A key observation to make is that when formulating these IBVPs as abstract evolution equations, the forcing terms in both of them is operator-like and, in some sense, can be viewed more generally as a mapping between function spaces. Specifically, for (5.202), consider the mapping $\mathfrak{F} : \mathbb{C}\left([0,T];\mathbb{R}^2\right) \to \mathbb{L}^1\left(0,T;\mathbb{R}^2\right)$ defined by

$$
\mathfrak{F}\left(\begin{bmatrix} y \\ z \end{bmatrix}\right)(t) = \begin{bmatrix} \int_0^t k(s) D(s,y(s),z(s)) ds \\ D(t,y(t),z(t)) \end{bmatrix}.
\tag{5.204}
$$

Exercise 5.9.1. Show that the mapping $\mathfrak{F}$ defined by (5.204) is globally Lipschitz on its domain.

Exercise 5.9.2. Suppose that we view the mapping in (5.204) as one that maps into the smaller space $\mathbb{C}\left([0,T];\mathbb{R}^2\right)$ instead of $\mathbb{L}^1\left(0,T;\mathbb{R}^2\right)$. Show that $\mathfrak{F}$ remains globally Lipschitz on its domain, but with a different Lipschitz constant.

Next, define the mapping $\mathfrak{H} : \left(\mathbb{C}\left([0,T];\mathbb{L}^2(0,\pi)\right)\right)^2 \to \left(\mathbb{L}^2\left(0,T;\mathbb{L}^2(0,\pi)\right)\right)^2$ by

$$
\mathfrak{H}\left(\begin{bmatrix} z \\ w \end{bmatrix}\right)(t,\cdot) = \begin{bmatrix} \int_0^t a_1(t-s) f_1(s,\cdot,z(s,\cdot),w(s,\cdot)) ds \\ \int_0^t a_2(t-s) f_2(s,\cdot,z(s,\cdot),w(s,\cdot)) ds \end{bmatrix}.
\tag{5.205}
$$

Exercise 5.9.3.
i.) Under what conditions on a_i and f_i can you conclude that $\mathfrak{H}$ is globally Lipschitz?
ii.) Under what conditions on a_i and f_i can you conclude that $\mathfrak{H}$ has sublinear growth (like (5.179))?

Loosely speaking, we are considering a modified version of (5.10) in which the semi-linear forcing term $f(t,u)$ is replaced by a more general form $\mathfrak{F}(u)(t)$. Can the theory developed for (5.10) be adapted to study an abstract evolution equation of the form

$$
\begin{cases}
u'(t) = Au(t) + \mathfrak{F}(u)(t), \, 0 < t < T, \\
u(0) = u_0,
\end{cases}
\tag{5.206}
$$

and if so, how?

5.10   Guidance for Exercises

5.10.1   Level 1: A Nudge in a Right Direction

5.1.1. Mimic (4.2).

5.1.2. Proceed as in the discussion of (4.56).

5.1.3. Combine the approaches used in Exer. 5.1.2 and the conversion of (4.59) to (4.60). The resulting system is comprised of how many equations?

5.1.4. Mimic Exer. 4.2.1 as it applies to (4.19). What are the differences?

5.1.5. This is a system of M ODEs. (Now what?)

5.1.6. Think about the classical heat equation.

5.2.1. Φ^{n_0} has a fixed point $x^{\star}$. (Why? So what?)

5.2.2. What is the definition of completeness? How does this help?

5.2.3. "$\{y_n\}$ converges to y in $L^2(a,b;\mathbb{R})$"0 means $\lim_{n\to\infty}\int_a^b |y_n(x) - y(x)|^2\, dx = 0$. What does it mean for $\{y_n\}$ to be Cauchy in $L^2(a,b;\mathbb{R})$? Compute $\left\|\cos\left(\frac{2n\pi\cdot}{b-a}\right)\right\|_{L^2(a,b;\mathbb{R})}$, $\forall n \in \mathbb{N}$.

5.2.4. <u>Proof of (i)</u>: Why is $\mathrm{cl}_{\mathscr{Y}}\left(B\left(\{x \in \mathscr{X} : \|x\|_{\mathscr{X}} = 1\}\right)\right)$ compact in $\mathscr{Y}$?
<u>Proof of (ii)</u>: Assume that $\mathrm{rng}(B)$ is finite-dimensional. Let $\{x_n\} \subset \mathscr{X}$ be bounded. Show that $\mathscr{K} = \{Bx_n : n \in \mathbb{N}\}$ is bounded. (Now what?)

Next, assume that $\mathrm{dom}(B)$ is finite-dimensional. Use the fact that the dimension of $\mathrm{rng}(B)$ is less than or equal to the dimension of $\mathrm{dom}(B)$ together with the fact that a linear operator defined on a finite-dimensional space must be bounded. (Why?)
<u>Proof of (iii)</u>: Use (ii). (How?)

5.2.5. Use the expression for $e^{At_0}[f]$ in (5.23) and determine an upper bound for $b_{k_1}^2$ in terms of $\|f\|_{L^2(0,a_1;\mathbb{R})}^2$.

5.2.6. Mimic the approach used to prove Thrm. 5.2.15. You will need to use other properties of continuous functions (cf. Section 1.8.2).

5.2.7. Note that $\left\|\frac{x^{\star}}{\|x^{\star}\|_{\mathscr{X}}} \cdot \frac{\delta}{2}\right\|_{\mathscr{X}} < \delta$. (So what? What else do we need?)

5.3.1. Suitably modify Def. 4.2.1.

5.3.2. Think about modifying Prop. 4.1.1.

5.3.3. i.) Clearly, $x = 0$ is a solution. For others, use separation of variables and construct a piecewise-defined function.

   ii.) (a) Verify this by direct differentiation.

   (b) For a given x_0, for what value(s) of t does the denominator of $x(t)$ equal zero?

5.4.1. Use Exer. 1.8.17, taking into account the regularity of f.

5.4.2. Evaluate the expression $\frac{f\left(\frac{1}{n}\right) - f(0)}{\frac{1}{n} - 0}$ and then compute its limit as $n \to \infty$.

5.4.3. The rate of growth of the graphs of all cross-sections in the t-direction is globally bounded above over the entire space $\mathscr{D}$.

5.4.4. i.) First think about how you would argue this for real-valued functions $f : \mathbb{R} \to \mathbb{R}$. Then, generalize.

ii.) Think of a function $f : \mathbb{R} \to \mathbb{R}$ for which you can construct a sequence of secant lines all passing through some specified point whose slopes become arbitrarily large.

5.4.5. i.) Consider $\int_y^x |f(z)| \, dz$.

ii.) Does this more general $\mathscr{X}$-valued integral satisfy the same properties needed to obtain an estimate as in (i)? (So what?)

5.4.6. i.) Must there exist positive constants M_1, M_2 for which

$$\int_{v(y)}^{v(x)} |f(t,z)| \, dz \leq M_1 |v(x) - v(y)| \leq M_2 |x - y|,$$

$\forall x, y \in (0, \infty)$ and $0 \leq t \leq T$? (Now what?)

ii.) Is this any different from (i)? Explain.

5.4.7. It is natural to begin by requiring f to be globally Lipschitz in its second and third variables (uniformly in t). Under this assumption, estimate

$$\left| f\left(t, x, \int_0^x g(z) dz\right) - f\left(t, y, \int_0^y g(z) dz\right) \right|.$$

5.4.8. Estimate $\left\| \sum_{i=1}^n a_i f_i(x) - \sum_{i=1}^n a_i f_i(y) \right\|_{\mathscr{X}}$, where $x, y \in \mathscr{D}$ and $\{a_i | i = 1, \ldots, n\} \subset \mathbb{R}$.

5.4.9. Local version of Def. 5.4.2: Can you simply demand that (5.57) hold uniformly in t?

(5.56): The ball is formed using elements from what space? Does anything else change?

5.4.10. Can such a function exhibit cusp-like behavior at any point?

5.4.11. Do constant functions belong to $\mathbb{C}([0,T];\mathbb{R})$? How about $\mathbb{L}^1(0,T;\mathbb{R})$? Conversely, consider the function $f : [0,T] \times \mathbb{R} \to \mathbb{R}$ defined by $f(t,x) = x^2$. Does it satisfy either (5.58) or (5.59)?

5.4.12. Proceed as in Exer. 5.4.11.

5.4.13. i.) Estimate the quantity $\|(\mathbf{A}x + \mathbf{B}) - (\mathbf{A}y + \mathbf{B})\|_{\mathbb{R}^N}$.

ii.) Is $\mathbf{f}$ linear? If so, use Prop. 5.2.12.

5.4.14. Let $\{x_n\} \subset \mathscr{X}$ be bounded and consider $\text{cl}_{\mathscr{Y}}(\{f(x_n) : n \in \mathbb{N}\})$.

5.4.15. i.) Each of (5.60) - (5.62) yields an estimate of the form

$$\sup_{0 \leq t \leq T} \sup_{x \in \mathscr{D}} \|f(t,x)\|_{\mathscr{Y}} \leq \left[\sup_{0 \leq t \leq T} M_1(t) \right] \sup_{x \in \mathscr{D}} \|x\|_{\mathscr{X}}^p + \left[\sup_{0 \leq t \leq T} M_2(t) \right].$$

Why must each of these be finite?

5.5.1. i.) Argue that $(\Phi u)(t) \in \mathscr{X}$, $\forall u \in \mathbb{C}([0,T_0];\mathscr{X})$ and $t \in [0,T_0]$.

ii.) Estimate $\|\Phi(x) - \Phi(y)\|_{\mathbb{C}}$ to determine a condition on f.

iii.) (a) Observe that $\forall x, y \in \mathbb{C}([0,T_0];\mathscr{X})$ and $t \in [0,T_0]$,

$$\left\| \Phi^2(x)(t) - \Phi^2(y)(t) \right\|_{\mathscr{X}} = \left\| \Phi(\Phi(x))(t) - \Phi(\Phi(y))(t) \right\|_{\mathscr{X}}$$

$$= \left\| \int_0^t e^{A(t-s)} \left[f(s, \Phi(x)(s)) - f(s, \Phi(y)(s)) \right] ds \right\|_{\mathscr{X}}.$$

(Now what?)

(b) Argue as in (a) and use the result of (b). (How?)

(c) Inductively argue that $\xi_n = \frac{(M_f \overline{M_A} T)^n}{n!}$, $\forall n \in \mathbb{N}$.

(d) Argue that $\lim_{n \to \infty} \xi_n = 0$. Why can we then conclude that $\exists n_0 \in \mathbb{N}$ such that Φ^{n_0} has a fixed point? (So what?)

5.5.2. i.) Use the change of variable $z = t - s$, $dz = -ds$. How do you compute

$$\frac{d}{dt} \int_0^t e^{Az} f(t - z, u(t - z)) \, dz?$$

5.5.3. Use the linearity of A. Explain why the following is true:

$$\mathbf{A} \int_0^t \mathbf{g}(s) ds = \int_0^t \mathbf{A} \mathbf{g}(s) ds$$

5.5.4. i.) Argue that (5.73) inductively using (5.71) and (5.72). For (5.74), use the Taylor representation formula for the exponential matrix.

ii.) The maximum difference between the limit function and the approximation.

5.5.5. Think of a critical property related to computing limits that the underlying space possesses.

5.5.6. i.) Argue that $\forall h > 0$, $\|u(t + h) - u(t)\|_{\mathscr{X}} \le e^{M_f t} \|u(h) - u_0\|_{\mathscr{X}}$ and use (5.11) to estimate the right-side. (Now what?) At what point do you need to use the fact that $u_0 \in \text{dom}(A)$?

ii.) Use (i) in $J_4(h)$ to argue that the integral term goes to zero as $h \to 0$.

5.5.7. The answer is different for parts (i) and (ii) of the theorem.

5.5.8. i.) Mimic the proof of Thrm. 5.5.2(i) that uses Approach 1.

ii.) Be careful. Such a function K need not be bounded on $[0, T]$.

5.5.9. Let $\mathscr{X} = \mathbb{L}^2(\mathbb{R}; \mathbb{R})$ in (5.10). Is f globally Lipschitz on $\mathscr{X}$ (uniformly in t)? Must it belong to $\mathbb{C}^1([0, T]; \mathbb{L}^2(\mathbb{R}; \mathbb{R}))$?

5.5.10. For the mild solution argument, how does (5.76) change? As for the classical solution portion of the argument, you must determine where the growth restriction (5.105) fits in amongst the various implications involving the growth conditions. Specifically, must f be continuously Frechet differentiable if (5.105) holds?

5.5.11. i.) The only change to (5.65) occurs in the forcing term.

ii.) Use norm and integral properties with the global Lipschitz condition on f.

iii.) Compute $\|\Phi^n(x)(t) - \Phi^n(y)(t)\|_{\mathscr{X}}$, $\forall n \in \mathbb{N}$. Does a nice explicit formula emerge?

5.5.12. Continuity plays various roles here.

5.5.13. The main changes center around the specific changes made to the space estimates and limits.

5.5.14. i.) Show that $t \mapsto g(t)h(u(t))$ is locally Lipschitz.

ii.) Choose $\{g_n\} \subset \mathbb{C}([0, T]; \mathscr{X})$ such that $\lim_{n \to \infty} \|g_n - g\|_{\mathbb{L}^1} = 0$ and apply (i) to (5.115) with $g_n(\cdot)$ in place of $g(\cdot)$. Call the mild solution u_n. (Now what?)

5.5.15. i.) Subtract the variation of parameters formulae and estimate using standard techniques. What new tool must you invoke in contrast to previous results?

ii.) Identify the space in which you are estimating f.

5.5.16. i.) Use the triangle inequality on the term $\left\| f(s, u(s)) - \widehat{f}(s, v(s)) \right\|_{\mathscr{X}}$.

ii.) Once the Lipschitz-type assumption is applied to $\left\| f(s,u(s)) - \widehat{f}(s,v(s)) \right\|_{\mathscr{X}}$, argue as in (i), making appropriate modifications to the estimate.

iii.) Is there a different version of Gronwall's lemma that would be useful?

5.5.17. i.) If the expression enclosed within brackets is continuous, the limit must exist. (So what?)

ii.) Use the definition of limit directly.

iii.) Yes. (Try to argue this.) You can then "paste" the two solutions together by way of a piecewise-defined mapping.

5.5.18. Use Exer. 5.5.17(ii) with the variation of parameters formula.

5.5.19. i.) Work through all results invoked in the proof of Prop. 5.5.8 assuming the more general condition. Does everything work?

ii.) Must such functions c_1, c_2 be bounded on a finite interval? Can this obstruct the argument?

5.5.20. i.) Use the Lipschitz condition with inputs $u(\cdot)$ and 0. (Now what?)

ii.) Compute $\lim\limits_{t \to \infty} \left\| \int_0^t e^{\omega(t-s)} \left[c_1 \|u(s)\|_{\mathscr{X}} + c_2 \right] ds \right\|_{\mathscr{X}}$.

iii.) Does Gronwall's lemma apply?

5.5.21. Repeat the arguments. In each case, the main difference arises in the final estimate. (How do you proceed?)

5.5.22. There are many answers. We simply need $\lim\limits_{t \to \infty} \left\| \int_0^t e^{\omega(t-s)} f(s,u(s)) ds \right\|_{\mathscr{X}}$ to exist.

5.5.23. Can you construct an IVP in $\mathbb{R}$ for which the solution is $u(t) = \frac{1}{t}$ for $t > 1$? (So what?)

5.5.24. i.) Use Exer. 5.5.7 once you have reformulated (5.1) abstractly.

ii.) (a) Use Exer. 5.5.7.

(b) Use the variation of parameters formula and compute $e^{\mathbf{A}t}$.

(c) Use (b) with (5.135). The continuity of $\mu \mapsto M(\mu)$ enables us to control the expression $|M(\mu) - M(\hat{\mu})|$ whenever μ and $\hat{\mu}$ are sufficiently close. (So what?)

(d) Use the definition of continuity with (c). (How?)

5.5.25. i.) First, convert to matrix form; the variation of parameters formula is standard. Then, argue that the forcing term is globally Lipschitz to verify existence. (So what?)

ii.) Is the forcing term continuously differentiable in its spatial variables?

iii.) Use the fact that all of the forcing terms are globally Lipschitz. (How?)

iv.) Yes, by completeness of $\mathbb{C}\left([0,T]; \mathbb{R}^2 \right)$. Determine the limit of each term of the variation of parameters formula.

5.5.26. i.) Proceed in a manner similar to (4.3).

ii.) This follows directly from the definition of an eigenvalue.

5.5.27. i.) Convert to matrix form and then argue that the forcing term is continuously differentiable in its spatial variables.

ii.) Periodicity is easily checked using the variation of parameters formula.

5.5.28. i.) & ii.) Use Thrm. 5.5.2.

5.5.29. Use the hint for Exer. 5.5.27.

5.5.30. i.) Convert to a system in a manner similar to Exer. 5.1.3.

ii.) You must impose conditions on the given forcing terms in such a way as to ensure that the corresponding forcing term in the abstract evolution equation satisfies the correct form of the condition in the Banach space.

iii.) Proceed in a manner analogous to Exer. 3.2.7 and 4.3.7.

iv.) Proceed as in Exer. 3.2.7 using $\mathscr{H} = \left(\mathbb{H}_0^1\left(\mathbb{R};\mathbb{R}\right) \times \mathbb{L}^2\left(\mathbb{R};\mathbb{R}\right)\right)^N$. Make appropriate modifications to the mapping and variable dependencies.

5.5.31. i.) Consider Exer. 5.1.4 for the abstract formulation. Argue that the forcing term is globally Lipschitz in $\mathbb{R}^2$ (uniformly in t). The remainder of the argument is standard.

ii.) Use Thrm. 5.5.2.

5.5.32. i.) Formulate (5.143) abstractly in $\mathbb{C}\left(\left[0, 2\pi\right];\mathbb{R}\right)$. Argue that f is globally Lipschitz on this space (uniformly in t).

ii.) Is f is globally Lipschitz on this space (uniformly in t)?

iii.) Identify the forcing term abstractly and prove it is globally Lipschitz.

iv.) Use the variation of parameters formula and proceed as in Exer. 5.5.15 taking into account this particular forcing term.

5.5.33. i.) Use Prop. 5.5.3.

ii.) Can you use either Prop. 5.5.7 or Prop. 5.5.8? What property of a locally Lipschitz function might be useful here?

5.5.34. i.) & ii.) Subtract the variation of parameters formulae. Use the continuity of the integral to compute the limit.

5.5.35. The mapping g is a composition which can be very badly behaved. One thought would be to try to approximate g by a sequence of continuous functions, but this depends on the regularity of g.

5.5.36. i.) Argue that $\forall \mathbf{w}, \mathbf{z} \in \mathfrak{B}_{\mathscr{H}}\left(0;R\right)$,

$$\|F(\mathbf{w}) - F(\mathbf{z})\|_{\mathscr{H}} \leq \beta_2 \|w_1 - z_1\|_{\mathbb{L}^2} + \|w_1^2\|_{\mathbb{L}^2} \|w_2 - z_2\|_{\mathbb{L}^2}$$
$$+ \|z_2\|_{\mathbb{L}^2} \|w_1 + z_1\|_{\mathbb{L}^2} \|w_2 - z_2\|_{\mathbb{L}^2}.$$

(Now what?)

ii.) Use Thrm. 5.5.2.

5.5.37. Argue that F is locally Lipschitz using standard estimates.

5.6.1. Consider f on the the interval $(0, \varepsilon)$, for arbitrarily small ε and consider appropriate secant lines.

5.6.2. i.) Use properties of absolute value, including Minkowski's inequality.

ii.) Use the fact that $\log(\cdot)$ is increasing and convex.

5.6.3. Use Exer. 5.6.2(i) with the given fact.

5.6.4. Lemma 5.6.2 only guarantees <u>local</u> existence. This must be strengthened.

5.7.1. Yes. Use Exer. 1.8.13.

5.7.2. There are many possibilities. We just need for **(H5.11)** to hold. Suppose that h is continuous. What condition(s) must g satisfy? As for (ii), h is bounded and continuous. (So what?)

5.7.3. i.) Let $M > 0$ be such that $\|u\|_{\mathscr{X}} \leq M, \forall u \in \mathscr{D}$. Estimate the quantity

$$\sup_{0 \leq t \leq T} \sup_{u \in \mathscr{D}} \|f(t, u)\|_{\mathscr{X}}.$$

ii.) Check the conditions of Thrm. 5.7.3.

5.7.4. i.) Does the mapping $(t,x,y) \mapsto \begin{bmatrix} f(t,x,y) \\ g(t,x,y) \end{bmatrix}$ satisfy **(H5.11)**? Use properties of compactness.

ii.) What do you know about the operator $B + D$ on the space $\mathscr{Y}$?

iii.) Express (5.201) in the form (5.7) and apply (i).

5.7.5. Reformulate (5.146) as an abstract evolution equation as in Exer. 5.5.32. Impose conditions like (5.179) on the forcing term. (Then what?)

5.9.1. Let $(y_1, z_1), (y_2, z_2) \in C([0,T]; \mathbb{R}^2)$ and estimate

$$\int_0^T \left[\left| \int_0^t k(s) \left[D(s, y_1(s), z_1(s)) - D(s, y_2(s), z_2(s)) \right] ds \right| + \left| \overline{D}(t, y_1(t), z_1(t)) - \overline{D}(t, y_2(t), z_2(t)) \right| \right] dt. \quad (5.207)$$

5.9.2. How does (5.207) change?

5.9.3. Set up the left-side of the inequality in the definition of *globally Lipschitz* for (i) and *sublinear* for (ii) for such a mapping.

5.10.2 Level 2: An Additional Thrust in a Right Direction

5.1.1. Mimic the approach in Exer. 4.1.1.

5.1.2. The right-side is now $F(t, z) = z(t, \cdot)$. (Now what?)

5.1.3. The result should consist of four equations. Define the operator A by

$$A \begin{bmatrix} z \\ \frac{\partial z}{\partial t} \\ w \\ \frac{\partial w}{\partial t} \end{bmatrix} = \begin{bmatrix} 0 & I & 0 & 0 \\ -c^2 \frac{\partial^2}{\partial x^2} & \alpha & 0 & 0 \\ 0 & 0 & 0 & I \\ 0 & 0 & -c^2 \frac{\partial^2}{\partial x^2} & \beta \end{bmatrix} \begin{bmatrix} z \\ \frac{\partial z}{\partial t} \\ w \\ \frac{\partial w}{\partial t} \end{bmatrix}.$$

(Now what?)

5.1.4. Use the same forcing term as in Exer. 5.1.2.

5.1.5. Use $\mathbf{A}$ as in (2.27) with diagonal entries $a_1, \ldots, a_M$ and define $\mathbf{F} : [0, T] \times \mathbb{R}^M \to \mathbb{R}^M$ by

$$\mathbf{F}\left(t, \begin{bmatrix} x_1 \\ \vdots \\ x_M \end{bmatrix}\right) = \begin{bmatrix} \sum_{j=1}^M \omega_{1j}(t) g_j(x_j(t)) \\ \vdots \\ \sum_{j=1}^M \omega_{Mj}(t) g_j(x_j(t)) \end{bmatrix}.$$

(Now what?)

5.1.6. Try $\mathscr{H} = \mathbb{L}^2(\Omega) \times \mathbb{L}^2(\Omega)$. (Why?) What is the inner product on this space?

5.2.1. Use $\Phi^{n_0}(\Phi(x^\star)) = \Phi(\Phi^{n_0}(x^\star)) = \Phi(x^\star)$. (Now what?)

5.2.2. Where was the fact that a Cauchy sequence in a complete space must converge in the space used?

5.2.3. Does any subsequence of $\left\{ \left\| \cos\left(\frac{2n\pi \cdot}{b-a}\right) \right\|_{\mathbb{L}^2(a,b;\mathbb{R})} : n \in \mathbb{N} \right\}$ converge in $\mathscr{H}$ in the $\mathbb{L}^2$-sense?

5.2.4. <u>Proof of (i)</u>: Compact sets in $\mathscr{X}$ must be closed and bounded. (So what?)
<u>Proof of (ii)</u>: Assuming that rng(B) is finite-dimensional, now argue that the set $\mathscr{K} = \{Bx_n : n \in \mathbb{N}\}$ is precompact. (How do you then conclude?)
Next, assuming that dom(B) is finite-dimensional, the result follows from the portion of the claim formed under the assumption that rng(B) is finite-dimensional. (Why?)
<u>Proof of (iii)</u>: The portion of (ii) used depends on which one of B_1 or B_2 is compact.
5.2.5. Why must the integral term in (5.24) converge?
5.2.7. Use the fact that $F'(x_0)$ is a linear operator.
5.3.1. Look ahead at Def. 5.5.1.
5.3.2. Use the notion of Frechet differentiability.
5.3.3. i.) Try $x(t) = \left(\frac{9}{8}t\right)^{8/3}$.
   ii.) (b) Use Def. 1.8.2.
5.4.1. Is $\int_0^t \frac{\partial}{\partial s}\left(e^{A(t-s)}f(s,u(s))\right)ds$ well-defined?
5.4.2. The difference quotient simplifies to $nf\left(\frac{1}{n}\right) = 3n^{3/8}$.
5.4.3. For simplicity, visualize the surface in $\mathbb{R}^3$ and interpret this as meaning that we cannot form a sequence of tangent planes to the surface that become asymptotically closer to a plane perpendicular to the xy-plane. Intuitively, there are bounds on the slopes of each of the three cross-sections of the set of all tangent planes, and so the plane whose slope is comprised of these bounds would serve as a boundary plane controlling the steepness of the surface throughout the domain. Of course, if the function is time-dependent, these bounds can change.
5.4.4. i.) Not in general, but if the derivative is bounded, then yes. Use (5.43). (How?)

.
ii.) Use $f(x) = \sqrt{x}$ on $[0,1]$ and argue as in Exer. 5.4.2.
5.4.5. i.) Is $f(x) = \frac{1}{\sqrt{x}}$ in $\mathbb{L}^1(0,1;\mathbb{R})$? If so, this suggests an additional condition to impose on f in order to ensure that g is globally Lipschitz.
ii.) Yes. Now use standard estimates.
5.4.6. i.) Neither of these constants M_1, M_2 must exist, in general. The function f must be bounded and v would need to satisfy what condition?
ii.) We can handle this as in (i). Alternatively, is it sufficient to assume that

$$|f(t,k(z))| \le M\,|k(z)|,$$

where $k(\cdot)$ is Lipschitz?
5.4.7. Now use Exer. 5.4.5. (How?) Identify a possible Lipschitz constant.
5.4.8. Yes. Use the triangle inequality and choose $M = \max\left\{\left|a_i M_{f_i}\right| : i = 1,\ldots,n\right\}$ as the Lipschitz constant.
5.4.9. Local version of Def. 5.4.2: Yes.
   (5.56): The space is $\mathscr{X}_1 \times \ldots \times \mathscr{X}_n$ and require it to hold uniformly in t.
5.4.10. Consider $f(x) = x^{2/3}$. Show that (5.57) fails on $(-\varepsilon, \varepsilon)$, $\forall \varepsilon > 0$. (So what?)
5.4.11. Def. 5.4.2 implies (5.58) and (5.59). The converse implications do not hold.
5.4.13. i.) This mapping is globally Lipschitz with $M_f = \|\mathbf{A}\|_{\mathbb{M}^N}$.
   ii.) Yes, f is linear.
5.4.14. Use (5.54) to determine if $\text{cl}_{\mathscr{Y}}\left(\{f(x_n) : n \in \mathbb{N}\}\right)$ is precompact in $\mathscr{Y}$. (How

do you conclude?)

5.4.15. i.) Use Prop. 1.8.6(iii).

ii.) No. Consult the discussion following Thrm. 5.2.4.

5.5.1. i.) $\mathscr{X}$ is a linear space and both terms in (5.65) are in $\mathscr{X}$.

ii.) The fact that f is globally Lipschitz in the second variable (uniformly in t) would be sufficient.

iii.) (a) Use the fact that f is globally Lipschitz in the second variable (uniformly in t) and the estimate from (ii).

(c) This is a consequence of iteration.

(d) Use Cor. 5.2.3.

5.5.2. i.) Apply the hint from Exer. 5.4.1.

5.5.3. Use the definition of the integral and the fact that $\mathbf{A}$, the limit operator, and finite sums are all linear operators.

5.5.4. i.) Observe that using (5.72),

$$\mathbf{U}_{m+1}(t) = \mathbf{U}_0 + \int_0^t \mathbf{A}\mathbf{U}_m(s)ds = \mathbf{U}_0 + \int_0^t \mathbf{A}\left(\sum_{k=0}^m \left(\frac{\mathbf{A}^k s^k}{k!}\right)\mathbf{U}_0\right)ds$$

Compute the integral and simplify. For (5.74), use the Taylor representation theorem.

ii.) The maximum difference between the limit function and the approximation.

5.5.5. How about completeness?

5.5.6. i.) Use Gronwall's lemma to help draw the conclusion. The term $\|Au_0\|_{\mathscr{X}}$ appears in the estimate.

5.5.7. Part (i) holds, but part (ii) no longer does, in general. Try to determine where f being continuously differentiable is used and argue that it cannot be weakened.

5.5.8. i.) The details are similar. See Hint 2 for Exer. 5.4.15(i).

ii.) A variant can be formulated if we impose the restriction $\overline{M}_A M_f \|K\|_{\mathbb{L}^1} < 1$. (Why?) The question of necessity of such a restriction is left as an open exercise.

5.5.9. The function $h(\cdot)$ must be globally Lipschitz on $\mathscr{X} = \mathbb{L}^2(\mathbb{R};\mathbb{R})$ (Why?) and $g(\cdot)$ is bounded on $[0,T]$. Hence, argue that f satisfies (5.55), $\forall x, y \in \mathbb{L}^2(\mathbb{R};\mathbb{R})$ (uniformly in t) and invoke Exer. 5.5.7. As for whether or not the mild solution must be classical, determine if f is continuously differentiable in its second variable (uniformly in t). If so, invoke Thrm. 5.5.2(ii).

5.5.10. Determine if you need to impose a restriction of the type suggested in Exer. 5.5.8. Determining if it must be a classical solution is left as an open exercise.

5.5.11. i.) Use $(\Phi u)(t) = e^{At}u_0 + \int_0^t e^{A(t-s)}f\left(s, u(s), \int_0^s a(\tau)u(\tau)d\tau\right)ds$.

ii.) Show that $c_1 = M_f \overline{M}_A$ and $c_2 = \frac{1}{2}M_f \overline{M}_A$ work.

iii.) Now, determine if you need to impose additional restrictions on the data in order to render Φ^{n_0} a strict contraction.

5.5.12. Choose $t_1 \in [0, t_0]$ such that $\forall y \in \mathscr{Y}$, $\|\Phi(y)(t) - x_0\|_{\mathscr{X}} < 1$. Then, $\Phi(\mathscr{Y}) \subset \mathscr{Y}$. Argue that the mild solution exists on $[0, t_1]$. (Now what?)

5.5.13. Use the Lipschitz constant to determine T_0. (Now what?)

5.5.14. i.) Invoke Prop. 5.5.3.

ii.) Make certain that the same Lipschitz constant can be used for all mappings $t \mapsto g_n(t)h(u_n(t))$, independent of $n \in \mathbb{N}$. Does it? If so, prove that $\|u_n - u\|_{\mathbb{C}} \longrightarrow 0$

as $n \to \infty$, where u is the mild solution from (i). If not, how can you modify the hypotheses imposed on $h(\cdot)$ to make it work?

5.5.15. i.) Use Gronwall's lemma.

5.5.16. i.) Use $\left\| f(s,u(s)) - f(s,v(s)) + f(s,v(s)) - \widehat{f}(s,v(s)) \right\|_{\mathscr{X}}$. (Now what?)

ii.) See Exer. 5.4.11 for a similar difference.

5.5.17. i.) Use the semigroup property and the FTC with Prop. 4.1.1 to handle g.

ii.) Use the strong continuity of the semigroup with an appropriate norm estimate on the integral term that employs the boundedness of f.

iii.) Look ahead at Prop. 5.5.7.

5.5.19. i.) c_1, c_2 are both bounded on all compact subintervals of $(0, \infty)$. So, all portions of the proof likely extend.

ii.) Redo the argument from scratch, employing previous results established under such conditions.

5.5.20. i.) Apply Gronwall's lemma to $\|u(t)\|_{\mathscr{X}} \le e^{\omega t} \|u_0\|_{\mathscr{X}} \int_0^t \|u(s)\|_{\mathscr{X}} \, ds$.

ii.) Argue in a manner similar to (i), although the constant on the right-side of the inequality in (i) will change.

iii.) Be careful because integrable functions need not be continuous. Can you argue as above?

5.5.21. Is there an alternative version of Gronwall's lemma that is useful?

5.5.22. i.) Try $g \in \mathbb{L}^1 (0, \infty; \mathbb{R})$ and h is globally Lipschitz. Verify that these work.

ii.) Try j globally Lipschitz and k bounded on $[0, \infty)$. Verify that these work.

5.5.23. Must such a solution have exponential decay on $[0, \infty)$? If yes, must it belong to $\mathbb{L}^1 (0, \infty; \mathbb{R})$? (Why?)

5.5.24. i.) Argue that the forcing term $t \mapsto \begin{bmatrix} D(\cdot, y(\cdot), z(\cdot)) \\ \overline{D}(\cdot, y(\cdot), z(\cdot)) \end{bmatrix}$ is globally Lipschitz on $\mathbb{R}^2$ (uniformly in t).

ii.) (b) $\exists \omega < 0$ such that $\left\| e^{\mathbf{A}t} \right\|_{\mathbb{M}^2} \le e^{\omega t}$. (Why?) Now, use a result from Section 5.4.

(d) An approximation result for semigroups and integral properties enter into the argument. (How?)

5.5.25. i.) Substitute into the variation of parameters formula (5.10). Simplifying this yields a vector in $\mathbb{R}^2$, both components of which resemble the variation of parameters formula for an ODE. (Now what?)

ii.) Yes. Now, use Thrm. 5.5.2(ii).

iii.) You need to use Gronwall's lemma as well. This is where $\alpha, \beta < 0$ enters into the argument. (How?) The sequence is Cauchy in $\mathbb{C}([0, T]; \mathbb{R}^2)$.

iv.) Consider the IVP

$$\begin{cases} \begin{bmatrix} y^\star \\ z^\star \end{bmatrix}'(t) = \begin{bmatrix} \alpha & 0 \\ 0 & \beta \end{bmatrix} \begin{bmatrix} y^\star \\ z^\star \end{bmatrix}(t) + \sin(2t) \begin{bmatrix} f(y^\star(t)) g(z^\star(t)) \\ \widehat{f}(y^\star(t)) \widehat{g}(z^\star(t)) \end{bmatrix}, t > 0, \\ \begin{bmatrix} y^\star \\ z^\star \end{bmatrix}(0) = \begin{bmatrix} y_0 \\ z_0 \end{bmatrix}. \end{cases}$$

5.5.26. i.) The matrix form is given by

$$\begin{cases} \mathbf{X}'(t) = \begin{bmatrix} 0 & 1 \\ \omega^2 & -\beta \end{bmatrix} \mathbf{X}(t), t > 0, \\ \mathbf{X}(0) = \begin{bmatrix} x_0 \\ x_1 \end{bmatrix}. \end{cases}$$

ii.) Both eigenvalues must have negative real parts in order for $\lim_{t \to \infty} \mathbf{X}(t)$ to exist. (Why?)

5.5.27. i.) Is the regularity of the IC sufficient?

ii.) Use the variation of parameters formula to determine if $x\left(t + \frac{2\pi}{\omega\Omega}\right) = x(t), \forall t$.

5.5.28. i.) Let $g(x) = 0$ and $h(t)$ be the right-side of (5.140).

ii.) Be careful when converting this to a system.

5.5.30. ii.) Impose a global Lipschitz condition on both f and g in all four spatial variables (uniformly in t).

iii.) Replace x by (x, y) throughout and make certain to account for the Neumann BCs correctly.

iv.) The operator A is an $N \times N$ matrix; what is $\text{dom}(A)$? The BCs are standard.

5.5.31. ii.) Keep in mind that the IC must also be sufficiently regular in addition to the regularity on the forcing term.

5.5.32. i.) Is f continuously differentiable? Does the IC belong to $\text{dom}(A)$?

ii.) If the solution exists on $[0, T]$, $\forall T > 0$, then yes.

iii.) It is easier to check if f is continuously differentiable.

5.5.33. i.) Argue that the forcing term is, in fact, locally Lipschitz.

ii.) Does the solution remain bounded as the input approaches T_0 obtained in (i)?

5.5.34. ii.) Observe that

$$\|\mathbf{U}_\varepsilon(t) - \mathbf{U}(t)\|_{\mathbb{R}^M} \leq \int_0^t \left\| e^{\mathbf{A}(t-s)} \right\|_{\mathbb{M}^M(\mathbb{R})} \cdot \left\| \sum_{j=1}^M \left[(\omega_{1j})_\varepsilon(t) g_j\left(x_j^\varepsilon(t)\right) - \omega_{1j}(t) g_j\left(x_j(t)\right) \right] \right. $$
$$\left. + \ldots + \sum_{j=1}^M \left[(\omega_{Mj})_\varepsilon(t) g_j\left(x_j^\varepsilon(t)\right) - \omega_{Mj}(t) g_j\left(x_j(t)\right) \right] \right\|_{\mathbb{R}^M} ds,$$

where $\mathbf{U}_\varepsilon(t) = \begin{bmatrix} x_1^\varepsilon(t) \\ \vdots \\ x_M^\varepsilon(t) \end{bmatrix}$. (Now what?)

5.5.36. i.) Use the fact that $\mathbf{w}, \mathbf{z} \in \mathfrak{B}_{\mathcal{H}}(0; R)$ to conclude.

ii.) The solution must be at least continuous. (So what?)

5.6.1. Can either hold uniformly $\forall t \in (0, \infty)$?

5.6.3. All functions involved are continuous and $|\cdot|^{5/3}$ is convex.

5.6.4. We must guarantee existence of a global solution so that when we show $\exists \lim_{t \to T_3^-} \|x(t)\|_{\mathcal{X}}$, we can use $x(T_3)$ as the new IC and reapply the result.

5.7.1. We know that $\exists 0 < \bar{\varepsilon} < \varepsilon$ such that $\int_{t_0 - \bar{\varepsilon}}^{t_0} g_k(s) ds < \frac{\varepsilon}{2}$. How can you use this?

5.7.2. There are various conditions that can be imposed on g. Does $g \in \mathbb{L}^1(0,T;\mathscr{X})$ work? Can you think of a weaker condition?

5.7.3. i.) $\exists M_h > 0$ such that

$$\left\| \frac{h(u)}{1 + \alpha \|h(u)\|_{\mathscr{X}}} \right\|_{\mathscr{X}} \leq \|h(u)\|_{\mathscr{X}} \leq M_h. \forall u \in \mathscr{D}. \text{ (Why?)}$$

ii.) Thrm. 5.7.3 does not apply directly. The other results in the section also do not apply. The function h would need to have sublinear growth. (Note that there are more general results in the literature that might apply.)

5.7.4. i.) This follows from Prop. 5.7.4. We can choose $g_k(t)$ in **(H5.11)** in a very particular manner.

ii.) Use Prop. 3.6.1.

5.7.5. Apply Thrm. 5.7.3. How could you apply Prop. 5.7.4?

5.9.1. Show that $T\left(\|k\|_{\mathbb{L}^1} M_D + M_{\bar{D}}\right)$ is a valid Lipschitz constant. Use standard computations involving properties of the integral and the sup norm.

5.9.2. Remove T from the Lipschitz constant in Exer. 5.9.1.

5.9.3. i.) Try assuming that $a_i \in \mathbb{L}^p$, for appropriate values of p, and f_i globally Lipschitz.

ii.) Weaken the Lipschitz growth assumption on f_i to sublinear growth. Does this work? Does the value of p change from (i)?

Chapter 6

Functional Evolution Equations

Overview

Incorporating more realistic assumptions into the formulation of a mathematical model leads to increased complexity in the resulting IBVP, which in turn must be accounted for in its formulation as an abstract evolution equation. The new forcing terms arising in these IBVPs can often be viewed as mappings between function spaces. This chapter focuses on the study of evolution equations equipped with such forcing terms.

6.1 Motivation by Models

We begin by revisiting two IBVPs that are now equipped with more complex forcing terms.

Model II.4 Pharmacokinetics with an Accumulation Effect

Consider the IVP

$$\begin{cases} \frac{dy}{dt} = -ay(t) + \int_0^t k(s)D(s,y(s),z(s))ds, t > 0, \\ \frac{dz}{dt} = ay(t) - bz(t) + \overline{D}(t,y(t),z(t)), t > 0, \\ y(0) = y_0, \ z(0) = 0, \end{cases} \tag{6.1}$$

where $k \in \mathbb{L}^1(0,T;(0,\infty))$ is a bounded function and $D, \overline{D}$ are globally Lipschitz. Recall that the mapping $\mathfrak{F}_1 : \mathbb{C}([0,T];\mathbb{R}^2) \to \mathbb{L}^1(0,T;\mathbb{R}^2)$ defined by (5.204) is globally Lipschitz (cf. Exer. 5.9.1). It effectively describes the forcing terms of this system when (6.1) is reformulated as an abstract evolution equation.

Model V.4 Two-dimensional Diffusion Equation Revisited

Consider the following generalization of IBVP (3.47):

$$\begin{cases} \frac{\partial}{\partial t}z(x,y,t) & = k\left(\frac{\partial^2}{\partial x^2}z(x,y,t) + \frac{\partial^2}{\partial y^2}z(x,y,t)\right) \\ & \quad + \int_0^t a(t,s)g\left(s,z(s,x,y),\int_0^s k(s,\tau,z(\tau,x,y))d\tau\right)ds, \\ z(x,y,0) & = z_0(x,y),\, 0 < x < a,\, 0 < y < b, \\ \frac{\partial z}{\partial x}(0,y,t) & = \frac{\partial z}{\partial x}(a,y,t) = 0,\, 0 < y < b, t > 0, \\ \frac{\partial z}{\partial y}(x,0,t) & = \frac{\partial z}{\partial y}(x,b,t) = 0,\, 0 < x < a, t > 0, \end{cases} \qquad (6.2)$$

where $z : [0,a] \times [0,b] \times [0,T] \to \mathbb{R}$, $g : [0,T] \times \mathbb{R} \times \mathbb{R} \to \mathbb{R}$, and $h : [0,T] \times [0,T] \times \mathbb{R} \to \mathbb{R}$. We can again view the forcing term in a somewhat more general manner as the mapping

$$\mathfrak{F}_2 : \mathbb{C}\left([0,T];\mathbb{L}^2\left((0,a) \times (0,b)\right)\right) \to \mathbb{L}^2\left(0,T;\mathbb{L}^2\left((0,a) \times (0,b)\right)\right)$$

defined by

$$\mathfrak{F}_2(z)(\cdot,\cdots,t) = \int_0^t a(t,s)g\left(s,z(s,\cdot,\cdots),\int_0^s k(s,\tau,z(\tau,\cdot,\cdots))d\tau\right)ds. \qquad (6.3)$$

Loosely speaking, for a given time t, the forcing term in (6.2) ought to be captured by $\mathfrak{F}_2(z)(t,\cdot,\cdots)$ when viewing the IBVP abstractly. The big difference here is that $\mathfrak{F}_2$ is itself a mapping between function spaces (similar to the solution map Φ in (5.65)), while in the previous chapter the domain and range spaces of the forcing terms were, in effect, one level lower in that they included terms that involved only $[0,T]$ and $\mathscr{X}$.

We can further improve the above diffusion model by accounting for advection. Doing so results in the alternate, somewhat more complex, version of IBVP (6.2):

$$\begin{cases} \frac{\partial z}{\partial t} = k\left(\frac{\partial^2 z}{\partial x^2} + \frac{\partial^2 z}{\partial y^2}\right) + \alpha_1 \frac{\partial z}{\partial x} + \alpha_2 \frac{\partial z}{\partial y} + \mathfrak{F}_2(z),\, 0 < x < a,\, 0 < y < b, t > 0, \\ z(x,y,0) = z_0(x,y),\, 0 < x < a,\, 0 < y < b, \\ \frac{\partial z}{\partial x}(0,y,t) = \frac{\partial z}{\partial x}(a,y,t) = 0,\, 0 < y < b, t > 0, \\ \frac{\partial z}{\partial y}(x,0,t) = \frac{\partial z}{\partial y}(x,b,t) = 0,\, 0 < x < a, t > 0, \end{cases}$$

$$(6.4)$$

where $z = z(x,y,t)$. The following symbolic substitutions are often used to simplify the notation:

$$\triangle z = \frac{\partial^2 z}{\partial x^2} + \frac{\partial^2 z}{\partial y^2}, \qquad (6.5)$$

$$\overrightarrow{\alpha} \cdot \nabla z = \alpha_1 \frac{\partial z}{\partial x} + \alpha_2 \frac{\partial z}{\partial y}. \qquad (6.6)$$

Substituting (6.5) and (6.6) into (6.4) yields the following equivalent, more succinct form of IBVP (6.4):

$$\begin{cases} \frac{\partial z}{\partial t} = k\triangle z + \overrightarrow{\alpha} \cdot \nabla z + \mathfrak{F}_2(z), \, 0 < x < a, \, 0 < y < b, t > 0, \\ z(x,y,0) = z_0(x,y), \, 0 < x < a, \, 0 < y < b, \\ \frac{\partial z}{\partial x}(0,y,t) = \frac{\partial z}{\partial x}(a,y,t) = 0, \, 0 < y < b, t > 0, \\ \frac{\partial z}{\partial y}(x,0,t) = \frac{\partial z}{\partial y}(x,b,t) = 0, \, 0 < x < a, t > 0, \end{cases} \quad (6.7)$$

where $z = z(x,y,t)$. The additional wrinkle we encounter when considering (6.7) is that the operator $k\triangle z$, which is known to generate a compact C_0−semigroup on $\mathbb{L}^2((0,a) \times (0,b))$ (see [**118, 267**]), is now perturbed by the first-order term $\overrightarrow{\alpha} \cdot \nabla z$). The question we will need to address is whether or not the operator $k\triangle z + \overrightarrow{\alpha} \cdot \nabla z$ still generates a C_0−semigroup on $\mathbb{L}^2((0,a) \times (0,b))$. If not, an alternative approach would be to determine if the mapping $\overline{\mathfrak{F}_2}(z) = \mathfrak{F}_2(z) + \overrightarrow{\alpha} \cdot \nabla z$ can be defined on convenient function spaces in a manner for which $\overline{\mathfrak{F}_2}$ satisfies appropriate growth conditions, and we can define a solution map to which a fixed-point will be a mild solution of (6.7) we seek. We shall revisit this problem later in the chapter.

The following model is an extension of (6.7) that arises when studying pollution [**187, 223**], chemotactic phenomena [**168, 169, 208, 254, 328**], population ecology [**119, 158, 298, 305**], and tumor growth [**153**]..

Model XI.1 A Reactive-Convective-Diffusive Model

We introduced a very elementary pollution model in Model IV.1. It was limited in scope since it did not account for the diffusive nature of a pollutant. As such, a pollution "cloud" would simply travel in a single direction at a constant speed and never disperse. A similar interpretation can be made if instead of a pollutant we consider a migratory herd or a travelling pathogen. We now consider a more general model that provides a more realistic description of such phenomena.

Introducing new elements into a local environment (such as a new species, pollution, etc.) can affect the evolution of populations present within the environment. We begin by considering a one-dimensional diffusive model involving two populations into which a convective term is introduced to account for the directed flow or spread of a new element affecting the system. Suppose that the region to which the population is confined is represented as the interval $[0,a]$ and denote by $P_i = P_i(x,t)$ $(i = 1,2)$ the population of the ith entity at position x in $[0,a]$ at time $t \geq 0$. A description of this phenomenon is given by the following IBVP:

$$\begin{cases} \frac{\partial P_1}{\partial t} = \alpha_{P_1} \frac{\partial^2 P_1}{\partial x^2} - \frac{\partial}{\partial x}(\beta_{P_1} P_1) + \sum_{i=1}^{N_1} \int_0^t k_1(t,s)g_{1,i}(s,P_1,P_2)\,ds, \\ \frac{\partial P_2}{\partial t} = \alpha_{P_2} \frac{\partial^2 P_2}{\partial x^2} - \frac{\partial}{\partial x}(\beta_{P_2} P_2) + \sum_{j=1}^{N_2} \int_0^t k_2(t,s)g_{2,j}(s,P_1,P_2)\,ds, \\ P_1(x,0) = P_1^0(x,0), \, P_2(x,0) = P_2^0(x,0), \, 0 < x < a, \\ \frac{\partial P_1}{\partial x}(0,t) = \frac{\partial P_1}{\partial x}(a,t) = \frac{\partial P_2}{\partial x}(0,t) = \frac{\partial P_2}{\partial x}(a,t) = 0, \, ,t > 0, \end{cases} \quad (6.8)$$

where $0 < x < a, t > 0$, $\alpha_{P_i}, \beta_{P_i}$ $(i = 1,2)$ are real constants, $k_i : \mathbb{R} \to \mathbb{R}$ $(i = 1,2)$ are sufficiently smooth functions, and $g_{1,i}(i = 1,\ldots,N_1)$ and $g_{2,j}(j = 1,\ldots,N_2)$ are globally Lipschitz on their domains. We can reformulate (6.8) as the system

$$
\begin{cases}
\dfrac{\partial}{\partial t}\begin{bmatrix} P_1 \\ P_2 \end{bmatrix} = \begin{bmatrix} \alpha_{P_1}\frac{\partial^2}{\partial x^2} - \beta_{P_1}\frac{\partial}{\partial x} & 0 \\ 0 & \alpha_{P_2}\frac{\partial^2}{\partial x^2} - \beta_{P_2}\frac{\partial}{\partial x} \end{bmatrix} \begin{bmatrix} P_1 \\ P_2 \end{bmatrix} \\
\qquad + \begin{bmatrix} \sum_{i=1}^{N_1}\int_0^t k_1(t,s)g_{1,i}(s,P_1,P_2)\,ds \\ \sum_{j=1}^{N_2}\int_0^t k_2(t,s)g_{2,j}(s,P_1,P_2)\,ds \end{bmatrix}, \ 0 < x < a, t > 0, \\
\begin{bmatrix} P_1 \\ P_2 \end{bmatrix}(x,0) = \begin{bmatrix} P_1^0(x,0) \\ P_2^0(x,0) \end{bmatrix}, \ 0 < x < a, \\
\dfrac{\partial}{\partial x}\begin{bmatrix} P_1 \\ P_2 \end{bmatrix}(0,t) = \dfrac{\partial}{\partial x}\begin{bmatrix} P_1 \\ P_2 \end{bmatrix}(a,t) = 0, \ t > 0.
\end{cases}
\tag{6.9}
$$

As with the other IBVPs, the forcing term can be described by the mapping

$$
\mathfrak{F}_3 : \mathbb{C}\left([0,T]; \left(\mathbb{L}^2(0,a)\right)^2\right) \to \mathbb{C}\left([0,T]; \left(\mathbb{L}^2(0,a)\right)^2\right)
$$

defined by

$$
\mathfrak{F}\left(\begin{bmatrix} P_1 \\ P_2 \end{bmatrix}\right)(t) = \begin{bmatrix} \sum_{i=1}^{N_1}\int_0^t k_1(t,s)g_{1,i}(s,P_1,P_2)\,ds \\ \sum_{j=1}^{N_2}\int_0^t k_2(t,s)g_{2,j}(s,P_1,P_2)\,ds \end{bmatrix}.
\tag{6.10}
$$

Exercise 6.1.1. Formulate an extension of this model to the case in which the region Ω to which the population is restricted is a bounded subset of $\mathbb{R}^3$ with smooth boundary $\partial\Omega$.

Remarks.
1. We encounter the same perturbation issue of the operator $\triangle$ as in (6.7).
2. The functions $g_{1,i}$, $g_{2,j}$ often possess logistic growth, as in (5.149).
3. Interesting questions regarding the dynamics of (6.9) (such as when a population will approach extinction or when the collection of all populations settles towards an equilibrium) are of particular interest and are explored in various references (see [306]).
4. A related model of genetically-engineered microbes is studied in [211]. This has particular utility in the agricultural community. With this context in mind, IBVP (6.9) can be improved by further subdividing the population into regions with which the microbes interact, such as top soil, deep soil, surface water, plants, etc. Each of these has a different mechanism of dispersal which should be accounted for in the model.
5. A related IBVP arises in the modeling of semiconductor technology in [138], although the underlying physical setting in this case requires that the IBVP be formulated in a function space consisting of more regular functions.
6. Other interpretations of IBVP (6.9) include an epidemiological model of the elementary spread of a viral infection through a community [131], or as a description of the damage incurred by a plague of locusts on grasslands [280].

Common Theme: The challenge with which we are now presented is to handle a forcing term whose description now involves the use of operators. We refer to such a mapping (from one function space into another) generically as a *functional*. Specifically, when reformulating the above IBVPs as abstract evolution equations, what *exactly* do we insert into the blank below?

$$\begin{cases} u'(t) & = Au(t) + \underline{\ ?\ }, \ t > 0, \\ u(0) & = u_0. \end{cases} \tag{6.11}$$

Ideally, we need a term that fills the above gap in a manner for which the fixed-point approach can be effectively used to study (6.11).

6.2 Functionals

Up to now, the forcing terms we have encountered in all IBVPs were, at worst, of the form $f : [0,T] \times \mathscr{X} \to \mathscr{X}$. The existence proof for the corresponding evolution equation involved the use of a mapping between function spaces when defining the solution map. We now wish to incorporate such a mapping as a term in the evolution equation itself, and investigate its properties. The purpose of this section is to expose you to different types of mappings arising in the mathematical modeling of various phenomena and to subsequently study their growth properties in preparation for the study of related IBVPs. The basic properties (e.g., Lipschitz growth conditions, sublinear growth conditions, etc.) are the same as those used in Chapter 5, but the concomitant calculations are more intricate. At the moment, we shall illustrate these computations outside the context of concrete models to help you to develop a working knowledge of them.

Example 6.2.1. Suppose that $c : [0,T] \times \mathbb{R} \to \mathbb{R}$ is a continuous function. Define $g : \mathbb{R} \to \mathbb{R}$ by $g(y) = y^2$ and consider the composition function $h : [0,T] \times \mathbb{R} \to \mathbb{R}$ given by

$$h(t,x) = [c(t,x)]^2. \tag{6.12}$$

Clearly, if such a function arose as the forcing term in an IBVP, then the abstract formulation of the IBVP would have fallen under the parlance of the theory developed in Chapter 5. That said, we shall show how it can be viewed in an arguably more general manner that will assist us in identifying the missing term in the abstract paradigm (6.11).

Let $\mathscr{X} = \mathbb{C}(\mathbb{R};\mathbb{R})$. By assumption, $c(t,\cdot) \in \mathscr{X}, \forall t \in [0,T]$. In Chapter 5, we would have defined a mapping $f : [0,T] \times \mathscr{X} \to \mathscr{X}$ by

$$f(t, c(t,\cdot)) = [c(t,\cdot)]^2. \tag{6.13}$$

We now approach this somewhat differently and define a mapping $\varphi : [0,T] \to \mathscr{X}$ by $\varphi(t)(x) = c(t,x)$. Observe that $\forall 0 \le t \le T$, $\varphi(t) \in \mathscr{X}$ and the mapping $\mathfrak{H}_1 :$

$\mathbb{C}\left([0,T];\mathscr{X}\right) \to \mathbb{C}\left([0,T];\mathscr{X}\right)$ given by $\mathfrak{H}_1(\varphi) = \varphi^2$ is well-defined Admittedly, this is not significantly different from (6.13) since, after all, $\forall 0 \leq t \leq T$,

$$\mathfrak{H}_1(\varphi)(t) = [c(t, \cdot)]^2. \tag{6.14}$$

But, the ability to relegate the intermittent dependence on t to "behind the scenes" is useful when establishing theoretical results for (6.11). This will become increasingly more evident as we progress through the chapter.

Exercise 6.2.1. Is the functional $\mathfrak{H}_1$ defined by (6.14) globally Lipschitz? If not, supply a sufficient condition to ensure that it is.

Example 6.2.2. Let $n \in \mathbb{N}$, $0 < t_1 < t_2 < \ldots < t_n < T$ be fixed times, and $\Omega = \left\{(u,v) \in \mathbb{R}^2 \,\middle|\, u^2 + v^2 \leq 1\right\}$. We make the following assumptions:
(H6.1) $\alpha \in \mathbb{L}^2\left(0,T;\mathbb{R}\right)$,
(H6.2) $f : [0,T] \times \mathbb{R} \to \mathbb{R}$ is continuous in the first variable and globally Lipschitz in the second variable (uniformly in t) with Lipschitz constant M_f,
(H6.3) $\beta_i \in \mathbb{C}\left(\Omega;\mathbb{R}\right)$, $i = 1,\ldots,n$.
Define the functional $\mathfrak{H}_2 : \mathbb{C}\left([0,T];\mathbb{L}^2\left(\Omega\right)\right) \to \mathbb{L}^2\left(\Omega\right)$ by

$$\mathfrak{H}_2(x)(\cdot) = \sum_{i=1}^{n} \beta_i(\cdot)x(t_i)(\cdot) + \int_0^T \alpha(s)f\left(s,x(s)(\cdot)\right)ds. \tag{6.15}$$

<u>Claim</u> 1: *If* **(H6.1)** - **(H6.3)** *hold, then $\mathfrak{H}_2$ is globally Lipschitz. For such a mapping, this means that $\exists M_{\mathfrak{H}_2} > 0$ such that $\forall x, y \in \mathbb{C}\left([0,T];\mathbb{L}^2\left(\Omega\right)\right)$,*

$$\left\|\mathfrak{H}_2(x) - \mathfrak{H}_2(y)\right\|_{\mathbb{L}^2(\Omega)} \leq M_{\mathfrak{H}_2}\left\|x - y\right\|_{\mathbb{C}\left([0,T];\mathbb{L}^2(\Omega)\right)}. \tag{6.16}$$

Proof: By definition,

$$
\begin{aligned}
\|w\|_{\mathbb{C}\left([0,T];\mathbb{L}^2(\Omega)\right)} &= \sup_{0 \leq t \leq T} \|w(t,\cdot)\|_{\mathbb{L}^2(\Omega)} \\
&= \sup_{0 \leq t \leq T}\left[\int_\Omega |w(t,z)|^2\, dz\right]^{1/2} \\
&\leq \left[\int_\Omega \sup_{0 \leq t \leq T} |w(t,z)|^2\, dz\right]^{1/2}.
\end{aligned}
\tag{6.17}
$$

The last inequality in (6.17) follows from standard properties of the integral. (Tell how.) Observe that $\forall x, y \in \mathbb{C}\left([0,T];\mathbb{L}^2\left(\Omega\right)\right)$, there exist positive constants K_1, K_2

for which the following string of inequalities holds, $\forall z \in \Omega$:

$$
|\mathfrak{H}_2(x)(z) - \mathfrak{H}_2(y)(z)|^2 \leq K_1 \left[\left| \sum_{i=1}^{n} \beta_i(z) \left[x(t_i)(z) - y(t_i)(z) \right] \right|^2 + \right.
$$

$$
\left. + \left| \int_0^T \alpha(s) \left[f(s, x(s)(z)) - f(s, y(s)(z)) \right] ds \right|^2 \right]
$$

$$
\leq K_1 \left[K_2 \sum_{i=1}^{n} |\beta_i(z)|^2 |x(t_i)(z) - y(t_i)(z)|^2 + \right.
$$

$$
\left. + \left[M_f \int_0^T |\alpha(s)| |x(s)(z) - y(s)(z)| ds \right]^2 \right]
$$

$$
\leq K_1 \left[K_2 \sum_{i=1}^{n} |\beta_i(z)|^2 |x(t_i)(z) - y(t_i)(z)|^2 + \qquad (6.18) \right.
$$

$$
\left. + M_f^2 \left(\sup_{0 \leq s \leq T} |x(s)(z) - y(s)(z)|^2 \right) \left(\int_0^T |\alpha(s)| ds \right)^2 \right]
$$

$$
\leq K_1 \left[K_2 \sum_{i=1}^{n} |\beta_i(z)|^2 |x(t_i)(z) - y(t_i)(z)|^2 + \qquad (6.19) \right.
$$

$$
\left. + M_f^2 \left(\sup_{0 \leq s \leq T} |x(s)(z) - y(s)(z)|^2 \right) T \, \|\alpha\|_{\mathbb{L}^2(0,T)}^2 \right].
$$

Note that in (6.18), we used the fact that

$$
\left(\sup_{0 \leq s \leq T} F(s) \right)^2 \leq \sup_{0 \leq s \leq T} (F(s))^2, \qquad (6.20)
$$

where $F : [0, T] \to [0, \infty)$ is a continuous function. Integrating both sides of (6.19) over Ω subsequently yields

$$
\|\mathfrak{H}_2(x) - \mathfrak{H}_2(y)\|_{\mathbb{L}^2(\Omega)}^2 \leq K_1 \left[K_2 \left(\sup_{1 \leq i \leq n} \left\{ \sup_{z \in \Omega} |\beta_i(z)|^2 \right\} \right) \cdot \right.
$$

$$
\sum_{i=1}^{n} \int_\Omega |x(t_i)(z) - y(t_i)(z)|^2 \, dz + M_f^2 T \, \|\alpha\|_{\mathbb{L}^2(0,T)}^2 \cdot
$$

$$
\int_\Omega \left(\sup_{0 \leq s \leq T} |x(s)(z) - y(s)(z)|^2 \right) dz \right] \qquad (6.21)
$$

$$
\leq K_1 \pi \left[K_2 n \left(\sup_{1 \leq i \leq n} \left\{ \sup_{z \in \Omega} |\beta_i(z)|^2 \right\} \right) + M_f^2 T \, \|\alpha\|_{\mathbb{L}^2(0,T)}^2 \right]
$$

$$
\times \|x - y\|_{\mathbb{C}([0,T];\mathbb{L}^2(\Omega))}^2 .
$$

The claim then follows by identifying the Lipschitz constant as

$$M_{\mathfrak{H}_2} = K_1 \pi \left[K_2 n \left(\sup_{1 \le i \le n} \left\{ \sup_{z \in \Omega} |\beta_i(z)|^2 \right\} \right) + M_f^2 T \, \|\alpha\|_{\mathbb{L}^2(0,T)}^2 \right]. \quad \square$$

Exercise 6.2.2. Identify the constants K_1, K_2 and justify all steps leading to (6.18) - (6.21).

Now, suppose we replace **(H6.2)** by the following weaker sublinear growth condition:

(H6.4) $f : [0,T] \times \mathbb{R} \to \mathbb{R}$ is a continuous mapping for which $\exists M_1 \in \mathbb{L}^1(0,T;(0,\infty))$ and $M_2 \in \mathbb{C}([0,T];(0,\infty))$ such that

$$|f(t,x)| \le M_1(t) + M_2(t) \, |x| , \forall t \in [0,T], x \in \mathbb{R}.$$

The following weaker claim holds:

<u>Claim 2</u>: *If* **(H6.1)**, **(H6.3)**, *and* **(H6.4)** *hold, then* $\mathfrak{H}_2$ *is a continuous mapping for which there exist positive constants* K_3, K_4 *such that*

$$\|\mathfrak{H}_2(x)\|_{\mathbb{L}^2(\Omega)} \le K_3 + K_4 \|x\|_{\mathbb{C}([0,T];\mathbb{L}^2(\Omega))} , \forall x \in \mathbb{C}\left([0,T];\mathbb{L}^2(\Omega)\right).$$

Exercise 6.2.3. Prove Claim 2.

Remark. The particular set Ω in this example can be replaced by any bounded subset of $\mathbb{R}^N$ with smooth boundary $\partial\Omega$ without any significant change to the argument. The "area" of Ω (namely π) enters into the calculation above, and would be replaced by $m(\Omega)$ (the measure of Ω).

Example 6.2.3. Let $0 < \varepsilon < T$ and Ω a bounded subset of $\mathbb{R}^N$ with smooth boundary. Define the functional $\mathfrak{H}_3 : \mathbb{C}\left([0,T];\mathbb{L}^2(\Omega)\right) \to \mathbb{L}^2(\Omega)$ by

$$\mathfrak{H}_3(x)(z) = \int_{T-\varepsilon}^{T} \left[1 + \sqrt{s}\right] \log\left(1 + x(s)(z)\right) ds. \quad (6.22)$$

Exercise 6.2.4.
i.) Show that $\exists M_{\mathfrak{H}_3} > 0$ and $p \in [0,1)$ such that $\forall x \in \mathbb{C}\left([0,T];\mathbb{L}^2(\Omega)\right)$,

$$\|\mathfrak{H}_3(x)\|_{\mathbb{L}^2(\Omega)} \le M_{\mathfrak{H}_3} \left[1 + \|x\|_{\mathbb{C}([0,T];\mathbb{L}^2(\Omega))}\right]^p. \quad (6.23)$$

ii.) Show that $\forall 0 < \varepsilon < T$, $\exists N \in \mathbb{N}$ such that

$$\|x\|_{\mathbb{C}([0,T];\mathbb{L}^2(\Omega))} \ge N \implies \frac{\|\mathfrak{H}_3(x)\|_{\mathbb{L}^2(\Omega)}}{\|x\|_{\mathbb{C}([0,T];\mathbb{L}^2(\Omega))}} < \varepsilon. \quad (6.24)$$

iii.) Do (i) and (ii) still hold (for possibly different values of N and $M_{\mathfrak{H}_3}$) if the term $[1 + \sqrt{s}]$ in (6.22) is replaced by a general $h \in \mathbb{C}([0,T];\mathbb{R})$? If so, can h be taken in an even larger space? Explain.

Example 6.2.4. Assume that
(H6.5) $a : [0,T] \times [0,T] \to (0,\infty)$ is such that $\sup\limits_{0 \le t \le T} \|a(t,\cdot)\|^2_{L^2(0,T)} < \infty$,
(H6.6) Let $\mathcal{U} = \left\{ (t,s) \in \mathbb{R}^2 \, | \, 0 \le s \le t \le T \right\}$ and assume that $k : \mathcal{U} \times \mathbb{R} \to \mathbb{R}$ is continuous in all variables and $\exists M_k > 0$ such that

$$|k(t,s,x_1) - k(t,s,x_2)| \le M_k |x_1 - x_2|, \forall (t,s) \in \mathcal{U}, x_1, x_2 \in \mathbb{R}. \tag{6.25}$$

(H6.7) $g : [0,T] \times \mathbb{R} \times \mathbb{R} \to \mathbb{R}$ is continuous in all three variables and satisfies
   (a) $g(\cdot, 0, 0) \in L^2(0,T)$, and
   (b) there exists $M_g > 0$ such that

$$|g(t,x_1,y_1) - g(t,x_2,y_2)| \le M_g [|x_1 - x_2| + |y_1 - y_2|], \tag{6.26}$$

$\forall (x_1,y_1), (x_2,y_2) \in \mathbb{R} \times \mathbb{R}$ (uniformly in t).

Define the functional $\mathfrak{H}_4 : \mathbb{C}\left([0,T]; L^2(\Omega)\right) \to L^2\left(0,T; L^2(\Omega)\right)$ by

$$\mathfrak{H}_4(x)(t)(\cdot) = \int_0^t a(t,s) g\left(s, x(s)(\cdot), \int_0^s k(s,\tau, x(\tau)(\cdot)) d\tau\right) ds. \tag{6.27}$$

Claim 3: $\mathfrak{H}_4$ *is globally Lipschitz.*
Outline of Proof: It is equivalent (and more convenient) to produce a Lipschitz constant $M_{\mathfrak{H}_4}$ for which

$$\|\mathfrak{H}_4(x) - \mathfrak{H}_4(y)\|^2_{L^2} \le M_{\mathfrak{H}_4} \|x - y\|^2_{\mathbb{C}}, \forall x,y \in \mathbb{C}\left([0,T]; L^2(\Omega)\right).$$

Using Holder's inequality and properties of the integral yields

$$
\begin{aligned}
|\mathfrak{H}_4(x)(t)(z) - \mathfrak{H}_4(y)(t)(z)|^2 &= \left| \int_0^t a(t,s) \left[g\left(s, x(s)(\cdot), \int_0^s k(s,\tau, x(\tau)(\cdot)) d\tau\right) \right. \right. \\
&\qquad \left. \left. - g\left(s, y(s)(\cdot), \int_0^s k(s,\tau, y(\tau)(\cdot)) d\tau\right) \right] ds \right|^2 \\
&\le \|a(t,\cdot)\|^2_{L^2(0,T)} \int_0^T \left| g\left(s, x(s)(\cdot), \int_0^s k(s,\tau, x(\tau)(\cdot)) d\tau\right) \right. \\
&\qquad \left. - g\left(s, y(s)(\cdot), \int_0^s k(s,\tau, y(\tau)(\cdot)) d\tau\right) \right|^2 ds.
\end{aligned}
$$

Now, how do you proceed? □

Exercise 6.2.5. Complete the proof of Claim 3.

Exercise 6.2.6. If **(H6.6)** and **(H6.7)** are replaced by appropriate sublinear growth conditions (like those used in **(H6.4)**), must there exist positive constants $M_{\mathfrak{H}_4}^1, M_{\mathfrak{H}_4}^2$ such that

$$\|\mathfrak{H}_4(x)\|_{\mathbb{L}^2} \leq M_{\mathfrak{H}_4}^1 \|x\|_{\mathbb{C}} + M_{\mathfrak{H}_4}^2, \forall x, y \in \mathbb{C}\left([0,T]; \mathbb{L}^2(\Omega)\right)? \tag{6.28}$$

If so, prove it; otherwise, explain why.

Exercise 6.2.7. Suppose we restrict the range of $\mathfrak{H}_4$ to be the subspace $\mathbb{C}\left([0,T]; \mathbb{L}^2(\Omega)\right)$ of $\mathbb{L}^2\left(0,T; \mathbb{L}^2(\Omega)\right)$. Do the conclusions of Claim 3 and Exer. 6.2.6 still hold? If so, how do the growth constants $M_{\mathfrak{H}_4}, M_{\mathfrak{H}_4}^1, M_{\mathfrak{H}_4}^2$ change?

Example 6.2.5. Let $\mathscr{X}$ be a Banach space, $a : [0,T] \to \mathbb{R}$ and $h : [0,T] \times \mathbb{R} \to \mathbb{R}$ be given, and define the functional $\mathfrak{H}_5 : \mathbb{L}^1(0,T; \mathscr{X}) \to \mathbb{L}^1(0,T; \mathscr{X})$ by

$$\mathfrak{H}_5(x)(t) = \int_0^t a(t-s)f(s,x(s))ds. \tag{6.29}$$

Exercise 6.2.8.
i.) Establish sufficient conditions on a and f that ensure that $\mathfrak{H}_5$ is a continuous, well-defined mapping that satisfies a sublinear growth condition of the type in (6.28).
ii.) Now, suppose the domain of $\mathfrak{H}_5$ is restricted to $\mathbb{C}([0,T]; \mathscr{X})$. Repeat (i), noting precisely where any necessary modifications occur.

Example 6.2.6. Assume $k_1, k_2 \in \mathbb{L}^1(0,T; \mathbb{R})$ and $f_i : [0,T] \times \mathscr{X} \times \mathscr{X} \to \mathscr{X}$ $(i=1,2)$ is continuous in the first variable and globally Lipschitz on $\mathscr{X} \times \mathscr{X}$ (uniformly in t). Define the functional $\mathfrak{H}_6 : (\mathbb{C}([0,T]; \mathscr{X}))^2 \to (\mathbb{C}([0,T]; \mathscr{X}))^2$ by

$$\mathfrak{H}_6 \begin{pmatrix} u \\ v \end{pmatrix} (t) = \begin{pmatrix} \int_0^t k_1(s)f_1(s,u(s),v(s))ds \\ \int_0^t k_2(s)f_2(s,u(s),v(s))ds \end{pmatrix}. \tag{6.30}$$

Exercise 6.2.9. Prove that $\mathfrak{H}_6$ is globally Lipschitz.

6.3 Theory in the Lipschitz Case

A comparison of the examples explored in Section 6.2 to the right-sides of the IVPs and IBVPs described in Section 6.1 suggests that the term inserted in (6.11) can be viewed as a functional defined on an appropriate space. As motivation for our discussion of the abstract theory, consider the following exercise.

Exercise 6.3.1. Show that each of the IVPs and IBVPs in Section 6.1 can be reformulated as an abstract evolution equation of the form

$$\begin{cases} u'(t) &= Au(t) + \mathfrak{F}(u)(t), \ 0 < t < T, \\ u(0) &= u_0, \end{cases} \tag{6.31}$$

in an appropriate Banach space $\mathscr{X}$. In each case, clearly identify the space $\mathscr{X}$, the operator $A : \text{dom}(A) \subset \mathscr{X} \to \mathscr{X}$, the functional $\mathfrak{F} : \mathscr{Y}_1 \to \mathscr{Y}_2$ including the spaces $\mathscr{Y}_1$ and $\mathscr{Y}_2$.

Although (6.31) is distinct from (5.10), the difference is not so great as to require us to formulate a drastically different notion of a mild solution for (6.31). Indeed, it is reasonable to slightly tweak Def. 5.5.1 to obtain the following definition:

Definition 6.3.1. A function $u : [0,T] \to \mathscr{X}$ is a *mild solution* of (6.31) on $[0,T]$ if u satisfies Def. 5.5.1(ii) with $f(t, u(t))$ replaced by $\mathfrak{F}(u)(t)$.

A natural first line of attack for treating (6.31) is to determine the extent to which the approach used in Chapter 5 is viable. More pointedly, how can we choose a function space $\mathscr{Z}$ such that, as a result of imposing appropriate growth conditions on $\mathfrak{F}$, the solution map $\Phi : \mathscr{Z} \to \mathscr{Z}$ defined by $\Phi(v) = u_v$, where $u_v : [0,T] \to \mathscr{X}$ is the unique mild solution of the IVP

$$\begin{cases} u_v'(t) &= Au_v(t) + \mathfrak{F}(v)(t), 0 < t < T, \\ u_v(0) &= u_0, \end{cases} \tag{6.32}$$

has a fixed-point which, in turn, satisfies Def. 6.3.1.

Exercise 6.3.2. Identify one natural choice of $\mathscr{Z}$ based on the experience you have acquired thus far.

The remainder of this chapter is devoted to forming an extension of the basic results from Chapter 5 to the present setting. Consider (6.31), where $\mathfrak{F}$ is a prescribed functional. The nature of how $\mathfrak{F}$ is defined suggests that the fixed-point theorems introduced in Section 5.2 might be useful tools. We begin with the Lipschitz case.

Theorem 6.3.2. *Assume* $(\mathbf{H_A})$, $u_0 \in \mathscr{X}$, *and*
$(\mathbf{H6.8})$ $\mathfrak{F} : \mathbb{C}([0,T];\mathscr{X}) \to \mathbb{C}([0,T];\mathscr{X})$ *is globally Lipschitz with Lipschitz constant* $M_{\mathfrak{F}}$.
If $\overline{M_A}M_{\mathfrak{F}}T < 1$, *then (6.31) has a unique mild solution on* $[0,T]$.

Proof. Let $v \in \mathbb{C}([0,T];\mathscr{X})$. Then, $\mathfrak{F}(v) \in \mathbb{C}([0,T];\mathscr{X})$, and so applying Prop. 4.2.3 (as it applies to guarantee the existence of a unique <u>mild</u> solution), we see that (6.32) has a unique mild solution on $[0,T]$. Define the solution map $\Phi : \mathbb{C}([0,T];\mathscr{X}) \to \mathbb{C}([0,T];\mathscr{X})$ by

$$(\Phi v)(t) = e^{At}u_0 + \int_0^t e^{A(t-s)}\mathfrak{F}(v)(s)ds, 0 < t < T. \tag{6.33}$$

Observe that $\forall v, w \in \mathbb{C}([0,T];\mathscr{X})$ and $0 < t < T$,

$$\|(\Phi v)(t) - (\Phi w)(t)\|_{\mathscr{X}} \leq \overline{M_A} \int_0^t \|(\mathfrak{F}v)(s) - (\mathfrak{F}w)(s)\|_{\mathscr{X}} ds,$$

and subsequently,

$$\|(\Phi v) - (\Phi w)\|_{\mathbb{C}} \leq \overline{M_A} M_{\mathfrak{F}} T \|v - w\|_{\mathbb{C}}.$$

(Why?) Thus, Φ is a contraction because $\overline{M_A} M_{\mathfrak{F}} T < 1$. So, Φ has a unique fixed-point by the Contraction Mapping Theorem. This completes the proof. $\square$

This result can be improved slightly by enlarging the range of the functional. Precisely, consider the following theorem:

Theorem 6.3.3. *Assume* ($\mathbf{H_A}$), $u_0 \in \mathscr{X}$, *and*
(H6.9) $\mathfrak{F} : \mathbb{C}([0,T];\mathscr{X}) \to \mathbb{L}^1(0,T;\mathscr{X})$ *is globally Lipschitz.*
If $\overline{M_A} M_{\mathfrak{F}} < 1$, *then (6.31) has a unique mild solution on* $[0,T]$.

Note that in the proofs of the above two theorems, we cannot simply iterate the solution map as we did in the proof of Thrm. 5.5.2(i) because the growth condition imposed on $\mathfrak{F}$ is expressed as an inequality involving a norm taken on the entire space $\mathbb{C}([0,T];\mathscr{X})$ rather than simply in $\mathscr{X}$. (Try doing so to see what happens.)

Exercise 6.3.3. Show that Thrm. 5.2.2(i) is a corollary of Thrm. 6.3.2.

Exercise 6.3.4. Prove Thrm. 6.3.3.

Exercise 6.3.5. Which is the more general result, Thrm. 6.3.2 or Thrm. 6.3.3?

Exercise 6.3.6. Define what is meant by a classical solution of (6.31).

Exercise 6.3.7. **(Continuous Dependence on Initial Data)**
i.) Let $\varepsilon > 0$ and consider the IVP

$$\begin{cases} u'_\varepsilon(t) & = A u_\varepsilon(t) + \mathfrak{F}(u_\varepsilon)(t), \, 0 < t < T, \\ u_\varepsilon(0) & = u_0 + \varepsilon, \end{cases} \tag{6.34}$$

where A, u_0 and $\mathfrak{F}$ are as in Thrm. 6.3.2. Show that $\lim\limits_{\varepsilon \to 0^+} \|u_\varepsilon - u\|_{\mathbb{C}} = 0$ under appropriate data restrictions.
ii.) If $\mathfrak{F}$ is defined as in Thrm. 6.3.3, does the same conclusion hold? Explain.

Proposition 6.3.4. *Assume* ($\mathbf{H_A}$), $u_0 \in \mathscr{X}$, *and* (**H6.8**). *For each* $n \in \mathbb{N}$, *consider the IVP*

$$\begin{cases} u'_n(t) & = A_n u_n(t) + \mathfrak{F}_n(u_n)(t), \, 0 < t < T, \\ u_n(0) & = u_0^n. \end{cases} \tag{6.35}$$

Assume that $\forall n \in \mathbb{N}$,
(H6.10) $A_n : \text{dom}(A_n) \subset \mathscr{X} \to \mathscr{X}$ *is such that* $\text{dom}(A_n) = \text{dom}(A)$ *and* A_n *generates a* C_0-*semigroup* $\{e^{tA_n} : t \geq 0\}$.

(H6.11) $\mathfrak{F}_n : \mathbb{C}([0,T];\mathscr{X}) \to \mathbb{C}([0,T];\mathscr{X})$ *is globally Lipschitz with Lipschitz constant* $M_{\mathfrak{F}_n} = M_{\mathfrak{F}}$.
(H6.12) $u_0^n \in \mathscr{X}$.
(H6.13) $\lim_{n\to\infty} \left\| e^{tA_n} - e^{tA} \right\|_{\mathbb{B}(\mathscr{X})} = 0$ *uniformly in* $t \in [0,T]$.
(H6.14) $\lim_{n\to\infty} \left\| \mathfrak{F}_n - \mathfrak{F} \right\|_{\mathbb{C}} = 0$.
(H6.15) $\lim_{n\to\infty} \left\| u_0^n - u_0 \right\|_{\mathscr{X}} = 0$.
Then, $\lim_{n\to\infty} \left\| u_n - u \right\|_{\mathbb{C}} = 0$, *where u is the unique mild solution of (6.31)*.

Exercise 6.3.8. Prove Prop. 6.3.4.

Exercise 6.3.9. If **(H6.8)** and **(H6.11)** are modified so that the ranges of $\mathfrak{F}_n$ and $\mathfrak{F}$ are enlarged to $\mathbb{L}^1(0,T;\mathscr{X})$, does the conclusion of Prop. 6.3.4 still hold? Explain.

Exercise 6.3.10. If $u_0^n \in \text{dom}(A^3)$, $\forall n \in \mathbb{N}$, and $\mathfrak{F} = 0$, what is the regularity of u? Or, can it not be determined? Explain.

Exercise 6.3.11. Consider IBVP (6.2).
i.) Impose conditions on a, g, k ensuring that (6.2) has a unique mild solution on $[0,T]$. Prove your result.
ii.) Formulate a convergence result for (6.2) in the spirit of Prop. 6.3.4. You need not form an approximation of the operator $\frac{\partial^2}{\partial x^2} + \frac{\partial^2}{\partial y^2}$.

Exercise 6.3.12. Repeat Exer. 6.3.11 for IVP (6.1).

Exercise 6.3.13. Repeat Exer. 6.3.11 for IBVP (6.8).

6.4   Theory under Compactness Assumptions

As expected, results for (6.31) in the spirit of those developed in Section 5.7 can be established. The differences are minimal and technical in nature, and arise when establishing the estimates and verifying precompactness (see **[11, 190]**). Consider, for instance, the following theorem.

Theorem 6.4.1. *Assume* $\left(\mathbf{H}_\mathbf{A}^\star \right)$ *and*
(H6.16) $\mathfrak{F} : \mathbb{C}([0,T];\mathscr{X}) \to \mathbb{C}([0,T];\mathscr{X})$ *is a continuous map for which there exist positive constants* d_1, d_2 *such that*

$$\left\| \mathfrak{F}x \right\|_{\mathbb{C}} \leq d_1 \left\| x \right\|_{\mathbb{C}} + d_2, \forall x \in \mathbb{C}([0,T];\mathscr{X}). \tag{6.36}$$

If $\overline{M_A} d_1 T < 1$, *then (6.31) has at least one mild solution on* $[0,T]$.

Exercise 6.4.1. Prove Thrm. 6.4.1 using Schaefer's fixed-point theorem. Where is the condition $M_A d_1 T < 1$ used?

Exercise 6.4.2. Can the conditions imposed on d_1, d_2 in Thrm. 6.4.1 be weakened in such a way that the same conclusion holds? Explain.

Exercise 6.4.3. If the range of the functional $\mathfrak{F}$ in Thrm. 6.4.1 is enlarged to $\mathbb{L}^1(0, T; \mathscr{X})$ and $\mathfrak{F}$ satisfies a suitably modified version of (6.36), use Schaefer's fixed-point theorem to prove that (6.31) has at least one mild solution on $[0, T]$.

6.5   Models – New and Old

We illustrate how the theory developed in this chapter can be applied to several different models.

Model VII.3 Functional Wave Equations

Consider the following generalization of IBVP (4.56):

$$\begin{cases} \frac{\partial^2 z}{\partial t^2} + \alpha \frac{\partial z}{\partial t} + c^2 \frac{\partial^2 z}{\partial x^2} = \int_0^t a(t, s) f\left(s, z, \frac{\partial z}{\partial s}\right) ds, \ 0 < x < L, t > 0, \\ z(x, 0) = z_0(x), \ \frac{\partial z}{\partial t}(x, 0) = z_1(x), \ 0 < x < L, \\ z(0, t) = z(L, t) = 0, t > 0, \end{cases} \tag{6.37}$$

where $z = z(x, t)$, $f : [0, T] \times \mathbb{R} \times \mathbb{R} \to \mathbb{R}$ and $a \in \mathbb{L}^1((0, T) \times (0, T))$. Related problems are investigated in **[54, 89, 90, 113, 238, 256, 263]**.

Exercise 6.5.1.
i.) Reformulate (6.37) abstractly as (6.31) in an appropriate space.
ii.) Assume that
(H6.17) $f : [0, T] \times \mathbb{R} \times \mathbb{R} \to \mathbb{R}$ is continuous in all three variables and is such that there exist positive constants m_1, m_2, m_3 for which

$$|f(t, x, y)| \leq m_1 + m_2 x + m_3 y, \forall t \in [0, T], \ x, y \in \mathbb{R}. \tag{6.38}$$

Show that if **(H6.17)** holds, then (6.37) has at least one mild solution on $[0, T]$.

Exercise 6.5.2.
i.) Formulate and prove an analog of Prop. 5.7.4 for (6.37).
ii.) Instead of assuming that **(H6.17)** holds, impose conditions on a and f sufficient to apply the result established in (i) to (6.37).

Exercise 6.5.3. Consider the following extension of IBVP (5.3):

$$\begin{cases} \frac{\partial^2 z}{\partial t^2} + \alpha \frac{\partial z}{\partial t} + c^2 \frac{\partial^2 z}{\partial x^2} = \int_0^t a_1(t-s)g_1(s,w,z)ds, \, 0 < x < L, t > 0, \\ \frac{\partial^2 w}{\partial t^2} + \beta \frac{\partial w}{\partial t} + c^2 \frac{\partial^2 w}{\partial x^2} = \int_0^t a_2(t-s)g_2(s,w,z)ds, \, 0 < x < L, t > 0, \\ z(x,0) = z_0(x), \, \frac{\partial z}{\partial t}(x,0) = z_1(x), \, 0 < x < L, \\ w(x,0) = w_0(x), \, \frac{\partial w}{\partial t}(x,0) = w_1(x), \, 0 < x < L, \\ \frac{\partial z}{\partial x}(0,t) = \frac{\partial z}{\partial x}(L,t) = \frac{\partial w}{\partial x}(0,t) = \frac{\partial w}{\partial x}(L,t) = 0, t > 0, \end{cases} \quad (6.39)$$

where $z = z(x,t)$, $w = w(x,t)$, $g_i : [0,T] \times \mathbb{R} \times \mathbb{R} \to \mathbb{R}$ $(i=1,2)$ and $a_1, a_2 \in \mathbb{L}^1((0,T))$.
i.) Reformulate (6.39) abstractly as (6.31) in an appropriate space.
ii.) Impose conditions on g_i, a_i that ensure (6.39) has a unique mild solution on $[0,T]$.
iii.) Impose conditions on g_i, a_i weaker than those in (i) that ensure merely the existence of at least one mild solution of (6.39) on $[0,T]$.

Model V.5 A Functional Diffusion-Advection Equation
Let $n \in \mathbb{N}$ and $0 < t_1 < t_2 < \ldots < t_n < T$ be fixed times. Consider the following IBVP governing a diffusive-advective process with accumulative external force:

$$\begin{cases} \frac{\partial z}{\partial t} + \alpha^2 \frac{\partial^2 z}{\partial x^2} + \gamma \frac{\partial z}{\partial x} = \sum_{i=1}^n \beta_i(x)z(x,t_i) + \int_0^T \zeta(s)f(s,z)ds, 0 < x < L, 0 < t < T, \\ \frac{\partial z}{\partial x}(0,t) = \frac{\partial z}{\partial x}(L,t) = 0, t > 0, \\ z(x,0) = z_0(x), \, 0 < x < L, \end{cases} \quad (6.40)$$

where $z = z(x,t)$. Such an IBVP arises, for instance, when describing atmospheric diffusion properties; see **[272]**.

Assume **(H6.1)** - **(H6.3)** (with Ω replaced by the interval $[0,L]$). When reformulating (6.40) abstractly as (6.31), we would like to identify the operator A as $\left(\alpha^2 \frac{\partial^2}{\partial x^2} + \gamma \frac{\partial}{\partial x} \right)$. We know that the operator $\alpha^2 \frac{\partial^2}{\partial x^2}$ generates a C_0-semigroup on $\mathbb{L}^2(0,L)$, but does adding the operator $\gamma \frac{\partial}{\partial x}$ to it prevent the sum operator from also generating a C_0-semigroup on $\mathbb{L}^2(0,L)$? We have encountered a similar notion in Section 3.6, but there the perturbing operator was bounded. Note that this is not true in the present scenario. (Why?) Even so, as long as the perturbation doesn't overpower $\alpha^2 \frac{\partial^2}{\partial x^2}$, we are okay. The answer to this question is given by the following theorem (see **[92, 142]** for a proof).

Theorem 6.5.1. *Assume that* $A : \text{dom}(A) \subset \mathscr{X} \to \mathscr{X}$ *generates a contractive* C_0- *semigroup on* $\mathscr{X}$ *and that* $B : \text{dom}(B) \subset \mathscr{X} \to \mathscr{X}$ *is a dissipative operator for which* $\text{dom}(A) \subset \text{dom}(B)$. *If there exist constants* $0 \le \delta_1 < 1$ *and* $\delta_2 \ge 0$ *for which*

$$\|Bg\|_{\mathscr{X}} \le \delta_1 \|Ag\|_{\mathscr{X}} + \delta_2 \|g\|_{\mathscr{X}}, \forall g \in \text{dom}(A), \quad (6.41)$$

then the operator $A+B : \text{dom}(A) \subset \mathscr{X} \to \mathscr{X}$ *generates a contractive* C_0-*semigroup on* $\mathscr{X}$.

Remark. If (6.41) holds, we say that B is *A-bounded*.

It can be shown that there exist $0 \le \delta_1 < 1$ and $\delta_2 \ge 0$ for which

$$\left\| \frac{\partial h}{\partial x} \right\|_{\mathbb{L}^2(0,a)} \le \delta_1 \left\| -\frac{\partial^2 h}{\partial x^2} \right\|_{\mathbb{L}^2(0,a)} + \delta_2 \|h\|_{\mathbb{L}^2(0,a)}, \, \forall h \in \mathbb{L}^2(0,a). \qquad (6.42)$$

(See **[192]**.) In light of (6.42) and Thrm. 6.5.1, we can conclude that the operator $C : \mathrm{dom}(C) \subset \mathbb{L}^2(0,a) \to \mathbb{L}^2(0,a)$ defined by

$$Cu = \alpha^2 \frac{d^2 u}{dx^2} + \gamma \frac{du}{dx} \qquad (6.43)$$

$$\mathrm{dom}(C) = \left\{ u \in \mathbb{L}^2(0,a) \,\Big|\, \exists \frac{du}{dx}, \frac{d^2 u}{dx^2}, \frac{d^2 f}{dx^2} \in \mathbb{L}^2(0,a) \wedge u(0) = u(a) = 0 \right\}$$

generates a C_0-semigroup on $\mathbb{L}^2(0,a)$. If the BCs are of Neumann or convective type (as defined in (3.28) and (3.29)), then the domain can be modified in order to draw the same conclusion. Use this fact to complete the following exercise.

Exercise 6.5.4.
i.) Reformulate (6.40) abstractly as (6.31) in $\mathscr{X} = \mathbb{L}^2(0,a)$.
ii.) Impose conditions to ensure that (6.40) has a unique mild solution on $[0,T]$.
iii.) Under what condition is the mild solution in (ii) continuable to $[0,\infty)$.

Exercise 6.5.5. Consider the following variant of (6.8) with logistic forcing:

$$\begin{cases} \frac{\partial P_1}{\partial t} = \alpha_{P_1} \frac{\partial^2 P_1}{\partial x^2} + \frac{\partial}{\partial x}(\beta_{P_1} P_1) + \frac{\gamma_1 g_1(P_1,P_2)}{1+\gamma_1 |g_1(P_1,P_2)|}, \, 0 < x < a, t > 0, \\ \frac{\partial P_2}{\partial t} = \alpha_{P_2} \frac{\partial^2 P_2}{\partial x^2} + \frac{\partial}{\partial x}(\beta_{P_2} P_2) + \frac{\gamma_2 g_2(P_1,P_2)}{1+\gamma_2 |g_2(P_1,P_2)|}, \, 0 < x < a, t > 0, \\ P_1(x,0) = P_1^0(x,0), \, P_2(x,0) = P_2^0(x,0), \, 0 < x < a, \\ P_1(0,t) = P_1(a,t) = \frac{\partial P_2}{\partial x}(0,t) = \frac{\partial P_2}{\partial x}(a,t) = 0, \, ,t > 0, \end{cases} \qquad (6.44)$$

where $P_i = P_i(x,t)$, $g_i : \mathbb{R} \times \mathbb{R} \to \mathbb{R}$, $\alpha_{P_i}, \beta_{P_i} \, (i = 1,2)$ are real constants, and γ_1, γ_2 are positive constants.
i.) Reformulate (6.44) abstractly as (6.31) in an appropriate space.
ii.) Impose conditions to ensure that (6.44) has a unique mild solution on $[0,T]$.
iii.) Formulate a continuity result with respect to the parameters γ_1, γ_2 from which you can deduce the limit function $\lim\limits_{\gamma_1,\gamma_2 \to 0^+} \begin{bmatrix} P_1 \\ P_2 \end{bmatrix}$.

Model III.3 Spring-Mass Systems with Logistic Forcing
   Logistic forcing can occur in the model of a spring-mass system. For instance, consider the following variant of (5.139):

$$\begin{cases} x''(t) + \beta x'(t) + \omega^2 x(t) = \frac{\alpha g(x)}{1+g(x)}, \, t > 0, \\ x(0) = x_0, \, x'(0) = x_1, \end{cases} \qquad (6.45)$$

where $g : \mathbb{R} \to (0,\infty)$ and $\alpha > 0$.

Exercise 6.5.6.

i.) Reformulate (6.45) abstractly as (6.31) in an appropriate space.

ii.) If $g \in C(\mathbb{R}; (0,\infty))$, must (6.45) have a unique mild solution on $[0,T]$?

iii.) Let $h : \mathbb{R} \times \mathbb{R} \to (0,\infty)$ and suppose that the right-side of (6.45) is replaced by the more general logistic term $\frac{\alpha h(x,x')}{1+h(x,x')}$. Impose sufficient conditions on h ensuring that this modified version of IVP (6.45) has a unique mild solution on $[0,T]$.

iv.) For each $n \in \mathbb{N}$, let $g_n \in C^1(\mathbb{R}; (0,\infty))$ and consider the variant of (6.45) obtained by replacing its right-side by $\frac{\alpha g_n(x)}{1+g_n(x)}$. Denote the mild solution of this IVP by x_n and assume that $g_n \to 0$ uniformly as $n \to \infty$.

    a.) Prove that $\exists x^\star \in C(\mathbb{R}; (0,\infty))$ such that $x_n \to x^\star$ uniformly as $n \to \infty$.

    b.) To which IVP is $x^\star$ a mild solution? Is it also a classical solution?

Exercise 6.5.7. Consider the following coupled system of two springs with accumulative forcing term where all constants are the same as in (2.9):

$$\begin{cases} m_A x_A''(t) + (k_A + k_B) x_A(t) + k_A x_B(t) = \int_0^t a_1(t-s) g_1(s, x_A(s), x_B(s)) ds, \\ m_B x_B''(t) - k_B x_A(t) + k_B x_B(t) = \int_0^t a_2(t-s) g_2(s, x_A(s), x_B(s)) ds, \\ x_A(0) = x_{0,A}, \ x_A'(0) = x_{1,A}, \\ x_B(0) = x_{0,B}, \ x_B'(0) = x_{1,B}. \end{cases}$$

$$(6.46)$$

i.) Reformulate (6.46) abstractly as (6.31) in an appropriate space.

ii.) Formulate and prove an existence-uniqueness result for (6.46).

Model XI.2 Pollution Model Revisited

A stochastic model of pollution is investigated in [187]. We presently consider a deterministic version of this model, but with different perturbative effects. Let Ω be a bounded region in $\mathbb{R}^N$ with smooth boundary $\partial\Omega$, and let $z(\mathbf{x},t)$ denote the pollution concentration at position $\mathbf{x} \in \Omega$ and time $t > 0$. Consider the following version of (6.7):

$$\begin{cases} \frac{\partial z}{\partial t} = k \triangle z + \overrightarrow{\alpha} \cdot \nabla z + \int_0^t a(t-s) g(s,z) ds, \ \mathbf{x} \in \Omega, t > 0, \\ z(\mathbf{x},0) = z_0(\mathbf{x}), \ \mathbf{x} \in \Omega, \\ \frac{\partial z}{\partial \mathbf{n}}(\mathbf{x},t) = 0, \ \mathbf{x} \in \partial\Omega, t > 0, \end{cases}$$

$$(6.47)$$

where $z = z(\mathbf{x},t)$ and $\frac{\partial z}{\partial \mathbf{n}}$ is the outward unit normal vector to $\partial\Omega$. Here, k is the dispersion coefficient, $\overrightarrow{\alpha} \cdot \nabla$ represents the water/air velocity (assumed, for simplicity, to be independent of spatial and temporal variables), and N is the dimension of the region under consideration. For instance, pollution in a river could be modeled by (6.47) with $N=1$, while describing the concentration of pollution across the surface of infected algae throughout a bay would require that we use $N=2$. The forcing term in (6.47) can be interpreted as an accumulation of concentration.

Exercise 6.5.8.

i.) Reformulate (6.47) abstractly as (6.31) in an appropriate space.

ii.) Formulate and prove an existence-uniqueness result for (6.47) in the spirit of Thrm. 6.3.3.

Exercise 6.5.9. Formulate a continuous dependence result, with respect to $\vec{\alpha}$ only, for (6.47). What happens to the mild solution of (6.47), as guaranteed to exist by Exer. 6.5.8 (ii), as $\left\| \vec{\alpha} \right\|_{\mathbb{R}^N} \longrightarrow 0$.

We now introduce three new models.

Model XII.1 Epidemiological Models

Diffusive phenomena occur in a wide variety of settings ranging from the spreading of ideas and rumors in a social setting **[94]**, to the spread of a virus through a region **[131]** or even a worm through the Internet, to the effects of predation on rain forests **[45]** and wetlands. The study of *epidemiology* is concerned with developing models that describe the evolution of such spreading. We consider such models below.

Many variables affect population density (e.g., demographics, environment, geographic elements, etc.). We begin with a version of a classical two-dimensional model explored in **[120]**. Let $\Omega = [0, L_1] \times [0, L_2]$ represent the spatial region of interest and $t > 0$. We consider two interrelated populations, given by

$$P_H = P_H(x,y,t) = \text{Host population at position } (x,y) \in \Omega \text{ at time } t > 0,$$
$$P_V = P_V(x,y,t) = \text{Viral population at position } (x,y) \in \Omega \text{ at time } t > 0.$$

Assume that:

(a) Both P_H and P_V are subject to diffusion (with diffusion constants α_H and α_V).

(b) The birth rate of the host population is the positive constant β_H.

(c) The rate at which the virus becomes inviable is the positive constant γ_V.

(d) The virus is transmitted via human interaction.

We account for nonlinearity in the transmission dynamics via the forcing term g and arrive at the IBVP

$$\begin{cases} \frac{\partial P_H}{\partial t} = \alpha_H \triangle P_H + \beta_H P_H - g\left(t, x, y, P_H, P_V\right) P_H P_V, \ (x,y) \in \Omega, t > 0, \\ \frac{\partial P_V}{\partial t} = \alpha_V \triangle P_V - \gamma_V P_V + g\left(t, x, y, P_H, P_V\right) P_H P_V, \ (x,y) \in \Omega, t > 0, \\ P_H(x,y,0) = P_H^0(x,y), \ P_V(x,y,0) = P_V^0(x,y), \ (x,y) \in \Omega, \\ \frac{\partial P_H}{\partial \mathbf{n}}(x,y,t) = \frac{\partial P_V}{\partial \mathbf{n}}(x,y,t) = 0, \ (x,y) \in \partial\Omega, t > 0, \end{cases}$$

which is equivalent to

$$\begin{cases} \dfrac{\partial}{\partial t}\begin{bmatrix} P_H \\ P_V \end{bmatrix} = \begin{bmatrix} \alpha_H \triangle + \beta_H I & 0 \\ 0 & \alpha_V \triangle - \gamma_V I \end{bmatrix}\begin{bmatrix} P_H \\ P_V \end{bmatrix} + \begin{bmatrix} -g\left(t,x,y,P_H,P_V\right)P_H P_V \\ g\left(t,x,y,P_H,P_V\right)P_H P_V \end{bmatrix}, \\ \begin{bmatrix} P_H \\ P_V \end{bmatrix}(x,y,0) = \begin{bmatrix} P_H^0 \\ P_V^0 \end{bmatrix}(x,y),\ (x,y) \in \Omega, \\ \dfrac{\partial}{\partial \mathbf{n}}\begin{bmatrix} P_H \\ P_V \end{bmatrix}(x,y,t) = 0,\ (x,y) \in \partial\Omega,\, t > 0, \end{cases}$$

(6.48)

where $g : [0,\infty) \times \Omega \times [0,\infty) \times [0,\infty) \rightarrow [0,\infty)$ is a continuous mapping.

Exercise 6.5.10.
i.) Reformulate (6.48) abstractly as (6.31) in an appropriate space.
ii.) Formulate and prove results analogous to those in Section 5.5 for (6.48).

Exercise 6.5.11. Assume that

$$g\left(t,x,y,P_H,P_V\right) = \int_0^t \frac{\eta P_H(x,y,s)P_V(x,y,s)}{\left(1 + P_H(x,y,s) + P_V(x,y,s)\right)^2}\,ds,$$

(6.49)

so that the rate of infection/transmission increases over time due to the fact that more people are being infected. Does the IBVP (6.48), where g is defined as in (6.49), have a unique mild solution on $[0,T]$, or must additional restrictions be imposed on the data in order to draw this conclusion?

Next, we incorporate a more general type of dispersion into (6.48) by adding appropriate integral terms. Doing so yields the modified forcing term

$$\begin{bmatrix} -g\left(t,x,y,P_H,P_V\right)P_H P_V + \int_\Omega a_1(x,y,w,z)P_H(w,z,t)dwdz \\ g\left(t,x,y,P_H,P_V\right)P_H P_V + \int_\Omega a_2(x,y,w,z)P_V(w,z,t)dwdz \end{bmatrix},$$

(6.50)

where $a_i : \Omega \times \Omega \rightarrow (0,\infty)$ $(i = 1,2)$ are continuous mappings and g is globally Lipschitz in its last four variables.

Exercise 6.5.12. Show that (6.50) can be expressed as a functional

$$\mathfrak{F} : \mathbb{C}\left([0,T];\left(\mathbb{L}^2(\Omega)\right)^2\right) \rightarrow \mathbb{L}^1\left(0,T;\left(\mathbb{L}^2(\Omega)\right)^2\right).$$

Is $\mathfrak{F}$ Lipschitz?

The complexity of the model increases if there exist N different strains of the virus, each of which attacks the population separately and is governed by its own dispersal and infection rates. Assuming no interaction among strains, the IBVP leading to

(6.48) becomes the following system of $N + 1$ equations:

$$\begin{cases} \frac{\partial P_H}{\partial t} = \alpha_H \triangle P_H + \beta_H P_H - \sum_{i=1}^N g_i(t,x,y,P_H,P_{V_1},\ldots,P_{V_N}) P_H P_{V_i}, \\ \frac{\partial P_{V_1}}{\partial t} = \alpha_{V_1} \triangle P_{V_1} - \sum_{i=1}^N \mathcal{W}_i P_{V_i} + \sum_{i=1}^N g_i(t,x,y,P_H,P_{V_1},\ldots,P_{V_N}) P_H P_{V_i}, \\ \qquad \vdots \\ \frac{\partial P_{V_N}}{\partial t} = \alpha_{V_N} \triangle P_{V_N} - \sum_{i=1}^N \mathcal{W}_i P_{V_i} + \sum_{i=1}^N g_i(t,x,y,P_H,P_{V_1},\ldots,P_{V_N}) P_H P_{V_i}, \\ P_H(x,y,0) = P_H^0(x,y), \ P_{V_i}(x,y,0) = P_{V_i}^0(x,y), \ (x,y) \in \Omega, i = 1,\ldots,N, \\ \frac{\partial P_H}{\partial \mathbf{n}}(x,y,t) = \frac{\partial P_{V_i}}{\partial \mathbf{n}}(x,y,t) = 0, \ (x,y) \in \partial\Omega, t > 0, i = 1,\ldots,N, \end{cases} \qquad (6.51)$$

where $P_H = P_H(x,y,t)$, $P_{V_i} = P_{V_i}(x,y,t)$, and $(x,y) \in \Omega$.

Exercise 6.5.13.
i.) Reformulate (6.51) abstractly as (6.31) in an appropriate space.
ii.) Formulate and prove an existence-uniqueness result for (6.51).

Exercise 6.5.14. Establish a continuous dependence result for (6.51) in terms of the dispersal and rate constants.

We could further subdivide the host population into subclasses based on susceptibility, age, and other factors to generate similar systems, albeit involving more equations and more complex nonlinearities.

Model XIII.1 Aeroelasticity - A Linear Approximation

Airplane wings are designed to bend and flap in a controlled manner during flight. Helicopter rotor blades undergo vibrations whose dynamics depend on the material with which the blades are composed, aerodynamic forces, etc. An important part of ensuring successful flight is controlling the vertical displacement of the wing/rotor blades in order to prevent *flutter*, an increasingly rapid and potentially destructive and uncontrollable vibration.

We consider a model introduced by Dowell in 1975 (see [136]) describing the deflection of a rectangular panel devoid of two-dimensional effects, meaning that we track only the cross-section of an edge of the panel. We further assume that the edges are supported so that there is no movement along them. If this panel is part of an aircraft moving through an airstream, it makes sense that it will deform as forces due to wind act on it; otherwise, it would snap. (If you have ever glanced out the window during a plane ride, you have undoubtedly noticed small bounce in the wing.) We also expect that the panel has a natural state to which it reverts in the absence of external forces, because otherwise the aircraft would permanently deform as a result of even the most insignificant of forces acting on it. Detailed discussions can be found in [35, 52, 136, 139, 236].

We model the panel as a one-dimensional segment, say $[0,a]$, since all cross-sections of the panel are assumed to be identical. (Accounting for twisting of the panel would lead to a more complicated nonlinear model, mentioned later in Chapter 9.)

Let $w = w(z,t)$ represent the panel deflection at (z,t), where $0 \le z \le a, t > 0$. Dowell considered the following second-order IBVP which serves as a linearized approximation of panel flutter:

$$\begin{cases} \frac{\partial^2 w}{\partial t^2} + \beta_1 \frac{\partial w}{\partial t} + \beta_2 \frac{\partial}{\partial t}\left(\frac{\partial^4 w}{\partial z^4}\right) + \frac{\partial^4 w}{\partial z^4} - \beta_3 \frac{\partial^2 w}{\partial z^2} + \beta_4 \frac{\partial w}{\partial z} = 0, \\ w(0,t) = w(a,t) = 0, t > 0, \\ \frac{\partial^2 w}{\partial z^2}(0,t) + \beta_2 \frac{\partial}{\partial t}\left(\frac{\partial^2 w}{\partial z^2}\right)(0,t) = 0, t > 0, \\ \frac{\partial^2 w}{\partial z^2}(a,t) + \beta_2 \frac{\partial}{\partial t}\left(\frac{\partial^2 w}{\partial z^2}\right)(a,t) = 0, t > 0, \\ w(z,0) = w_0(z), \frac{\partial w}{\partial t}(z,0) = w_1(z), 0 \le z \le a. \end{cases} \qquad (6.52)$$

The physical parameters $\beta_1, \beta_2, \beta_3, \beta_4$ are assumed to be positive constants and represent the measures of viscoelastic structural damping, aerodynamic pressure, inplane tensile load, and aerodynamic damping, respectively.

We want to reformulate (6.52) as an abstract evolution equation. But, which of the forms studied thus far is most appropriate? The equation involves several more terms than previous models, and it is not initially clear which terms should be used to form the operator A and which should be subsumed into the forcing term. We outline one possible reformulation of (6.52), developed in [236], below.

First, the presence of a second-order time derivative suggests that viewing (6.52) as a system of two equations, as we did in the study of the classical wave equation, might be a prudent first step. To this end, we use the Hilbert space $\mathscr{H} = \mathbb{H}^2(0,a) \times \mathbb{L}^2(0,a)$ equipped with the following inner product and norm:

$$\left\langle \begin{bmatrix} f \\ \frac{\partial f}{\partial t} \end{bmatrix}, \begin{bmatrix} g \\ \frac{\partial g}{\partial t} \end{bmatrix} \right\rangle_{\mathscr{H}} \equiv \left\langle \frac{\partial^2 f}{\partial z^2}, \frac{\partial^2 g}{\partial z^2} \right\rangle_{\mathbb{L}^2(0,a)} + \left\langle \frac{\partial f}{\partial t}, \frac{\partial g}{\partial t} \right\rangle_{\mathbb{L}^2(0,a)} \qquad (6.53)$$

$$\left\| \begin{bmatrix} f \\ \frac{\partial f}{\partial t} \end{bmatrix} \right\|_{\mathscr{H}}^2 \equiv \left\| \frac{\partial^2 f}{\partial z^2} \right\|_{\mathbb{L}^2(0,a)}^2 + \left\| \frac{\partial f}{\partial t} \right\|_{\mathbb{L}^2(0,a)}^2 \qquad (6.54)$$

$$= \int_0^a \left[\left| \frac{\partial^2 f}{\partial z^2}(y, \cdot) \right|^2 + \left| \frac{\partial f}{\partial t}(y, \cdot) \right|^2 \right] dy.$$

Exercise 6.5.15. Show that (6.53) is a well-defined inner product on $\mathscr{H}$.

The equation in (6.52) can be written equivalently as

$$\frac{\partial}{\partial t}\begin{bmatrix} w \\ \frac{\partial w}{\partial t} \end{bmatrix} = \begin{bmatrix} 0 & I \\ -\frac{\partial^4}{\partial z^4} & -\beta_2 \frac{\partial^4}{\partial z^4} \end{bmatrix}\begin{bmatrix} w \\ \frac{\partial w}{\partial t} \end{bmatrix} + \begin{bmatrix} 0 & 0 \\ \beta_3 \frac{\partial^2}{\partial z^2} - \beta_4 \frac{\partial}{\partial z} & -\beta_1 I \end{bmatrix}\begin{bmatrix} w \\ \frac{\partial w}{\partial t} \end{bmatrix}. \qquad (6.55)$$

Symbolically, it is reasonable to identify the unknown $U(t) = \begin{bmatrix} w \\ \frac{\partial w}{\partial t} \end{bmatrix}$ and the operators $A : \text{dom}(A) \subset \mathcal{H} \to \mathcal{H}$ and $B : \text{dom}(B) \subset \mathcal{H} \to \mathcal{H}$ by

$$A \begin{bmatrix} w \\ \frac{\partial w}{\partial t} \end{bmatrix} = \begin{bmatrix} 0 & I \\ -\frac{\partial^4}{\partial z^4} & -\beta_2 \frac{\partial^4}{\partial z^4} \end{bmatrix} \begin{bmatrix} w \\ \frac{\partial w}{\partial t} \end{bmatrix}, \tag{6.56}$$

$$B \begin{bmatrix} w \\ \frac{\partial w}{\partial t} \end{bmatrix} = \begin{bmatrix} 0 & 0 \\ \beta_3 \frac{\partial^2}{\partial z^2} - \beta_4 \frac{\partial}{\partial z} & -\beta_1 I \end{bmatrix} \begin{bmatrix} w \\ \frac{\partial w}{\partial t} \end{bmatrix}. \tag{6.57}$$

These identfications would render (6.52) as an abstract <u>homogenous</u> evolution equation similar to (3.87)!

Exercise 6.5.16. Keeping in mind the space and the BCs, determine $\text{dom}(A)$ and $\text{dom}(B)$. Why are the operators A and B well-defined?

Exercise 6.5.17.
i.) Prove the following:
    a.) The operator B in (6.57) is bounded.
    b.) The operator A in (6.56) is dissipative.
    c.) For all $\lambda > 0$, $\text{rng}(\lambda I - A)$ is closed.
    d.) $\text{rng}(\lambda I - A)$ is dense in $\mathcal{H}$.
ii.) Conclude that (6.52) has a unique global mild solution on $[0, \infty)$.
iii.) Identify sufficient conditions that guarantee the existence of $\lim_{t \to \infty} w(\cdot, t)$.

Exercise 6.5.18. Consider the sequence of IBVPs obtained by replacing β_3 and β_4 by β_3^n and β_4^n, where $\beta_3^n, \beta_4^n \to 0$ as $n \to \infty$. Argue that the limit of the sequence of mild solutions of these IBVPs is a mild solution of the IBVP obtained by taking $\beta_3 = \beta_4 = 0$ in (6.52).

As with the other models explored in this chapter, we can account for an accumulation effect of external forces by adding the following forcing term on the right-side of the equation in (6.52):

$$g \left(t, z, \int_0^t a_1(t,s) w(s,z) ds, \int_0^t a_2(t,s) \frac{\partial w}{\partial s}(s,z) ds \right). \tag{6.58}$$

Exercise 6.5.19. Consider the IBVP obtained by adding the forcing term (6.58) to the right-side of (6.52) and keeping all other conditions the same.
i.) Reformulate (6.52) abstractly as (6.31) in an appropriate space.
ii.) Formulate and prove an existence-uniqueness result for (6.52).

Model XIV.1 Transverse Vibrations of Extensible Beams
We now consider a model similar to the one used to describe waves in a vibrating elastic string, but now we account for a different kind of vibration, namely transverse vibrations of a uniform bar (see **[32, 36, 132, 226, 265]**).

Let $w = w(z,t)$ represent the deflection of the beam at position $0 \leq z \leq a$ at time $t > 0$. The most rudimentary equation, coupled with its initial profile, involved in describing such a phenomenon is given by

$$\begin{cases} \frac{\partial^2 w}{\partial t^2} + \alpha \frac{\partial^4 w}{\partial z^4} = 0, \ 0 < z < a, t > 0, \\ w(z,0) = w_0(z), \ \frac{\partial w}{\partial t}(z,0) = w_1(z), \ 0 < z < a, \end{cases} \tag{6.59}$$

where $\alpha > 0$ describes the bending stiffness. A complete description of this phenomenon requires that we prescribe what happens at both ends of the rod. This can be done in many natural ways, two of which are described below:

1. Clamped at $z = a$:

$$\frac{\partial w}{\partial z}(a,t) = w(a,t) = 0, t > 0, \tag{6.60}$$

2. Hinged at $z = a$:

$$\frac{\partial^2 w}{\partial z^2}(a,t) = w(a,t) = 0, t > 0. \tag{6.61}$$

Exercise 6.5.20. Consider (6.59) equipped with the following 4 BCs:

$$\frac{\partial w}{\partial z}(0,t) = \frac{\partial^2 w}{\partial z^2}(0,t) = \frac{\partial w}{\partial z}(a,t) = \frac{\partial^2 w}{\partial z^2}(a,t) = 0, t > 0 \tag{6.62}$$

i.) Solve (6.59) coupled with (6.62) using the separation of variables method.
ii.) Using the form of the solution obtained in (i), conjecture the form of the semigroup as we did in Section 3.3.1.

Now, as in our study of the wave equation, we can view (6.59) abstractly by making the following change of variable (in the spirit of (4.42)):

$$v_1 = w, \ v_2 = \frac{\partial w}{\partial t}, \ \frac{\partial v_1}{\partial t} = v_2, \ \frac{\partial v_2}{\partial t} = -\alpha \frac{\partial^4 v_1}{\partial z^4} \tag{6.63}$$

Then, (6.59) becomes

$$\begin{cases} \frac{\partial}{\partial t} \begin{bmatrix} v_1 \\ v_2 \end{bmatrix}(z,t) = \begin{bmatrix} 0 & I \\ -\alpha \frac{\partial^4}{\partial z^4} & 0 \end{bmatrix} \begin{bmatrix} v_1 \\ v_2 \end{bmatrix}(z,t), \ 0 < z < a, t > 0, \\ \begin{bmatrix} v_1 \\ v_2 \end{bmatrix}(z,0) = \begin{bmatrix} w_0 \\ w_1 \end{bmatrix}(z,0), \ 0 < z < a, \\ \frac{\partial w}{\partial z}(0,t) = \frac{\partial^2 w}{\partial z^2}(0,t) = \frac{\partial w}{\partial z}(a,t) = \frac{\partial^2 w}{\partial z^2}(a,t) = 0, t > 0. \end{cases} \tag{6.64}$$

Guided by our discussion of the wave equation in Section 4.3, we shall reformulate (6.59) as an abstract evolution equation in the Hilbert space

$$\mathscr{H} = \text{dom}\left(-\frac{d^2}{dz^2}\right) \times \mathbb{L}^2(0,a) \tag{6.65}$$

$$\left\langle \begin{bmatrix} v_1 \\ v_2 \end{bmatrix}, \begin{bmatrix} v_1^\star \\ v_2^\star \end{bmatrix} \right\rangle_{\mathscr{H}} \equiv \int_0^a \left[\frac{\partial^2 v_1}{\partial z^2} \frac{\partial^2 v_1^\star}{\partial z^2} + v_2 v_2^\star \right] dz \tag{6.66}$$

by defining the operator $A : \text{dom}(A) \subset \mathscr{H} \to \mathscr{H}$ by

$$A \begin{bmatrix} v_1 \\ v_2 \end{bmatrix} = \begin{bmatrix} 0 & I \\ -\alpha \frac{\partial^4}{\partial z^4} & 0 \end{bmatrix} \begin{bmatrix} v_1 \\ v_2 \end{bmatrix} = \begin{bmatrix} v_1 \\ -\alpha \frac{\partial^4 v_2}{\partial z^4} \end{bmatrix}, \tag{6.67}$$

$$\text{dom}(A) = \text{dom}\left(-\frac{d^4}{dz^4}\right) \times \text{dom}\left(-\frac{d^2}{dz^2}\right),$$

where

$$\text{dom}\left(-\frac{d^4}{dz^4}\right) = \left\{ w \in \mathbb{L}^2(0,a) \,\middle|\, w, \frac{\partial w}{\partial z}, \frac{\partial^2 w}{\partial z^2}, \frac{\partial^3 w}{\partial z^3} \text{ are AC}, \frac{\partial^4 w}{\partial z^4} \in \mathbb{L}^2(0,a), \wedge \right.$$
$$\left. \frac{\partial w}{\partial z}(0,\cdot) = \frac{\partial^2 w}{\partial z^2}(0,\cdot) = \frac{\partial w}{\partial z}(a,\cdot) = \frac{\partial^2 w}{\partial z^2}(a,\cdot) = 0 \right\} \tag{6.68}$$

and

$$\text{dom}\left(-\frac{d^2}{dz^2}\right) = \left\{ w \in \mathbb{L}^2(0,a) \,\middle|\, w, \frac{\partial w}{\partial z} \text{ are AC}, \frac{\partial^2 w}{\partial z^2} \in \mathbb{L}^2(0,a), \wedge \right.$$
$$\left. \frac{\partial w}{\partial z}(0,\cdot) = \frac{\partial^2 w}{\partial z^2}(0,\cdot) = \frac{\partial w}{\partial z}(a,\cdot) = \frac{\partial^2 w}{\partial z^2}(a,\cdot) = 0 \right\}. \tag{6.69}$$

Arguing that $(\mathscr{H}, \langle \cdot, \cdot \rangle_{\mathscr{H}})$ is a Hilbert space is standard (see [320]) and we can show that A generates a C_0–semigroup on $\mathscr{H}$ by proceeding in a manner similar to Exer. 6.5.17. As such, we see that the IBVP (6.59) coupled with the BCs (6.62) can be reformulated as the abstract evolution equation (3.87).

Exercise 6.5.21. Supply the details in the above discussion.

As such, we recover the unique classical solution of (6.59) coupled with (6.62) as $U(t) = e^{At} \begin{pmatrix} w_0 \\ w_1 \end{pmatrix}$ which coincides with the classical solution from Exer. 6.5.20(i).

A more general principle at work behind the scenes is being used to show that the operators arising in second-order problems, such as the wave equation and the beam equation, generate C_0–semigroups (actually, C_0–groups). Before stating this result, we introduce the following notion of a square root of an operator.

Definition 6.5.2. Let $(\mathscr{H}, \langle \cdot, \cdot \rangle_{\mathscr{H}})$ be a real Hilbert space and $A : \text{dom}(A) \subset \mathscr{H} \to \mathscr{H}$. A positive operator $B : \text{dom}(B) \subset \mathscr{H} \to \mathscr{H}$ such that $B^2 = A$ is called a *positive square root of* A, and is denoted by $A^{1/2}$.

The domain of a square root operator can be equipped with the usual graph norm. We consider general powers of operators in the next chapter. The following result is fundamental in the theory of second-order evolution equations, and applies to our study of the wave and beam equations, as well as the equations arising in aeroelasticity models. (See [236] for a proof.)

Theorem 6.5.3. *Let* $(\mathscr{H}, \langle \cdot, \cdot \rangle_{\mathscr{H}})$ *be a real Hilbert space and* $A : \mathrm{dom}(A) \subset \mathscr{H} \to$ $\mathscr{H}$ *a self-adjoint operator for which* $\exists c > 0$ *such that* $\langle Ah, h \rangle \geq c \, \|h\|_{\mathscr{H}}^2$, $\forall h \in \mathrm{dom}(A)$. *The operator* $\mathscr{A} : \mathrm{dom}(\mathscr{A}) \subset \mathrm{dom}\left(A^{1/2}\right) \times \mathscr{H} \to \mathrm{dom}\left(A^{1/2}\right) \times \mathscr{H}$ *defined by* $\mathscr{A} = \begin{bmatrix} 0 & I \\ -A & 0 \end{bmatrix}$ *generates a* C_0*–group* $\left\{ e^{t\mathscr{A}} : t \in \mathbb{R} \right\}$ *on* $\mathrm{dom}\left(A^{1/2}\right) \times \mathscr{H}$*. Moreover,* $U(t) =$ $e^{t\mathscr{A}} \begin{bmatrix} u_0 \\ u_1 \end{bmatrix}$ *is a classical solution of the second-order abstract evolution equation*

$$\begin{cases} U''(t) = -AU(t), \, t \in \mathbb{R}, \\ U(0) = u_0, \, U'(0) = u_1. \end{cases}$$

We illustrate the applicability of this theorem in the present model following the approach used in **[132]**.. Define the operators $B_1, B_2 : \mathbb{L}^2(0,a) \to \mathbb{L}^2(0,a)$ by

$$B_1[f] = \alpha \frac{d^4 f}{dz^4}, \tag{6.70}$$

$$B_2[f] = \sqrt{\alpha} \frac{d^2 f}{dz^2}. \tag{6.71}$$

Their domains are given by (6.68) and (6.69), respectively. Using a standard Fourier-type basis $\{ e_n : n \in \mathbb{N} \}$ for $\mathbb{L}^2(0,a)$, these operators can be expressed as follows:

$$B_1[f] = \alpha \sum_{n=1}^{\infty} \left(\frac{n\pi}{\alpha} \right)^4 \langle f, e_n \rangle_{\mathbb{L}^2(0,a)} e_n, \tag{6.72}$$

$$B_2[f] = \sqrt{\alpha} \sum_{n=1}^{\infty} \left(\frac{n\pi}{\alpha} \right)^2 \langle f, e_n \rangle_{\mathbb{L}^2(0,a)} e_n. \tag{6.73}$$

Exercise 6.5.22. Try to verify that B_2 is the square root of B_1. Then, describe its connection to the abstract formulation of the classical wave equation.

Exercise 6.5.23. Let $\beta > 0$. Prove that the following IBVP has a unique classical solution:

$$\begin{cases} \frac{\partial^2 w}{\partial t^2} + \alpha \frac{\partial^4 w}{\partial z^4} + \beta w = 0, \, 0 < z < a, t > 0, \\ w(z,0) = w_0(z), \, \frac{\partial w}{\partial t}(z,0) = w_1(z), \, 0 \leq z \leq a, \\ \frac{\partial^2 w}{\partial z^2}(0,t) = w(0,t) = 0 = \frac{\partial^2 w}{\partial z^2}(a,t) = w(a,t), t > 0. \end{cases} \tag{6.74}$$

Now, we incorporate additional physical terms into the IBVP to improve the model. For instance, consider the following more general IBVP:

$$\begin{cases} \frac{\partial^2 w}{\partial t^2} + \alpha \frac{\partial^4 w}{\partial z^4} + \beta w = g\left(t, z, w(z,t)\right), \, 0 < z < a, 0 < t < T, \\ w(z,0) = w_0(z), \, \frac{\partial w}{\partial t}(z,0) = w_1(z), \, 0 < z < a, \\ \frac{\partial^2 w}{\partial z^2}(0,t) = w(0,t) = 0 = \frac{\partial^2 w}{\partial z^2}(a,t) = w(a,t), \, 0 < t < T, \end{cases} \tag{6.75}$$

where $g : [0,T] \times [0,a] \times \mathbb{R} \to \mathbb{R}$ is continuous on $[0,T] \times [0,a]$ and globally Lipschitz in the third variable (uniformly in (t,z)) with Lipschitz constant M_g. We can reformulate (6.75) abstractly as (5.10), where the forcing term $f : [0,T] \times \mathscr{H} \to \mathscr{H}$ (where $\mathscr{H}$ is given by (6.65)) is defined by

$$f\left(t, \begin{pmatrix} v_1 \\ v_2 \end{pmatrix}\right)(\cdot) = \begin{bmatrix} 0 \\ g(t,\cdot,v_1) \end{bmatrix}. \tag{6.76}$$

To see that f is globally Lipschitz on $\mathscr{H}$, observe that

$$\left\| f\left(t, \begin{pmatrix} v_1 \\ v_2 \end{pmatrix}\right) - f\left(t, \begin{pmatrix} v_1^{\star} \\ v_2^{\star} \end{pmatrix}\right) \right\|_{\mathscr{H}}^2 = \left\| \begin{bmatrix} 0 \\ g(t,\cdot,v_1) \end{bmatrix} - \begin{bmatrix} 0 \\ g(t,\cdot,v_1^{\star}) \end{bmatrix} \right\|_{\mathscr{H}}^2$$

$$= \int_0^a |g(t,v_1(t,z)) - g(t,v_1^{\star}(t,z))|^2 \, dz$$

$$\leq M_g^2 \int_0^a |v_1(t,z) - v_1^{\star}(t,z)|^2 \, dz \tag{6.77}$$

$$\leq M_g^2 \left\| \begin{bmatrix} v_1 \\ v_2 \end{bmatrix} - \begin{bmatrix} v_1^{\star} \\ v_2^{\star} \end{bmatrix} \right\|_{\mathscr{H}}^2.$$

As such, we can invoke Thrm. 5.5.2 to conclude that (6.75) has a unique mild solution on $[0,a]$.

Exercise 6.5.24. Assume that g satisfies **(H5.7)** - **(H5.9)**. Show that (6.75) has a unique mild solution on $[0,a]$ by appealing to Thrm. 5.6.3.

Exercise 6.5.25. Let $\gamma > 0$. Modify the above reasoning to deduce that the following IBVP has a unique mild solution on $[0,a]$:

$$\begin{cases} \frac{\partial^2 w}{\partial t^2} + \alpha \frac{\partial^4 w}{\partial z^4} + \beta w - \gamma \frac{\partial^2 w}{\partial z^2} = 0, \ 0 < z < a, t > 0, \\ w(z,0) = w_0(z), \ \frac{\partial w}{\partial t}(z,0) = w_1(z), \ 0 < z < a, \\ \frac{\partial w}{\partial z}(0,t) = w(0,t) = 0 = \frac{\partial w}{\partial z}(a,t) = w(a,t), t > 0, \end{cases} \tag{6.78}$$

Exercise 6.5.26.
i.) Let $\delta > 0$. Show that the following IBVP has a unique mild solution on $[0,a]$:

$$\begin{cases} \frac{\partial^2 w}{\partial t^2} + \delta \frac{\partial w}{\partial t} + \alpha \frac{\partial^4 w}{\partial z^4} + \beta w - \gamma \frac{\partial^2 w}{\partial z^2} = 0, \ 0 < z < a, t > 0, \\ w(z,0) = w_0(z), \ \frac{\partial w}{\partial t}(z,0) = w_1(z), \ 0 < z < a, \\ \frac{\partial w}{\partial z}(0,t) = w(0,t) = 0 = \frac{\partial w}{\partial z}(a,t) = w(a,t), t > 0, \end{cases} \tag{6.79}$$

ii.) More generally, suppose that the term $\delta \frac{\partial w}{\partial t}$ is replaced by $\left(\frac{\partial w}{\partial t}\right)^2$. Argue that the same conclusion as in (i) holds.
iii.) Argue that if $J : \mathbb{R} \to \mathbb{R}$ is an increasing globally Lipschitz function, then the IBVP (6.79) with $\delta \frac{\partial w}{\partial t}$ replaced by $J\left(\frac{\partial w}{\partial t}\right)$ has a unique mild solution on $[0,a]$.

Going one step further, we incorporate a more general forcing term into the above IBVPs to obtain

$$
\begin{cases}
\frac{\partial^2 w}{\partial t^2} + \alpha \frac{\partial^4 w}{\partial z^4} + \beta w = \int_0^t a(t,s) f\left(s,z,w,\frac{\partial^2 w}{\partial z^2},\frac{\partial w}{\partial s}\right) ds, \ 0 < z < a, \ 0 < t < T, \\
w(z,0) = w_0(z), \ \frac{\partial w}{\partial t}(z,0) = w_1(z), \ 0 < z < a, \\
\frac{\partial^2 w}{\partial z^2}(0,t) = w(0,t) = 0 = \frac{\partial^2 w}{\partial z^2}(a,t) = w(a,t), \ t > 0,
\end{cases}
$$
(6.80)

where $w = w(z,t)$, $f : [0,T] \times [0,a] \times \mathbb{R} \times \mathbb{R} \times \mathbb{R} \to \mathbb{R}$ is continuous on $[0,T] \times [0,a]$ and globally Lipschitz in the last three variables (uniformly in (t,z)), and $a : [0,T] \times [0,T] \to \mathbb{R}$ is continuous. Of course, this is only one of many different types of possible forcing terms.

Observe that (6.80) is a particular case of the abstract functional evolution equation

$$
\begin{cases}
U'(t) + AU(t) + BU(t) = \mathfrak{F}(U)(t), \ 0 < t < T, \\
U(0) = U_0,
\end{cases}
$$
(6.81)

in a Banach space $\mathscr{X}$, where $U(\cdot) = \begin{bmatrix} v_1(\cdot) \\ v_2(\cdot) \end{bmatrix}$, $A : \mathrm{dom}(A) \subset \mathscr{X} \to \mathscr{X}$ and $B : \mathrm{dom}(B) \subset \mathscr{X} \to \mathscr{X}$ are operators, and $\mathfrak{F} : \mathbb{C}([0,T];\mathscr{X}) \to \mathbb{C}([0,T];\mathscr{X})$ is a Lipschitz functional.

Exercise 6.5.27. Prove that (6.80) has a unique mild solution on $[0,T]$ by reformulating it as (6.81) and then applying the appropriate theorem.

Remark. The study of fiber dynamics (or elastodynamics) is related to that of beam dynamics, but for a material that has different inherent characteristics. The goal is to model the motion of long flexible fibers in a moving "fluid," such as an airstream or liquid. There have been numerous articles written on this subject (see [170, 233]). The following simple linearized model of the horizontal component of the fiber is very similar to the beam models discussed above:

$$
\begin{cases}
\frac{\partial^2 w}{\partial t^2} + \alpha \frac{\partial^4 w}{\partial z^4} - \beta \frac{\partial}{\partial z}\left(c(z)\frac{\partial w}{\partial z}\right) = F(z,t), \ 0 < z < a, \ 0 < t < T, \\
w(z,0) = w_0(z), \ \frac{\partial w}{\partial t}(z,0) = w_1(z), \ 0 < z < a, \\
\frac{\partial^2 w}{\partial z^2}(0,t) = w(0,t) = 0 = \frac{\partial^2 w}{\partial z^2}(a,t) = w(a,t), \ t > 0,
\end{cases}
$$
(6.82)

where $w = w(z,t)$. The external forcing term can be due to aerodynamic drag.

6.6 Looking Ahead

The IBVPs considered thus far have all been reformulated as abstract evolution equations in which one could solve for $u'(t)$ explicitly. But, this is not the case for all

IBVPs arising in practice. Indeed, consider, for instance, the following IBVP arising in the study of soil mechanics.

$$\begin{cases} \frac{\partial}{\partial t}\left(z - \frac{\partial^2 z}{\partial x^2}\right) + \frac{\partial^2 z}{\partial x^2} = \int_0^t k(s)f(s,z)ds, \, 0 < x < \pi, 0 < t < T, \\ z(x,0) = z_0(x), \, 0 < x < \pi, \\ z(0,t) = z(\pi,t) = 0, \, t > 0, \end{cases} \tag{6.83}$$

where $z = z(x,t)$, $k : [0,T] \to \mathbb{R}$ is integrable, and $f : [0,T] \times \mathbb{R} \to \mathbb{R}$ is a mapping. In order to reformulate (6.83) as an abstract evolution equation, let $\mathscr{X} = \mathbb{L}^2(0,\pi;\mathbb{R})$ and define the operators $A : \text{dom}(A) \subset \mathscr{X} \to \mathscr{X}$ and $B : \text{dom}(B) \subset \mathscr{X} \to \mathscr{X}$ as in (3.57). Also, define the functional $\mathfrak{F} : \mathbb{C}([0,T];\mathscr{X}) \to \mathbb{C}([0,T];\mathscr{X})$ by

$$\mathfrak{F}(z)(\cdot,t) = \int_0^t k(s)f(s,z(\cdot,s))ds.$$

These identifications enable us to reformulate (6.83) as the abstract evolution equation

$$\begin{cases} (Bu)'(t) + Au(t) = \mathfrak{F}(u)(t), \, 0 < t < T, \\ u(0) = u_0. \end{cases} \tag{6.84}$$

Ideally, we would like to further express (6.84) in the form (6.3). The only way, symbolically, to do so is to use the formal substitution $v(t) = Bu(t)$ in (6.84). While making this substitution does create an isolated term $v'(t)$, it comes at the expense of requiring the operator B be at least invertible. (Why?)

Exercise 6.6.1. What other technical complications arise as a result of making this substitution? Specifically, what compatibility requirements must exist between the operators A and B?

Despite the apparent shortcomings of the suggested substitution, our discussion in the next chapter reveals that it constitutes a viable approach.

6.7   Guidance for Exercises

6.7.1   Level 1: A Nudge in a Right Direction

6.1.1. The obvious change is to replace "$x \in [0,a]$" by "$(x,y,z) \in \Omega$." The second BC in (6.8) needs to be imposed on $\partial\Omega$. What other changes should be made?

6.2.1. Not necessarily. (Why?)

6.2.2. (6.18): Use Minkowski's inequality with standard properties of the integral.

(6.19): Use Holder's inequality and the definition of the $\mathbb{L}^2$−norm.

(6.20): Perform typical calculations and use $|f(t)| \le \|f\|_{\mathbb{C}([0,T];\mathbb{R})}, \forall t \in [0,T]$. Also use Exer. 1.4.4 for K_1, K_2.

6.2.3. The estimates are similar to those in Exer. 6.2.2. How do you verify the continuity?

6.2.4. i.) Apply Holder's inequality on the inner integral term of

$$\int_\Omega \left[\int_{T-\varepsilon}^T [1+\sqrt{s}] \log\left(1+x(s)(z)\right) ds\right]^2 dz. \text{ (Now what?)}$$

ii.) Show that (6.23) implies that

$$\frac{\|\mathfrak{H}_3(x)\|_{L^2(\Omega)}}{\|x\|_{C([0,T];L^2(\Omega))}} \longrightarrow 0 \text{ as } \|x\|_C \to 0. \tag{6.85}$$

iii.) Work through the calculations closely. What essential fact about the mapping $s \mapsto (1+\sqrt{s})$ is used?

6.2.5. Apply (6.26), followed by (6.25), together with Prop. 1.8.13(vi)(a). Be certain to correctly account for the square.

6.2.6. Yes. Argue as in Exer. 6.2.5.

6.2.7. Yes. Use the sup norm in the last step of the calculation instead of integrating from 0 to T. (How does the estimate change as a result?)

6.2.8. i.) There are different ways to proceed. For well-definedness, you must ensure that $\forall x \in L^1(0,T;\mathscr{X})$, the mapping $t \mapsto \int_0^t a(t-s)f(s,x(s))ds$ belongs to $L^1(0,T;\mathscr{X})$. The regularity of $\mathfrak{H}_5$ depends on that of a and f. We must estimate $\int_0^T \int_0^t a(t-s)f(s,x(s))dsdt$.

ii.) Compute the sup over $[0,T]$ in (6.29) rather than integrating over this interval.

6.2.9. Let $\begin{pmatrix} x \\ y \end{pmatrix} \in (C([0,T];\mathscr{X}))^2$. Note that

$$\left\|\begin{pmatrix} x \\ y \end{pmatrix}\right\|_{(C([0,T];\mathscr{X}))^2} = \sup_{0 \le t \le T} \left(\|x(t)\|_{\mathscr{X}} + \|y(t)\|_{\mathscr{Y}}\right).$$

Apply this with calculations analogous to those used to verify Claims 2 and 3.

6.3.1. In each case, use the previous version of the model with the same space $\mathscr{X}$ and operator A. The functional in each case is specified in Section 6.1 and $\mathscr{Y}_1 = \mathscr{Y}_2$ in all instances (and they are Sobolev spaces comprised of certain $\mathscr{X}$-valued functions).

6.3.2. Try $C([0,T];\mathscr{X})$.

6.3.3. Define a functional in a manner similar to (6.14), though a little more general. What must be true about f under the hypotheses of Thrm. 5.2.2?

6.3.4. Refer to Prop. 4.1.1. Does the fact that $\mathfrak{F}(v) \in L^1(0,T;\mathscr{X})$ provide sufficient regularity in order to be able to proceed with the proof used in Thrm. 6.3.2?

6.3.5. Note that $C([0,T];\mathscr{X}) \subset L^1(0,T;\mathscr{X})$. (So what?)

6.3.6. Suitably modify Def. 5.5.1(i).

6.3.7. Observe that

$$u_\varepsilon(t) = e^{At}(u_0+\varepsilon) + \int_0^t e^{A(t-s)}\mathfrak{F}(u_\varepsilon)(s)ds \tag{6.86}$$

and

$$u(t) = e^{At}u_0 + \int_0^t e^{A(t-s)}\mathfrak{F}(u)(s)ds. \tag{6.87}$$

Estimate $\|u_\varepsilon - u\|_{\mathbb{C}}$. (Now what?)

6.3.8. Subtract the variation of parameters formula (6.87) and

$$u_n(t) = e^{A_n t}u_0^n + \int_0^t e^{A_n(t-s)}\mathfrak{F}_n(u_n)(s)ds. \tag{6.88}$$

Then estimate $\|u_n - u\|_{\mathbb{C}}$. You will need to use the triangle inequality.

6.3.9. Consider the term in the second hint for Exer. 6.3.8. Apply the $\mathbb{L}^1$−norm instead of the sup norm.

6.3.10. Is Prop. 3.4.7 useful?

6.3.11. i.) Identify the space $\mathscr{X}$, operator A, and the IC in the same manner as we did when reformulating (3.47) abstractly. The functional $\mathfrak{F}$ is given by $\mathfrak{F}_2$ defined by (6.3). We need for $\mathfrak{F}_2$ to be globally Lipschitz. Showing this will involve computations similar to those used in Example 6.2.2. Provide a precise restriction on the data involving a, g, and k.

ii.) You need to consider sequences $\{a_n\}$, $\{g_n\}$, and $\{k_n\}$ defined on the same spaces as and which are convergent (in appropriate senses) to a, g, and k, respectively. Moreover, they must be sufficiently regular as to ensure that the sequence of functionals $\{\mathfrak{F}_n\}$ given by

$$\mathfrak{F}_n(z_n)(\cdot,\cdots,t) = \int_0^t a_n(t,s)g_n\left(s,z_n(s,\cdot,\cdots),\int_0^s k_n(s,\tau,z_n(\tau,\cdot,\cdots))d\tau\right)ds$$

converges to the functional $\mathfrak{F}_2$ defined in (6.3).

6.3.12. i.) Let $\mathbf{X} = \begin{bmatrix} y \\ z \end{bmatrix}$. Assume that $k \in \mathbb{L}^1(0,T;(0,\infty))$ is bounded and that D and $\overline{D}$ are globally Lipschitz in their state variables. Define $\mathfrak{F} : \mathbb{C}([0,T];\mathbb{R}^2) \to \mathbb{L}^1(0,T;\mathbb{R}^2)$ by (5.204). Prove that $\mathfrak{F}$ is well-defined and globally Lipschitz.

ii.) Approximate k, D and $\overline{D}$ by k_n, D_n and $\overline{D}_n$, respectively, and assume that each sequence converges in an appropriate sense and space to the mapping that it approximates. (Now what?)

6.3.13. i.) Use (6.9) and (6.10) to aid in reformulating (6.8) as (6.31). Use the notion of A-boundedness to argue that the operator A behaves appropriately (cf. Thrm. 6.5.1). Alternatively, we dealt with a similar operator earlier; mimic that approach here. In order to ensure that the functional given by (6.10) is globally Lipschitz, modify the assumptions used in Exer. 6.3.12 for all mappings involved.

ii.) This is the same as Exer. 6.3.12.

6.4.1. Argue as in the proof (using Approach 1) of Thrm. 5.7.3. Be careful to make appropriate modifications since $\mathfrak{F}$ is now a functional.

6.4.2. Yes. In fact, one possibility was used in Thrm. 5.7.3.

6.4.3. The proof is similar to that of Thrm. 6.4.1 (cf. Exer. 6.4.1). But, be especially careful when verifying precompactness.

6.5.1. i.) Use Exer. 4.3.5 (iv). How do you define the functional $\mathfrak{F}$?

ii.) Prove that $\mathfrak{F}$ has sublinear growth and that A satisfies $(\mathbf{H}_{\mathbf{A}}^{\star})$.

6.5.2. i.) Modify (5.195) for the function $f(t,x,y)$ and define a new ball $\mathfrak{B}_n$. Do the proofs of the continuity and compactness change? (If so, how?)

ii.) Assume the modified hypothesis used to formulate the theorem in (i) and assume that a is continuous on $[0,L]^2$. Can this be weakened?

6.5.3. i.) Refer to the hint for Exer. 5.5.30(i). How do you handle the forcing term? Note that the same space $\mathscr{H}$ works.

ii.) Do the conditions in Exer. 5.5.30(ii) imply that the functional is globally Lipschitz? What else must be assumed?

iii.) Impose conditions like **(H6.17)** and apply the analog of Prop. 5.7.4 established in Exer. 6.5.2(i).

6.5.4. i.) Combine the approach used in Exer. 5.5.32(i) with the operator C defined by (6.43). How do you handle the functional?

ii.) Consider Example 6.2.2.

iii.) Can you somehow modify Prop. 5.5.8?

6.5.5. i.) Use the argument for (6.8) and (6.9) to handle everything except for the forcing term.

ii.) Refer to Exer. 5.5.37. Does assuming that g_i is globally Lipschitz work?

iii.) To what functional does $\mathfrak{F}$ converge as $\gamma_1, \gamma_2 \to 0$?

6.5.6. i.) Mimic Exer. 5.5.26(i).

ii.) Can you apply Thrm. 6.3.2?

iii.) If (ii) worked, then assuming that h is continuous on its domain would yield similar results. Make certain to define the corresponding functional on the appropriate space.

iv.) To what function $g \in C(\mathbb{R};[0,\infty))$ does the sequence $\left\{ \frac{\alpha g_n(\cdot)}{1+g_n(\cdot)} \right\}$ converge (uniformly) as $n \to \infty$? (Now what?)

6.5.7. i.) Use Exer. 2.1.1 for everything except the forcing term.

ii.) Modify Exer. 6.5.3(ii) for this setting.

6.5.8. i.) & ii.) While more general, the structure of (6.47) is NOT any different from (6.8). Use this to guide your thinking. Use $\mathscr{X} = \mathbb{L}^2(\Omega)$.

6.5.9. $\vec{\alpha} \cdot \nabla z(x,t) \to 0$ uniformly for (x,t) as $\|\vec{\alpha}\|_{\mathbb{R}^N} \to 0$. (So what?)

6.5.10. i.) Use $\mathscr{X} = \left(\mathbb{L}^2(\Omega)\right)^2$. (Why?) Notice that $\beta_H I$ and $-\gamma_V I$ are bounded perturbations of $\alpha_H \triangle$ and $\alpha_V \triangle$, respectively. So, the natural identification of A as the matrix operator is sensible. (Why?) How about the functional?

6.5.11. Yes, as long as a standard data restriction is imposed. Argue that the functional defined using (6.49) is globally Lipschitz.

6.5.12. Simply compute the $\mathbb{L}^1\left(0,T;\left(\mathbb{L}^2(\Omega)\right)^2\right)$ −norm of the expression in (6.50).

Remember that for $\mathbf{j} = \begin{bmatrix} j_1(t,x,y) \\ j_2(t,x,y) \end{bmatrix}$,

$$\|\mathbf{j}\|_{\mathbb{L}^1\left(0,T;(\mathbb{L}^2(\Omega))^2\right)} = \int_0^T \|\mathbf{j}(t,\cdot)\|_{(\mathbb{L}^2(\Omega))^2} \, dt$$

$$= \int_0^T \left[\|j_1(t,\cdot)\|_{\mathbb{L}^2(\Omega)} + \|j_2(t,\cdot)\|_{\mathbb{L}^2(\Omega)}\right] dt.$$

6.5.13. i.) Mimic (6.48) and Exer. 6.5.10. The operator A should be a matrix with what dimensions? The forcing term consists of the terms $\sum_{i=1}^N g_i$. (Now what?)
ii.) Modify Exer. 6.5.10 appropriately.
6.5.14. Consider two versions of the IBVP (6.51), namely as it presently reads and also with all of the dispersal rate constants replaced by the same symbol with an overscore; for instance, replace α_H by $\overline{\alpha_H}$, etc. Let $\varepsilon > 0$. Assume that the corresponding rate constants are within ε of each other. Subtract the corresponding variation of parameters formulae. (Now what?)
6.5.15. Verify the properties of Def. 1.7.10. Linearity of the differential operators and the well-definedness of the $\mathbb{L}^2(0,a)$ inner product should make this easy.
6.5.16. The easier one to identify is

$$\text{dom}(B) = \left\{ \begin{pmatrix} w \\ \frac{\partial w}{\partial t} \end{pmatrix} \in \mathbb{H}^2(0,a) \times \mathbb{L}^2(0,a) \, \middle| \, w(0,t) = w(a,t) = 0 \right\}.$$

Explain why this makes sense.
6.5.17. i.) (a) Verify that Def. 3.1.1(i) and make certain to use the $\mathscr{H}$−norm.
(b) Note that

$$\left\langle A \begin{pmatrix} w \\ \frac{\partial w}{\partial t} \end{pmatrix}, \begin{pmatrix} w \\ \frac{\partial w}{\partial t} \end{pmatrix} \right\rangle_{\mathscr{H}} = \left\langle \frac{\partial^2}{\partial x^2}\left(\frac{\partial w}{\partial t}\right), \frac{\partial^2}{\partial x^2}(w) \right\rangle_{\mathbb{L}^2(0,a)} +$$
$$\left\langle \alpha \frac{\partial^4}{\partial x^4}\left(\frac{\partial w}{\partial t}\right) + \frac{\partial^4}{\partial x^4}(w), \frac{\partial w}{\partial t} \right\rangle_{\mathbb{L}^2(0,a)}$$

(Why? So what?)
(c) Let $\left\{ \begin{pmatrix} w_n \\ \frac{\partial w_n}{\partial t} \end{pmatrix} \right\} \subset \text{dom}(A)$ and assume that $(\lambda I - A)\begin{pmatrix} w_n \\ \frac{\partial w_n}{\partial t} \end{pmatrix} \longrightarrow y$ in $\mathscr{H}$.
Show that $\exists x \in \text{dom}(\lambda I - A)$ such that $(\lambda I - A)x = y$.
(d) Try it!
ii.) Apply Prop. 3.4.2.
iii.) Is the semigroup contractive? Why does this matter?
6.5.18. Refer to Exer. 6.5.5(iii) and Exer. 6.3.8 for related ideas.
6.5.19. i.) Suitably modify (6.27) to handle (6.58), assuming that the range of the functional is $\mathbb{C}([0,T];\mathscr{H})$.
ii.) Assume $a_i \in \mathbb{C}\left([0,T]^2;(0,\infty)\right)$ and show the functional is globally Lipschitz.
6.5.20. i.) Assume $w(z,t) = Z(z)T(t)$. Substituting this into (6.59) yields

$$\frac{Z^{(4)}(z)}{Z(z)} = -\frac{T''(t)}{\alpha T(t)} = \lambda.$$

Take into account the BCs and solve the resulting BVP for Z, which only has non-trivial solutions when $\lambda > 0$, say $\lambda = c^2$. (Why?) Now, continue...

ii.) Identify this in a manner similar to the example following Def. 3.3.1 and in the discussion of the classical wave equation.

6.5.22. Show that $B_2^2 f = B_1 f$. Remember that $B_2^2 f = B_2 (B_2(f))$.

6.5.23. Modifying the operator used in (6.64) yields

$$\begin{bmatrix} 0 & I \\ -\alpha\frac{\partial^4}{\partial z^4} - \beta I & 0 \end{bmatrix} = \begin{bmatrix} 0 & I \\ -\alpha\frac{\partial^4}{\partial z^4} & 0 \end{bmatrix} + \begin{bmatrix} 0 & 0 \\ -\beta I & 0 \end{bmatrix}.$$

What can be said about the operator $U \mapsto \begin{bmatrix} 0 & 0 \\ -\beta I & 0 \end{bmatrix} U$?

6.5.24. Verify directly that the forcing term satisfies the hypotheses of Thrm. 5.6.3. Do not reinvent the wheel!

6.5.25. Adjust the setting in Exer. 6.5.23. Be careful how you interpret the term $-\gamma\frac{\partial^2 w}{\partial z^2}$ in the abstract formulation.

6.5.26. i.) & ii.) Modify Exer. 6.5.25 appropriately.

6.5.27. This follows from using Exer. 6.5.23, together with a slightly modified version of the approach used in Exer. 6.5.18.

6.6.1. The operators A and B must be compatible. Their domains are related. (How?) They both ought to be linear in order for us to have any hope of applying the theory developed thus far. And, we need somehow, upon making the indicated change of variable, to identify a generator of a linear C_0-semigroup.

6.7.2 Level 2: An Additional Thrust in a Right Direction

6.1.1. The operator $\begin{bmatrix} \alpha_{P_1}\frac{\partial^2}{\partial x^2} - \beta_{P_1}\frac{\partial}{\partial x} & 0 \\ 0 & \alpha_{P_2}\frac{\partial^2}{\partial x^2} - \beta_{P_2}\frac{\partial}{\partial x} \end{bmatrix}$ must account for the replacement of x by (x,y,z). Use the general Laplacian $\triangle$ and $\nabla \cdot \overrightarrow{F} P_i$, where $\overrightarrow{F}$ is a constant vector in $\mathbb{R}^3$. (How?)

6.2.1. We must ensure that $\|c\|_{C([0,T]\times\mathbb{R};\mathbb{R})} < \infty$.

6.2.2. K_1, K_2 are powers of 2.

6.2.3. Let $\{x_n\} \subset C([0,T];L^2(\Omega))$ be such that $\|x_n - x\|_C \to 0$. Use Prop. 1.8.16 to show that $\|\mathfrak{H}_2(x_n) - \mathfrak{H}_2(x)\|_{L^2(\Omega)} \to 0$.

6.2.4. i.) The expression provided in Hint 1 is $\leq M(T,\varepsilon)\|1 + \log(T,\cdot)\|_{L^2(\Omega)}^2$, where $M(T,\varepsilon) = \left(\frac{(1+\sqrt{T})^3 - (1+\sqrt{T-\varepsilon})^3}{3}\right)\varepsilon$. Now use the fact that $z \mapsto (1+\log(T,z))$ is dominated by Cz, for some $C > 0$. (How? So what?)

ii.) Note that (6.85) implies (6.24) by definition. (Why?)

iii.) It holds and you can enlarge the space to at least $L^2(0,T;\mathbb{R})$.

6.2.5. Continuing the inequality in the proof of Claim 3, we see that

$$|\mathfrak{H}_4(x)(t)(z) - \mathfrak{H}_4(y)(t)(z)|^2 \le \left(2M_g^2 M_k^2 T \sup_{0 \le t \le T} \|a(t, \cdot)\|_{\mathbb{L}^2(0,T)}\right) \cdot$$
$$\sup_{0 \le t \le T} |x(t)(z) - y(t)(z)|^2.$$

Now, integrate both sides over Ω, and then over $[0, T]$. Use the fact that

$$\|x - y\|_{\mathbb{C}([0,T];\mathbb{L}^2(\Omega))}^2 = \int_\Omega \sup_{0 \le t \le T} |x(t)(z) - y(t)(z)|^2 \, dz.$$

6.2.6. Argue as in Hint 2 of Exer. 6.2.5. The estimate will involve more constants, but the approach is essentially the same.

6.2.7. The constants involve multiples of T because we are now taking sups and hence can invoke the linearity of the integral.

6.2.8. i.) Assume that $a \in \mathbb{C}([0,T];(0,\infty))$ and that f satisfies **(H6.4)** with $M_i \in \mathbb{C}([0,T];(0,\infty))$. Now, use standard computations to show that $\forall x \in \mathbb{L}^1(0,T;\mathscr{X})$,

$$\|\mathfrak{H}_5(x)\|_{\mathbb{L}^1} \le \|a\|_{\mathbb{C}} \left[T^2 \|M_1\|_{\mathbb{C}} + T \|M_2\|_{\mathbb{C}} \|x\|_{\mathbb{L}^1}\right].$$

ii.) The same conditions used in (i) work here. (Why?) The estimate, however, changes slightly to

$$\|\mathfrak{H}_5(x)\|_{\mathbb{L}^1} \le T \|a\|_{\mathbb{C}} \left[\|M_1\|_{\mathbb{C}} + \|M_2\|_{\mathbb{C}} \|x\|_{\mathbb{L}^1}\right].$$

6.2.9. Observe that

$$\left\|\mathfrak{H}_6\begin{pmatrix}u_1 \\ v_1\end{pmatrix} - \mathfrak{H}_6\begin{pmatrix}u_2 \\ v_2\end{pmatrix}\right\|_{\mathbb{C}} \le \left(M_{f_1} \|k_1\|_{\mathbb{L}^1} + M_{f_2} \|k_2\|_{\mathbb{L}^1}\right) \cdot$$
$$\left[\|u_1 - u_2\|_{\mathbb{C}} + \|v_1 - v_2\|_{\mathbb{C}}\right].$$

(So what?)

6.3.3. Define $\mathfrak{F} : \mathbb{C} \to \mathbb{C}$ by $\mathfrak{F}(x)(\cdot) = f(\cdot, x)$. Note that f is globally Lipschitz in its second variable. (So what?)

6.3.4. Yes. In fact, if $\mathfrak{F}(v) \in \mathbb{C}$, we might be able to improve the regularity of the mild solution of (6.32) in Thrm. 6.3.2.

6.3.5. The functional $\mathfrak{F}$ satisfying **(H6.8)** must also satisfy **(H6.9)**, but not vice versa. As such, Thrm. 6.3.2 implies Thrm. 6.3.3, but not conversely. So, Thrm. 6.3.3 is applicable to a wider variety of problems.

6.3.6. This closely resembles Def. 4.2.1(i). A good question is how to formulate an existence/uniqueness result for classical solutions of (6.31).

6.3.7. Observe that $\|\mathfrak{F}(u_\varepsilon) - \mathfrak{F}(u)\|_{\mathbb{C}} \longrightarrow 0$ as $\varepsilon \to 0^+$ using **(H6.8)**. (Why?) Use this fact with standard integral properties.

6.3.8. The trickiest term to handle is $\left\|e^{A_n(t-s)}\mathfrak{F}_n(u_n)(s) - e^{A(t-s)}\mathfrak{F}(u)(s)\right\|_{\mathscr{X}}$. This is equivalent to

$$\left\|e^{A_n(t-s)}\mathfrak{F}_n(u_n)(s) - e^{A(t-s)}\mathfrak{F}_n(u_n)(s) + e^{A(t-s)}\mathfrak{F}_n(u_n)(s)\right.$$
$$\left. - e^{A(t-s)}\mathfrak{F}_n(u)(s) + e^{A(t-s)}\mathfrak{F}_n(u)(s) - e^{A(t-s)}\mathfrak{F}(u)(s)\right\|_{\mathscr{X}}.$$

(Now what?)

6.3.9. The main concern is the boundedness of the set $\left\{\int_0^T \mathfrak{F}_n(u_n)(s)ds \mid n \in \mathbb{N}\right\}$. (Why?) Is it bounded? How do the computations differ from those in Exer. 6.3.8?

6.3.10. Since dom (A^3) is closed (Why?) and $u_0^n \to u_0$, we know that $u_0 \in$ dom (A^3). Now apply Prop. 3.4.7.

6.3.11. i.) Assume that $a \in \left(\mathbb{L}^2(0,T)\right)^2$ and that both g and k are globally Lipschitz in their spatial variables. The computations are a bit tedious, but are comparable to those in Example 6.2.2.

ii.) Assume they are globally Lipschitz in their spatial variables with the same Lipschitz constants as g and k, and that $\|a_n - a\|_{\left(\mathbb{L}^2(0,T)\right)^2} \to 0$. Now, the calculations are similar to those in (i).

6.3.12. i.) Use the fact that

$$|D(t,y(t),z(t))| = |D(t,y(t),z(t)) - D(t,0,0) + D(t,0,0)|,$$

and the same for $\overline{D}$, to argue that $\|\mathfrak{F}(\mathbf{X})\|_{\mathbb{L}^1} < \infty$. In order to verify the Lipschitz continuity, observe that

$$\|\mathfrak{F}(\mathbf{X}) - \mathfrak{F}(\mathbf{Y})\|_{\mathbb{L}^1} \leq \int_0^T \max\left\{ \int_0^s |k(\tau)| |D(\tau,y(\tau),z(\tau)) - D(\tau,\overline{y}(\tau),\overline{z}(\tau))| d\tau, \right.$$
$$\left. \left|\overline{D}(\tau,y(\tau),z(\tau)) - \overline{D}(\tau,\overline{y}(\tau),\overline{z}(\tau))\right| \right\} ds$$
$$\leq T \max\left\{ M_D \|k\|_{\mathbb{L}^1}, M_{\overline{D}} \right\} \|\mathbf{X} - \mathbf{Y}\|_{\mathbb{C}},$$

where M_D and $M_{\overline{D}}$ are the Lipschitz constants. When are we assured that the solution map is a contraction?

ii.) Assume that $\|k_n - k\|_{\mathbb{L}^1} \to 0$, $|D_n - D| \to 0$, and $\left|\overline{D}_n - \overline{D}\right| \to 0$.

6.4.1. The data restriction is needed in Step 3 because you have an inequality of the form

$$\|x(t)\|_{\mathscr{X}} \leq M(t) + M_S d_1 T \|x\|_{\mathbb{C}}, \forall t \in [0,T],$$

and you eventually need to find $C^\star > 0$ such that

$$\|x\|_{\mathbb{C}} \leq C^\star, \forall x \in \xi(\Phi).$$

6.4.2. Certainly, assuming that $d_1, d_2 \in \mathbb{C}([0,T];(0,\infty))$ works. (Show why.) Does assuming that $d_1, d_2 \in \mathbb{L}^1(0,T;(0,\infty))$ also work?

6.4.3. You need a revised version of the Arzela-Ascoli theorem that works for families of functions in $\mathbb{L}^1(0,T;\mathscr{X})$.

6.5.1. i.) Modify the functional $\mathfrak{H}_4$ given in (6.27) to define $\mathfrak{F}$. Be certain to incorporate the space $\mathscr{H}$ appropriately.

ii.) Apply Thrm. 6.4.1.

6.5.2. i.) The computations are slightly altered because of the presence of two variables. The data restriction that arises when showing that $\xi(\Phi)$ is bounded (so that Schaefer's fixed-point theorem can be applied) will, of course, change accordingly.

ii.) Does assuming that $a \in \left(\mathbb{L}^1(0,T)\right)^2$ work? Alternatively, you can establish a version of Prop. 5.7.4 for (6.31) directly and apply it.

6.5.3. i.) Mimic (6.30) for the functional.

ii.) You also need to impose a data restriction to ensure that the solution map is a contraction.

iii.) Alternatively, apply Thrm. 6.4.1 directly. Interpret the data restriction that arises.

6.5.4. i.) Can you adapt (6.15) to handle this functional? If so, tell how.

ii.) The standard contraction mapping argument works here.

iii.) If you proceed directly without reformulating the forcing term as a functional, show that the solution given by the variation of parameters formula is bounded on $[0, T_{\max})$. The conclusion then follows. (Why?)

6.5.5. i.) The forcing term is transformed into a functional $\mathfrak{F} : \mathbb{C}\left([0,T] ; \left(\mathbb{L}^2(0,a)\right)^2\right) \to$ $\mathbb{C}\left([0,T] ; \left(\mathbb{L}^2(0,a)\right)^2\right)$ by replacing the components in (6.10) with the two logistic terms in (6.44).

ii.) Note that

$$\left| \frac{\gamma_i g_i (P_1, P_2)}{1 + \gamma_i |g_i (P_1, P_2)|} \right| \le \gamma_i |g_i (P_1, P_2)| . \text{ (So what?)}$$

iii.) The functional approaches $\begin{bmatrix} 0 \\ 0 \end{bmatrix}$ as $\gamma_1, \gamma_2 \to 0$. So, the sequence of mild solutions $\begin{bmatrix} P_1 \\ P_2 \end{bmatrix}$ of (6.44) corresponding to the parameters γ_1, γ_2 approaches the mild solution of the homogenous version of (6.44).

6.5.6. i.) What is the functional in this case? On what space is it defined?

ii.) Is the mapping $\begin{bmatrix} x \\ x' \end{bmatrix} \mapsto \begin{bmatrix} 0 \\ \frac{\alpha g(x)}{1+g(x)} \end{bmatrix}$ globally Lipschitz as a functional from $\mathbb{C}\left([0,T] ; \mathbb{R}^2\right)$ into itself? If so, you can use Thrm. 6.3.2 (see **[49]**).

iii.) Determine if the mapping $\begin{bmatrix} x \\ x' \end{bmatrix} \mapsto \begin{bmatrix} 0 \\ \frac{\alpha h(x,x')}{1+h(x,x')} \end{bmatrix}$ is globally Lipschitz.

iv.) The limit function $g : \mathbb{R} \to \mathbb{R}$ is $g(x) = \alpha$. (Why?) Consider (6.45) with the right-side replaced by α. Denote the mild solution of this IVP by x^* (Why does it exist?) and show that $\|x_n - x^*\|_\mathbb{C} \to 0$ by appealing to the variation of parameters formula. Also, given the simplicity of the forcing term of the limiting IVP, x^* is also a classical solution. (Why?)

6.5.7. i.) Use the functional from Exer. 6.5.3, albeit on a different space.

6.5.8. i.) & ii.) You do not need the matrix operator, but an entry of it is close to what the operator A ought to be. Treat the functional in a manner similar to (6.29).

6.5.9. Consider (6.47) without the term $\overrightarrow{\alpha} \cdot \nabla z(x,t)$. Denote the mild solution of this IVP by z^*. The mild solution of (6.47) corresponding to $\overrightarrow{\alpha}$ approaches z^* uniformly for (x,t) as $\left\|\overrightarrow{\alpha}\right\|_{\mathbb{R}^N} \to 0$.

6.5.10. Define $\mathfrak{F} : \mathbb{C}\left([0,T]; \left(\mathbb{L}^2(0,a)\right)^2\right) \to \mathbb{C}\left([0,T]; \left(\mathbb{L}^2(0,a)\right)^2\right)$ by

$$\mathfrak{F}\begin{bmatrix} P_H \\ P_V \end{bmatrix}(x,y)(t) = \begin{bmatrix} -g\left(t,x,y,P_H,P_V\right)P_H P_V \\ g\left(t,x,y,P_H,P_V\right)P_H P_V \end{bmatrix}.$$

The standard growth conditions can be imposed on the components to apply the theory. (Try it!)

6.5.11. You need to bound the integrand above by an expression that can be subsequently bounded above by $M_g\left[\left\|P_H - \overline{P_H}\right\|_{\mathbb{C}} + \left\|P_V - \overline{P_V}\right\|_{\mathbb{C}}\right]$. To this end, note that $\frac{\eta P_H P_V}{(1+P_H+P_V)^2} \le \eta P_H P_V$. (So what?)

6.5.12. The functional is globally Lipschitz and the Lipschitz constant involves the quantities $\|a_i\|_{\mathbb{C}}, m(\Omega)$, and M_g.

6.5.13. i.) The operator A is the $(N+1) \times (N+1)$ matrix operator whose value at $\begin{bmatrix} P_H & P_{V_1} & \cdots & P_{V_N} \end{bmatrix}^T$ is defined by

$$\begin{bmatrix} \alpha_H I + \beta_H I & 0 & 0 & \cdots & 0 \\ 0 & \alpha_{V_1}\triangle - \sum_{i=1}^{N} \mathcal{W}_i I & 0 & \cdots & 0 \\ 0 & 0 & \alpha_{V_2}\triangle - \sum_{i=1}^{N} \mathcal{W}_{2i} I & \cdots & \\ \vdots & \vdots & \vdots & 0 & \vdots \\ 0 & 0 & 0 & \cdots & \alpha_{V_N}\triangle - \sum_{i=1}^{N} \mathcal{W}_N I \end{bmatrix} \begin{bmatrix} P_H \\ P_{V_1} \\ \vdots \\ P_{V_N} \end{bmatrix}.$$

Make certain to incorporate the BCs into dom(A). Does this operator generate a C_0−semigroup on $\left(\mathbb{L}^2(\Omega)\right)^{N+1}$? Next, define

$$\mathfrak{F} : \mathbb{C}\left([0,T]; \left(\mathbb{L}^2(\Omega)\right)^{N+1}\right) \to \mathbb{C}\left([0,T]; \left(\mathbb{L}^2(\Omega)\right)^{N+1}\right)$$

by

$$\mathfrak{F}\begin{bmatrix} P_H \\ P_{V_1} \\ \vdots \\ P_{V_N} \end{bmatrix}(t) = \begin{bmatrix} -\sum_{i=1}^{N} g_i\left(t,x,y,P_H,P_{V_1},\ldots,P_{V_N}\right)P_H P_{V_i} \\ \sum_{i=1}^{N} g_i\left(t,x,y,P_H,P_{V_1},\ldots,P_{V_N}\right)P_H P_{V_i} \\ \vdots \\ \sum_{i=1}^{N} g_i\left(t,x,y,P_H,P_{V_1},\ldots,P_{V_N}\right)P_H P_{V_i} \end{bmatrix}.$$

(Now what?)

6.5.14. The rest is standard estimation. A related result could be established assuming that all constants approach zero and arguing that the sequences of mild solutions generated by these sequences of constants tends towards a continuous function that is a mild solution of a related IBVP. Are these results the same?

6.5.16. The easier one to identify is

$$\text{dom}(A) = \left\{ \begin{pmatrix} w \\ \frac{\partial w}{\partial t} \end{pmatrix} \in \mathbb{H}^2(0,a) \times \mathbb{L}^2(0,a) \,|\, w \text{ satisfies } \textbf{(i)} - \textbf{(vi)} \right\},$$

where conditions (i) - (vi) are:

(i) $w(0,t) = w(a,t) = 0$, **(ii)** $\frac{\partial w}{\partial t}(0,t) = \frac{\partial w}{\partial t}(a,t) = 0$, **(iii)** $\frac{\partial^2 w}{\partial x^2}(0,t) + \beta_2 \frac{\partial^2 w}{\partial x^2}(0,t) = 0$, **(iv)** $\frac{\partial^2 w}{\partial x^2}(a,t) + \beta_2 \frac{\partial^2 w}{\partial x^2}(a,t) = 0$, **(v)** $w + \beta_2 \frac{\partial w}{\partial t} \in \mathbb{H}^4(0,a)$, and **(vi)** $\frac{\partial w}{\partial t} \in \mathbb{H}^2(0,a)$.

6.5.17. i.) (b) Continuing the inequality from Hint 1, we have

$$\left\langle \frac{\partial^2}{\partial x^2}\left(\frac{\partial w}{\partial t}\right), \frac{\partial^2 w}{\partial x^2}\right\rangle_{\mathbb{L}^2(0,a)} - \left\langle \alpha \frac{\partial^2}{\partial x^2}\left(\frac{\partial w}{\partial t}\right) + \frac{\partial^2 w}{\partial x^2}, \frac{\partial^2}{\partial x^2}\left(\frac{\partial w}{\partial t}\right)\right\rangle_{\mathbb{L}^2(0,a)} = $$

$$-\left\langle \alpha \frac{\partial^2}{\partial x^2}\left(\frac{\partial w}{\partial t}\right), \frac{\partial^2}{\partial x^2}\left(\frac{\partial w}{\partial t}\right)\right\rangle_{\mathbb{L}^2(0,a)} \le 0.$$

(Why?)

(c) We conclude that $\exists \lim\limits_{n\to\infty}\begin{pmatrix} w_n \\ \frac{\partial w_n}{\partial t} \end{pmatrix} = \begin{pmatrix} w_1^\star \\ w_2^\star \end{pmatrix}$ and $\exists \lim\limits_{n\to\infty} A\begin{pmatrix} w_n \\ \frac{\partial w_n}{\partial t} \end{pmatrix}$. **(So what?)**

(d) Refer to **[236]**.

6.5.19. ii.) Make certain to account for the $\mathscr{H}$-norm when showing the functional is globally Lipschitz.

6.5.20. Continue as we did in the examples discussed in Chapters 3 and 4.

6.5.22. Use the continuity and basic properties of the inner product to argue that

$$B_1[f] = \sqrt{\alpha} \sum_{n=1}^{\infty} \left(\frac{n\pi}{\alpha}\right)^2 \langle B_2[f], e_n\rangle_{\mathbb{L}^2} e_n.$$

6.5.23. It must be bounded. So, use Prop. 3.6.1, followed by Thrm. 3.4.2.

6.5.24. Verify directly that the forcing term satisfying the hypotheses of Thrm. 5.6.3. Do not reinvent the wheel!

6.5.25. We cannot simply incorporate the term $-\gamma\frac{\partial^2 w}{\partial z^2}$ as part of the operator in Exer. 6.5.23 since then it would no longer generate a C_0−semigroup on $\mathscr{H}$. Rather, use the same operator as in Exer. 6.5.23 and define the forcing term $f(t,w) = -\gamma\frac{\partial^2 w}{\partial z^2}$. To prove that this mapping is globally Lipschitz, show that

$$\left\| f\left(t, \begin{pmatrix} w \\ \frac{\partial w}{\partial t} \end{pmatrix}\right) - f\left(t, \begin{pmatrix} w^\star \\ \frac{\partial w^\star}{\partial t} \end{pmatrix}\right)\right\|_{\mathscr{H}} = \gamma\left\|\frac{\partial^2 w}{\partial z^2} - \frac{\partial^2 w^\star}{\partial z^2}\right\|_{\mathscr{H}}$$

$$\le \left\|\begin{pmatrix} w \\ \frac{\partial w}{\partial t} \end{pmatrix} - \begin{pmatrix} w^\star \\ \frac{\partial w^\star}{\partial t} \end{pmatrix}\right\|_{\mathscr{H}}.$$

Then, invoke Thrm. 5.5.2(i).

6.5.26. i.) Modify the forcing term as

$$f\left(t, \begin{pmatrix} w \\ \frac{\partial w}{\partial t} \end{pmatrix}\right) = \begin{bmatrix} 0 \\ \delta\frac{\partial w}{\partial t} - \gamma\frac{\partial^2 w}{\partial z^2} \end{bmatrix}. \qquad (6.89)$$

Show this mapping is globally Lipschitz on $\mathscr{H}$.

ii.) Replace $\delta\frac{\partial w}{\partial t}$ in (6.89) by $\delta\left(\frac{\partial w}{\partial t}\right)^2$ and argue that the new forcing term is globally Lipschitz on $\mathscr{H}$.

iii.) Extract the underlying principle that makes (i) and (ii) work to establish this result.

6.6.1. Look ahead to Section 7.1.

Chapter 7

Implicit Evolution Equations

Overview

We consider two special classes of evolution equations in which the time derivative of the unknown is defined implicitly in the equation, but for which we can still generate a variation of parameters formula for a mild solution. Such equations arise in a vast assortment of fields, including soil mechanics, thermodynamics, civil engineering, and non-Newtonian fluids.

7.1   Sobolev-Type Equations

7.1.1   Motivation by Models

The intention of the following discussion is to focus only on the forms of the equations arising in many different models rather than to provide a rigorous derivation of them. As such, references are provided throughout the section to facilitate further study of the underlying detail in the development of these models.

Model XV.1 Soil Mechanics and Clay Consolidation

The erosion of beaches and grasslands is an ongoing environmental concern for various species of wildlife and human development. Avalanches occur due to the movement and changing of soil. Understanding such phenomena has important environmental ramifications. *Hypoplasticity* is an area of study that examines the behavior of granular solids, such as soil, sand, and clay. The IBVPs that arise in the modeling of this phenomena are complicated, mainly due to the presence of phase changes that the material undergoes. Refer to **[149, 197, 251, 303, 327, 330]** for further study.

We examine a particular form of a system of equations discussed in **[292, 293, 294, 297]** relating the fluid pressure and structural displacement, ignoring the physical meaning of the constants involved. Let $\Omega \subset \mathbb{R}^3$ be a bounded domain with smooth boundary $\partial\Omega$. The fluid pressure is denoted by $p(\mathbf{x},t)$ and the (three-dimensional) structural displacement by $\mathbf{w}(\mathbf{x},t)$ at position $\mathbf{x} = (x,y,z)$ in the soil at time t. Consider the following IBVP:

$$\begin{cases} -\beta_1 \nabla \left(\nabla \cdot \mathbf{w}(\mathbf{x},t) \right) - \beta_2 \triangle \mathbf{w}(\mathbf{x},t) + \beta_3 \nabla p(\mathbf{x},t) = \mathbf{f}(\mathbf{x},t), \ \mathbf{x} \in \Omega, t > 0, \\ \frac{\partial}{\partial t} \left(\beta_4 p(\mathbf{x},t) + \beta_3 \nabla \cdot \mathbf{w}(\mathbf{x},t) \right) - \nabla \cdot \beta_5 \nabla p(\mathbf{x},t) = h(\mathbf{x},t), \ \mathbf{x} \in \Omega, t > 0, \\ p(\mathbf{x},0) = p_0(\mathbf{x}), \ w(\mathbf{x},0) = w_0(\mathbf{x}), \ \mathbf{x} \in \Omega, \\ \frac{\partial}{\partial \mathbf{n}} p(\mathbf{x},t) = \frac{\partial}{\partial \mathbf{n}} \mathbf{w}(\mathbf{x},t) = 0, \ x \in \partial \Omega, t > 0. \end{cases} \quad (7.1)$$

It can be shown that solving the first PDE in (7.1) for $\mathbf{w}(\mathbf{x},t)$ and then substituting this into the second PDE yields an abstract evolution equation in the space $\left(\mathbb{L}^2(\Omega) \right)^3$ of the form

$$\begin{cases} \frac{d}{dt} \left(\alpha p(t) - \nabla \cdot v^{-1} \left(\nabla p(t) \right) \right) + \triangle p(t) = h(t), \ t > 0, \\ p(0) = p_0, \end{cases} \quad (7.2)$$

where

$$v(\mathbf{w}) = -\beta_1 \nabla \left(\nabla \cdot \mathbf{w} \right) - \beta_2 \triangle \mathbf{w}. \quad (7.3)$$

An overly simplified, yet comprehensible, one-dimensional version of this evolution equation is given by

$$\begin{cases} \frac{\partial}{\partial t} \left(z(x,t) - \frac{\partial^2}{\partial x^2} z(x,t) \right) + \alpha \frac{\partial^2}{\partial x^2} z(x,t) = f(x,t), \ 0 < x < a, t > 0, \\ z(x,0) = z_0(x), \ 0 < x < a, \\ \frac{\partial z}{\partial x}(0,t) = \frac{\partial z}{\partial x}(a,t) = 0, \ t > 0, \end{cases} \quad (7.4)$$

where $\alpha > 0$. The following more general version of this model in $\Omega \subset \mathbb{R}^3$ is of neutral-type (discussed more extensively in Section 7.2):

$$\begin{cases} \frac{\partial}{\partial t} \left(z(\mathbf{x},t) - \triangle z(\mathbf{x},t) \right) + \alpha \triangle z(\mathbf{x},t) = f(\mathbf{x},t), \ \mathbf{x} \in \Omega, t > 0, \\ z(\mathbf{x},0) = z_0(\mathbf{x}), \ \mathbf{x} \in \Omega, \\ \frac{\partial z}{\partial \mathbf{n}}(\mathbf{x},t) = 0, \ \mathbf{x} \in \partial \Omega, t > 0. \end{cases} \quad (7.5)$$

A detailed discussion of (7.5), including some subtle complications, is given in **[22]**. A variant of (7.5) in a two-dimensional domain with a more general functional forcing term, related to those discussed in **[30, 31, 33, 188, 215, 309]**, is as follows:

$$\begin{cases} \frac{\partial}{\partial t} \left(z(x,y,t) - \frac{\partial^2}{\partial x^2} z(x,y,t) - \frac{\partial^2}{\partial y^2} z(x,y,t) \right) + \frac{\partial^2}{\partial x^2} z(x,y,t) + \frac{\partial^2}{\partial y^2} z(x,y,t) \\ \quad = \int_0^{a_2} \int_0^{a_1} k(t,x,y) f\left(t, z(x,y,t) \right) dx dy, \ 0 < x < a_1, 0 < y < a_2, 0 < t < T, \\ z(x,y,0) = z_0(x,y), \ 0 < x < a_1, 0 < y < a_2, \\ z(0,y,t) = z(a_1,y,t) = 0, \ 0 < x < a_1, 0 < t < T, \\ z(x,0,t) = z(x,a_2,t) = 0, \ 0 < y < a_2, 0 < t < T. \end{cases} \quad (7.6)$$

Exercise 7.1.1. Try to reformulate (7.6) as an abstract evolution equation.

Model XVI.1 Seepage of Fluid Through Fissured Rocks

Try to visualize a sizeable stack of rocks separated by a network of mini-cracks or fissures. Liquid flows along the arteries of this network, but also through tiny pores in the rocks themselves. The modeling of this situation in a bounded domain $\Omega \subset \mathbb{R}^3$ involves PDEs governing the pressure of the liquid in the fissures. A simplified version of one such IBVP is as follows:

$$
\begin{cases}
\frac{\partial}{\partial t} p(x,y,z,t) - \alpha \frac{\partial}{\partial t} \left(\triangle p(x,y,z,t) \right) = \beta \triangle p(x,y,z,t), \ (x,y,z) \in \Omega, t > 0, \\
p(x,y,z,0) = p_0(x,y,z), \ (x,y,z) \in \Omega, \\
\frac{\partial p}{\partial \mathbf{n}}(x,y,z,,t) = 0, \ (x,y,z) \in \partial\Omega, t > 0.
\end{cases}
\tag{7.7}
$$

The parameters α and β are dependent on the characteristics of the rocks (e.g., porosity and permeability). A detailed discussion of such models can be found in [41].

Exercise 7.1.2. Thinking ahead, if it can be shown that $\forall \alpha > 0$, (7.7) has a unique mild solution p_α, must $\exists p \in C([0,T]; \mathscr{X})$ for which $\lim\limits_{\alpha \to 0^+} p_\alpha = p$? To what IBVP is p a mild solution?

Model XVII.1 Second-order Fluids

Non-Newtonian fluids are fluids characterized by a variable viscosity (e.g., some oils and grease, shampoo, blood, and polymer melts). The dynamics of such fluids have been investigated extensively (see [124, 177, 308]). Assuming a unidirectional, nonsteady flow, a model of the velocity field $w(x,t)$ for the flow over a wall can be characterized by

$$
\begin{cases}
\frac{\partial w}{\partial t}(x,t) = \alpha \frac{\partial^2 w}{\partial x^2}(x,t) + \beta \frac{\partial^3 w}{\partial x^2 \partial t}(x,t), \ x > 0, t > 0, \\
w(0,t) = 0, \ t > 0, \\
w(x,0) = w_0(x), \ x > 0.
\end{cases}
\tag{7.8}
$$

Model VII.4 Wave Equations of Sobolev-type

Consider our earlier discussion of classical wave equations. The various characteristics that we incorporated into the model (e.g., dissipation, diffusion, advection, etc.) manifested as distinct differential terms being incorporated into the PDE portion of the IBVP. In some cases, depending on the term being added, the resulting PDE is of Sobolev-type. For instance, consider the following IBVP:

$$
\begin{cases}
\frac{\partial}{\partial t} \left(\frac{\partial z}{\partial t}(x,t) + \frac{\partial^2 z}{\partial x^2}(x,t) \right) + \frac{\partial^2 z}{\partial x^2}(x,t) = g(x,t), \ 0 < x < L, t > 0, \\
z(x,0) = z_0(x), \ \frac{\partial z}{\partial t}(x,0) = z_1(x), \ 0 < x < L, \\
z(0,t) = z(L,t) = 0, \ t > 0.
\end{cases}
$$

Exercise 7.1.3. How would you convert this second-order IBVP into an abstract evolution equation?

Remark. There are other applications in which such equations arise. We refer you to **[186]** for a discussion of such a model in thermodynamics and **[4]** for one arising in civil engineering.

7.1.2 The Abstract Framework

The IBVPs in Section 7.1.1 can be reformulated as an abstract evolution equation of the form

$$\begin{cases} (Bu)'(t) = Au(t) + f(t, u(t)), \ 0 < t < T, \\ u(0) = u_0, \end{cases} \tag{7.9}$$

for any given $T > 0$ in an appropriate Banach space $\mathscr{X}$. Here, $u : [0, T] \to \mathrm{dom}(B) \subset \mathscr{X}$, $A : \mathrm{dom}(A) \subset \mathscr{X} \to \mathscr{X}$, $B : \mathrm{dom}(B) \subset \mathscr{X} \to \mathscr{X}$, $f : [0, T] \times \mathscr{X} \to \mathscr{X}$, and $u_0 \in \mathscr{X}$. Such an evolution equation is said to be of *Sobolev-type*. The main difference from (5.10), of course, is the presence of the operator B.

What can we do to transform (7.9) into an equivalent evolution equation of the form (5.10) so that the approach used in Chapter 5 is applicable? Different applications require that different choices for the operators A and B be used in the abstract formulation of the problem, and these choices naturally lead to different relationships between A and B. As such, we can be assured that a single approach will not handle all possibilities. We shall focus in this section on one particular scenario guided by a change of variable suggested in Section 6.6. This is the approach adopted in **[64, 293, 309]**. Doing so requires that we impose certain assumptions on the operators A and B, the first one of which is:

(H7.1) $A : \mathrm{dom}(A) \subset \mathscr{X} \to \mathscr{X}$ and $B : \mathrm{dom}(B) \subset \mathscr{X} \to \mathscr{X}$ are linear operators.

Of course, additional restrictions must be imposed in order to express (7.9) as (5.10). To this end, a natural approach is to define the new function $v : [0, T] \to \mathrm{rng}(B)$ by

$$v(t) = Bu(t), \ 0 \le t \le T. \tag{7.10}$$

Keep in mind that the goal of the substitution is to produce an evolution equation equivalent to (7.9), but for which the time-derivative term is not obstructed by an operator. If we substitute (7.10) into (7.9), we need to replace each occurrence of $u(t)$ by an equivalent term involving $v(t)$. As such, we would like to further say that (7.10) is equivalent to

$$B^{-1}v(t) = u(t), \ 0 \le t \le T. \tag{7.11}$$

This leads to the second assumption:

(H7.2) $B : \mathrm{dom}(B) \subset \mathscr{X} \to \mathscr{X}$ is invertible.

Now, substituting (7.11) into (7.9) yields

$$\begin{cases} v'(t) = A\left(B^{-1}v(t)\right)v(t) + f(t, B^{-1}v(t)), \ 0 < t < T, \\ B^{-1}v(0) = u_0. \end{cases} \tag{7.12}$$

Since $B^{-1}v(t) \in \mathrm{dom}(B)$, $\forall 0 \le t \le T$, (Why?) we need to further impose the following two assumptions:

(H7.3) $u_0 \in \mathrm{dom}(B)$.
(H7.4) $\mathrm{dom}(B) \subset \mathrm{dom}(A)$.

It is now meaningful to rewrite (7.12) as

$$\begin{cases} v'(t) = \left(AB^{-1}\right)v(t) + f(t, B^{-1}v(t)), \ 0 < t < T, \\ v(0) = Bu_0. \end{cases} \tag{7.13}$$

In order to apply the existence results from Chapter 5, we need for the operator $AB^{-1} : \mathrm{dom}(AB^{-1}) \subset \mathscr{X} \to \mathscr{X}$ to generate a C_0-semigroup $\left\{ e^{(AB^{-1})t} : t \ge 0 \right\}$ on $\mathscr{X}$. The assumptions imposed up to now merely guarantee that AB^{-1} is a well-defined, linear operator on $\mathscr{X}$. In light of Thrm. 3.3.4, if $AB^{-1} \in \mathbb{B}(\mathscr{X})$, then it must generate a C_0-semigroup on $\mathscr{X}$. But, can we impose natural assumptions on A and/or B in order to guarantee this? After all, individually, each of them can be an unbounded operator.

This takes a bit of creativity and foresight. We need to impose conditions on A and B that are reasonable from a practical viewpoint in the sense that there exist nontrivial IBVPs to which the theory under these assumptions would apply, and these conditions should be readily verifiable. It turns out that the Closed Graph Theorem (cf. Thrm. 3.1.15) is especially useful. (Before reading further, why do you suppose this is the case?) With this in mind, we shall scout ahead to determine conditions to impose on A and B that enable us to infer from the Closed Graph Theorem that $AB^{-1} \in \mathbb{B}(\mathscr{X})$. We will then revisit the models to make certain the theory applies to them. Consider the following claim.

Proposition 7.1.1. *Assume that (H7.1) - (H7.4) hold, as well as*
(H7.5) *A and B are closed operators, and*
(H7.6) $B^{-1} : \mathrm{rng}(B) \subset \mathscr{X} \to \mathscr{X}$ *is a compact operator.*
Then, $AB^{-1} \in \mathbb{B}(\mathscr{X})$.

Proof. Let $\{x_n : n \in \mathbb{N}\} \subset \mathrm{dom}(AB^{-1})$ be such that

$$\lim_{n \to \infty} x_n = x \text{ and } \lim_{n \to \infty} \left(AB^{-1}\right)x_n = y. \tag{7.14}$$

Clearly, $x \in \mathscr{X}$, so we need only to show that $\left(AB^{-1}\right)x = y$. We infer from **(H7.6)** and Prop. 5.2.12(i) that B^{-1} is bounded and hence continuous by Exer. 3.1.7(i). Thus,

$$\lim_{n \to \infty} B^{-1}x_n = B^{-1}x. \tag{7.15}$$

Using the fact that B is closed, together with **(H7.4)**, yields

$$\left\{ B^{-1}(x_n) : n \in \mathbb{N} \right\} \subset \operatorname{rng}\left(B^{-1} \right) = \operatorname{dom}(B) \subset \operatorname{dom}(A),$$

so that $B^{-1}x \in \operatorname{dom}(B) \subset \operatorname{dom}(A)$. Hence, AB^{-1} is well-defined.

Now, we infer from the fact that A is closed and (7.14) that

$$\lim_{n \to \infty} A\left(B^{-1}x_n \right) = A\left(B^{-1}x \right) = \left(AB^{-1} \right)x.$$

The uniqueness of limits ensures that $\left(AB^{-1} \right)x = y$. Thus, AB^{-1} is a closed operator. Moreover, it is not difficult to see that $\operatorname{dom}(AB^{-1})$ is closed in $\mathscr{X}$. (Tell why.) Hence, we conclude using the Closed Graph Theorem that $AB^{-1} \in \mathbb{B}(\mathscr{X})$, as desired. $\qquad\square$

Summarizing, we have shown that (7.9) is equivalent to (7.12) via the substitution (7.10) provided that **(H7.1)** - **(H7.6)** hold. Moreover, under these assumptions, it follows that $AB^{-1} : \operatorname{dom}(AB^{-1}) \subset \mathscr{X} \to \mathscr{X}$ generates a C_0−semigroup $\left\{ e^{(AB^{-1})t} : t \geq 0 \right\}$ on $\mathscr{X}$.

Of course, this is very convenient from a theoretical perspective, but is it applicable in the various applied settings introduced in Section 7.1.1? Thankfully, yes. It has been shown that the models in Section 7.1.1 can be treated within this framework. For instance, consider the following exercise.

Exercise 7.1.4. Show directly that IBVP (7.4) can be transformed into a first-order semi-linear abstract evolution equation by verifying the hypotheses formulated in the above discussion.

7.1.3   Main Results

Imposing the assumptions **(H7.1)** - **(H7.6)** essentially renders (7.9) as an abstract evolution equation of the form (5.10), assuming that $u_0 \in \operatorname{dom}(B)$. As such, it should not be surprising that the results and proofs of Chapter 5 carry over with minimal changes.

Exercise 7.1.5. Before proceeding, try to formulate the definitions of mild and classical solutions for (7.9) as well as the theoretical results by suitably modifying the theory developed in Chapter 5. Pay particular attention to the spaces to which various terms must belong.

The definition of a mild solution for (7.13) is a simple extension of Def. 5.5.1. Indeed, we have

Definition 7.1.2. The function $v : [0,T] \to \operatorname{rng}(B)$ is a *mild solution* of (7.13) if v is continuous and

$$v(t) = e^{(AB^{-1})t} B u_0 + \int_0^t e^{(AB^{-1})(t-s)} f(s, B^{-1}v(s)) ds, \ 0 \leq t \leq T. \qquad (7.16)$$

We would like to say that v is a mild solution of (7.13) if and only if $u = B^{-1}v$ is a mild solution of (7.9). For if this were the case, then substituting (7.10) into (7.16) would yield

$$Bu(t) = e^{(AB^{-1})t}Bu_0 + \int_0^t e^{(AB^{-1})(t-s)}f(s,\underbrace{B^{-1}(Bu(s))}_{=u(s)})ds$$

so that

$$u(t) = B^{-1}e^{(AB^{-1})t}Bu_0 + B^{-1}\int_0^t e^{(AB^{-1})(t-s)}f(s,u(s))ds$$

$$= B^{-1}e^{(AB^{-1})t}Bu_0 + \int_0^t B^{-1}e^{(AB^{-1})(t-s)}f(s,u(s))ds.$$

Exercise 7.1.6. Explain why B^{-1} can be brought inside the integral sign in the last equality above.

This suggests that the following definition of a mild solution of (7.9) is practical.

Definition 7.1.3. A function $u : [0,T] \to \mathscr{X}$ is a *mild solution* of (7.9) if u is continuous and

$$u(t) = B^{-1}e^{(AB^{-1})t}Bu_0 + \int_0^t B^{-1}e^{(AB^{-1})(t-s)}f(s,u(s))ds, \ 0 \le t \le T. \quad (7.17)$$

Consequently, we define the solution map $\Phi : \mathbb{C}([0,T];\mathscr{X}) \to \mathbb{C}([0,T];\mathscr{X})$ by

$$(\Phi u)(t) = B^{-1}e^{(AB^{-1})t}Bu_0 + \int_0^t B^{-1}e^{(AB^{-1})(t-s)}f(s,u(s))ds. \quad (7.18)$$

Remark. It is very tempting to first reduce $B^{-1}e^{(AB^{-1})t}Bu_0$ to $e^{(AB^{-1})t}B^{-1}Bu_0$ and then to $e^{(AB^{-1})t}u_0$. However, recall that operators do not commute in general (even in the finite-dimensional case). Specifically, $B^{-1}e^{(AB^{-1})t} \ne e^{(AB^{-1})t}B^{-1}$. So, we must leave (7.18) as is. Nevertheless, the compactness of B^{-1} enables us to establish a reasonable estimate on this term. Indeed, $\forall x \in \text{dom}(B)$,

$$\left\|B^{-1}e^{(AB^{-1})t}Bx\right\|_{\mathscr{X}} \le \left\|B^{-1}\right\|_{\mathbb{B}(\mathscr{X})}\left\|e^{(AB^{-1})t}Bx\right\|_{\mathscr{X}}$$

$$\le \left\|B^{-1}\right\|_{\mathbb{B}(\mathscr{X})}\left\|e^{(AB^{-1})t}\right\|_{\mathbb{B}(\mathscr{X})}\|Bx\|_{\mathscr{X}}$$

$$\le \left\|B^{-1}\right\|_{\mathbb{B}(\mathscr{X})}\left(\overline{M_{AB^{-1}}}\right)\|Bx\|_{\mathscr{X}},$$

where $\overline{M_{AB^{-1}}}$ is defined as in (5.63).

The following definition of a classical solution of (7.9) is natural (see **[64]**).

Definition 7.1.4. A function $u : [0,T] \to \mathscr{X}$ is a *classical solution* of (7.9) if
  a.) $u(t) \in \mathrm{dom}(A)$, $\forall t \in [0,T]$,
  b.) $u'(t) \in \mathrm{dom}(B)$, $\forall t \in (0,T)$,
  c.) $u(\cdot)$ is continuous on $[0,T]$,
  d.) $u(\cdot)$ is differentiable on $(0,T)$,
  e.) (7.9) is satisfied.

The formulation and proofs of the standard results from Chapter 5 carry over to the present setting without issue. For instance, the following is the analog of Thrm. 5.5.2(i). The existence portion follows from a straightforward application of the Contraction Mapping Theorem, while the continuous dependence portion is a direct consequence of Gronwall's lemma.

Proposition 7.1.5. *Assume that (H7.1) - (H7.6) hold and that $f : [0,T] \times \mathscr{X} \to \mathscr{X}$ is continuous in the first variable and globally Lipschitz in the second variable (uniformly in t). Then, (7.9) has a unique mild solution on $[0,T]$. Moreover, if $u_0, v_0 \in \mathrm{dom}(B)$ and u and v are the corresponding mild solutions of (7.9), then $\forall 0 \le t \le T$,*

$$\|u(t) - v(t)\|_{\mathscr{X}} \le \|B^{-1}\|_{\mathbb{B}(\mathscr{X})} \left(\overline{M_{AB^{-1}}}\right) \|B(u_0 - v_0)\|_{\mathscr{X}} \, e^{\left(\|B^{-1}\|_{\mathbb{B}(\mathscr{X})} M_f\right)t}, \quad (7.19)$$

where $\overline{M_{AB^{-1}}}$ is defined as in (5.63).

Exercise 7.1.7. Prove Prop. 7.1.5.

Exercise 7.1.8. Formulate sufficient conditions ensuring that (7.9) has a unique classical solution. Clearly mention any modifications that must be made to the proof of Thrm. 5.5.2(ii) in order to verify it.

Exercise 7.1.9. Consider IBVP (7.4). Identify conditions guaranteeing that (7.4) has a classical solution.

Exercise 7.1.10. Formulate existence results for (7.9) analogous to the results developed in Sections 5.5 - 5.7. Carefully indicate the necessary modifications to the hypotheses in the statements, as well as to the proofs of the results. Pay particular attention to how the computations change due to the presence of B^{-1} and B.

Exercise 7.1.11.
i.) Apply the appropriate existence result established under compactness assumptions in Exer. 7.1.10 to argue that IBVP (7.4) has at least one local mild solution.
ii.) Must the mild solution from (i) be globally defined? Why or why not? If not, indicate how to strengthen the hypotheses to ensure that it is globally defined.

We now turn our attention to the more general functional Sobolev-type evolution equation

$$\begin{cases} (Bu)'(t) = Au(t) + \mathfrak{F}(u)(t), \ 0 < t < T, \\ u(0) = u_0, \end{cases} \quad (7.20)$$

under the same hypotheses **(H7.1)** - **(H7.6)**, where $\mathfrak{F}: \mathbb{C}([0,T];\mathscr{X}) \to \mathbb{C}([0,T];\mathscr{X})$. We say that $u:[0,T] \to \mathscr{X}$ is a *mild* (respectively, *classical*) *solution* of (7.20) if u satisfies Def. 7.1.3 (respectively, Def. 7.1.4) with $f(t,u(t))$ replaced by $\mathfrak{F}(u)(t)$.

We must define a solution map to aid in our investigation of (7.20). Guided by (6.32), let $v \in \mathbb{C}([0,T];\mathscr{X})$ and define $\Phi: \mathbb{C}([0,T];\mathscr{X}) \to \mathbb{C}([0,T];\mathscr{X})$ by $\Phi(v) = u_v$, where u_v is the unique mild solution of the IVP

$$\begin{cases} (Bu_v)'(t) = Au_v(t) + \mathfrak{F}(v)(t),\ 0 < t < T, \\ u_v(0) = u_0. \end{cases} \tag{7.21}$$

Define the mapping $f:[0,T] \times \mathscr{X} \to \mathscr{X}$ by $f(t,u_v(t)) = \mathfrak{F}(v)(t)$. As long as $\mathfrak{F}$ is continuous, we can invoke Prop. 7.1.5 to conclude that Φ is well-defined. (Why?)

We apply the theory developed in Chapter 6 to establish results for (7.20). Particular versions of (7.20) corresponding to specific choices for the functional $\mathfrak{F}$ have been studied in the literature under a variety of assumptions (see, for instance, **[193]**). The theory outlined below encompasses many of these results as special cases. We begin with the following extension of Prop. 7.1.5.

Proposition 7.1.6. *Assume that (H7.1) - (H7.6) hold and that $\mathfrak{F}$ satisfies (H6.8). Then, (7.20) has a unique mild solution on $[0,T]$ provided that*

$$\left\|B^{-1}\right\|_{\mathbb{B}(\mathscr{X})} \left(\overline{M_{AB^{-1}}}\right) M_{\mathfrak{F}} T < 1.$$

Moreover, if $u_0, v_0 \in \mathrm{dom}(B)$ and u and v are the corresponding mild solutions of (7.20), then

$$\|u - v\|_{\mathbb{C}} \le \frac{\left\|B^{-1}\right\|_{\mathbb{B}(\mathscr{X})} \left(\overline{M_{AB^{-1}}}\right) \|B(u_0 - v_0)\|_{\mathscr{X}}}{1 - \left\|B^{-1}\right\|_{\mathbb{B}(\mathscr{X})} \left(\overline{M_{AB^{-1}}}\right) M_{\mathfrak{F}} T}. \tag{7.22}$$

Proof. Let $v \in \mathbb{C}([0,T];\mathscr{X})$. Then, $\mathfrak{F}(v) \in \mathbb{C}([0,T];\mathscr{X})$, so that Prop. 7.1.5 ensures that (7.20) has a unique mild solution on $[0,T]$. (Tell why carefully.) Since

$$(\Phi v)(t) = B^{-1}e^{\left(AB^{-1}\right)t}Bu_0 + \int_0^t B^{-1}e^{\left(AB^{-1}\right)(t-s)}\mathfrak{F}(v)(s)ds,$$

we know that $\forall v, w \in \mathbb{C}([0,T];\mathscr{X})$ and $0 < t < T$,

$$\|(\Phi v)(t) - (\Phi w)(t)\|_{\mathscr{X}} \le \left\|B^{-1}\right\|_{\mathbb{B}(\mathscr{X})} \left(\overline{M_{AB^{-1}}}\right) \int_0^t \underbrace{\|(\mathfrak{F}v)(s) - (\mathfrak{F}w)(s)\|_{\mathscr{X}}}_{\le \|\mathfrak{F}v - \mathfrak{F}w\|_{\mathbb{C}}} ds$$

$$\le \left\|B^{-1}\right\|_{\mathbb{B}(\mathscr{X})} \left(\overline{M_{AB^{-1}}}\right) M_{\mathfrak{F}} \int_0^t \|v - w\|_{\mathbb{C}} ds. \tag{7.23}$$

Taking the supremum over $[0,T]$ in (7.23) subsequently yields

$$\|(\Phi v) - (\Phi w)\|_{\mathbb{C}} \le \left\|B^{-1}\right\|_{\mathbb{B}(\mathscr{X})} \left(\overline{M_{AB^{-1}}}\right) M_{\mathfrak{F}} T \|v - w\|_{\mathbb{C}} < \|v - w\|_{\mathbb{C}}.$$

Thus, Φ is a contraction and so, has a unique fixed-point by the Contraction Mapping Theorem. This fixed-point coincides with a mild solution of (7.20). (Why?) The continuous dependence result (7.22) follows easily. (Tell how.) This completes the proof. $\qquad\square$

A similar result formulated under a slightly different data restriction can be established if the range space of $\mathfrak{F}$ is enlarged to $\mathbb{L}^1(0,T;\mathscr{X})$ and the semigroup is required to be contractive. Consider the following exercise.

Exercise 7.1.12. Assume in Prop. 7.1.6 that $\mathfrak{F}: \mathbb{C}([0,T];\mathscr{X}) \to \mathbb{L}^1(0,T;\mathscr{X})$ satisfies **(H6.9)** and that A generates a contractive C_0-semigroup on $\mathscr{X}$. Show that (7.20) has a global mild solution on $[0,T]$ provided that $\left\|B^{-1}\right\|_{\mathbb{B}(\mathscr{X})} M_{\mathfrak{F}} < 1$.

Corollary 7.1.7. *Consider the evolution equation*

$$\begin{cases} (Bu)'(t) = Au(t) + f\left(t,u(t),\int_0^t a(t-s)g(s,u(s))ds\right), \ 0 < t < T, \\ u(0) = u_0, \end{cases} \quad (7.24)$$

where $f: [0,T] \times \mathscr{X} \times \mathscr{X} \to \mathscr{X}$ and $g: [0,T] \times \mathscr{X} \to \mathscr{X}$ are continuous in the first variable and globally Lipschitz in the remaining variables (uniformly in t) and $a \in \mathbb{C}(\mathbb{R};\mathbb{R})$. Then, (7.24) has a unique global mild solution.

Exercise 7.1.13.
i.) Prove Cor. 7.1.7 directly without first reformulating (7.24) as (7.20).
ii.) Alternatively, reformulate (7.24) as (7.20) in an appropriate space and recover the result directly from Prop. 7.1.6.

Exercise 7.1.14. Formulate and prove an existence-uniqueness result for IBVP (7.6).

Exercise 7.1.15. Define what is meant by a classical solution of (7.24). Then, formulate and prove an existence result in the spirit of Thrm. 5.5.2(ii).

Consider the evolution equation

$$\begin{cases} (Bu)'(t) = Au(t) + Cu(t) + \mathfrak{F}(u)(t), \ 0 < t < T, \\ u(0) = u_0, \end{cases} \quad (7.25)$$

in $\mathscr{X}$ under hypotheses **(H7.1)** - **(H7.6)**, where $C \in \mathbb{B}(\mathscr{X})$ and $\mathfrak{F}: \mathbb{C}([0,T];\mathscr{X}) \to \mathbb{C}([0,T];\mathscr{X})$.

Exercise 7.1.16.
i.) Explain why the operator $(A+C)B^{-1}$ generates a C_0-semigroup on $\mathscr{X}$.
ii.) Assume that $\mathfrak{F}$ satisfies **(H6.8)**. Show that (7.25) has a unique global mild solution.
iii.) Formulate and prove a continuous dependence result for (7.25).

As an application of (7.25), consider the following system of Sobolev PDEs governing the behavior of $z = z(x,t)$ and $w = w(x,t)$ for $0 < x < a, 0 < t < T$:

$$\begin{cases} \frac{\partial}{\partial t}\left(z - \frac{\partial^2 z}{\partial x^2}\right) + \alpha_1 \frac{\partial^2 z}{\partial x^2} + \beta_1 z = h_1(z,w) \int_0^t a_1(t-s) f_1(s,x,z,w) ds, \\ \frac{\partial}{\partial t}\left(w - \frac{\partial^2 w}{\partial x^2}\right) + \alpha_2 \frac{\partial^2 w}{\partial x^2} + \beta_2 w = h_2(z,w) \int_0^t a_2(t-s) f_2(s,x,z,w) ds, \\ z(x,0) = z_0(x), w(x,0) = w_0(x), \ 0 < x < a, \\ \frac{\partial z}{\partial x}(0,t) = \frac{\partial z}{\partial x}(a,t) = 0 = \frac{\partial w}{\partial x}(0,t) = \frac{\partial w}{\partial x}(a,t), \ 0 < t < T, \end{cases} \tag{7.26}$$

where α_i, β_i $(i = 1,2)$ are real constants, $f_i : [0,T] \times [0,a] \times \mathbb{R}^2 \to \mathbb{R}$, $h_i : \mathbb{R}^2 \to \mathbb{R}$, and $a_i \in \mathbb{C}(\mathbb{R};(0,\infty))$ $(i = 1,2)$.

Exercise 7.1.17.
i.) Reformulate (7.26) abstractly as (7.25) in an appropriate space.
ii.) Verify that **(H7.1)** - **(H7.6)** are satisfied.
iii.) Impose appropriate growth and/or regularity restrictions on a_i, f_i, h_i $(i = 1,2)$ that ensure (7.26) has a unique global mild solution on $[0,T]$.
iv.) If h_i is continuous and globally bounded and f_i has sublinear growth $(i = 1,2)$, must (7.26) still have a mild solution on $[0,T]$?

Sometimes, an operator is too complicated to work with directly. In such case it is beneficial to approximate it by a sequence of nicer operators. We want to be assured that the sequence of mild solutions of the IVPs formed using the sequence of approximate operators converges (in a suitable sense) to the mild solution of the original IVP. For instance, $\forall n \in \mathbb{N}$, consider the IVP

$$\begin{cases} (B_n u_n)'(t) = A u_n(t) + \mathfrak{F}(u_n)(t), \ 0 < t < T, \\ u_n(0) = u_0, \end{cases} \tag{7.27}$$

in $\mathscr{X}$, where the sequence of operators $\{B_n : n \in \mathbb{N}\}$ converges in norm to an operator $B : \text{dom}(B) \subset \mathscr{X} \to \mathscr{X}$, and A and B satisfy **(H7.1)** - **(H7.6)**. We would like to say that if "B_n converges to B (in an appropriate sense)," then $\exists u \in \mathbb{C}([0,T];\mathscr{X})$ such that $\lim_{n \to \infty} \|u_n - u\|_{\mathbb{C}} = 0$. Several assumptions must be imposed to ensure that the operators involved are "compatible" and that we can apply earlier results.

Exercise 7.1.18. Before moving on to Prop. 7.1.8, think about what needs to be imposed on B_n, B, and A so that the problem is well-defined.

Proposition 7.1.8. *Assume that the following hold in addition to **(H7.1)** - **(H7.6)**:*
(H7.7) *For every $n \in \mathbb{N}$, $B_n : \text{dom}(B_n) \subset \mathscr{X} \to \mathscr{X}$ is a closed, linear, bijective operator such that $\text{cl}_{\mathscr{X}}(\text{dom}(B_n)) = \text{dom}(B)$ and*

$$\lim_{n \to \infty} \|B_n x - Bx\|_{\mathscr{X}} = 0, \forall x \in \mathscr{Y} = \text{dom}(B) \cap \bigcap_{n=1}^{\infty} \text{dom}(B_n).$$

(H7.8) $\forall n \in \mathbb{N}, B_n^{-1} : \mathscr{X} \to \text{dom}(B_n)$ *is compact and* $\lim_{n \to \infty} \left\| B_n^{-1} - B^{-1} \right\|_{\mathbb{B}(\mathscr{Y})} = 0$.

(H7.9) $u_0 \in \mathscr{Y}$.

(H7.10) $\forall n \in \mathbb{N}, \text{dom}(B_n) \subset \text{dom}(A)$.

(H7.11) $1 - T M_{\mathfrak{F}} \sup_{0 \le t \le T} \left\{ \sup_{n \in \mathbb{N}} \left\| B_n^{-1} e^{(AB_n^{-1})t} \right\|_{\mathbb{B}(\mathscr{Y})} \right\} > 0$.

Then, $\lim_{n \to \infty} \left\| u_n - u \right\|_{\mathbb{C}} = 0$.

Proof. For every $n \in \mathbb{N}$, we know that

$$AB_n^{-1} \text{ generates a } C_0 - \text{semigroup } \left\{ e^{(AB_n^{-1})t} : t \ge 0 \right\} \text{ on } \mathscr{X},$$

$$AB^{-1} \text{ generates a } C_0 - \text{semigroup } \left\{ e^{(AB^{-1})t} : t \ge 0 \right\} \text{ on } \mathscr{X}.$$

Using **(H7.8)** and the facts that A is closed and B_n, B are surjective, we infer that

$$\lim_{n \to \infty} \left\| A(B_n^{-1}x) - A(B^{-1}x) \right\|_{\mathscr{X}} = 0, \forall x \in \mathscr{X}.$$

(Tell why.) Consequently, Thrm. 3.7.3 implies that $\forall x \in \mathscr{X}, \lim_{n \to \infty} e^{(AB_n^{-1})t}x = e^{(AB^{-1})t}x$ uniformly on $[0, T]$.

Observe that

$$\|u_n(t) - u(t)\|_{\mathscr{X}} \le \left\| \left[B_n^{-1} e^{(AB_n^{-1})t} B_n - B^{-1} e^{(AB^{-1})t} B \right] u_0 \right\|_{\mathscr{X}} +$$
$$\int_0^t \left\| B_n^{-1} e^{(AB_n^{-1})(t-s)} \mathfrak{F}(u_n)(s) - B^{-1} e^{(AB^{-1})(t-s)} \mathfrak{F}(u)(s) \right\|_{\mathscr{X}} ds$$
$$= J_1(t) + J_2(t). \tag{7.28}$$

We estimate each term separately. First, the triangle inequality yields

$$J_1(t) \le \left\| \left[B_n^{-1} e^{(AB_n^{-1})t} B_n - B_n^{-1} e^{(AB^{-1})t} B_n \right] u_0 \right\|_{\mathscr{X}} +$$
$$\left\| \left[B_n^{-1} e^{(AB^{-1})t} B_n - B_n^{-1} e^{(AB^{-1})t} B \right] u_0 \right\|_{\mathscr{X}} + \tag{7.29}$$
$$\left\| \left[B_n^{-1} e^{(AB^{-1})t} B - B^{-1} e^{(AB^{-1})t} B \right] u_0 \right\|_{\mathscr{X}}.$$

Each term on the right-side of (7.29) goes to zero as $n \to \infty$. (Why?) Next, we estimate $J_2(t)$ to obtain

$$J_2(t) \le \int_0^t \left\| B_n^{-1} e^{(AB_n^{-1})(t-s)} \mathfrak{F}(u_n)(s) - B_n^{-1} e^{(AB_n^{-1})(t-s)} \mathfrak{F}(u)(s) \right\|_{\mathscr{X}} ds +$$
$$\int_0^t \left\| B_n^{-1} e^{(AB_n^{-1})(t-s)} \mathfrak{F}(u)(s) - B_n^{-1} e^{(AB^{-1})(t-s)} \mathfrak{F}(u)(s) \right\|_{\mathscr{X}} ds + \tag{7.30}$$
$$\int_0^t \left\| B_n^{-1} e^{(AB^{-1})(t-s)} \mathfrak{F}(u)(s) - B^{-1} e^{(AB^{-1})(t-s)} \mathfrak{F}(u)(s) \right\|_{\mathscr{X}} ds$$
$$= J_3(t) + J_4(t) + J_5(t).$$

We estimate each term in (7.30) separately. Observe that

$$J_3(t) \leq \int_0^t \left\| B_n^{-1} e^{\left(AB_n^{-1}\right)(t-s)} \right\|_{\mathbb{B}(\mathscr{Y})} \left\| \mathfrak{F}(u_n)(s) - \mathfrak{F}(u)(s) \right\|_{\mathscr{X}} ds$$

$$\leq \sup_{0 \leq s \leq t \leq T} \left\{ \sup_{n \in \mathbb{N}} \left\| B_n^{-1} e^{\left(AB_n^{-1}\right)(t-s)} \right\|_{\mathbb{B}(\mathscr{Y})} T M_{\mathfrak{F}} \left\| u_n - u \right\|_{\mathbb{C}} \right\} \qquad (7.31)$$

$$= \overline{M_1} \left\| u_n - u \right\|_{\mathbb{C}}.$$

Estimating $J_4(t)$ leads to

$$J_4(t) \leq \left(\sup_{n \in \mathbb{N}} \left\| B_n^{-1} \right\|_{\mathbb{B}(\mathscr{Y})} \right) \cdot \qquad (7.32)$$

$$\int_0^t \left\| e^{\left(AB_n^{-1}\right)(t-s)} - e^{\left(AB^{-1}\right)(t-s)} \right\|_{\mathbb{B}(\mathscr{X})} \left\| \mathfrak{F}(u)(s) \right\|_{\mathscr{X}} ds,$$

and $\forall 0 \leq s \leq t$,

$$\left\| \mathfrak{F}(u)(s) \right\|_{\mathscr{X}} \leq \left\| \mathfrak{F}(u)(s) - \mathfrak{F}(0)(s) \right\|_{\mathscr{X}} + \left\| \mathfrak{F}(0)(s) \right\|_{\mathscr{X}}$$

$$\leq \sup_{0 \leq s \leq T} \left\| \mathfrak{F}(u)(s) - \mathfrak{F}(0)(s) \right\|_{\mathscr{X}} + \sup_{0 \leq s \leq T} \left\| \mathfrak{F}(0)(s) \right\|_{\mathscr{X}}$$

$$\leq M_{\mathfrak{F}} \left\| u \right\|_{\mathbb{C}} + \left\| \mathfrak{F}(0) \right\|_{\mathbb{C}}.$$

The last expression in the above string of inequalities is finite because u is a mild solution to (7.20) on $[0, T]$ and hence, is continuous on $[0, T]$. Consequently, since

$$\lim_{n \to \infty} \left\| e^{\left(AB_n^{-1}\right)(t-s)} - e^{\left(AB^{-1}\right)(t-s)} \right\|_{\mathbb{B}(\mathscr{X})} = 0 \text{ uniformly on } [0, T],$$

we conclude that the right-side of (7.32) goes to zero as $n \to \infty$.

Next, **(H7.8)** ensures that the right-side of the following inequality goes to zero as $n \to \infty$:

$$J_5(t) \leq \int_0^t \left\| B_n^{-1} - B^{-1} \right\|_{\mathbb{B}(\mathscr{Y})} \left\| e^{\left(AB^{-1}\right)(t-s)} \mathfrak{F}(u)(s) \right\|_{\mathscr{X}} ds. \qquad (7.33)$$

Summarizing the above discussion, we have shown that

$$\left\| u_n(t) - u(t) \right\|_{\mathscr{X}} \leq \varepsilon^\star(n) + \overline{M_1} \left\| u_n - u \right\|_{\mathbb{C}}, \qquad (7.34)$$

where $\lim_{n \to \infty} \varepsilon^\star(n) = 0$ (uniformly in t). As such, (7.34) implies that

$$\left(1 - \overline{M_1} \right) \left\| u_n - u \right\|_{\mathbb{C}} \leq \varepsilon^\star(n). \qquad (7.35)$$

Thus, the result follows from (7.35) due to **(H7.11)**. This completes the proof. $\square$

Exercise 7.1.18. Explain why $\displaystyle \sup_{0 \leq t \leq T} \left\{ \sup_{n \in \mathbb{N}} \left\| B_n^{-1} e^{\left(AB_n^{-1}\right)t} \right\|_{\mathbb{B}(\mathscr{Y})} \right\} < \infty.$

Remarks.
1. The restriction on the data given by **(H7.11)** is the price that is paid when dealing with the type of <u>functional</u> evolution equation under consideration. The restriction is due to a technical complication and sometimes can be avoided by directly considering individual examples of the functional $\mathfrak{F}$. (For instance, consider Exer. 7.1.12(i).) Integral inequalities are often more directly applicable when studying such problems since the supremum on the right-side of the estimate is not computed over the entire interval $[0,T]$ prior to applying the integral inequality.

2. Hypothesis **(H7.11)** simplifies slightly if $\forall n \in \mathbb{N}$, the semigroup $\left\{ e^{\left(AB_n^{-1}\right)t} : t \geq 0 \right\}$ is contractive. Indeed, the restriction becomes

$$1 - \sup_{n \in \mathbb{N}} \left\| B_n^{-1} \right\|_{\mathbb{B}(\mathscr{X})} TM_{\mathfrak{F}} > 0.$$

3. Prop. 7.1.8 is a general result and has straightforward practical applications in which some of the hypotheses simplify. For instance, consider IBVPs (7.6) and (7.7). Observe that

$$\lim_{\varepsilon_n \to 0} \left(\triangle + \varepsilon_n I \right) = \triangle \text{ in } \mathbb{B} \left(\mathbb{L}^2(\Omega) \right). \tag{7.36}$$

Further, we point out that $\text{dom} \left(\triangle + \varepsilon_n I \right) = \text{dom} \left(\triangle \right)$, $\forall n \in \mathbb{N}$, which greatly simplifies **(H7.7)** and **(H7.9)**. Also, one can explicitly compute $\left(\triangle + \varepsilon_n I \right)^{-1}$ and its norm, which simplifies the computations.

Exercise 7.1.20. For every $n \in \mathbb{N}$, consider the Sobolev functional evolution equation

$$\begin{cases} (B_n u_n)'(t) = A_n u_n(t) + \mathfrak{F}_n(u_n)(t), \ 0 < t < T, \\ u_n(0) = u_0, \end{cases} \tag{7.37}$$

in $\mathscr{X}$, in which the sequence of operators $\{B_n : n \in \mathbb{N}\}$ converges in norm to some operator $B : \text{dom}(B) \subset \mathscr{X} \to \mathscr{X}$, $\{A_n : n \in \mathbb{N}\}$ converges in norm to some operator $A : \text{dom}(A) \subset \mathscr{X} \to \mathscr{X}$, and the operators A and B satisfy **(H7.1)** - **(H7.6)**. Formulate and prove a result in the spirit of Prop. 7.1.8 for (7.37).

Next, we consider (7.20) under compactness conditions. The proofs of the following results are nearly identical to the proofs in Section 5.7 with the exception of some necessary modifications involving $\left\| B^{-1} \right\|_{\mathbb{B}(\mathscr{X})}$ to the estimates.

Proposition 7.1.9. *Assume that A and B satisfy* **(H7.1)** *-* **(H7.6)** *and are such that AB^{-1} generates a <u>compact</u> C_0-semigroup on $\mathscr{X}$. If either*
i.) $\mathfrak{F} : \mathbb{C}([0,T];\mathscr{X}) \to \mathbb{C}([0,T];\mathscr{X})$ *is a continuous map for which there exist positive constants d_1, d_2 such that*

$$\left\| \mathfrak{F}(x) \right\|_{\mathbb{C}} \leq d_1 \left\| x \right\|_{\mathbb{C}} + d_2, \forall x \in \mathbb{C}([0,T];\mathscr{X});$$

or
ii.) $\mathfrak{F} : \mathbb{C}([0,T];\mathscr{X}) \to \mathbb{L}^1(0,T;\mathscr{X})$ *is a continuous map for which there exist positive constants d_1, d_2 such that*

$$\left\| \mathfrak{F}(x) \right\|_{\mathbb{L}^1} \leq d_1 \left\| x \right\|_{\mathbb{C}} + d_2, \forall x \in \mathbb{C}([0,T];\mathscr{X});$$

then (7.20) has at least one global mild solution on $[0,T]$ *provided that the growth constants are sufficiently small.*

Exercise 7.1.21.
i.) Prove Prop. 7.1.9. Make certain to give precise meaning to the restriction "the growth constants are sufficiently small."
ii.) Can the constants d_1, d_2 be replaced by more general quantities in such a way as to ensure the same conclusion holds? Explain.

Remark. Verifying that AB^{-1} generates a compact C_0-semigroup on $\mathscr{X}$ can be difficult. As such, we look for a more readily verifiable condition that ensures that this occurs. One such condition is:

$$\exists \lambda_0 \in \rho\left(AB^{-1}\right) \text{ such that } R_{\lambda_0}\left(AB^{-1}\right) \text{ is a compact operator.} \qquad (7.38)$$

For instance, consider Exer. 7.1.16. We can argue as in **[64]** to verify (7.38).

Exercise 7.1.22. Formulate a result for (7.4) in the spirit of Prop. 7.1.9. Assume the functions involved satisfy appropriate sublinear growth conditions.

Exercise 7.1.23. Formulate a result for (7.26) in the spirit of Prop. 7.1.9.

7.2   Neutral Evolution Equations

A different class of implicit evolution equations arising in applications is given by

$$\begin{cases} [u(t) + g(t, u(t))]' = Au(t) + f(t, u(t)), \ 0 < t < T, \\ u(0) = u_0, \end{cases} \qquad (7.39)$$

in $\mathscr{X}$, where $f : [0,T] \times \mathscr{X} \to \mathscr{X}$ and $g : [0,T] \times \mathscr{X} \to \mathscr{X}$ are given mappings. This is an abstract formulation of models that arise in electric circuit theory **[47]** and describing oscillatory behavior **[23, 60, 331]**, among others. Note that, once again, $u'(t)$ is described implicitly in (7.39), but the situation is different from (7.9). (Why?) This requires an alternate approach.

7.2.1   Finite-Dimensional Case

We begin by considering the IVP (7.39) with $\mathscr{X} = \mathbb{R}^N$, so that $\mathbf{A} \in \mathbb{M}^N(\mathbb{R})$. We proceed as in Chapters 4 and 5 to develop a variation of parameters formula that will serve as the form of a mild solution that we seek. As we progress through its development, keep track of any additional criteria that must be satisfied so that all computations are well-defined.

Beginning with (7.39), we have $\forall 0 \le s \le T$,

$$e^{-\mathbf{A}s}\left[\mathbf{u}'(s) - \mathbf{A}\mathbf{u}(s)\right] + e^{-\mathbf{A}s}\frac{d}{ds}\mathbf{g}(s,\mathbf{u}(s)) = e^{-\mathbf{A}s}\mathbf{f}(s,\mathbf{u}(s))$$

$$\frac{d}{ds}\left[e^{-\mathbf{A}s}\mathbf{u}(s)\right] + e^{-\mathbf{A}s}\frac{d}{ds}\mathbf{g}(s,\mathbf{u}(s)) = e^{-\mathbf{A}s}\mathbf{f}(s,\mathbf{u}(s)). \qquad (7.40)$$

Now, let $0 \le s \le t \le T$ and integrate (7.40) over $(0,t)$ to obtain

$$e^{-\mathbf{A}t}\mathbf{u}(t) - \underbrace{e^{-\mathbf{A}0}\mathbf{u}_0}_{=\mathbf{u}_0} + \int_0^t e^{-\mathbf{A}s}\frac{d}{ds}\mathbf{g}(s,\mathbf{u}(s))ds = \int_0^t e^{-\mathbf{A}s}\mathbf{f}(s,\mathbf{u}(s))ds. \qquad (7.41)$$

Computing the integral on the left-side of (7.41) using integration by parts yields

$$e^{-\mathbf{A}t}\mathbf{u}(t) = \mathbf{u}_0 - e^{-\mathbf{A}t}\mathbf{g}(t,\mathbf{u}(t)) + \mathbf{g}(0,\mathbf{u}_0) - \int_0^t \mathbf{A}e^{-\mathbf{A}s}\mathbf{g}(s,\mathbf{u}(s))ds$$

$$+ \int_0^t e^{-\mathbf{A}s}\mathbf{f}(s,\mathbf{u}(s))ds. \qquad (7.42)$$

Finally, multiplying (7.42) by $e^{\mathbf{A}t}$ yields the formula

$$\mathbf{u}(t) = e^{\mathbf{A}t}\left(\mathbf{u}_0 + \mathbf{g}(0,\mathbf{u}_0)\right) - \mathbf{g}(t,\mathbf{u}(t)) - \int_0^t \mathbf{A}e^{\mathbf{A}(t-s)}\mathbf{g}(s,\mathbf{u}(s))ds$$

$$+ \int_0^t e^{\mathbf{A}(t-s)}\mathbf{f}(s,\mathbf{u}(s))ds. \qquad (7.43)$$

Assuming sufficient regularity on $\mathbf{f}$, this suggests a natural form of a mild solution of (7.39). Indeed, we have

Definition 7.2.1. A function $\mathbf{u} : [0,T] \to \mathbb{R}^N$ is a *mild solution* of (7.39) if
i.) $\mathbf{u}(0) = \mathbf{u}_0$,
ii.) The mapping $s \mapsto \mathbf{A}e^{\mathbf{A}(t-s)}\mathbf{g}(s,\mathbf{u}(s))$ is integrable on $[0,t]$, $\forall t \in [0,T]$,
iii.) $\mathbf{u}$ satisfies (7.43).

Compare (7.43) to the variation of parameters formulae (4.22) and (5.65). Obviously, the main difference is the presence of $-\mathbf{g}(t,\mathbf{u}(t)) - \int_0^t \mathbf{A}e^{\mathbf{A}(t-s)}\mathbf{g}(s,\mathbf{u}(s))ds$. Since $\mathbf{A} \in \mathbb{M}^N(\mathbb{R})$, we know that

$$\sup_{0 \le t \le T}\left\{\left\|\mathbf{A}e^{\mathbf{A}(t-s)}\right\|_{\mathbb{M}^N(\mathbb{R})} : 0 \le t \le s \le T\right\} < \infty.$$

(Why?) This suggests that establishing results for (7.39) analogous to those in Section 5.2 should boil down to appropriately controlling the growth and regularity of $\mathbf{g}$. With this in mind, complete the following exercises.

Exercise 7.2.1. What is the weakest assumption on $\mathbf{g}$ you can think of that ensures the integrability of $s \mapsto \mathbf{A}e^{\mathbf{A}(t-s)}\mathbf{g}(s, \mathbf{u}(s))$ over $(0, T)$?

Exercise 7.2.2.
i.) Formulate and prove the existence and uniqueness of a mild solution, as well as continuous dependence, for (7.39) under Lipschitz conditions.
ii.) Let $\{\mathbf{g}_n : n \in \mathbb{N}\} \subset \mathbb{C}\left([0, T]; \mathbb{R}^N\right)$ be such that $\mathbf{g}_n(\cdot, \cdots) \longrightarrow 0$ uniformly as $n \to \infty$. For every $n \in \mathbb{N}$, consider the IVP

$$\begin{cases} [\mathbf{u}_n(t) + \mathbf{g}_n(t, \mathbf{u}_n(t))]' = \mathbf{A}\mathbf{u}_n(t) + \mathbf{f}(t, \mathbf{u}_n(t)), \ 0 < t < T, \\ \mathbf{u}_n(0) = \mathbf{u}_0. \end{cases} \tag{7.44}$$

a.) Assuming the same conditions as in (i), argue $\exists \mathbf{u} \in \mathbb{C}\left([0, T]; \mathbb{R}^N\right)$ such that $\lim\limits_{n \to \infty} \|\mathbf{u}_n - \mathbf{u}\|_{\mathbb{C}} = 0$. To which IVP is $\mathbf{u}$ a mild solution?
b.) Identify a sufficient condition that guarantees that $\mathbf{u} \in \mathbb{C}^1\left((0, T); \mathbb{R}^N\right)$.
iii.) Formulate hypotheses that guarantee that $\mathbf{u} \in \mathbb{C}\left([0, \infty); \mathbb{R}^N\right)$ satisfies (7.43) and that $\exists \lim\limits_{t \to \infty} \mathbf{u}(t)$. Prove your assertion.

Open Exercise 7.2.3. (Non-Lipschitz Case)
Track through the details of the proof of Thrm. 5.6.3 and indicate any changes to the hypotheses and computations needed to establish such a result for (7.39).

The existence of mild solutions of (7.39) can be established under weaker growth conditions as in Chapter 5, but not without some notable changes. Indeed, we had to argue that the solution map was compact in the proofs of all such results. Presently, $\mathbf{A} \in \mathbb{B}\left(\mathbb{R}^N\right)$ and so $\{e^{\mathbf{A}t} : t \geq 0\}$ is automatically a compact semigroup on $\mathbb{R}^N$. (Why?) As such, if it were not for the presence of the term $\mathbf{g}(t, \mathbf{u}(t))$, the proofs of Thrm. 5.7.3 and Prop. 5.7.4 with minimal modifications could be used to establish analogous results for (7.39).

Exercise 7.2.4. What complications does the term $\mathbf{g}(t, \mathbf{u}(t))$ create in the proof? How might they be overcome?

An approach used in the literature involves decomposing the solution map Φ_0 as a sum of two operators Φ_1 and Φ_2, where Φ_1 is comprised of the new problematic terms and Φ_2 resembles the solution map used in Chapter 5. The standard sublinear growth conditions (cf. (**H5.10**) and (**H5.11**)) are imposed on f (so that the result coincides with the corresponding result in Chapter 5 when $\mathbf{g} = 0$), while it is customary to impose a global Lipschitz-type condition on $\mathbf{g}$. The motivation for doing so is that *Krasnoselskii's fixed-point theorem*, stated below and proven in [66], will be applied to Φ_0 to obtain the desired fixed-point.

Theorem 7.2.2. (Krasnoselskii's Fixed-Point Theorem) *Let $\mathscr{D} \subset \mathbb{C}\left([0, T]; \mathscr{X}\right)$ be a closed, convex set and suppose that Φ_1 and Φ_2 are operators on $\mathscr{D}$ such that*
i.) *Φ_1 is a contraction,*

ii.) Φ_2 *is a continuous, compact map.*
Then, $\Phi_1 + \Phi_2$ *has at least one fixed-point in* $\mathscr{D}$.

We use this fixed-point theorem to prove the following result. A similar approach is used in **[127]**.

Proposition 7.2.3. *Assume that*
(H7.12) $\mathbf{g} : [0,T] \times \mathbb{R}^N \to \mathbb{R}^N$ *is a continuous map for which there exist positive constants* $M_{\mathbf{g}}, \overline{M}_{\mathbf{g}_1}, \overline{M}_{\mathbf{g}_2}$ *such that*

$$\|\mathbf{g}(t,\mathbf{x}) - \mathbf{g}(t,\mathbf{y})\|_{\mathbb{R}^N} \leq M_{\mathbf{g}} \|\mathbf{x} - \mathbf{y}\|_{\mathbb{R}^N}, \, \forall \mathbf{x}, \mathbf{y} \in \mathbb{R}^N, t > 0,$$

$$\|\mathbf{g}(t,\mathbf{x})\|_{\mathbb{R}^N} \leq \overline{M}_{\mathbf{g}_1} \|\mathbf{x}\|_{\mathbb{R}^N} + \overline{M}_{\mathbf{g}_2}, \, \forall \mathbf{x}, \mathbf{y} \in \mathbb{R}^N, t > 0.$$

(H7.13) $\mathbf{f} : [0,T] \times \mathbb{R}^N \to \mathbb{R}^N$ *is a continuous mapping such that*
 i.) *For every* $k \in \mathbb{N}$, $\exists h_k \in \mathbb{C}([0,T];(0,\infty))$ *such that*

$$\sup_{\|\mathbf{x}\|_{\mathbb{R}^N} \leq k} \|\mathbf{f}(t,\mathbf{x})\|_{\mathbb{R}^N} \leq h_k(t), \forall t > 0,$$

ii.) $\lim\limits_{k \to \infty} \frac{1}{k} \int_0^T h_k(s) ds = \alpha < \infty.$
If

$$\overline{M}_{\mathbf{g}_1} \left(1 + \|\mathbf{A}\|_{\mathbb{M}^N(\mathbb{R})} M^\star T \right) + M^\star \alpha < 1 \tag{7.45}$$

$$M_{\mathbf{g}} \left(1 + TM^\star \|\mathbf{A}\|_{\mathbb{M}^N(\mathbb{R})} \right) < 1, \tag{7.46}$$

where $M^\star = \sup\limits_{0 \leq t \leq T} \left\| e^{\mathbf{A}t} \right\|_{\mathbb{M}^N(\mathbb{R})}$, *then (7.39) has at least one global mild solution.*

Proof. Define $\Phi : \mathbb{C}([0,T];\mathbb{R}^N) \to \mathbb{C}([0,T];\mathbb{R}^N)$ by

$$\Phi(\mathbf{u})(t) = e^{\mathbf{A}t} (\mathbf{u}_0 + \mathbf{g}(0,\mathbf{u}_0)) - \mathbf{g}(t,\mathbf{u}(t)) - \int_0^t \mathbf{A} e^{\mathbf{A}(t-s)} \mathbf{g}(s,\mathbf{u}(s)) ds$$

$$+ \int_0^t e^{\mathbf{A}(t-s)} \mathbf{f}(s,\mathbf{u}(s)) ds. \tag{7.47}$$

The continuity of $\mathbf{f}$ and $\mathbf{g}$ ensures that Φ is well-defined. (Why?) Let $n \in \mathbb{N}$ and define $\mathfrak{B}_n = \left\{ \mathbf{x} \in \mathbb{C}([0,T];\mathbb{R}^N) \, | \, \|\mathbf{x}\|_{\mathbb{C}} \leq n \right\}$. The existence of $N_0 \in \mathbb{N}$ such that $\Phi(\mathfrak{B}_{N_0}) \subset \mathfrak{B}_{N_0}$ can be established as in the proof of Prop. 5.7.4. (Show this.)
Define the operators $\Phi_1, \Phi_2 : \mathfrak{B}_{N_0} \to \mathfrak{B}_{N_0}$ by

$$\Phi_1(\mathbf{u})(t) = e^{\mathbf{A}t} \mathbf{g}(0,\mathbf{u}_0) - \mathbf{g}(t,\mathbf{u}(t)) - \int_0^t \mathbf{A} e^{\mathbf{A}(t-s)} \mathbf{g}(s,\mathbf{u}(s)) ds, \tag{7.48}$$

$$\Phi_2(\mathbf{u})(t) = e^{\mathbf{A}t} \mathbf{u}_0 + \int_0^t e^{\mathbf{A}(t-s)} \mathbf{f}(s,\mathbf{u}(s)) ds. \tag{7.49}$$

Observe that $\Phi = \Phi_1 + \Phi_2$. We verify the conditions of Thrm. 7.2.2.

<u>Claim 1</u>: Φ_1 is a contraction.

Proof: Let $\mathbf{x}, \mathbf{y} \in \mathfrak{B}_{N_0}$. Observe that

$$\|\Phi_1(\mathbf{x})(t) - \Phi_1(\mathbf{y})(t)\|_{\mathbb{R}^N} \leq \|\mathbf{g}(t, \mathbf{x}(t)) - \mathbf{g}(t, \mathbf{y}(t))\|_{\mathbb{R}^N}$$
$$+ M^\star \|\mathbf{A}\|_{\mathbb{M}^N(\mathbb{R})} \int_0^t \|\mathbf{g}(s, \mathbf{x}(s)) - \mathbf{g}(s, \mathbf{y}(s))\|_{\mathbb{R}^N} \, ds$$
$$\leq M_{\mathbf{g}} \left(1 + T M^\star \|\mathbf{A}\|_{\mathbb{M}^N(\mathbb{R})} \right) \|\mathbf{x}(t) - \mathbf{y}(t)\|_{\mathbb{R}^N}$$

so that taking the supremum over $0 \leq t \leq T$ and applying (7.46) shows that Φ_1 is a contraction, as claimed. $\diamondsuit$

<u>Claim 2</u>: Φ_2 is continuous and compact.

Proof: This is argued as in the proof of Thrm. 5.7.3 because $\displaystyle\sup_{0 \leq t \leq T} \|\mathbf{A}e^{\mathbf{A}t}\|_{\mathbb{M}^N(\mathbb{R})} < \infty$

when $\mathbf{A} \in \mathbb{M}^N(\mathbb{R})$. $\diamondsuit$

Thus, we conclude from Thrm. 7.2.2 that Φ has at least one fixed point in $\mathfrak{B}_{N_0}$, which is a mild solution of (7.39) on $[0,T]$. This completes the proof. $\square$

Exercise 7.2.5. Assume that **(H7.12)** holds and replace **(H7.13)** by
(H7.14) $\mathbf{f} : [0,T] \times \mathbb{R}^N \to \mathbb{R}^N$ is a continuous map that satisfies **(H5.10)**.
Prove that (7.39) has at least one global mild solution on $[0,T]$. (Note: Imposing a data restriction on the growth constants is typical.)

We now consider a more general functional neutral evolution equation of the form

$$\begin{cases} [\mathbf{u}(t) + \mathbf{g}(t, \mathbf{u}(t))]' = \mathbf{A}\mathbf{u}(t) + \mathfrak{F}(\mathbf{u})(t), \ 0 < t < T, \\ \mathbf{u}(0) = \mathbf{u}_0, \end{cases} \tag{7.50}$$

in $\mathbb{R}^N$, where $\mathfrak{F} : \mathbb{C}\left([0,T]; \mathbb{R}^N\right) \to \mathbb{C}\left([0,T]; \mathbb{R}^N\right)$. A *mild solution* of (7.50) is a continuous function $\mathbf{u} : [0,T] \to \mathbb{R}^N$ satisfying Def. 7.2.1 with $\mathbf{f}(t, \mathbf{u}(t))$ replaced by $\mathfrak{F}(\mathbf{u})(t)$.

Proposition 7.2.4. *Assume that **(H6.8)** and **(H7.12)** hold. Then, (7.50) has a unique global mild solution, provided that* $M_{\mathbf{g}}\left(1 + T M^\star \|\mathbf{A}\|_{\mathbb{M}^N(\mathbb{R})}\right) + M^\star M_{\mathfrak{F}} T < 1$.

Proof. Let $\mathbf{x}, \mathbf{y} \in \mathbb{C}\left([0,T]; \mathbb{R}^N\right)$. Observe that

$$\|\Phi(\mathbf{x})(t) - \Phi(\mathbf{y})(t)\|_{\mathbb{R}^N} \leq \|\mathbf{g}(t, \mathbf{x}(t)) - \mathbf{g}(t, \mathbf{y}(t))\|_{\mathbb{R}^N}$$
$$+ M^\star \int_0^t \|\mathfrak{F}(\mathbf{x})(s) - \mathfrak{F}(\mathbf{y})(s)\|_{\mathbb{R}^N} \, ds$$
$$+ M^\star \|\mathbf{A}\|_{\mathbb{M}^N(\mathbb{R})} \int_0^t \|\mathbf{g}(s, \mathbf{x}(s)) - \mathbf{g}(s, \mathbf{y}(s))\|_{\mathbb{R}^N} \, ds$$
$$\leq M_{\mathbf{g}} \|\mathbf{x}(t) - \mathbf{y}(t)\|_{\mathbb{R}^N} + M^\star M_{\mathfrak{F}} T \|\mathbf{x} - \mathbf{y}\|_{\mathbb{C}}$$
$$+ M^\star M_{\mathbf{g}} \|\mathbf{A}\|_{\mathbb{M}^N(\mathbb{R})} \int_0^t \|\mathbf{x}(s) - \mathbf{y}(s)\|_{\mathbb{R}^N} \, ds$$

Hence, taking the supremum over $[0,T]$ yields

$$\|\Phi \mathbf{x} - \Phi \mathbf{y}\|_{\mathbb{C}} \leq \left[M_{\mathbf{g}}\left(1 + TM^{\star}\|\mathbf{A}\|_{\mathbb{M}^N(\mathbb{R})}\right) + M^{\star}M_{\mathfrak{F}}T\right]\|\mathbf{x} - \mathbf{y}\|_{\mathbb{C}} < \|\mathbf{x} - \mathbf{y}\|_{\mathbb{C}}.$$

So, Φ is a strict contraction and hence, has a unique fixed-point by the Contraction Mapping Theorem. This fixed-point coincides with the mild solution we seek. $\qquad\square$

Exercise 7.2.6. How does the restriction on the data change if we enlarge the range space of $\mathfrak{F}$ to $\mathbb{L}^1\left(0,T;\mathbb{R}^N\right)$?

Exercise 7.2.7. Consider the evolution equation

$$\begin{cases} [\mathbf{u}(t) + \mathbf{g}(t,\mathbf{u}(t))]' = \mathbf{A}\mathbf{u}(t) + \mathbf{f}\left(t,\mathbf{u}(t),\int_0^t a(t-s)\mathbf{h}(s,\mathbf{j}(s,\mathbf{u}(s)))ds\right), \ 0 < t < T, \\ \mathbf{u}(0) = \mathbf{u}_0, \end{cases}$$
(7.51)

where $\mathbf{f}: [0,T] \times \mathbb{R}^N \times \mathbb{R}^N \to \mathbb{R}^N$, $\mathbf{g}: [0,T] \times \mathbb{R}^N \to \mathbb{R}^N$, $\mathbf{h}: [0,T] \times \mathbb{R}^N \to \mathbb{R}^N$, $\mathbf{j}: [0,T] \times \mathbb{R}^N \to \mathbb{R}^N$, and $a: [0,T] \to (0,\infty)$. Impose appropriate restrictions so that Prop. 7.2.4 can be invoked to conclude that (7.51) has a unique global mild solution.

Exercise 7.2.8. Formulate and prove an analog of Prop. 6.3.4 for (7.50).

7.2.2   Infinite-Dimensional Case

Certain IBVPs can be reformulated abstractly as (7.50), but where the operator A is unbounded and $\mathscr{X}$ is an infinite-dimensional Banach space. By comparison, it is certainly reasonable to define a mild solution in the same manner as in the finite-dimensional case. But, unfortunately this is the extent to which the approach used in Section 7.2.1 is applicable. This might be somewhat surprising since the evolution equations *look* exactly the same. So, where does the approach break down?

The main obstacle is that for a general unbounded generator A of a C_0−semigroup $\left\{e^{At} : t \geq 0\right\}$ on $\mathscr{X}$, it need not be the case that

$$\sup_{0 \leq t \leq T} \sup_{x \in \mathscr{X}} \left\|Ae^{At}x\right\|_{\mathscr{X}} < \infty.$$
(7.52)

(See [228] for an example.) Naturally, the question arises as to whether or not there are specific operators A for which (7.52) *does* hold. One thought might be that since compact semigroups possess nice properties, they might satisfy (7.52). Alas, no. In fact, the combination of compactness of the semigroup and condition (7.52) necessarily implies that $\mathscr{X}$ is finite dimensional!

Exercise 7.2.9. Prove the last assertion in the above paragraph.

So, we have encountered another dead end. Naively, with these technical problems in view, the question is whether we can somehow express the operator A in a different manner that is amenable to forming such estimates as to avoid this obstacle. The short answer is yes, but the story requires a considerable amount of technical discussion to unfold completely. We shall provide only a broad stroke discussion below and mention that there are two key components that, when combined, help us to overcome this hurdle.

The first component is to consider operators A for which the parameter t in the semigroup $\{e^{At} : t \geq 0\}$ is now allowed to be a complex number rather than simply a nonnegative real number and which possesses somewhat greater regularity. Precisely, we restrict our attention to operators that generate *analytic semigroups* in the following sense:

Definition 7.2.5. Let $\mathscr{Z} = \left\{ re^{i\theta} \,|\, \theta_1 < \theta < \theta_2 \wedge r \geq 0 \right\} \subset C$, where $\theta_1 < 0 < \theta_2$ and $\mathscr{C}$ is the set of complex numbers. The family of operators $\{T(z) : z \in \mathscr{Z}\} \subset \mathbb{B}(\mathscr{X})$ is an *analytic semigroup* on $\mathscr{X}$ if
i.) $z \mapsto T(z)$ is analytic in $\mathscr{Z}$,
ii.) $T(0) = I$,
iii.) $\lim\limits_{z \to 0} T(z)x = x$, $\forall x \in \mathscr{X}$,
iv.) $T(z_1 + z_2)x = T(z_1)T(z_2)x$, $\forall x \in \mathscr{X}$, $z_1, z_2 \in \mathscr{Z}$.

Remarks.
1. Since $(0, \infty) \subset \mathscr{Z}$, the restriction of an analytic semigroup to $(0, \infty)$ is a C_0−semigroup. But, not all C_0−semigroups are analytic due to its increased regularity. See **[246, 267]** for particular examples.
2. Assumption (i) is strong because analytic functions are very regular. Also, the limit in (iii) is taken in the complex sense, which is more restrictive than in the real-valued setting. Refer to **[16]** for a discussion of complex limits and analyticity.
3. Consult **[123, 246, 267]** for a rigorous development of analytic semigroup theory.

The second key component runs parallel to the development of analytic semigroup theory, although it plays a critical role in establishing some of the estimates. To begin, we define the notion of a *fractional power of a closed operator* as follows:

Definition 7.2.6. Assume that $A : \mathrm{dom}(A) \subset \mathscr{X} \to \mathscr{X}$ is a closed operator and generates a C_0−semigroup $\{e^{At} : t \geq 0\}$ on $\mathscr{X}$ satisfying (3.75).
i.) For $\alpha \in (-\infty, 0)$, the *fractional power* α of A, denoted $A^\alpha : \mathrm{dom}(A^\alpha) \subset \mathscr{X} \to \mathscr{X}$, is defined by

$$A^\alpha x = \frac{1}{\Gamma(-\alpha)} \int_0^\infty t^{-\alpha-1} e^{At} x \, dt, \, \forall x \in \mathscr{X},$$

where $\Gamma(-\alpha) = \int_0^\infty y^{-\alpha-1} e^{-y} dy$ is the Gamma function, well-known from statistical theory.
ii.) For $\alpha \in (0, \infty)$, $A^\alpha : \mathrm{dom}(A^\alpha) \subset \mathscr{X} \to \mathscr{X}$, is defined by the inverse of $A^{-\alpha}$.

The well-definedness of (ii) of Def. 7.2.6 follows from the invertibility of the negative fractional powers of A, which is not difficult to verify. These operators possess many important properties listed below without proof. (See **[198, 246]** for details.) Most of them are actually easily verified. (Try it!)

Proposition 7.2.7. (Properties of Fractional Powers of a Closed Operator)
Let $A : \mathrm{dom}(A) \subset \mathscr{X} \to \mathscr{X}$ *be a closed operator that generates a* C_0-*semigroup* $\{e^{At} : t \geq 0\}$ *on* $\mathscr{X}$.
i.) $A^{\alpha} \in \mathbb{B}(\mathscr{X})$, $\forall \alpha \in (-\infty, 0)$.
ii.) $\mathrm{dom}(A^{\alpha})$ *is dense in* $\mathscr{X}$, $\forall \alpha \in (-\infty, 0)$.
iii.) *Let* $\alpha \in (-\infty, 0)$, $\mathscr{H}$ *be a Hilbert space with a complete orthonormal basis* $\{\phi_n : n \in \mathbb{N}\}$, *and* $\{\lambda_n : n \in \mathbb{N}\} \subset (0, \infty)$ *be such that* $\inf_{n \in \mathbb{N}} \lambda_n > 0$. *If* A *is expressed by* (3.120), *then*

$$A^{\alpha} z = \sum_{n=1}^{\infty} \lambda_n^{\alpha} \langle z, \phi_n \rangle_{\mathscr{H}} \, \phi_n, \, \forall z \in \mathrm{dom}(A^{\alpha}).$$

iv.) $A^{\alpha} A^{\beta} = A^{\alpha + \beta}$, $\forall \alpha, \beta \in (-\infty, 0)$.
v.) *The operator* $A^{-\alpha}$ *also satisfies (ii) - (iv).*
vi.) $A^{\alpha} e^{At} z = e^{At} A^{\alpha} z$, $\forall \alpha \in \mathbb{R}$, $z \in \mathrm{dom}(A^{\alpha})$, $t \geq 0$.
vii.) *For each* $\alpha \in [0, 1]$, *there exists a positive constant* C_{α},

$$\|A^{\alpha} z\|_{\mathscr{X}} \leq C_{\alpha} \|Az\|_{\mathscr{X}}^{\alpha} \|z\|_{\mathscr{X}}^{1-\alpha}, \, \forall z \in \mathrm{dom}(A^{\alpha}).$$

viii.) $\alpha < \beta \implies \mathrm{dom}(A^{\alpha}) \supset \mathrm{dom}(A^{\beta})$.

The key to establishing existence results for (7.39) and (7.50) lies in combining the above two components. Indeed, if A generates an analytic semigroup on $\mathscr{X}$, then the following properties hold and constitute precisely what we need to overcome the obstacles that presently prevent a direct extension of the theory from the finite-dimensional case. (See [**198, 246, 267**] for a proof.)

Proposition 7.2.8. *Assume that* $A : \mathrm{dom}(A) \subset \mathscr{X} \to \mathscr{X}$ *generates an analytic semigroup* $\{e^{At} : t \geq 0\}$ *on* $\mathscr{X}$ *and that* $0 \in \rho(A)$. *Then,* $\forall 0 < \alpha \leq 1$,
i.) $(\mathrm{dom}(A^{\alpha}), \|\cdot\|_{\alpha})$ *is a Banach space, where* $\|x\|_{\alpha} \equiv \|A^{\alpha} x\|_{\mathscr{X}}$.
ii.) *There exists a positive constant* C_{α} *such that*

$$\sup_{x \in \mathscr{X}} \|A^{\alpha} e^{At} x\|_{\mathscr{X}} \leq \frac{C_{\alpha}}{t^{\alpha}}, \, \forall t > 0.$$

In particular, $\forall t > 0$, *the operator* $A^{\alpha} e^{At}$ *is bounded!*

How does all of this really help? Consider the following result and pay particular attention to the hypotheses concerning g and how they are used in the proof.

Proposition 7.2.9. *Assume that*
(H7.15) $A : \mathrm{dom}(A) \subset \mathscr{X} \to \mathscr{X}$ *is a closed operator that generates an analytic semigroup* $\{e^{At} : t \geq 0\}$ *on* $\mathscr{X}$.
(H7.16) $\mathfrak{F} : \mathbb{C}([0, T]; \mathscr{X}) \to \mathbb{L}^1(0, T; \mathscr{X})$ *satisfies* **(H6.9)**.
(H7.17) $g : [0, T] \times \mathscr{X} \to \mathscr{X}$ *is a mapping for which* $\exists 0 < \beta < 1$ *such that*
   a.) $\mathrm{rng}(g) \subset \mathrm{dom}(A^{\beta})$;
   b.) $A^{\beta} g$ *is a continuous mapping;*

c.) *There exists $M_g > 0$ such that*

$$\left\| A^{\beta} g(t,x) - A^{\beta} g(t,y) \right\|_{\mathscr{X}} \leq M_g \left\| x - y \right\|_{\mathscr{X}}, \forall x, y \in \mathscr{X}, t \geq 0.$$

If $M_g \left\| A^{1-\beta} \right\|_{\mathbb{B}(\mathscr{X})} + M^{\star} M_{\mathfrak{F}} + M_g C_{1-\beta} \frac{T^{2-\beta}}{2-\beta} < 1$, then (7.50) has a unique global mild solution on $[0,T]$.

Proof. Define the solution map $\Phi : \mathbb{C}([0,T];\mathscr{X}) \to \mathbb{C}([0,T];\mathscr{X})$ by

$$\Phi(u)(t) = e^{At}\left(u_0 + g(0,u_0)\right) - g(t,u(t)) - \int_0^t Ae^{A(t-s)} g(s,u(s)) ds$$

$$+ \int_0^t e^{A(t-s)} \mathfrak{F}(u)(s) ds. \tag{7.53}$$

We begin by verifying the continuity of Φ. Let $t_0 \in (0,T)$ and $|h|$ sufficiently small. Observe that

$$\Phi(u)(t_0 + h) - \Phi(u)(t_0) = \left(e^{A(t_0+h)} - e^{At_0}\right)\left(u_0 + g(0,u_0)\right)$$

$$+ \left(g(t_0 + h, u(t_0+h)) - g(t_0, u(t_0))\right)$$

$$+ \left(\int_0^{t_0+h} Ae^{A(t_0+h-s)} g(s,u(s)) ds \right.$$

$$\left. - \int_0^{t_0} Ae^{A(t_0-s)} g(s,u(s)) ds \right) \tag{7.54}$$

$$+ \left(\int_0^{t_0+h} e^{A(t_0+h-s)} \mathfrak{F}(u)(s) ds - \int_0^{t_0} e^{A(t_0-s)} \mathfrak{F}(u)(s) ds \right)$$

$$= \sum_{i=1}^4 J_i(t_0,h).$$

We shall show that $\lim_{h \to 0} J_i(t_0,h) = 0$, $i = 1,2,3,4$. To begin, note that

$$\|J_1(t_0,h)\|_{\mathscr{X}} = \left\| \left(e^{Ah}e^{At_0} - e^{At_0}\right)\left(u_0 + g(0,u_0)\right) \right\|_{\mathscr{X}}. \tag{7.55}$$

The strong continuity of the semigroup guarantees that the right-side of (7.55) goes to zero as $|h| \to 0$. (Why?) Next, observe that

$$\|J_2(t_0,h)\|_{\mathscr{X}} \leq \|g(t_0 + h, u(t_0+h)) - g(t_0 + h, u(t_0))\|_{\mathscr{X}}$$

$$+ \|g(t_0 + h, u(t_0)) - g(t_0, u(t_0))\|_{\mathscr{X}} \tag{7.56}$$

$$\leq M_g \|u(t_0 + h) - u(t_0)\|_X + \|g(t_0 + h, u(t_0)) - g(t_0, u(t_0))\|_{\mathscr{X}}.$$

The continuity of u and g guarantees that the right-side of (7.56) goes to zero as $|h| \to 0$. (Tell how.)

Next, observe that

$$
\begin{aligned}
\|J_3(t_0,h)\|_{\mathscr{X}} &= \left\| \int_0^{t_0} \left[A e^{A(t_0+h-s)} g(s,u(s)) - A e^{A(t_0-s)} g(s,u(s)) \right] ds \right. \\
&\quad \left. + \int_{t_0}^{t_0+h} A e^{A(t_0+h-s)} g(s,u(s)) ds \right\|_{\mathscr{X}} \\
&\leq \int_0^{t_0} \left\| \left(e^{Ah} - I \right) A^{1-\beta} e^{A(t_0-s)} A^{\beta} g(s,u(s)) \right\|_{\mathscr{X}} ds \\
&\quad + \int_{t_0}^{t_0+h} \left\| A^{1-\beta} e^{A(t_0+h-s)} A^{\beta} g(s,u(s)) \right\|_{\mathscr{X}} ds \qquad (7.57) \\
&\leq \int_0^{t_0} \left\| e^{Ah} - I \right\|_{\mathbb{B}(\mathscr{X})} \left\| A^{1-\beta} e^{A(t_0-s)} \right\|_{\mathscr{X}} \left\| A^{\beta} g(s,u(s)) \right\|_{\mathscr{X}} ds \\
&\quad + \int_{t_0}^{t_0+h} \left\| A^{1-\beta} e^{A(t_0+h-s)} \right\|_{\mathbb{B}(\mathscr{X})} \left\| A^{\beta} g(s,u(s)) \right\|_{\mathscr{X}} ds \\
&= \int_0^{t_0} \left\| e^{Ah} - I \right\|_{\mathbb{B}(\mathscr{X})} J_5(s) J_6(s) ds + \int_{t_0}^{t_0+h} J_7(s) \left\| A^{\beta} g(s,u(s)) \right\|_{\mathscr{X}} ds
\end{aligned}
$$

Using Prop. 7.2.7, we see that $\forall s \in [0,t_0]$,

$$
J_5(s) \leq \frac{C_{1-\beta}}{(t_0-s)^{1-\beta}} \qquad (7.58)
$$

$$
\begin{aligned}
J_6(s) &\leq \left\| A^{\beta} g(s,u(s)) - A^{\beta} g(s,0) + A^{\beta} g(s,0) \right\|_{\mathscr{X}} \\
&\leq M_g \|u(s)\|_{\mathscr{X}} + \left\| A^{\beta} g(s,0) \right\|_{\mathscr{X}} \qquad (7.59)
\end{aligned}
$$

$$
J_7(s) \leq \frac{C_{1-\beta}}{(t_0+h-s)^{1-\beta}}. \qquad (7.60)
$$

Since $u \in \mathbb{C}([0,T];\mathscr{X})$ and the continuity of $A^{\beta} g$ ensures that

$$
\sup \left\{ \left\| A^{\beta} g(s,0) \right\|_{\mathscr{X}} : 0 \leq s \leq T \right\} < \infty,
$$

we conclude that the right-side of (7.57) is dominated by an integrable function. As such, Prop. 1.8.16, together with the strong continuity of the semigroup, implies that the right-side of (7.57) goes to zero as $|h| \to 0$. Finally, it is not difficult to argue that $\|J_4(t_0,h)\|_{\mathscr{X}}$ goes to zero as as $|h| \to 0$. (Tell how.) Thus, we conclude that Φ is continuous.

Finally, observe that $\forall x, y \in \mathbb{C}\left([0,T];\mathscr{X}\right)$,

$$\|\Phi(x)(t) - \Phi(y)(t)\|_{\mathscr{X}} \leq \|g(t,x(t)) - g(t,y(t))\|_{\mathscr{X}}$$
$$+ \int_0^t \left\|Ae^{A(t-s)}\left(g(s,x(s)) - g(s,y(s))\right)\right\|_{\mathscr{X}} ds$$
$$+ \int_0^t \left\|e^{A(t-s)}\left(\mathfrak{F}(x)(s) - \mathfrak{F}(y)(s)\right)\right\|_{\mathscr{X}} ds$$
$$\leq \left\|A^{1-\beta}\right\|_{\mathbb{B}(\mathscr{X})}\left\|A^{\beta}g(t,x(t)) - A^{\beta}g(t,y(t))\right\|_{\mathscr{X}} \tag{7.61}$$
$$+ \int_0^t \left\|A^{1-\beta}e^{A(t-s)}A^{\beta}\left(g(s,x(s)) - g(s,y(s))\right)\right\|_{\mathscr{X}} ds$$
$$+ M^{\star}\|\mathfrak{F}(x) - \mathfrak{F}(y)\|_{\mathbb{L}^1}$$
$$\leq M_g \left\|A^{1-\beta}\right\|_{\mathbb{B}(\mathscr{X})}\|x(t) - y(t)\|_{\mathscr{X}}$$
$$+ \int_0^t \frac{C_{1-\beta}}{(t-s)^{1-\beta}}M_g\|x(s) - y(s)\|_{\mathscr{X}} ds + M^{\star}M_{\mathfrak{F}}\|x-y\|_{\mathbb{C}}.$$

Now, taking the supremum over $[0,T]$ on both sides of (7.61) yields

$$\|\Phi x - \Phi y\|_{\mathbb{C}} \leq \left(M_g\left\|A^{1-\beta}\right\| + M^{\star}M_{\mathfrak{F}} + M_gC_{1-\beta}\int_0^T (T-s)^{1-\beta}ds\right)\|x-y\|_{\mathbb{C}}$$
$$\leq \left(M_g\left\|A^{1-\beta}\right\| + M^{\star}M_{\mathfrak{F}} + M_gC_{1-\beta}\frac{T^{2-\beta}}{2-\beta}\right)\|x-y\|_{\mathbb{C}} < \|x-y\|_{\mathbb{C}}.$$

Thus, we conclude that Φ is a strict contraction and so, has a unique fixed-point which is a mild solution of (7.50). This completes the proof. $\qquad\square$

Exercise 7.2.10. Verify the details in (7.61).

Similar estimates can be used to establish results for infinite dimensional neutral evolution equations of the form (7.39) under non-Lipschitz conditions of the type used in Section 5.6, as well as compactness assumptions of the form used in Section 5.7. Some work in this direction can be found in [17, 21, 106, 147], but open problems remain. Refer to [29] for related work on second-order problems.

Consider the following Sobolev neutral functional evolution equation:

$$\begin{cases} [Bu(t) + g(t,u(t))]' = Au(t) + \mathfrak{F}(u)(t), \ 0 < t < T, \\ u(0) = u_0. \end{cases} \tag{7.62}$$

Open Exercise 7.2.11. Formulate an existence theory for (7.62) under Lipschitz, non-Lipschitz, and compactness assumptions. Also, develop related convergence and continuous dependence results. Such theory would provide an abstract formulation of more general IBVPs arising in the modeling of soil mechanics and fluid flow through fissured rocks.

7.3 Looking Ahead

Some external influences on a system take effect only after a certain amount of time has elapsed, resulting in a delayed reaction. This can be true for any of the models we have encountered thus far. For instance, when one ingests medication, the effect is not felt immediately. We can easily incorporate such a time delay in the IVP (6.1) as follows:

$$
\begin{cases}
\frac{dy}{dt} = -ay(t) + \int_0^t k(s)D(s,y(s-\sigma_1),z(s-\sigma_2))ds,\, t > 0, \\
\frac{dz}{dt} = ay(t) - bz(t) + \overline{D}(t,y(t-\sigma_1),z(t-\sigma_2)),\, t > 0, \\
y(t) = \theta_1(t),\ -\sigma_1 \le t \le 0, \\
z(t) = \theta_2(t),\ -\sigma_2 \le t \le 0,
\end{cases}
\tag{7.63}
$$

where σ_1 and σ_2 are positive constants. This creates various technical complications. For example, consider the equation at $t = 0$. The inputs into $y(\cdot)$ and $z(\cdot)$ are now negative. As such, in order for the equation to make sense, information about the state of the phenomenon is required for a certain length of time prior to the beginning of the experiment. This will naturally require us to modify the form of the initial condition. We handle this and other technical difficulties in the next chapter.

7.4 Guidance for Exercises

7.4.1 Level 1: A Nudge in a Right Direction

7.1.1. Define $A : \mathrm{dom}(A) \subset \mathbb{L}^2(\Omega) \to \mathbb{L}^2(\Omega)$ and $B : \mathrm{dom}(B) \subset \mathbb{L}^2(\Omega) \to \mathbb{L}^2(\Omega)$ by

$$
A[z] = \left(\frac{\partial^2}{\partial x^2} + \frac{\partial^2}{\partial y^2} \right)[z],
$$

$$
B[z] = (I - A)[z].
$$

Identify $\mathrm{dom}(A)$ and $\mathrm{dom}(B)$. How do you handle the forcing term?

7.1.2. To what operator must $I - \alpha \frac{\partial}{\partial t}$ converge as $\alpha \to 0^+$?

7.1.3. Does a change of variable similar to the one used to convert a classical wave equation into a system of first-order PDEs work? (Refer to (4.57).)

7.1.4. Define $A : \mathrm{dom}(A) \subset \mathbb{L}^2(0,a) \to \mathbb{L}^2(0,a)$ and $B : \mathrm{dom}(B) \subset \mathbb{L}^2(0,a) \to \mathbb{L}^2(0,a)$ by

$$
A[u] = \alpha \frac{\partial^2}{\partial x^2}[u],
$$

$$
B[u] = (I - \frac{1}{\alpha}A)[u],
$$

where $u(t) = z(\cdot,t)$. Then, loosely speaking, (7.4) can be reformulated as (7.9) in $\mathscr{X} = \mathbb{L}^2(0,a)$ with $f(t,u(t)) = f(t)$. Now, verify **(H7.1)** - **(H7.6)**.

7.1.5. Keep trying!

7.1.6. Since B^{-1} is a linear operator, it commutes with certain limit operators and behaves well when applied to inputs in the forms of finite sums. (So what?)

7.1.7. Modify the proof of Thrm. 5.5.2(i).

7.1.8. Does it suffice to impose hypothesis **(H5.1)**?

7.1.9. Impose/check the hypotheses determined in Exer. 7.1.8.

7.1.10. Keep trying!

7.1.11. i.) Use the analog of either Thrm. 5.7.3 or Prop. 5.7.4, assuming the same conditions on f.

ii.) No, because even in the non-Sobolev framework, it need not be globally defined. How did we ensure it was global in the easier case?

7.1.12. Argue well-definedness (i.e., the first step in the proof of Prop. 7.1.6) as we did in the proof of Thrm. 6.3.3. (How? What else?)

7.1.13. i.) Modify the solution map in (7.18) appropriately and then argue it is a strict contraction in the usual manner. Of course, the estimate involves more terms and this affects the data restriction needed to ensure that the solution map is a contraction.

ii.) View the forcing term in a manner similar to (6.3). Then, the result follows immediately.

7.1.14. Technically, this follows from Cor. 7.1.7 if you define f appropriately.

7.1.15. Use Exer. 6.3.6 and Exer. 7.1.8.

7.1.16. i.) Use Prop. 3.6.1.

ii.) Apply Prop. 7.1.6 directly. (How?)

iii.) Modify (7.22).

7.1.17. i.) Use $\mathscr{X} = \left(\mathbb{L}^2(0,a)\right)^2$. The functional $\mathfrak{F}$ will consist of two components, while each of the operators A, B, and C are 2×2 matrices. (Now what?)

ii.) See Exer. 7.1.4.

iii.) Be careful here. The functional must be globally Lipschitz, however the initial thought of simply imposing Lipschitz conditions on all of the functions involved might need to be reconsidered.

iv.) Check to see whether the functional satisfies **(H6.16)**. If so, adapt that argument to establish a similar existence result for (7.20).

7.1.19. Both sequences $\left\{B_n^{-1}\right\}$ and $\left\{e^{(AB_n^{-1})t}\right\}$ converge in appropriate spaces, the latter uniformly in t. (Now what?)

7.1.20. While (7.37) is very similar to (7.27), the fact that A_n and $\mathfrak{F}_n$ now vary with n complicates matters. The most crucial modification concerns **(H7.7)**, **(H7.10)**, and **(H7.11)**, which deal with the compatibility of A, B, A_n, and B_n.

7.1.21. i.) Modify the proof of Thrm. 6.4.1.

ii.) Yes, see Exer. 6.4.2.

7.1.22 & 7.1.23. Appeal to Exercises 5.7.2 - 5.7.5 for both of these IBVPs.

7.2.1. Certainly assuming that g is continuous works, but this is not necessary.

7.2.2. i.) In order to obtain the desired estimates and data restriction, define the solution map $\Phi : \mathbb{C}\left([0,T];\mathbb{R}^N\right) \to \mathbb{C}\left([0,T];\mathbb{R}^N\right)$ by

$$\Phi(\mathbf{u})(t) = \text{right} - \text{side of } \textbf{(??)}.$$

ii.) (a) It is evident that the limit function $\mathbf{u}$ is a mild solution of (5.10). (Why?) Subtract the variation of parameters formulae to show that $\lim_{n\to\infty} \|\mathbf{u}_n - \mathbf{u}\|_{\mathbb{C}} = 0$.

  (b) We need for f and g to be sufficiently smooth. Trace back through the development of (7.43) to determine these conditions.

iii.) Consider the theory developed in Section 5.5.3.

7.2.4. Compare (7.43) to (5.49), for instance. When $\left\{e^{At}|t \geq 0\right\}$ is compact, the sets $\left\{e^{At}u_0|0 \leq t \leq T\right\}$ and $\left\{e^{At}g(0,u_0)|0 \leq t \leq T\right\}$ are automatically precompact, but the set $\{g(t,u(t))|0 \leq t \leq T\}$ will not be.

7.2.5. Adapt the proof of Thrm. 5.7.3 that uses Schaefer's fixed-point theorem. Step 3 requires the most modification.

7.2.6. See Exercises 6.4.2 and 6.4.3.

7.2.7. Assume **(H7.12)** for g. What conditions should be imposed on f,h,a, and j to ensure that the forcing term is globally Lipschitz?

7.2.8. Consider the hypotheses **(H6.10)** - **(H6.15)**. Some of these work in their present form, but some will need to change and new ones will need to be added.

7.2.9. Note that $e^{At} - I = \int_0^t Ae^{As}ds$. So, $\left\{e^{At}|t \geq 0\right\}$ is uniformly continuous. But, since e^{At} is compact, $\forall t > 0$, what can be said about I? (So what?)

7.2.10. Use Prop. 7.2.7(vi). (How? Then what?)

7.4.2   Level 2: An Additional Thrust in a Right Direction

7.1.1. Make certain to incorporate the BCs into dom(A). Define the functional $\mathfrak{F}$: $\mathbb{C}\left([0,T];\mathbb{L}^2(\Omega)\right) \to \mathbb{C}\left([0,T];\mathbb{L}^2(\Omega)\right)$ by

$$\mathfrak{F}(z)(t) = \int_0^{a_2}\int_0^{a_1} k(t,x,y)f\left(t,z(x,y,t)\right)dxdy.$$

Is $\mathfrak{F}$ well-defined? What form is the abstract evolution equation? (Look ahead to (7.20).)

7.1.2. It can be shown that the limit operator is I. As such, p is a mild solution to a classical diffusion equation.

7.1.3. It is tempting to try to use a change of variable similar to the one used to convert the wave equation to a system of first-order PDEs. However, mimicking that approach results in having to make the following identification:

$$A\begin{bmatrix} z \\ \frac{\partial z}{\partial t} \end{bmatrix} = \begin{bmatrix} 0 & I \\ \frac{\partial^2}{\partial x^2} & \frac{\partial^2}{\partial x^2} \end{bmatrix}\begin{bmatrix} z \\ \frac{\partial z}{\partial t} \end{bmatrix},$$

which does not generate a C_0-semigroup on $\mathscr{H}$. As such, we can at best reformulate the given IBVP as the abstract evolution equation

$$\begin{cases} u''(t) + (Au)'(t) + Au(t) = f(t,u(t)), \ 0 < t < T, \\ u(0) = u_0, \ u'(0) = u_1. \end{cases} \tag{7.64}$$

It remains to be seen if such a second-order evolution equation can be transformed into a system of first-order PDEs that can be handled using the theory developed thus far.

7.1.4. This is studied in **[215]**. Here are the highlights:
Define $\text{dom}(A) = \text{dom}(B)$ to be

$$\left\{ f \in \mathbb{L}^2(0,a) | f, \frac{df}{dx} \text{ are AC, } \frac{d^2 f}{dx^2} \in \mathbb{L}^2(0,a), \wedge f(0) = f(a) = 0 \right\}.$$

Then, **(H7.4)** holds. Moreover, A and B can be expressed in the form (3.120) as

$$Az = \sum_{n=1}^{\infty} n^2 \langle z, e_n \rangle_{\mathbb{L}^2} e_n,$$

$$Bz = \sum_{n=1}^{\infty} \left(1 + n^2\right) \langle z, e_n \rangle_{\mathbb{L}^2} e_n,$$

where $e_n = \sqrt{\frac{2}{a}} \sin\left(\frac{2n\pi x}{a}\right)$, $n \in \mathbb{N}$. The linearity of A implies the linearity of B, so that **(H7.1)** holds. Also, **(H7.5)** holds by Exer. 3.1.14, and the fact that B^{-1} is closed follows from Exer. 3.1.15. You can check directly that B is invertible with

$$B^{-1}z = \sum_{n=1}^{\infty} \frac{1}{1+n^2} \langle z, e_n \rangle_{\mathbb{L}^2} e_n,$$

so that **(H7.2)** holds. (Do so!) The fact that **(H7.3)** holds is an assumption that must be imposed on z_0. Finally, try to verify **(H7.6)** directly using the definition.

7.1.5. Look ahead in the section now to check your work.

7.1.6. This implies that $B^{-1}\left(\sum_{i=1}^{n} H\left(x_i^{\star}\right)\triangle x_i\right) = \sum_{i=1}^{n} B^{-1}\left(H\left(x_i^{\star}\right)\triangle x_i\right)$, for all partitions of $(0,t)$. Hence, taking $\lim_{n \to \infty}$ of both sides yields equal results. By definition of the integral, this suggests that B^{-1} commutes with the integral operator.

7.1.7. Argue that the solution map defined by (7.18) is a contraction. Make certain to take into account the estimate in the Remark directly following (7.18).

7.1.8. Work through the details of Thrm. 5.5.2(i) for (7.9) directly and check to see that **(H5.1)** is sufficient without imposing additional restrictions on the data. In addition, you need to make certain that $u'(t) \in \text{dom}(B)$, $\forall 0 \le t \le T$, and $Bu' \in \mathbb{C}([0,T];\mathscr{X})$. (Why?)

7.1.9. Most of this is straightforward. Why must $Bu' \in \mathbb{C}\left([0,T];\mathbb{L}^2(0,a)\right)$?

7.1.10. Refer to **[38, 122, 229]** for results obtained by imposing non-Lipschitz conditions. For existence without uniqueness, use the fact that AB^{-1} generates a compact, C_0-semigroup under **(H7.1)** - **(H7.6)**. (Why is this important?) Aside from this, the proofs closely resemble those in the nonSobolev semi-linear setting.

7.1.11. i.) Note also that for (7.4), the forcing term is identified with an abstract function $\mathbf{f}(t)$, not $\mathbf{f}(t,u(t))$ which simplifies the situation.

ii.) Impose a restriction on the data similar to the one used to prove Φ was a contraction in Exer. 7.1.7. This will likely involve growth constants and certain norm estimates involving B^{-1}. (Why?)

7.1.12. You must also modify the estimate (7.23) using the $\mathbb{L}^1-$norm instead of the $\mathbb{C}-$norm.

7.1.13. i.) & ii.) Proving that the forcing term is globally Lipschitz is the crucial part of this approach. When viewing it as a functional, make certain to adapt $\mathfrak{F}_2$ in (6.3) so that it maps into the smaller space $\mathbb{C}([0,T];\mathscr{X})$.

7.1.14. Alternatively, define the functional

$$\mathfrak{F}: \mathbb{C}\left([0,T];\mathbb{L}^2\left((0,a_1)\times(0,a_2)\right)\right) \to \mathbb{C}\left([0,T];\mathbb{L}^2\left((0,a_1)\times(0,a_2)\right)\right)$$

as in Exer. 7.1.1. Assuming that k is continuous is sufficient (but might be overly strong) and that f is globally Lipschitz.

7.1.15. To prove the existence of a classical solution, you need for f to be Frechet differentiable in its second and third variables. What must we assume about a and g to ensure differentiability of the composition?

7.1.16. ii.) Treat the operator A in (7.20) as $A+C$ in (7.25). Then, the result is immediate.

ii.) The only modification is that $(A+C)B^{-1}$ is the generator of interest, so we change the notation in the constant $\overline{M_{AB^{-1}}}$ in (7.22) accordingly to $\overline{M_{(A+C)B^{-1}}}$.

7.1.17. i.) Symbolically,

$$A\begin{bmatrix} z \\ w \end{bmatrix} = \begin{bmatrix} \alpha_1 \frac{\partial^2}{\partial x^2} & 0 \\ 0 & \alpha_2 \frac{\partial^2}{\partial x^2} \end{bmatrix}\begin{bmatrix} z \\ w \end{bmatrix},$$

$$B\begin{bmatrix} z \\ w \end{bmatrix} = \begin{bmatrix} I - \frac{\partial^2}{\partial x^2} & 0 \\ 0 & I - \frac{\partial^2}{\partial x^2} \end{bmatrix}\begin{bmatrix} z \\ w \end{bmatrix},$$

$$C\begin{bmatrix} z \\ w \end{bmatrix} = \begin{bmatrix} \beta_1 I & 0 \\ 0 & \beta_2 I \end{bmatrix}\begin{bmatrix} z \\ w \end{bmatrix}.$$

Make certain to identify their domains and incorporate the BCs appropriately. Define the functional $\mathfrak{F}: \mathbb{C}([0,T];\mathscr{X}) \to \mathbb{C}([0,T];\mathscr{X})$ by

$$\mathfrak{F}\begin{bmatrix} z \\ w \end{bmatrix}(t) = \begin{bmatrix} h_1(z,w)\int_0^t a_1(t-s)f_1(s,x,z,w)ds \\ h_2(z,w)\int_0^t a_2(t-s)f_2(s,x,z,w)ds \end{bmatrix}.$$

iii.) Suppose that h_1 and f_1 are globally Lipschitz. How would you argue that

$$\left|\mathfrak{F}\begin{bmatrix} z \\ w \end{bmatrix}(t) - \mathfrak{F}\begin{bmatrix} \overline{z} \\ \overline{w} \end{bmatrix}(t)\right| \le M\left[|z-\overline{z}| + |w-\overline{w}|\right]?$$

The triangle inequality can help only so much. Alternatively, you might consider assuming that h_i continuous and globally bounded and that f_i are globally Lipschitz. Can these be weakened?

iv.) Yes, but prove it.

7.1.19. Apply Thrm. 3.1.7. (Tell how.)

7.1.20. I will leave this as an open exercise. Be careful!

7.1.21. The proofs are basically the same, with the exception of a change to the data restriction.

7.2.1. Using a more general measure-theoretic setting, assuming continuity a.e. with measurability also works.

7.2.2. i.) The continuous dependence result is with respect to the IC $\mathbf{u}_0$ and $\mathbf{v}_0$. The only additional wrinkle is the presence of $\mathbf{g}(0, \mathbf{u}_0)$. But, assuming that $\mathbf{g}$ is globally Lipschitz in its second variable easily takes care of this.

   ii.) (b) Assuming that both f and g are continuously differentiable works. (Why?)

   iii.) Consider (7.43) and determine conditions on f and g that ensure global existence using a suitably modified result in Section 5.5.3.

7.2.4. Assuming that g is a compact mapping works. Is there anything weaker?

7.2.7. Refer to the discussion of (6.27).

7.2.8. You need to impose a condition governing the convergence of $\{g_n\}$ and adapt **(H6.13)** to account for the presence of the terms $A_n e^{A_n t}$ and $A e^{At}$. Refer to Exer. 6.3.8 to aid in the proof.

7.2.9. $I = e^{A(0)}$ must be compact, which renders $\mathscr{X}$ finite-dimensional. (Why?)

7.2.10. Now apply Prop. 7.2.8(ii) to get the final step in the inequality.

Chapter 8

Delay Evolution Equations

Overview

Many different external forces can be accounted for when formulating a mathematical model governing the behavior of any phenomenon. While some of these forces can have an impact on the system that is realized immediately (e.g., a sharp blow or electrical surge), the effects of the others are noticeable only after a certain time delay. Extending our theoretical development to account for such a delay is the focus of the present chapter.

8.1 Motivation by Models

Below are several IBVPs formed by incorporating various types of time delays into the unknowns of IBVPs explored earlier in the text. At the moment, we do not provide an extensive discussion of them and we do not impose any conditions on the data. Rather, we shall revisit them throughout the chapter as we develop the theory necessary to study them rigorously.

Model III.4 Spring Mass System with Finite Time Delay

A possible semi-linear counterpart of IVP (5.139) with time delays is given by

$$\begin{cases} x''(t) + \beta x'(t) + \omega^2 x(t) = f\left(x(t - \sigma_1), x'(t - \sigma_2)\right), \ 0 < t < T, \\ x(t) = \theta_1(t), \ -\sigma_1 \leq t \leq 0, \\ x'(t) = \theta_2(t), \ -\sigma_2 \leq t \leq 0, \end{cases} \tag{8.1}$$

where $\sigma_1 > \sigma_2 > 0$. Try to interpret this physically.

Model II.5 Pharmacokinetics with delayed concentration-dependent dosage

Suppose that the effects of a dosage at time t, $D(t)$, begin to noticeably affect the concentration level of the drug in the bloodstream and GI tract after a $\frac{1}{4}$-hour time delay. A possible extension of IVP (5.1) is as follows:

$$\begin{cases} \frac{dy}{dt} = -ay(t) + \int_0^t k(s)D(s, y(s-\sigma_1), z(s-\sigma_2))ds, \, 0 < t < T, \\ \frac{dz}{dt} = ay(t) - bz(t) + \overline{D}(t, y(t-\sigma_1), z(t-\sigma_2)), \, 0 < t < T, \\ y(t) = \theta_1(t), \, -\sigma_1 \le t \le 0, \\ z(t) = \theta_2(t), \, -\sigma_2 \le t \le 0, \end{cases} \tag{8.2}$$

where θ_1 and θ_2 describe the concentration levels of the drug for a particular time interval prior to ingesting the drug.

Model IX.2 Neural Networks with Variable Delayed-Forcing

Incorporating a variable time delay into the various components of the output signal yields this extension of IVP (5.5):

$$\begin{cases} \frac{d}{dt}x_1(t) = a_1 x_1(t) + \sum_{j=1}^M \omega_{1j}(t)g_j\left(x_j(t-\sigma_1(t))\right), t > 0, \\ \vdots \\ \frac{d}{dt}x_M(t) = a_M x_M(t) + \sum_{j=1}^M \omega_{Mj}(t)g_j\left(x_j(t-\sigma_M(t))\right), t > 0, \\ x_1(t) = \theta_1(t), \, -R_1 \le t \le 0, \\ \vdots \\ x_M(t) = \theta_M(t), \, -R_M \le t \le 0, \end{cases} \tag{8.3}$$

where R_i $(i = 1,\dots,M)$ depends on $\sigma_1, \sigma_2, \dots, \sigma_M$ (in a manner specified later).

Model VII.5 Wave Equations with Dispersion and Time Delay

A two-dimensional wave equation on a bounded circular domain Ω involving dispersion and a finite time-delay can be described by the following IBVP:

$$\begin{cases} \frac{\partial^2 z}{\partial t^2} + \alpha\frac{\partial z}{\partial t} + \beta^2\left(\frac{\partial^2 z}{\partial x^2} + \frac{\partial^2 z}{\partial y^2}\right) = \sum_{i=1}^M \gamma_i g_i\left(t-\sigma_i, x, y, z(t-\sigma_i, x, y)\right), \\ \frac{\partial z}{\partial \mathbf{n}}(x, y, t) = 0, \, (x,y) \in \Omega, t > 0, \\ z(x,y,t) = \theta(x,y,t), \, (x,y) \in \partial\Omega = \left\{(x,y) : x^2 + y^2 = 1\right\}, \, -R \le t \le 0, \end{cases} \tag{8.4}$$

where $(x,y) \in \Omega$, $0 \le t \le T$, $\gamma_i > 0$ and $\sigma_i > 0$, $(i = 1,\dots,M)$ and R is chosen appropriately.

Model XIV.2 Extensible Beam Model with Delay Effects

Multiple time delays can occur within the same IBVP. For instance, consider the following model of transverse vibrations of an extensible beam that is subject to a forcing term whose effects are realized in waves. This can be viewed as a generalized version of (6.80):

$$\begin{cases} \frac{\partial^2 w}{\partial t^2}(z,t) + \alpha \frac{\partial^4 w}{\partial z^4}(z,t) + \beta w(z,t) \\ = f\left(z,t,w(z,t-\sigma_1), \frac{\partial^2 w}{\partial z^2}(z,t-\sigma_2(t)), \frac{\partial w}{\partial t}(z,t-\sigma_3)\right), 0 \le z \le a, 0 < t < T, \\ w(z,t) = w_0(z,t), \frac{\partial w}{\partial t}(z,t) = w_1(z,t), 0 \le z \le a, -R \le t \le 0, \\ \frac{\partial^2 w}{\partial z^2}(0,t) = w(0,t) = 0 = \frac{\partial^2 w}{\partial z^2}(a,t) = w(a,t), 0 < t < T, \end{cases}$$

(8.5)

where R is chosen appropriately.

Model XV.2 Sobolev-Type Models with Delays

Various types of time delays arise naturally in the modeling of soil mechanics, second-order fluids, etc. The following two-dimensional model is a generalization of IBVP (7.6):

$$\begin{cases} \frac{\partial}{\partial t}\left(z(x,y,t) - \frac{\partial^2}{\partial x^2}z(x,y,t) - \frac{\partial^2}{\partial y^2}z(x,y,t)\right) + \frac{\partial^2}{\partial x^2}z(x,y,t) + \frac{\partial^2}{\partial y^2}z(x,y,t) \\ = \int_{-r}^{0} a(\tau)z(x,y,t+\tau)d\tau, 0 < x < a_1, 0 < y < a_2, 0 < t < T, \\ z(x,y,t) = z_0(x,y,t), 0 < x < a_1, 0 < y < a_2, -R \le t \le 0, \\ z(0,y,t) = z(a_1,y,t) = 0, 0 < x < a_1, 0 < t < T, \\ z(x,0,t) = z(x,a_2,t) = 0, 0 < y < a_2, 0 < t < T, \end{cases}$$

(8.6)

where R is chosen appropriately. Note that the time-delay occurs within an integral term this time, rather than simply as a delayed temporal input in the unknown.

Model XII.2 Time-Delay Epidemiological Model

The following is a generalization of IBVP (6.51); see [102] for related work.

$$\begin{cases} \frac{\partial P_H}{\partial t} = \alpha_H \triangle P_H + \beta_H P_H - \left(\int_{t-h(t)}^{t} \sum_{i=1}^{N} g_i\left(s,x,y,P_H,P_{V_1},\ldots,P_{V_N}\right)ds\right) P_H P_{V_i}, \\ \frac{\partial P_{V_1}}{\partial t} = \alpha_{V_1} \triangle P_{V_1} - \sum_{i=1}^{N} \mathcal{W}_i P_{V_i} + \left(\int_{t-h(t)}^{t} \sum_{i=1}^{N} g_i\left(s,x,y,P_H,P_{V_1},\ldots,P_{V_N}\right)ds\right) P_H P_{V_i}, \\ \vdots \\ \frac{\partial P_{V_N}}{\partial t} = \alpha_{V_N} \triangle P_{V_N} - \sum_{i=1}^{N} \mathcal{W}_i P_{V_i} + \left(\int_{t-h(t)}^{t} \sum_{i=1}^{N} g_i\left(s,x,y,P_H,P_{V_1},\ldots,P_{V_N}\right)ds\right) P_H P_{V_i}, \\ P_H(x,y,t) = P_H^0(x,y,t), P_{V_i}(x,y,t) = P_{V_i}^0(x,y,t), (x,y) \in \Omega, i = 1,\ldots,N, -R \le t \le 0 \\ \frac{\partial P_H}{\partial \mathbf{n}}(x,y,t) = \frac{\partial P_{V_i}}{\partial \mathbf{n}}(x,y,t) = 0, (x,y) \in \partial\Omega, t > 0, i = 1,\ldots,N, \end{cases}$$

(8.7)

where $P_H = P_H(x,y,t)$, $P_{V_i} = P_{V_i}(x,y,t)$, and $h_i : [0,T] \to [0,\infty)(i = 1,\ldots,n)$ are continuous. The applicability of such IBVPs in a biological setting is discussed in [67].

Exercise 8.1.1. Try to reformulate the IBVPs (8.1) - (8.7) as abstract evolution equations. What hurdles do you encounter?

Exercise 8.1.2. Think about the above models and try to interpret why taking into account the past history in each case leads to an improvement of the description of the phenomenon.

Common Theme: A critical observation to make is that the forcing term in all of the above IBVPs depends temporally on more than just the time t at which the state process is being computed. Some involve a single constant delay, while for other models the actual delay experienced at a particular time t itself depends on t. And, there can be multiple delays in a single equation. Several other more complicated forms of delay (of so-called *Volterra type*) were also used (cf. (8.6) and (8.7)). We seek a common umbrella under which to treat all of these delay IBVPs.

8.2 Setting and Formulation of the Problem

Up to now, we have assumed that at any time t, the system under investigation was determined only by its state <u>at that time</u> rather than on a subset of states occurring prior to that time (called *past history*). Accounting for past history in the description of a phenomenon is important because most systems experience some sort of delayed reaction to an external influence, even if it is considered to be negligible. We seek the formal description of such forcing terms, which is somewhat more general than what we have encountered thus far.

By way of motivation, we first consider some particular forms of forcing terms out of context and extract some common salient features. We begin by considering the following simple ODE with <u>constant delay</u>:

$$x'(t) = bx(t - \alpha), \ 0 \leq t \leq T, \tag{8.8}$$

where $b \neq 0$ and $\alpha > 0$. Note that at $t = 0$, we need to know the value of the state x at $t = -\alpha$, and as we move forward in time we need *a priori* to know the values of $x(t), \forall t \in [-\alpha, 0]$. Without this information, (8.8) would be not well-posed since the right-side would be undefined. As such, the standard initial condition $x(0) = x_0$ must be replaced by a condition of the form

$$x(t) = \varphi(t), \ -\alpha \leq t \leq 0. \tag{8.9}$$

The same would be true if we added a function $f(t, x(t))$ to the right-side of (8.8). (Why?)

Remark. The type of delay incorporated into (8.8) is called *constant delay* since at any time $0 \leq t \leq T$, information about the state is needed for the interval $[t - \alpha, t]$ in order for the equation to be well-defined. All such intervals have the same length α; ergo, they are referred to as *constant* delays.

Exercise 8.2.1. Consider the equation

$$x'(t) + ax(t) = \sum_{i=1}^{n} b_i x(t - \alpha_i) + f(t), \ 0 \le t \le T, \tag{8.10}$$

where $\alpha_i > 0$, $i = 1, \ldots, n$, and at least one b_i does not equal zero.
i.) With what IC should (8.10) be coupled to guarantee that the resulting IVP is well-defined?
ii.) How should a mild solution of (8.10), coupled with the IC from (i), be defined?

Time delays could conceivably change with time since the actual system in which they arise is concurrently changing. For instance, if the time necessary for an external force to be realized by a system decreases whenever the state surpasses a certain threshold, it might be more sensible from a modeling perspective to allow for the time delays α_i in (8.8) and (8.10) to depend on t. This would result in a so-called *variable time delay*.

Exercise 8.2.2. Consider (8.10) with α_i replaced by a continuous function $\alpha_i : [0, T] \rightarrow [0, \infty)$.
i.) What conditions are sufficient to impose on $\alpha_i(t)$ to ensure that the value of the state at any time does not depend on knowledge of the state at future times?
ii.) Redo Exer. 8.2.1 for this more general form of time delay.

Extending our discussion to analogous systems in $\mathbb{R}^N$, namely

$$\mathbf{x}'(t) + \mathbf{A}\mathbf{x}(t) = \sum_{i=1}^{N} \mathbf{B}_i \mathbf{x}(t - \alpha_i) + \mathbf{f}(t), \ 0 \le t \le T, \tag{8.11}$$

where $\mathbf{x}(t) = [x_1(t), \ldots, x_N(t)]^T$, $\mathbf{A}, \mathbf{B}_i \in \mathbb{M}^N(\mathbb{R})$, $\mathbf{f} : [0, T] \rightarrow \mathbb{R}^N$, and $\alpha_i > 0$, $(i = 1, \ldots, N)$, can be done without any significant difficulty. (Why?) Hence, each component $x_i(t)$ of the unknown state vector $\mathbf{x}(t)$ resembles (8.10). As such, it is reasonable that a mild solution to an IVP formed using (8.11) would be formulated using a variation of parameters formula. Indeed, as in Exer. 8.2.1, a candidate for a mild solution $\mathbf{x} : [0, T] \rightarrow \mathbb{R}^N$ of (8.11) coupled with the initial condition

$$\mathbf{x}(t) = \varphi(t), \ -\alpha \le t \le 0, \tag{8.12}$$

where $\alpha = \max\{\alpha_i : i = 1, \ldots, N\}$ and $\varphi : [0, T] \rightarrow \mathbb{R}^N$, is as follows:

$$\mathbf{x}(t) = \begin{cases} \varphi(t), & -\alpha \le t \le 0, \\ e^{\mathbf{A}t}\varphi(0) + \int_0^t e^{\mathbf{A}(t-s)} \left(\sum_{i=1}^N \mathbf{B}_i \mathbf{x}(s - \alpha_i) + \mathbf{f}(s) \right) ds, & 0 \le t \le T. \end{cases} \tag{8.13}$$

Exercise 8.2.3. Assume that $\varphi \in \mathbb{C}\left([-\alpha, 0]; \mathbb{R}^N\right)$ and $\mathbf{f} \in \mathbb{C}\left([0, T]; \mathbb{R}^N\right)$.
i.) Why is $\mathbf{x}$, as defined by (8.13), continuous on $[-\alpha, T]$?
ii.) Does $\mathbf{x}$ actually satisfy (8.11) and (8.12) in the classical sense? Explain.

iii.) Can there be more than one solution to the IVP (8.11)-(8.12)? Explain.
iv.) If α_i were replaced by $\alpha_i(t)$ (as in Exer. 8.2.2), explain how your responses to (i)-(iii) would change.

The examples considered thus far have involved a "discrete delay" in the sense that information about x at only finitely many previous times was used to determine x' at the current time. But, IBVPs (8.6) and (8.7) involve a "continuous delay " in the sense that information about x for an entire interval of previous times is needed in order to accurately describe x' at the current time. For simplicity, consider the scalar ODE

$$x'(t) + ax(t) = \int_{t-\eta}^{t} f(s,x(s))ds. \tag{8.14}$$

Note that the integrand on the right-side of (8.14) involves $x(s)$. As such, at $t = 0$, values of x on the entire interval $[-\eta, 0]$ must be prescribed to describe the system. Making modifications to (8.14) similar to those previously discussed yields similar results, although the interval of times required to describe x at a particular time t can itself evolve with t (i.e., $\eta = \eta(t)$).

It would be convenient from a theoretical viewpoint to find a succinct way of subsuming the forcing terms corresponding to different types of delay as special cases of a more general *delay functional*. It is natural to expect the generic form of such a functional to be

$$F\left(t, \underline{(\text{past history})}\right). \tag{8.15}$$

But, what terms can we use to capture these different types of delays?

We begin with a simple case. Let $0 \le t_0 \le T$ be fixed and consider the following term from the right-side of the second equation in (8.13):

$$\int_{0}^{t_0} e^{\mathbf{A}(t_0-s)} \mathbf{B}_1 \mathbf{x}(s - \alpha_1)ds. \tag{8.16}$$

Note that

$$0 \le s \le t_0 \implies \underbrace{-\alpha_1 \le s - \alpha_1 \le t_0 - \alpha_1}_{(\star)} < t_0 \le T.$$

Observe that $(\star)$ is the range of inputs that enters into forming (8.16). As s moves from 0 to t_0, $s - \alpha_1$ moves from $-\alpha_1$ to $t_0 - \alpha_1$. Since the delay is constant, the set of inputs that influence the behavior of $\mathbf{x}$ also shifts to the right as the value of t_0 increases.

Let us define a new function as follows. For a given $s_0 \in [0, t_0]$, define $\mathbf{x}_{s_0} : [-\alpha_1, 0] \to \mathbb{R}^N$ by

$$\mathbf{x}_{s_0}(\theta) = \mathbf{x}(s_0 + \theta), \quad -\alpha_1 \le \theta \le 0. \tag{8.17}$$

This function embodies <u>all</u> possible history values needed in order to calculate (8.16). But, since the delay is constant in (8.16), we only need these values when $\theta = -\alpha_1$. (Why?) As such, we could express (8.16) equivalently as

$$\int_{0}^{t_0} e^{\mathbf{A}(t_0-s)} \mathbf{B}_1 \mathbf{x}_s(-\alpha_1)ds. \tag{8.18}$$

With (8.18) in mind, if we expand the domain to $[-\alpha_1, 0]$ we could express the second equation in (8.13) as

$$\int_0^t e^{\mathbf{A}(t-s)} \left(\sum_{i=1}^N \mathbf{B}_i \mathbf{x}_s(-\alpha_i) + \mathbf{f}(s) \right) ds. \tag{8.19}$$

Exercise 8.2.4. How could you modify the function $\mathbf{x}_s$ to handle a variable delay in which α_i is replaced by $\alpha_i(s)$ in (8.13)?

Expressing the delay portion in the previous examples required that the function $\mathbf{x}_s$ be calculated at a particular input. However, this is not the only case. Indeed, note that a delay of the form $\int_{-r}^0 a(\theta) x(t+\theta) d\theta$, called a Volterra delay, can be reformulated as

$$\int_{-r}^0 a(\theta) x_t d\theta, \tag{8.20}$$

where x_t is evaluated at all values in its domain are needed in the expression (8.20).

The following functional is the common link among all cases in formally expressing the history.

Definition 8.2.1. Let $\beta > 0$ and $\mathbf{x} \in \mathbb{C}\left([-\beta, T]; \mathbb{R}^N\right)$. For every $0 \le t \le T$, define the *past history functional* $\mathbf{x}_t : [-\beta, 0] \to \mathbb{R}^N$ by

$$\mathbf{x}_t(\theta) = \mathbf{x}(t+\theta), -\beta \le \theta \le 0. \tag{8.21}$$

Proposition 8.2.2. *The following are true.*
i.) *For every fixed* $0 \le t \le T$, $\mathbf{x}_t \in \mathbb{C}\left([-\beta, 0]; \mathbb{R}^N\right)$.
ii.) *The function* $\mathbf{z} : [0, T] \to \mathbb{C}\left([-\beta, 0]; \mathbb{R}^N\right)$ *defined by* $\mathbf{z}(t) = \mathbf{x}_t$ *is continuous.*

Exercise 8.2.5. Prove Prop. 8.2.2.

Notation: We write $\mathbb{C}_\beta = \mathbb{C}\left([-\beta, 0]; \mathbb{R}^N\right)$.

We can formally express the delay perturbations (8.18) - (8.20) under the common umbrella of a mapping $\mathbf{F} : [0, T] \times \mathbb{C}_\beta \to \mathbb{R}^N$ as follows:

(8.18) becomes	$\mathbf{F}(s, \mathbf{x}_s) = \mathbf{B}_1 \mathbf{x}_s(-\alpha_1)$
(8.19) becomes	$\mathbf{F}(s, \mathbf{x}_s) = \sum_{i=1}^N \mathbf{B}_i \mathbf{x}_s(-\alpha_i) + \mathbf{f}(s)$
(8.20) becomes	$\mathbf{F}(s, \mathbf{x}_s) = \int_{-r}^0 a(\theta) x_s d\theta$

Exercise 8.2.6. How would you modify the above identifications in the case of a variable delay?

The form of our finite-dimensional IVPs with finite delay can therefore be formulated abstractly as

$$\begin{cases} \mathbf{u}'(t) = \mathbf{A}\mathbf{u}(t) + \mathbf{F}(t, \mathbf{u}_t), 0 \le t \le T, \\ \mathbf{u}(t) = \varphi(t), -\beta \le t \le 0, \end{cases} \tag{8.22}$$

where $\mathbf{A} \in \mathbb{M}^N(\mathbb{R})$, $\mathbf{F} : [0,T] \times \mathbb{C}_\beta \to \mathbb{R}^N$, $\varphi : [-\beta, 0] \to \mathbb{R}^N$, and $\mathbf{u} : [-\beta, T] \to \mathbb{R}^N$.

Definition 8.2.3. A continuous function $\mathbf{u} : [-\beta, T] \to \mathbb{R}^N$ is a
i.) *mild solution* of (8.22) if
    a.) $\mathbf{u}(t) = \varphi(t)$, $-\beta \le t \le 0$,
    b.) $\mathbf{u}(t) = e^{\mathbf{A}t}\varphi(0) + \int_0^t e^{\mathbf{A}(t-s)}\mathbf{F}(s,\mathbf{u}_s)\,ds$, $0 \le t \le T$.
ii.) *classical solution* of (8.22) if $\mathbf{u} \in \mathbb{C}^1\left((-\beta, T); \mathbb{R}^N\right)$ and the formula for $\mathbf{u}(t)$
given by (i)(b) satisfies (8.22).

Guided by the above discussion, replacing $\mathbb{R}^N$ by a general Banach space $\mathscr{X}$ throughout and matrix $\mathbf{A}$ by a linear operator $A : \mathrm{dom}(A) \subset \mathscr{X} \to \mathscr{X}$ yields a natural, more general abstract framework that can be used to reformulate the IBVPs in Section 8.1 abstractly. All details carry over without issue. (Convince yourself!) (See [43, 156] for more details regarding the framework for delay equations.)

Exercise 8.2.7. Before moving on to the next section, try to formulate the usual existence-uniqueness results under Lipschitz and non-Lipschitz growth conditions. Pay particular attention to the modifications necessary to establish the results, as well as any notable similarities.

8.3   Theory for Lipschitz-Type Forcing Terms

We begin by considering the abstract delay evolution equation

$$\begin{cases} u'(t) = Au(t) + F(t,u_t), \, 0 \le t \le T, \\ u(t) = \varphi(t), \, -\beta \le t \le 0, \end{cases} \tag{8.23}$$

in $\mathscr{X}$. Here, $u : [-\beta, T] \to \mathscr{X}$, $A : \mathrm{dom}(A) \subset \mathscr{X} \to \mathscr{X}$ is a linear (possibly unbounded) operator, $\varphi : [-\beta, 0] \to \mathscr{X}$, and $F : [0,T] \times \mathbb{C}_\beta \to \mathscr{X}$. Throughout this section, unless otherwise specified, we shall assume $(\mathbf{H_A})$ and

$(\mathbf{H_\varphi})$ $\varphi \in \mathbb{C}_\beta$.

Definition 8.3.1. A *mild solution* of (8.23) is a continuous function $u : [-\beta, T] \to \mathscr{X}$ satisfying Def. 8.2.3 with $\mathbb{R}^N$ replaced by $\mathscr{X}$.

We begin with:

Theorem 8.3.2. *Assume that*
(H8.1) $F : [0,T] \times \mathbb{C}_\beta \to \mathscr{X}$ *is continuous in the first variable and* $\exists M_f > 0$ *such that*

$$\|F(t,u_t) - F(t,v_t)\|_{\mathscr{X}} \le M_f \|u_t - v_t\|_{\mathbb{C}_\beta} \tag{8.24}$$

globally on $\mathbb{C}_\beta$ *(uniformly in t).*

Then, (8.23) has a unique global mild solution on $[-\beta, T]$. *Moreover, let* $\varphi, \psi \in \mathbb{C}_\beta$ *and denote the corresponding mild solutions of (8.23) by* u_φ *and* u_ψ, *respectively. Then,* $\exists M^* > 0$ *such that*

$$\left\| u_\varphi - u_\psi \right\|_{\mathbb{C}} \leq M^* e^{(\alpha + M_f)T} \left\| \varphi - \psi \right\|_{\mathbb{C}_\beta}. \tag{8.25}$$

Proof. Define the solution map $\Phi : \mathbb{C}\left([-\beta, T]; \mathscr{X}\right) \to \mathbb{C}\left([-\beta, T]; \mathscr{X}\right)$ by

$$(\Phi x)(t) = \begin{cases} e^{At}\varphi(0) + \int_0^t e^{A(t-s)} F(s, x_s)\, ds, & 0 \leq t \leq T, \\ \varphi(t), & -\beta \leq t \leq 0. \end{cases} \tag{8.26}$$

We apply the Contraction Mapping Theorem to prove that Φ has a unique fixed-point which is a mild solution of (8.23).

Let $x \in \mathbb{C}\left([-\beta, T]; \mathscr{X}\right)$. We argue that $\Phi x \in \mathbb{C}\left([-\beta, T]; \mathscr{X}\right)$ by showing that

$$\sup_{-\beta \leq t \leq T} \left\| (\Phi x)(t) \right\|_{\mathscr{X}} < \infty. \tag{8.27}$$

Since $\varphi \in \mathbb{C}_\beta$,

$$\sup_{-\beta \leq t \leq 0} \left\| (\Phi x)(t) \right\|_{\mathscr{X}} = \sup_{-\beta \leq t \leq 0} \left\| \varphi(t) \right\|_{\mathscr{X}} < \infty. \tag{8.28}$$

Also, $\forall 0 \leq t \leq T$,

$$\begin{aligned} \left\| (\Phi x)(t) \right\|_{\mathscr{X}} &\leq \left\| e^{At}\varphi(0) \right\|_{\mathscr{X}} + \left\| \int_0^t e^{A(t-s)} F(s, x_s)\, ds \right\|_{\mathscr{X}} \\ &\leq \overline{M_A} \left[\left\| \varphi(0) \right\|_{\mathscr{X}} + \int_0^t \left[M_f \left\| x_s \right\|_{\mathbb{C}_\beta} + \left\| F(s, 0) \right\|_{\mathscr{X}} \right] ds \right] < \infty. \end{aligned}$$

Since Φ itself is clearly continuous, we conclude that Φ is well-defined.

Next, $\forall x, y \in \mathbb{C}\left([-\beta, T]; \mathscr{X}\right)$, observe that

$$\begin{aligned} \left\| (\Phi x)(t) - (\Phi y)(t) \right\|_{\mathscr{X}} &\leq \overline{M_A} \int_0^t \left\| F(s, x_s) - F(s, y_s) \right\|_{\mathscr{X}} ds \\ &\leq \overline{M_A} M_f \int_0^t \left\| x_s - y_s \right\|_{\mathbb{C}_\beta} ds. \end{aligned} \tag{8.29}$$

Now, applying (8.29), we see that

$$\left\|(\Phi^2 x)(t) - (\Phi^2 y)(t)\right\|_{\mathscr{X}} \leq \overline{M_A} M_f \int_0^t \left\|(\Phi x)_s - (\Phi y)_s\right\|_{\mathbb{C}_\beta} ds$$

$$\leq \overline{M_A} M_f \int_0^t \sup_{-\beta \leq s \leq T} \left\|(\Phi x)(s) - (\Phi y)(s)\right\|_{\mathscr{X}} ds$$

$$\leq \overline{M_A} M_f \int_0^t \sup_{-\beta \leq s \leq T} \left\{ \overline{M_A} M_f \int_0^s \left\|x_\tau - y_\tau\right\|_{\mathbb{C}_\beta} d\tau \right\} ds$$

$$\leq \left(\overline{M_A} M_f\right)^2 \left(\int_0^t \int_0^s 1 d\tau ds \right) \|x - y\|_{\mathbb{C}([-\beta,t];\mathscr{X})} \quad (8.30)$$

$$= \left(\overline{M_A} M_f\right)^2 \frac{t^2}{2} \|x - y\|_{\mathbb{C}([-\beta,t];\mathscr{X})}.$$

Continuing as in (8.30), we obtain for every $n \in \mathbb{N}$,

$$\left\|(\Phi^n x)(t) - (\Phi^n y)(t)\right\|_{\mathscr{X}} \leq \frac{\left(t \overline{M_A} M_f\right)^n}{n!} \|x - y\|_{\mathbb{C}([-\beta,t];\mathscr{X})}. \quad (8.31)$$

Taking the supremum over $[-\beta, T]$ on both sides of (8.31) yields

$$\left\|\Phi^n x - \Phi^n y\right\|_{\mathbb{C}([-\beta,T];\mathscr{X})} \leq \frac{\left(T \overline{M_A} M_f\right)^n}{n!} \|x - y\|_{\mathbb{C}([-\beta,T];\mathscr{X})}. \quad (8.32)$$

There exists a sufficiently large $n_0 \in \mathbb{N}$ such that $\frac{\left(T \overline{M_A} M_f\right)^{n_0}}{n_0!} < 1$ (Why?), so that Φ^{n_0} is a strict contraction. Thus, Cor. 5.2.3 ensures that Φ^{n_0} has a unique fixed-point x which is a mild solution of (8.23).

Finally, we establish the continuous dependence estimate. Let $\varphi, \psi \in \mathbb{C}_\beta$ and denote the corresponding mild solutions of (8.23) by u_φ and u_ψ, respectively. Then,

$$x_\varphi(t) - x_\psi(t) = \begin{cases} e^{At}\left[\varphi(0) - \psi(0)\right] + \\ \int_0^t e^{A(t-s)}\left[F\left(s, (x_\varphi)_s\right) - F\left(s, (x_\psi)_s\right)\right] ds, \ 0 \leq t \leq T, \\ \varphi(t) - \psi(t), \ -\beta \leq t \leq 0. \end{cases} \quad (8.33)$$

Hence,

$$\left\|x_\varphi(t) - x_\psi(t)\right\|_{\mathscr{X}} \leq \begin{cases} \overline{M_A}\left[\|\varphi(0) - \psi(0)\|_{\mathscr{X}} \\ + M_f \int_0^t \left\|(x_\varphi)_s - (x_\psi)_s\right\|_{\mathbb{C}_\beta} ds\right], 0 \leq t \leq T, \\ \|\varphi(t) - \psi(t)\|_{\mathscr{X}}, \ -\beta \leq t \leq 0. \end{cases} \quad (8.34)$$

For every $0 \leq t \leq T$, define

$$\left\|x_\varphi - x_\psi\right\|_t = \sup_{-\beta \leq \theta \leq t} \left\|x_\varphi(\theta) - x_\psi(\theta)\right\|_{\mathscr{X}}. \quad (8.35)$$

We deduce from (8.34) and Gronwall's lemma that

$$
\|x_\varphi - x_\psi\|_t \leq
\begin{cases}
\overline{M_A}\left[\|\varphi(0) - \psi(0)\|_{\mathscr{X}} \\
+ M_f \int_0^t \|x_\varphi - x_\psi\|_s \, ds\right], 0 \leq t \leq T, \\
\|\varphi - \psi\|_t, \, -\beta \leq t \leq 0,
\end{cases}
$$

$$
\leq
\begin{cases}
\overline{M_A}\left[\|\varphi(0) - \psi(0)\|_{\mathscr{X}} \, e^{(M_f)t}, 0 \leq t \leq T, \\
\|\varphi - \psi\|_t, \, -\beta \leq t \leq 0.
\end{cases}
\tag{8.36}
$$

(Tell how.) Hence, (8.25) follows, as desired. $\qquad\square$

Exercise 8.3.1. Obtain an estimate for $\left\|(x_\varphi)_t - (x_\psi)_t\right\|_{\mathbb{C}_\beta}$.

The following two corollaries are useful in practice.

Corollary 8.3.3. *Let* $g : [0,T] \times \mathscr{X}^N \to \mathscr{X}$ *and* $B \in \mathbb{B}(\mathscr{X})$, *and assume that A and* φ *satisfy* $(\mathbf{H_A})$ *and* $(\mathbf{H_\varphi})$, *respectively. Consider the abstract evolution equation*

$$
\begin{cases}
u'(t) = (A+B)u(t) + g\left(t, u(t-\beta_1), \ldots, u(t-\beta_N)\right), 0 \leq t \leq T, \\
u(t) = \varphi(t), \, -\max\{\beta_1, \ldots, \beta_N\} \leq t \leq 0.
\end{cases}
\tag{8.37}
$$

Assume that
(H8.2) $g : [0,T] \times \mathscr{X}^N \to \mathscr{X}$ *is continuous in the first variable and* $\exists M_g > 0$ *such that*

$$
\|g(t, y_1, \ldots, y_N) - g(t, \overline{y_1}, \ldots, \overline{y_N})\|_{\mathscr{X}} \leq M_g \sum_{i=1}^N \|y_i - \overline{y_i}\|_{\mathscr{X}}
\tag{8.38}
$$

globally on $\mathscr{X}^N$ *(uniformly in t).*
Then, (8.37) *has a unique global mild solution.*

Exercise 8.3.2. Prove Cor. 8.3.3. Explain how the statement and proof change for variable delay.

Corollary 8.3.4. *Assume that* $B : \mathrm{dom}(B) \subset \mathscr{X} \to \mathscr{X}$ *is an A-bounded perturbation (cf. Thrm. 6.5.1),* $\{B_j : 1 \leq j \leq n\} \subset \mathbb{B}(\mathscr{X})$, *and that* $(\mathbf{H_A})$ *and* $(\mathbf{H_\varphi})$ *hold. Then,*

$$
\begin{cases}
u'(t) = \left(A + B + \sum_{j=1}^n B_j\right) u(t) + F(t, u_t), 0 \leq t \leq T, \\
u(t) = \varphi(t), \, -\beta \leq t \leq 0,
\end{cases}
\tag{8.39}
$$

has a unique global mild solution on $[-\beta, T]$.

Exercise 8.3.3. Prove Cor. 8.3.4. How does it change for variable delay?

Exercise 8.3.4. Impose appropriate growth conditions so that Cor. 8.3.3 can be invoked to conclude that (8.1) has a unique mild solution.

Exercise 8.3.5. Impose appropriate growth conditions so that Cor. 8.3.3 can be invoked to conclude that (8.4) has a unique mild solution.

Exercise 8.3.6. Impose appropriate growth conditions so that Cor. 8.3.3 can be invoked to conclude that (8.5) has a unique mild solution.

Exercise 8.3.7. Impose appropriate growth conditions so that Cor. 8.3.3 can be invoked to conclude that (8.7) has a unique mild solution.

The extent to which a time delay is incorporated into a model can vary. The following convergence result illustrates the effect of a delay perturbation on the following nondelay IVP:

$$\begin{cases} y'(t) & = Ay(t) + f(t, y(t)), \ 0 \le t \le T, \\ y(0) & = \varphi(0), \end{cases} \tag{8.40}$$

in $\mathscr{X}$, where $f : [0,T] \times \mathscr{X} \to \mathscr{X}$ and $\varphi \in \mathbb{C}_\beta$.

Proposition 8.3.5. (Delay Perturbation Convergence Result)
For every $n \in \mathbb{N}$, consider the IVP

$$\begin{cases} z_n'(t) + Az_n(t) = f(t, z_n(t)) + \overline{f_n}(t, (z_n)_t), \ 0 \le t \le T, \\ z_n(t) = \varphi(t), \ -\beta \le t \le 0. \end{cases} \tag{8.41}$$

Assume that $(\mathbf{H_A})$ *and* $(\mathbf{H_\varphi})$ *hold, as well as*
(H8.3) $f : [0,T] \times \mathscr{X} \to \mathscr{X}$ *is continuous in the first variable and* $\exists M_f > 0$ *such that*

$$\|f(t,x_1) - f(t,x_2)\|_{\mathscr{X}} \le M_f \|x_1 - x_2\|_{\mathscr{X}}$$

globally on $\mathscr{X}$ (uniformly in t);
(H8.4) $\lim\limits_{n\to\infty} \left\|\overline{f_n}(t,x_t)\right\|_{\mathscr{X}} = 0$ *uniformly in $[0,T] \times \mathbb{C}_\beta$.*
Then, $\lim\limits_{n\to\infty} \|z_n - y\|_{\mathbb{C}([0,T];\mathscr{X})} = 0$.

Exercise 8.3.8. Prove Prop. 8.3.5.

Exercise 8.3.9. Establish an analog of Prop. 8.3.5 for the variable delay case.

Exercise 8.3.10. Formulate and prove a convergence result for (8.23) in the spirit of Prop. 7.1.8 and Exer. 7.1.20 (with $B_n = 0, \forall n \in \mathbb{N}$).

The right-side of a delay evolution equation can be more complicated than what appears in (8.23). For instance, consider

$$\begin{cases} u'(t) = Au(t) + g\left(t, u_t, \int_0^t k(s, u_s)ds\right), \ 0 \le t \le T, \\ u(t) = \varphi(t), \ -\beta \le t \le 0, \end{cases} \tag{8.42}$$

where $k : [0,T] \times \mathbb{C}_\beta \to \mathscr{X}$ and $g : [0,T] \times \mathbb{C}_\beta \times \mathscr{X} \to \mathscr{X}$. The following result can be established in a manner similar to Thrm. 8.3.2.

Proposition 8.3.6. *Assume that* $(\mathbf{H_A})$ *and* $(\mathbf{H_\varphi})$ *hold, and that both k and g are continuous in the first variable and globally Lipschitz in their spatial variables (uniformly in t). Then, (8.42) has a unique global mild solution on* $[-\beta, T]$.

Exercise 8.3.11. Use Thrm. 8.3.2 to prove Prop. 8.3.6.

Exercise 8.3.12. Consider the evolution equation

$$\begin{cases} u'(t) = (A+B)u(t) + f(t)\int_{-\beta}^{0} a(\tau)u_\tau d\tau, \, 0 \le t \le T, \\ u(t) = \varphi(t), \, -\beta \le t \le 0, \end{cases} \tag{8.43}$$

in $\mathscr{X}$, where $(\mathbf{H_A})$ and $(\mathbf{H_\varphi})$ hold, $B \in \mathbb{B}(\mathscr{X})$, $f : [0,T] \to \mathscr{X}$ is continuous, and $a \in \mathbb{L}^1(-\beta, 0; \mathbb{R})$.
i.) Show that (8.43) has a unique global mild solution on $[-\beta, T]$.
ii.) Apply the result in (i) to IBVP (8.6).

Since T was arbitrary in our existence results, the mild solution in each case actually exists on $[0,\infty)$. In general, various types of behavior can occur; the $\mathbb{C}$-norm of this solution can tend to infinity, or the solution can be time-periodic or possess no predictable long-term behavior whatsoever. To study such long-term behavior, we extend our setting to $[0,\infty)$ in a natural way:

Consider IVP (8.23) on $[0,\infty)$, where $F : [0,\infty) \times \mathbb{C}_\beta \to \mathscr{X}$ and $u : [-\beta, \infty) \to \mathscr{X}$ is a mild solution of (8.23). We have the following result in the spirit of Prop. 4.2.4:

Proposition 8.3.7. *If* $\{e^{At} : t \ge 0\}$ *is contractive, then* $\lim_{t\to\infty}\|u(t)\|_{\mathscr{X}} = 0$.

Exercise 8.3.13. Prove Prop. 8.3.7.

Exercise 8.3.14.
i.) Formulate and prove a version of Prop. 8.3.7 for the variable delay case.
ii.) Under what conditions would the result in (i) apply to IVP (8.3)?

Related problems are studied in [**2, 231, 232, 241, 264, 304**].

8.4   Theory for Non-Lipschitz-Type Forcing Terms

We now consider (8.23) under non-Lipschitz type conditions of the type introduced in Section 5.6.

Exercise 8.4.1. Before moving on, look back at Section 5.6 and try to determine precisely what changes would need to be made to formulate and prove a version of Thrm. 5.6.3 for (8.23).

Assume that $(\mathbf{H_A})$ and $(\mathbf{H_\varphi})$ hold, as well as the following slightly modified versions of **(H5.7)** - **(H5.9)**:

(H8.5) There exists $K_1 : [0,T] \times [0,\infty) \to [0,\infty)$ such that

    a.) $K_1(\cdot,x)$ is integrable, $\forall x \in [0,\infty)$,

    b.) $K_1(t,\cdot\cdot)$ is continuous, nondecreasing, and concave, $\forall t \in [0,T]$,

    c.) $\|f(t,x_t)\|_{\mathscr{X}} \leq K_1(t,\|x\|_t)$, $\forall t \in [0,T]$, $x_t \in \mathbb{C}_\beta$.

(H8.6) There exists $K_2 : [0,T] \times [0,\infty) \to [0,\infty)$ such that

    a.) $K_2(\cdot,x)$ is integrable, $\forall x \in [0,\infty)$,

    b.) $K_2(t,\cdot\cdot)$ is continuous, nondecreasing, and $K_2(t,0) = 0$, $\forall t \in [0,T]$,

    c.) $\|f(t,x_t) - f(t,y_t)\|_{\mathscr{X}} \leq K_2(t,\|x-y\|_t)$, $\forall t \in [0,T]$, $x_t, y_t \in \mathbb{C}_\beta$.

(H8.7) Any function $w : [0,T] \to [0,\infty)$ which is continuous, $w(0) = 0$, and

$$w(t) \leq \overline{M_A} \int_0^t K_2(s,w(s))\,ds, \forall 0 \leq t \leq T^\star \leq T,$$

must be identically 0 on $[0,T^\star]$.

(See (8.35) for the notation $\|\cdot\|_t$.) A version of the following theorem was proven in **[144, 145]**. We point out the main differences that occur in the proof as compared to the proof of Thrm. 5.6.3.

Theorem 8.4.1. *If $(\mathbf{H_A})$, $(\mathbf{H_\varphi})$, and (H8.5) - (H8.7) hold, then (8.23) has a unique local mild solution on $[-\beta, T^\star)$, for some $0 < T^\star \leq T$.*

Outline of Proof: Let $n \in \mathbb{N}$ and consider the following sequence of successive approximations:

$$u_0(t) = \begin{cases} e^{At}\varphi(0), \ 0 \leq t \leq T, \\ \varphi(t), \ -\beta \leq t \leq 0, \end{cases} \tag{8.44}$$

$$u_n(t) = \begin{cases} e^{At}\varphi(0) + \int_0^t e^{A(t-s)}F(s,(u_n)_s)\,ds, \ 0 \leq t \leq T, \\ \varphi(t), \ -\beta \leq t \leq 0. \end{cases}$$

For any $\beta_1 > \overline{M_A}\|\varphi(0)\|_{\mathscr{X}}$, Lemma 5.6.2 guarantees the existence of $0 < T_1 \leq T$ for which the equation

$$z(t) = \beta_1 + \overline{M_A} \int_0^t K_1(s,z(s))\,ds$$

has a unique solution $z : [0,T_1) \to [0,\infty)$. The main changes in the proof occur in Claims 2 and 3 because they involve deriving estimates of terms involving the delay part. These claims are reformulated in the present setting as follows:

<u>Claim 2:</u> For every $\delta_0 > 0$, $\exists 0 < T_2 \leq T_1$ (independent of n) such that $\forall n \in \mathbb{N}$,

$$\|u_n - u_0\|_t \leq \delta_0, \forall 0 \leq t < T_2 \leq T_1 \leq T.$$

<u>Claim 3</u>: For every $n, m \in \mathbb{N}$,

$$\|u_{n+m} - u_n\|_t \leq \overline{M_A} \int_0^t K_2(s, 2\delta_0) \, ds, \forall 0 \leq t < T_2.$$

Now, $\forall n, m \in \mathbb{N}$, define $\gamma_n : [0, T_2] \to (0, \infty)$ as in (5.167) and $\theta_{m,n} : [0, T_2] \to (0, \infty)$ by

$$\theta_{m,n}(t) = \|u_{n+m} - u_n\|_t.$$

Using **(H8.5)(b)**, the continuity of K_2 guarantees the existence of $0 < T_3 \leq T_2$ such that

$$\gamma_1(t) \leq 2\delta_0, \forall 0 \leq t < T_3.$$

The remainder of the argument proceeds as in the proof of Thrm. 5.6.3. □

Exercise 8.4.2. Provide the details in the proof of Thrm. 8.4.1. Under what conditions is the mild solution globally defined?

Exercise 8.4.3. Formulate and prove a result in the spirit of Thrm. 8.4.1 for

$$\begin{cases} u'(t) = Au(t) + F(t, u(\rho(t))), 0 \leq t \leq T, \\ u(t) = \varphi(t), -\beta \leq t \leq 0, \end{cases} \tag{8.45}$$

where $\rho : [0, T] \to [0, T]$ is increasing and such that $\rho(t) \leq t, \forall t \in [0, T]$. (A stochastic version of this result studied in **[144, 145]**.)

Exercise 8.4.4. Consider (8.7) and assume that f satisfies the appropriate modified form of **(H8.5)** - **(H8.7)**. Use Exer. 8.4.3 to argue that (8.7) has a unique global mild solution.

8.5   Implicit Delay Evolution Equations

We now turn our attention to a class of Sobolev-type evolution equations with delay of the form

$$\begin{cases} (Bu)'(t) = Au(t) + F(t, u_t), 0 \leq t \leq T, \\ u(t) = \varphi(t), -\beta \leq t \leq 0, \end{cases} \tag{8.46}$$

where $A : \text{dom}(A) \subset \mathscr{X} \to \mathscr{X}$ and $B : \text{dom}(B) \subset \mathscr{X} \to \mathscr{X}$ satisfy **(H7.1)** - **(H7.6)**, and $F : [0, T] \times \mathbb{C}_\beta \to \mathscr{X}$ and $\varphi \in \mathbb{C}([-\beta, 0]; \text{dom}(B))$ are given. The theory established in this chapter thus far can be easily combined with the development in Chapter 7 to establish the standard results for (8.46) concerning existence-uniqueness, continuous dependence, etc. We outline the main results, but leave it to you to supply the details as an instructive exercise.

Definition 8.5.1. A *mild solution* of (8.46) is a continuous function $u : [-\beta, T] \to \mathscr{X}$ satisfying

$$u(t) = \begin{cases} B^{-1} e^{(AB^{-1})t} \varphi(0) + \int_0^t B^{-1} e^{(AB^{-1})(t-s)} F(s, u_s) \, ds, \ 0 \le t \le T, \\ \varphi(t), \ -\beta \le t \le 0. \end{cases} \tag{8.47}$$

Proposition 8.5.2. *If (H7.1) - (H7.6) hold and F satisfies (8.24), then (8.46) has a unique global mild solution. Moreover, let $\varphi, \psi \in \mathbb{C}_\beta$ and denote the corresponding mild solutions of (8.46) by u_φ and u_ψ, respectively. Then, there exists a positive constant C^*(depending on $M_f, \alpha, T, \varphi, \psi, \|B\|$, and $\|B^{-1}\|$) such that*

$$\|u_\varphi - u_\psi\|_{\mathbb{C}} \le C^* \|\varphi - \psi\|_{\mathbb{C}_\beta}.$$

Corollary 8.5.3. *Assume that (H7.1) - (H7.6) hold. Consider the evolution equation*

$$\begin{cases} (Bu)'(t) = (A+C)u(t) + g(t, u(t-\beta_1), \ldots, u(t-\beta_N)) \ 0 \le t \le T, \\ u(t) = \varphi(t), \ -\max\{\beta_1, \ldots, \beta_N\} \le t \le 0, \end{cases} \tag{8.48}$$

where $g : [0, T] \times \mathscr{X}^N \to \mathscr{X}$ satisfies (8.38) and $C \in \mathbb{B}(\mathscr{X})$. Then, (8.48) has a unique global mild solution.

Exercise 8.5.1. Prove Prop. 8.5.2 and Cor. 8.5.3.

Exercise 8.5.2. Let $\Omega \subset \mathbb{R}^3$ and consider the following generalization of IBVP (7.5), now equipped with delay:

$$\begin{cases} \frac{\partial}{\partial t}(z(\mathbf{x}, t) - \triangle z(\mathbf{x}, t)) + \alpha \triangle z(\mathbf{x}, t) - \gamma z(\mathbf{x}, t) = \\ \quad f\left(\mathbf{x}, t, z_t, \int_0^t k(t, s, z_s) ds\right), \ \mathbf{x} \in \Omega, 0 \le t \le T, \\ z(\mathbf{x}, t) = \phi(\mathbf{x}, t), \ \mathbf{x} \in \Omega, -\beta \le t \le 0, \\ \frac{\partial z}{\partial \mathbf{n}}(\mathbf{x}, t) = 0, \ \mathbf{x} \in \partial \Omega, 0 \le t \le T, \end{cases} \tag{8.49}$$

where $\alpha, \beta, \gamma > 0$, f and k are continuous in their temporal variables and globally Lipschitz in their spatial variables (uniformly in t, s), and $\phi \in \mathbb{C}(\Omega \times [-\beta, 0]; \mathbb{R})$.
i.) Prove that (8.49) has a unique global mild solution.
ii.) Let $\{\phi_\varepsilon : \varepsilon > 0\} \subset \mathbb{C}(\Omega \times [-\beta, 0]; \mathbb{R})$ be such that $\lim_{\varepsilon \to 0^+} |\phi_\varepsilon(x, t) - \phi(x, t)| = 0$ uniformly on $\Omega \times [-\beta, 0]$, and denote the solution of the IBVP obtained by replacing ϕ by ϕ_ε in (8.49) by z_ε. Prove that $z_\varepsilon \longrightarrow z$ in $\mathbb{C}(\Omega \times [-\beta, 0]; \mathbb{R})$ as $\varepsilon \to 0^+$.

Exercise 8.5.3. Consider the following generalization of IBVP (7.8) now equipped with a Volterra-type delay:

$$\begin{cases} \frac{\partial w}{\partial t}(x, t) = \alpha \frac{\partial^2 w}{\partial x^2}(x, t) + \beta \frac{\partial^3 w}{\partial x^2 \partial t}(x, t) \\ \quad + g(t) \int_{-\sigma}^0 a(\tau) w_\tau(x, \cdot) d\tau, \ x > 0, 0 < t < T, \\ w(0, t) = 0, \ 0 < t < T, \\ w(x, t) = \phi(x, t), \ x > 0, -\sigma \le t \le 0, \end{cases} \tag{8.50}$$

where $\alpha, \beta, \sigma > 0$ and $\phi \in \mathbb{C}((0, \infty) \times [-\sigma, 0]; \mathbb{R})$.

i.) Specify appropriate growth restrictions under which (8.50) has a unique global mild solution. Prove your result.

ii.) Repeat Exer. 8.5.2 (ii) for IBVP (8.50).

Exercise 8.5.4. Formulate and prove a result in the spirit of Prop. 8.3.5 for (8.46).

Exercise 8.5.5. For every $n \in \mathbb{N}$, consider the following Sobolev-type evolution equation with Volterra delay:

$$\begin{cases} (B_n u_n)'(t) = A u_n(t) + g_n(t) \int_{-\beta}^{0} a(\tau)(u_n)_\tau d\tau, \ 0 \le t \le T, \\ u_n(t) = \varphi_n(t), \ -\beta \le t \le 0. \end{cases} \tag{8.51}$$

Impose sufficient conditions on the operators and mappings in (8.51) in a manner that ensures the existence of $u \in \mathbb{C}([-\beta, T]; \mathscr{X})$ for which $\lim_{n \to \infty} \|u_n - u\|_{\mathbb{C}} = 0$. To what IVP is u a mild solution? Prove your assertions.

Next, consider the following generalization of IBVP (7.6):

$$\begin{cases} \frac{\partial}{\partial t}\left(z(x,y,t) - \frac{\partial^2}{\partial x^2} z(x,y,t) - \frac{\partial^2}{\partial y^2} z(x,y,t) \right) + \frac{\partial^2}{\partial x^2} z(x,y,t) + \frac{\partial^2}{\partial y^2} z(x,y,t) \\ \quad + \alpha z(x,y,t) = f(t, z(x,y,t-\rho(t))), \ 0 < x < a_1, 0 < y < a_2, 0 < t < T, \\ z(x,y,t) = \phi(x,y,t), \ 0 < x < a_1, 0 < y < a_2, -\beta \le t \le 0, \\ z(0,y,t) = z(a_1,y,t) = 0, \ 0 < x < a_1, 0 < t < T, \\ z(x,0,t) = z(x,a_2,t) = 0, \ 0 < y < a_2, 0 < t < T, \end{cases} \tag{8.52}$$

where f is assumed to satisfy an appropriately modified version of **(H8.5)** - **(H8.7)**. Does (8.52) have a unique global mild solution?

When faced with such a problem, try to identify an abstract evolution equation of which it is a special case. Guided by the approach we have applied throughout the text, the following approach might be useful when studying (8.52).

1. Reformulate (8.52) as an abstract evolution equation.

2. Specify conditions under which to study the equation from Step 1.

3. Establish an existence-uniqueness result.

4. Apply the abstract result from Step 3 to (8.52).

Exercise 8.5.6. Argue that (8.52) has a unique local mild solution. When is it globally defined?

Similar results can be established in the same manner for neutral equations of the abstract form

$$\begin{cases} [u(t) + g(t, u_{\rho(t)})]' = A u(t) + f(t, u_{\rho(t)}), \ 0 < t < T, \\ u(t) = \varphi(t), \ -\beta \le t \le 0. \end{cases} \tag{8.53}$$

Open Exercise 8.5.7. Establish the analogs of the theoretical results from Sections 8.3 and 8.4 for (8.53) in the case when $\mathscr{X} = \mathbb{R}^N$. (The infinite-dimensional case for first and second-order equations is studied in [**20, 30, 163, 164**].)

___

8.6   Other Forms of Delay

Our discussion has merely scratched the surface when it comes to the study of delay equations. Aside from changing the assumptions imposed on φ (e.g., assuming that it belongs to a larger space, like $\mathbb{L}^2(-\beta,T;\mathscr{X})$), the system may depend differently on the past history than described above. We briefly mention some such possibilities here. You are encouraged to consult the cited references to gain a deeper understanding.

8.6.1   Unbounded Delay

Consider the following IBVP related to those discussed in [**165, 166**]:

$$\begin{cases} \frac{\partial}{\partial t}\left[u(z,t) + \int_{-\infty}^{t}\int_{0}^{L}k_1(y,z,s-t)f(u(y,s)dyds\right] = \frac{\partial^2}{\partial z^2}u(z,t) + k_2(z)u(z,t) \\ \quad + \int_{-\infty}^{t}k_3(s-t)u(z,s)ds + k_4(z,t), 0 < t < T, 0 < z < L, \\ u(0,t) = u(L,t) = 0, \ 0 < t < T, \\ u(z,t) = \phi(z,t), \ t \leq 0, 0 < z < L, \end{cases}$$

(8.54)

where k_i $(i = 1,2,3,4)$ are prescribed functions. The formulation of this IBVP as an abstract evolution equation in $\mathbb{L}^2(0,L)$ is very similar to (8.23), with the exception that the dependence on the past history is not limited to a finite time interval. Indeed, consider the abstract IVP

$$\begin{cases} u'(t) = Au(t) + F(t,u_t), 0 \leq t \leq T, \\ u(t) = \varphi(t), \ -\infty < t \leq 0. \end{cases}$$

(8.55)

Naturally, we seek a function $u \in C((-\infty,T];\mathscr{X})$ that satisfies the variation of parameters formula and $u_t \in C((-\infty,0];\mathscr{X})$. The idea is not to allow ϕ to grow "too large" since its influence will simply overwhelm the state $u(t)$. Indeed, in response, we restrict our attention to a particular subspace of $C((-\infty,0];\mathscr{X})$ (called a *phase space*) to which our IC must belong and in which the growth of $\phi(t)$ is controlled as $t \to -\infty$. There are different ways of choosing such a subspace. We mention two common ones below.

1. For any continuous function $h : (-\infty,0] \to (0,\infty)$, the space

$$\mathfrak{C}_h = \left\{ \phi \in C((-\infty,0];\mathscr{X}) \,\middle|\, \lim_{s \to -\infty}\frac{\|\phi(s)\|_{\mathscr{X}}}{h(s)} = 0 \right\}$$

(8.56)

equipped with the norm

$$\|\phi\|_{\mathfrak{C}_h} = \sup_{-\infty < s \leq 0} \frac{\|\phi(s)\|_{\mathcal{X}}}{h(s)} \qquad (8.57)$$

is complete. (See [171] for details.)

2. For any $\lambda \geq 0$, the space

$$\mathfrak{C}_\lambda = \left\{ \phi \in \mathcal{C}((-\infty, 0]; \mathcal{X}) \,\Big|\, \exists \lim_{s \to -\infty} e^{\lambda s} \|\phi(s)\|_{\mathcal{X}} \right\} \qquad (8.58)$$

equipped with the norm

$$\|\phi\|_{\mathfrak{C}_h} = \sup_{-\infty < s \leq 0} e^{\lambda s} \|\phi(s)\|_{\mathcal{X}} \qquad (8.59)$$

is complete. (See [282] for details.)

Consider the following analog of Thrm. 8.3.2.

Proposition 8.6.1. *Assume* $(\mathbf{H_A})$, $\phi \in \mathfrak{C}_\lambda$, *and*
$(\mathbf{H8.8})$ $f : [0, T] \times \mathfrak{C}_\lambda \to \mathcal{X}$ *is continuous in the first variable and*

$$\|f(t, x) - f(t, y)\|_{\mathcal{X}} \leq e^{\lambda t} \|x - y\|_{\mathfrak{C}_\lambda}$$

globally on $\mathfrak{C}_\lambda$ *(uniformly in* t*).*
Then, (8.55) has a unique global mild solution on $(-\infty, T]$.

Exercise 8.6.1. Prove Prop. 8.6.1.

Exercise 8.6.2. Establish a result analogous to Prop. 8.6.1 in which $\mathfrak{C}_\lambda$ is replaced by $\mathfrak{C}_h$.

Exercise 8.6.3. Formulate and prove a result for (8.55) in which non-Lipschitz conditions are imposed on f.

Exercise 8.6.4. Formulate and prove results analogous to Prop. 8.6.1 and Exer. 8.6.3 in the case of a variable delay.

The applicability of such theoretical results to IBVPs like (8.54) is discussed in several articles and books, including [20, 58, 157, 159, 171, 181, 218].

8.6.2   State-Dependent Delay

Consider the following IBVP discussed in [**165, 166**]:

$$
\begin{cases}
\frac{\partial}{\partial t}\left[u(z,t) + \int_{-\infty}^{t} k_1(z,s-t)u(z,s)ds\right] = \frac{\partial^2}{\partial z^2}\left[u(z,t) + \int_{-\infty}^{t} k_1(z,s-t)u(z,s)ds\right] \\
\quad + \int_{-\infty}^{t} k_2(s-t)u(z,s-\rho_1(t)\rho_2(|u(z,t)|))ds,\, 0 < t < T, 0 < z < L, \\
u(0,t) = u(L,t) = 0,\, 0 < t < T, \\
u(z,t) = \phi(z,t),\, t \le 0, 0 < z < L,
\end{cases}
$$

$$(8.60)$$

where k_1, k_2, ρ_1, ρ_2 are given functions. This problem can be reformulated as the abstract evolution equation

$$
\begin{cases}
u'(t) = Au(t) + F(t, u(t - u(t))),\, 0 \le t \le T, \\
u(t) = \varphi(t),\, -\infty < t \le 0.
\end{cases}
$$

$$(8.61)$$

In this case the state process itself depends on a time delay which, in turn, depends on the current value of the state. (Interpret this in terms of an application.) Here, we cannot determine *a priori* a particular value of β for which the history dependence requires knowledge of only $[-\beta, 0]$. (Why?) As such, we automatically consider the problem to be one with unbounded delay. Recent related work is established in [**46, 165, 166**].

8.7   Models – New and Old

We now revisit some earlier models and introduce some new ones.

Model XI.3 Pollution Model with Variable Delay

Let Ω be a bounded region in $\mathbb{R}^N$ with smooth boundary $\partial\Omega$, and let $w(\mathbf{x}, t)$ denote the pollution concentration at position $\mathbf{x} \in \Omega$ and time $t > 0$. We consider the following generalization of (6.47):

$$
\begin{cases}
\frac{\partial z}{\partial t}(\mathbf{x}, t) = k\triangle z(\mathbf{x}, t) + \overrightarrow{\alpha} \cdot \nabla z(\mathbf{x}, t) \\
\quad + \sum_{i=1}^{n} \int_{0}^{t} (t-s)^2 \sin(s, \mathbf{x}, z(\mathbf{x}, s - \sigma_i(s))) ds,\, \mathbf{x} \in \Omega, 0 < t < T, \\
z(\mathbf{x}, t) = \phi(\mathbf{x}, t),\, \mathbf{x} \in \Omega, -\beta^\star \le t \le 0, \\
\frac{\partial z}{\partial \mathbf{n}}(\mathbf{x}, t) = 0,\, \mathbf{x} \in \partial\Omega, t > 0,
\end{cases}
$$

$$(8.62)$$

where $\beta^\star > 0$ is completely determined by $\{\sigma_i : i, \ldots, n\}$. (See Model XI.2 for other contextual details.)

Exercise 8.7.1.
i.) Formulate and prove an existence-uniqueness result for (8.62).

ii.) Establish a continuous dependence result for (8.62).

Model XII.2 Epidemiological Model with Volterra Delay

It is natural to account for delay in an epidemiological model for many reasons. Consider the following generalization of IBVP (6.48):

$$
\begin{cases}
\frac{\partial}{\partial t}\begin{bmatrix} P_H \\ P_V \end{bmatrix} = \begin{bmatrix} \alpha_H \triangle + \beta_H & 0 \\ 0 & \alpha_V \triangle - \gamma_V \end{bmatrix}\begin{bmatrix} P_H \\ P_V \end{bmatrix} \\
\quad + \begin{bmatrix} -(P_H)_t\,(P_V)_t \int_{-\sigma}^{0} \frac{\eta\,(P_H)_s(x,y,\cdot)(P_V)_s(x,y,\cdot)}{((P_H)_s(x,y,\cdot)+(P_V)_s(x,y,\cdot))^2}ds \\ (P_H)_t\,(P_V)_t \int_{-\sigma}^{0} \frac{\eta\,(P_H)_s(x,y,\cdot)(P_V)_s(x,y,\cdot)}{((P_H)_s(x,y,\cdot)+(P_V)_s(x,y,\cdot))^2}ds \end{bmatrix}, (x,y) \in \Omega, 0 < t < T, \\
\begin{bmatrix} P_H \\ P_V \end{bmatrix}(x,y,t) = \begin{bmatrix} \phi_1 \\ \phi_2 \end{bmatrix}(x,y,t),\ (x,y) \in \Omega, -\sigma \le t \le 0, \\
\frac{\partial}{\partial \mathbf{n}}\begin{bmatrix} P_H \\ P_V \end{bmatrix}(x,y,t) = 0,\ (x,y) \in \partial\Omega, t > 0.
\end{cases}
$$

(8.63)

(See Model XII.1 for contextual details.)

Exercise 8.7.2.
i.) Formulate and prove an existence-uniqueness result for (8.63).
ii.) Establish a continuous dependence result for (8.63).

Model XVII.1 Diffusion Revisited - Effects of Random Motility on a Bacteria Population

Lauffenburger, in 1981, suggested a model for the effects of random motility on a bacterial population that consume a diffusible substrate (see **[208]**). A three-dimensional version of this model in a bounded region $\Omega \subset \mathbb{R}^3$ with smooth boundary $\partial\Omega$ is described by the following IBVP:

$$
\begin{cases}
\frac{\partial B}{\partial t} = \alpha \triangle B + (G(S) - K_D)\,B(x,y,z,t - \sigma(t)),\ (x,y,z) \in \Omega,\ 0 < t < T, \\
\frac{\partial S}{\partial t} = \beta \triangle S - \frac{G(S)}{M}B(x,y,z,t - \sigma(t)),\ (x,y,z) \in \Omega,\ 0 < t < T, \\
B(x,y,z,t) = S(x,y,z,t) = 0,\ (x,y,z) \in \partial\Omega,\ 0 < t < T, \\
B(x,y,z,t) = \phi_1(x,y,z,t),\ (x,y,z) \in \Omega,\ -\beta \le t \le 0, \\
S(x,y,z,t) = \phi_2(x,y,z,t),\ (x,y,z) \in \Omega,\ -\beta \le t \le 0,
\end{cases}
$$

(8.64)

where $\sigma : [0,T] \to [0,T]$ is a continuous, increasing function. Here,
$B = B(x,y,z,t) =$ density of bacteria population at position $(x,y,z) \in \Omega$ at time t,
$S = S(x,y,z,t) =$ substrate concentration at position $(x,y,z) \in \Omega$ at time t,
$M =$ mass of bacteria per unit mass of nutrient,
$K_D =$ bacterial death rate in the absence of the substrate,
$G(S) =$ substrate-dependent growth rate.

We can rewrite (8.64) equivalently as

$$
\begin{cases}
\frac{\partial}{\partial t}\begin{bmatrix} B \\ S \end{bmatrix} = \begin{bmatrix} \alpha\triangle & -K_D \\ 0 & \beta\triangle \end{bmatrix}\begin{bmatrix} B \\ S \end{bmatrix} + \begin{bmatrix} G(S)B(x,y,z,t-\sigma(t)) \\ -\frac{G(S)}{M}B(x,y,z,t-\sigma(t)) \end{bmatrix}, \\[2mm]
\begin{bmatrix} B \\ S \end{bmatrix}(x,y,z,t) = \begin{bmatrix} \phi_1 \\ \phi_2 \end{bmatrix}(x,y,z,t),\ (x,y,z)\in\Omega,\ -\beta\le t\le 0, \\[2mm]
\begin{bmatrix} B \\ S \end{bmatrix}(x,y,z,t) = 0,\ (x,y,z)\in\partial\Omega,\ 0<t<T.
\end{cases}
\tag{8.65}
$$

where $(x,y,z)\in\Omega$ and $0<t<T$. In order to reformulate (8.65) as the abstract evolution equation (8.39) in the space $\mathscr{X} = \mathbb{L}^2(\Omega)\times\mathbb{L}^2(\Omega)$, identify the unknown as $U = \begin{bmatrix} B \\ S \end{bmatrix}$, let $\Phi = \begin{bmatrix} \phi_1 \\ \phi_2 \end{bmatrix}$, and define the forcing term $F : [0,T]\times\mathbb{C}_\beta \to \mathscr{X}$ by

$$
F\left(t,\begin{bmatrix} B \\ S \end{bmatrix}\right) = \begin{bmatrix} G(S)B(x,y,z,t-\sigma(t)) \\ -\frac{G(S)}{M}B(x,y,z,t-\sigma(t)) \end{bmatrix}.
\tag{8.66}
$$

Also, define the operator $A : \mathrm{dom}(A)\subset\mathscr{X}\to\mathscr{X}$ by

$$
A = \begin{bmatrix} \alpha\triangle & 0 \\ 0 & \beta\triangle \end{bmatrix},\quad \mathrm{dom}(A) = \mathrm{dom}(\triangle)\times\mathrm{dom}(\triangle),
\tag{8.67}
$$

and $B : \mathrm{dom}(B)\subset\mathscr{X}\to\mathscr{X}$ by

$$
B = \begin{bmatrix} 0 & -K_D \\ 0 & 0 \end{bmatrix},\quad \mathrm{dom}(B) = \mathscr{X}.
\tag{8.68}
$$

It is known that A generates a C_0−semigroup on $\mathscr{X}$ and that $B\in\mathbb{B}(\mathscr{X})$. (Why?) In view of the above identifications, (8.65) is a special case of (8.39).

Exercise 8.7.3. Assume that $G : [0,\infty)\to[0,\infty)$ is continuous.
i.) Use Cor. 8.3.4 to show that (8.65) has a unique global mild solution.
ii.) If $G\in\mathbb{C}^2((0,\infty);[0,\infty))$, must the mild solution be classical?
iii.) If $G(S)$ were replaced by $G(S_{\rho(t)})$, how would the results of (i) and (ii) change?

Exercise 8.7.4. Consider a sequence of IBVPs of the form (8.65) in which σ is replaced by σ_n. Assume that $G : [0,\infty)\to[0,\infty)$ is continuous and $\forall n\in\mathbb{N}$, denote the mild solution by $\begin{bmatrix} B_n \\ S_n \end{bmatrix}$. If $\lim\limits_{n\to\infty}\|\sigma_n - \sigma\|_C = 0$, prove that $\exists\begin{bmatrix} B^\star \\ S^\star \end{bmatrix}\in\mathbb{C}_\beta$ such that

$$
\lim_{n\to\infty}\left\|\begin{bmatrix} B_n \\ S_n \end{bmatrix} - \begin{bmatrix} B^\star \\ S^\star \end{bmatrix}\right\|_{\mathbb{C}_\beta} = 0.
$$

Model XIX.1 Crab Population Model

Suppose that the following conditions are taken into account when modeling the crab population $C = C(x,y,z,t)$ in a local bay, represented by $\Omega\subset\mathbb{R}^3$:

1. Logistic growth with delay: $\beta C_{t_1}\left(1 - \frac{C_{t_1}}{K}\right)$, where $\beta > 0$, K is the carrying capacity, and $t_1 = t - r_1$;

2. Diffusive dispersion throughout the bay: $\alpha \triangle C$, where $\alpha > 0$;

3. Constant harvesting with success rate h: $-hC_{t_2}$, where $h > 0$ and $t_2 = t - r_2$;

4. Time-dependent ill-environmental seasonal effects (e.g., algae): $g(t)C_{t_3}$, where $t_3 = t - r_3$.

Accounting for all of these effects and assuming that there is no change in bacteria population along the boundary of the bay leads to the following IBVP describing the crab population:

$$\begin{cases} \frac{\partial C}{\partial t} = \alpha \triangle C + (g(t) - h)C_t + \beta C_t\left(1 - \frac{C_t}{K}\right), \ (x,y,z) \in \Omega, 0 < t < T, \\ C(x,y,z,t) = \phi(x,y,z,t), \ (x,y,z) \in \Omega, \ -\min\{r_1, r_2, r_3\} \le t \le 0, \\ \frac{\partial C}{\partial \mathbf{n}}(x,y,z,t) = 0, \ (x,y,z) \in \Omega, 0 < t < T. \end{cases} \qquad (8.69)$$

Exercise 8.7.5. Assume that ϕ is continuous and g is twice continuously differentiable. Show that (8.69) has a unique global classical solution.

Exercise 8.7.6. Establish a continuous dependence result for (8.69) in terms of the parameters h, α, β, K.

Model XX.1 The Sunflower Equation

Somolinas, in 1978, developed a model to describe the periodic movement of the tip of growing sunflowers as affected by the plant's growth factors along the stem (see **[84, 212, 224]**). Other researchers have studied the existence of periodic solutions and the stability of solutions of the resulting IVP. We shall focus on the second-order IVP

$$\begin{cases} \theta''(t) + \frac{a}{\sigma}\theta'(t) + \frac{b}{\sigma}\sin\left(\theta(t - \sigma)\right) = 0, t > 0, \\ \theta(t) = \phi_1(t), \ -\sigma \le t \le 0, \\ \theta'(t) = \phi_2(t), \ -\sigma \le t \le 0, \end{cases} \qquad (8.70)$$

where $a \ge b > 0$ and $\theta(t) = $ the angle between the stem and the vertical axis at time t.

Exercise 8.7.7.
i.) Prove that (8.70) has a unique global mild solution.
ii.) Establish a continuous dependence result for (8.70).
iii.) Describe the behavior of the solutions of (8.70) $\theta(\cdot)$ as

$$\sup\left[\{|\phi_1(t)| : -\sigma \le t \le 0\} \cup \{|\phi_2(t)| : -\sigma \le t \le 0\}\right] \to 0.$$

Model XXI.1 Chemo Taxis

Substances, organisms, and collections of animals (such as bacteria and schools of fish) tend to travel in the direction dictated by a specific gradient (chemically-driven or otherwise) which coincides with the direction of maximum increase in food supply, optimal temperature, pheromone concentration, etc. The equation governing the population concentration of such substances or organisms cannot be simply diffusion-based due to the existence of such directive gradients. However, the concentration of the attractant is subject to diffusion. As such, the description of this scenario requires two equations. For simplicity, we begin our study of a one-dimensional version without delay. To this end, let

$p(x,t)$ =density of the underlying population being attracted (e.g., by a pheromone),

$c(x,t)$ = concentration level of the pheromone.

Consider the following IBVP:

$$
\begin{cases}
\frac{\partial p}{\partial t}(x,t) = \alpha_p \frac{\partial^2 p}{\partial x^2}(x,t) - \beta \frac{\partial}{\partial x}\left(p(x,t)\frac{\partial c}{\partial x}(x,t)\right), 0 < x < L, t > 0, \\
\frac{\partial c}{\partial t}(x,t) = \alpha_c \frac{\partial^2 c}{\partial x^2}(x,t) + f(t,x,c(x,t)), 0 < x < L, t > 0, \\
p(x,0) = p_0(x), c(x,0) = c_0(x), 0 < x < L, \\
\frac{\partial p}{\partial x}(0,t) = \frac{\partial p}{\partial x}(L,t) = \frac{\partial c}{\partial x}(0,t) = \frac{\partial c}{\partial x}(L,t) = 0, t > 0.
\end{cases}
\tag{8.71}
$$

Remarks.

1. The diffusivity constants α_p, α_c are assumed to be positive. They can be replaced by more realistic nonlinear versions $\frac{\partial}{\partial x}\left(\alpha_1(p)\frac{\partial p}{\partial x}\right)$ and $\frac{\partial}{\partial x}\left(\alpha_2(c)\frac{\partial c}{\partial x}\right)$, respectively. This is explored in Chapter 9.

2. The parameter $\beta > 0$ is a measure of attractivity of c and generally depends on c.

3. The second equation in (8.71) governs the signal emitted by the pheromone. Usually, it also involves a perturbation involving $p(x,t)$, the presence of which renders the two equations more strongly coupled. For simplicity, we assume no such dependence upon p here, but refer you to **[168, 169, 254, 328]** for other analyses in this case.

4. The forcing term f describes the kinetics of the process.

How do we attack this problem? Without using any context-specific knowledge, the space $\mathbb{L}^2(0,L)$ would seem to be a reasonable space to use for both p and c when reformulating the problem as an abstract evolution equation. (Why?) Further, the independence of the second equation of p suggests that a viable approach might be to first solve the IBVP consisting of the second equation in (8.71), together with its IC and BC, and then substitute the result into the first equation in (8.71) and solve the resulting IBVP. Seems reasonable, right? Well, note that even if we have $c(x,t)$, we must be able to prove the existence of a mild solution of an IBVP of the form

$$
\begin{cases}
\frac{\partial p}{\partial t}(x,t) = \alpha_p \frac{\partial^2 p}{\partial x^2}(x,t) + a(x,t)\frac{\partial p}{\partial x}(x,t) + b(x,t)p(x,t), 0 < x < L, t > 0, \\
p(x,0) = p_0(x), 0 < x < L, \\
\frac{\partial p}{\partial x}(0,t) = \frac{\partial p}{\partial x}(L,t) = 0, t > 0,
\end{cases}
$$

$$\tag{8.72}$$

which can be reformulated abstractly as

$$
\begin{cases}
X'(t) = AX(t), t > 0, \\
X(0) = X_0,
\end{cases}
\tag{8.73}
$$

in a Hilbert space $\mathscr{H}$, where the operator $A : \text{dom}(A) \subset \mathscr{H} \to \mathscr{H}$ is given by

$$A = c_1 \frac{\partial^2}{\partial x^2} + \underbrace{c_2(x)\frac{\partial}{\partial x} + c_3(x)}_{\text{Call this } \mathscr{M}(x)} = c_1 \frac{\partial^2}{\partial x^2} + \mathscr{M}(x). \tag{8.74}$$

It can be shown that the operator $\mathscr{M} : \text{dom}(A) \subset \mathscr{H} \to \mathscr{H}$ is A-bounded whenever c_2, c_3 are continuous functions (see **[142]**). Hence, we can conclude that A satisfies $(\mathbf{H_A})$, so that (8.73) has a unique global mild solution. (Why?)

We can incorporate a variable delay naturally into (8.71), as well as impose non-Lipschitz conditions (in the spirit of Exer. 8.4.3) on the forcing term f. For instance, consider the following IBVP:

$$\begin{cases} \frac{\partial p}{\partial t}(x,t) = \alpha_p \frac{\partial^2 p}{\partial x^2}(x,t) - \beta \frac{\partial}{\partial x}\left(p(x,t)\frac{\partial c}{\partial x}(x,t)\right) \\ \quad + g\left(t,x,c(x,t-\sigma(t)),p(x,t-\sigma(t))\right), 0 < x < L, t > 0, \\ \frac{\partial c}{\partial t}(x,t) = \alpha_c \frac{\partial^2 c}{\partial x^2}(x,t) + f\left(t,x,c(x,t-\sigma(t))\right), 0 < x < L, t > 0, \\ p(x,0) = p_0(x), c(x,0) = c_0(x), 0 < x < L, \\ \frac{\partial p}{\partial x}(0,t) = \frac{\partial p}{\partial x}(L,t) = \frac{\partial c}{\partial x}(0,t) = \frac{\partial c}{\partial x}(L,t) = 0, t > 0. \end{cases} \tag{8.75}$$

Exercise 8.7.8. Formulate and prove an existence-uniqueness result for mild solutions of (8.75).

Model XXII.1 Some Important Equations From Mathematical Physics

PDEs arise in the study of physical phenomena, many of which can be studied under the parlance of the theory developed in this text. We provide a brief encounter with five such equations. More general perturbations of them, with and without delay, can be studied using the techniques developed thus far. Indeed, the versions mentioned below can all be reformulated abstractly as (6.31). We refer you to **[13, 19, 56, 91, 104, 117, 255, 279, 306, 342]** for detailed analyses of these, and related, equations of mathematical physics.

1. Burger's Equation

The following equation is an elementary quasilinear diffusion equation arising in the mathematical modeling of fluid dynamics, magneto-hydrodynamics, and traffic flow.

$$\begin{cases} \frac{\partial u}{\partial t}(x,t) + u(x,t)\frac{\partial u}{\partial x}(x,t) - \gamma \frac{\partial^2 u}{\partial x^2}(x,t) = 0, 0 < x < L, t > 0, \\ u(x,0) = \cos(2x), 0 < x < L, \\ u(0,t) = u(L,t) = 0, t > 0, \end{cases} \tag{8.76}$$

where $\gamma > 0$.

Exercise 8.7.9. Use the elementary substitution $u = -2\gamma\frac{\partial z/\partial x}{z}$ in (8.76) to transform the IBVP into a basic diffusion equation. Then, show that the transformed IBVP has a unique global mild solution, and subsequently determine an explicit representation formula for it.

2. Schrodinger Equations

These complex-valued PDEs arise in the study of quantum physics, specifically in the modeling of the dynamics of free particles in a bounded region Ω with smooth boundary $\partial\Omega$. The most basic form is given by

$$\begin{cases} \frac{\partial \psi}{\partial t}(\mathbf{x},t) - i\triangle\psi(\mathbf{x},t) = 0, \ \mathbf{x} \in \Omega, t > 0, \\ \psi(\mathbf{x},t) = 0, \ \mathbf{x} \in \partial\Omega, t > 0, \\ \psi(\mathbf{x},0) = \Psi_0(\mathbf{x}), \ \mathbf{x} \in \Omega. \end{cases} \tag{8.77}$$

Here, ψ is a complex-valued function expressed as $\psi = \psi_1 + i\psi_2$, where ψ_1 and ψ_2 are real-valued functions. Using the fact that

$$a + bi = c + di \text{ iff } a = c \text{ and } b = d,$$

we can formulate (8.77) as a system of two real-valued PDEs. (Tell how.) We can apply that same process to the following perturbed version of (8.77):

$$\begin{cases} \frac{\partial \psi}{\partial t}(\mathbf{x},t) - i\triangle\psi(\mathbf{x},t) = \int_0^t a(t-s)\psi(\mathbf{x},s)ds, \ \mathbf{x} \in \Omega, t > 0, \\ \psi(\mathbf{x},t) = 0, \ \mathbf{x} \in \partial\Omega, t > 0, \\ \psi(\mathbf{x},0) = \Psi_0(\mathbf{x}), \ \mathbf{x} \in \Omega, \end{cases} \tag{8.78}$$

where $a : [0,T] \to [0,\infty)$ is a continuous function.

Exercise 8.7.10.
i.) Prove that (8.78) has a unique mild solution on $[0,T]$, $\forall T > 0$.
ii.) Establish a continuous dependence result for (8.78).
iii.) Suppose that the right-side of (8.78) is replaced by $f(\mathbf{x},t, \psi(\mathbf{x},t))$, where f satisfied the non-Lipschitz conditions. Clearly indicate the modifications needed in the proof of Thrm. 5.6.3 to obtain analogous result for (8.78).
iv.) Repeat (iii) for the more general forcing term $f(\mathbf{x},t, \psi(\mathbf{x},t), \int_0^t a(t-s)\psi(\mathbf{x},s)ds)$?

Exercise 8.7.11. Incorporate a variable delay into the forcing term f in Exer. 8.7.10(iii) and formulate a result analogous to the one in Exer. 8.4.3 for (8.78).

3. Sine-Gordon Equation

This second-order PDE arises in the theory of semiconductors, lasers, and particle physics. Its most basic form in bounded region Ω with smooth boundary $\partial\Omega$ is given by:

$$\begin{cases} \frac{\partial^2 u}{\partial t^2}(\mathbf{x},t) + \alpha\frac{\partial u}{\partial t}(\mathbf{x},t) - \beta\triangle u(\mathbf{x},t) + \gamma\sin(u(\mathbf{x},t)) = 0, \ \mathbf{x} \in \Omega, t > 0, \\ u(\mathbf{x},t) = 0, \ \mathbf{x} \in \partial\Omega, t > 0, \\ u(\mathbf{x},0) = u_0(\mathbf{x}), \ \frac{\partial u}{\partial t}(\mathbf{x},0) = u_1(\mathbf{x}), \ \mathbf{x} \in \Omega. \end{cases} \tag{8.79}$$

More generally, we can view (8.79) as a special case of the semi-linear wave equation

$$\begin{cases} \frac{\partial^2 u}{\partial t^2}(\mathbf{x},t) + \alpha \frac{\partial u}{\partial t}(\mathbf{x},t) - \beta \triangle u(\mathbf{x},t) = F(\mathbf{x},t,u(\mathbf{x},t)), \ \mathbf{x} \in \Omega, \ t > 0, \\ u(\mathbf{x},t) = 0, \ \mathbf{x} \in \partial\Omega, \ t > 0, \\ u(\mathbf{x},0) = u_0(\mathbf{x}), \ \frac{\partial u}{\partial t}(\mathbf{x},0) = u_1(\mathbf{x}), \ \mathbf{x} \in \Omega. \end{cases} \tag{8.80}$$

IBVP (8.80) can be expressed as an equivalent system via the following substitution:

$$v = \frac{\partial u}{\partial t}, \ \frac{\partial v}{\partial t} = -\alpha v + \beta \triangle u - F(u). \tag{8.81}$$

Indeed, using (8.81) then enables us to rewrite (8.80) as

$$\begin{cases} \frac{\partial}{\partial t} \begin{bmatrix} u \\ v \end{bmatrix} = \begin{bmatrix} v \\ -\alpha v + \beta \triangle u - F(u) \end{bmatrix} = \begin{bmatrix} 0 & I \\ \beta\triangle & -\alpha I \end{bmatrix} \begin{bmatrix} u \\ v \end{bmatrix} + \begin{bmatrix} 0 \\ F(u) \end{bmatrix} \\ \begin{bmatrix} u \\ v \end{bmatrix}(0) = \begin{bmatrix} u_0 \\ u_1 \end{bmatrix} \end{cases} \tag{8.82}$$

in the space $\mathbb{L}^2(\Omega) \times \mathbb{L}^2(\Omega)$.

Exercise 8.7.12. Prove that if F is globally Lipschitz, then (8.82), and hence (8.79), has a global mild solution on $[0,T]$, $\forall T > 0$.

4. Klein-Gordon Equation
This second-order complex-valued PDE arises in quantum theory and the study of nonlinear dispersion. Its most basic form is given by

$$\begin{cases} \frac{\partial^2 u}{\partial t^2}(\mathbf{x},t) - \alpha \triangle u(\mathbf{x},t) + \beta u(\mathbf{x},t) = 0, \ \mathbf{x} \in \Omega, \ t > 0, \\ u(\mathbf{x},t) = 0, \ \mathbf{x} \in \partial\Omega, \ t > 0, \\ u(\mathbf{x},0) = u_0(\mathbf{x}), \ \frac{\partial u}{\partial t}(\mathbf{x},0) = u_1(\mathbf{x}), \ \mathbf{x} \in \Omega, \end{cases} \tag{8.83}$$

where $u = u_1 + iu_2$. The term βu is usually taken be the derivative of the potential function and so takes on a more complicated form in general.

Exercise 8.7.13.
i.) Express the second-order PDE (8.83) as a system of two real-valued PDEs, and rewrite the ICs and BCs in a suitable manner. Then, prove that (8.83) has a unique global mild solution on $[0,T]$, $\forall T > 0$.
ii.) Equip (8.83) with a forcing term $F(\mathbf{x},t,u(\mathbf{x},t), \frac{\partial u}{\partial t}(\mathbf{x},t))$. Formulate existence-uniqueness results for the newly-formed IBVP under Lipschitz and non-Lipschitz growth conditions.
iii.) Now, incorporate a constant delay into the forcing term in (ii), and formulate corresponding existence-uniqueness results for the resulting IBVP.
iv.) How do the results in (iii) change if a variable delay is incorporated instead?

5. Cahn-Hillard Equation

This PDE arises in the study of pattern formation in materials undergoing phase transitions, especially in alloys and glass. A version of the model is given by

$$\begin{cases} \frac{\partial u}{\partial t}(\mathbf{x},t) - \triangle \left(\alpha \triangle u(\mathbf{x},t)\right) + \triangle \left(\beta u(\mathbf{x},t) + \gamma u^3(\mathbf{x},t)\right) = 0, \mathbf{x} \in \Omega, t > 0, \\ \frac{\partial u}{\partial \mathbf{n}}(\mathbf{x},t) = 0, \mathbf{x} \in \partial\Omega, t > 0, \\ u(\mathbf{x},0) = u_0(\mathbf{x}), \mathbf{x} \in \Omega, \end{cases} \tag{8.84}$$

where $\alpha, \beta, \gamma > 0$.

Open Exercise 8.7.14.
i.) Use the techniques used in the study of the beam equation (cf. Model XIV.1) to reformulate (8.84) as an abstract evolution equation.
ii.) Prove the existence and uniqueness of a mild solution on $[0,T]$, $\forall T > 0$.
iii.) Try to formulate and prove results for the various delay versions of (8.84).

Remark. We have presented several standard models in the text, but there are many, many more. We mention some additional topics, with references, below. The theory developed thus far can be used to study some models arising in these areas. You are strongly encouraged to explore them at will!

1. Forced elongation (like spider web emission): **[154, 155, 239, 268]**
2. Fluid dynamics: **[40, 99]**
3. Fragmentation and mass loss: **[18]**
4. Population dynamics: **[85, 189, 243, 326]**
5. Oceanography and geotropic movement: **[184]**
6. Solitary waves: **[269]**
7. Lateral inhibition in sensory systems: **[180]**

8.8 An Important Look Back!

Before moving onto the last leg of our journey, we pause to take a critical look at the theoretical development thus far and examine the necessity of the crucial underlying hypotheses. The following exercises serve as preparation for our study of nonlinear evolution equations in the next chapter.

Exercise 8.8.1. (A Look Back - The Assumption of Linearity)
Revisit Chapters 2 - 8 and identify those places where the linearity of either the operator A or the semigroup it generated was used. Also, make particular note of those instances in which the commutativity of a semigroup and its generator was needed.

Exercise 8.8.2. (A Look Back - Model Development)

It is instructive to trace back through the development of the various models discussed thus far. Pay particular attention to how the theory developed in each new chapter enabled us to incorporate additional complexity into the models, thereby rendering more accurate descriptions of the underlying phenomena.

8.9 Looking Ahead

The linearity of the operator A is a central assumption in the theory of abstract linear evolution equations. All models considered thus far were able to be reformulated abstractly as such an evolution equation. In many ways, the theory presented is an idealized view of the true behavior of the phenomena under investigation. Indeed, one can take into account more realistic features of the phenomena which result in severe changes to the IBVP. For instance, consider the following IBVP which is a generalization of the pollution model (6.47):

$$\begin{cases} \frac{\partial z}{\partial t}(\mathbf{x},t) = k(z(\mathbf{x},t))\triangle z(\mathbf{x},t) + \overrightarrow{\alpha} \cdot \nabla z(\mathbf{x},t) + \int_0^t a(t-s)g(s,z(\mathbf{x},s))ds, \ \mathbf{x} \in \Omega, t > 0, \\ z(\mathbf{x},0) = z_0(\mathbf{x}), \ \mathbf{x} \in \Omega, \\ \frac{\partial z}{\partial \mathbf{n}}(\mathbf{x},t) = 0, \ \mathbf{x} \in \partial\Omega, t > 0. \end{cases}$$

$$(8.85)$$

Reformulating this IBVP as the abstract functional evolution equation (6.31) requires that we identify the operator $A : \text{dom}(A) \subset \mathscr{X} \to \mathscr{X}$ as

$$A = k(\cdot)\triangle(\cdot) + \overrightarrow{\alpha} \cdot \nabla(\cdot).$$

The only change is that the diffusion operator $k\triangle z$ used in (6.47) has been replaced by the more general operator $k(z)\triangle z$, where k is a continuous function. But, one very notable problem is that this operator is NOT linear! As such, the theory developed thus far cannot be used directly to conclude that the corresponding evolution equation (6.31) into which (8.85) is transformed is well-defined, and it certainly does not help us to conclude the existence of a mild solution of (8.85). Now what?

The question is whether or not it is possible to develop a generalization of the theory that can handle abstract evolution equations of the form

$$\begin{cases} u'(t) = \mathscr{A}u(t) + \mathfrak{F}(u)(t), 0 < t < T, \\ u(0) = u_0, \end{cases}$$

$$(8.86)$$

in a Banach space $\mathscr{X}$, where the operator $\mathscr{A} : \text{dom}(\mathscr{A}) \subset \mathscr{X} \to \mathscr{X}$ is nonlinear. This turns out to be a very deep question, and the theory developed in response is much richer than what we have already experienced. We shall take a very brief tour of this world in the next chapter.

8.10   Guidance for Exercises

8.10.1   Level 1: A Nudge in a Right Direction

8.1.1. Compare each model to its previous version. Nearly the same identifications work, with the notable exception of the forcing term and IC. What new hurdles do you encounter?

8.1.2. This is more realistic because a system rarely reacts to an external force, or changes therein, instantaneously. There is usually a time delay before such effects are realized.

8.2.1. i.) What is the earliest time prior to $t = 0$ at which the state process must be defined?

ii.) You still want a variation of parameters formula, but you must also account for an IC of the form presented in (i).

8.2.2. i.) Do not remove its time dependence. Bound it above appropriately.

ii.) For (i), the same general idea applies. The variation of parameters formula will be a special case of one developed in this section.

8.2.3. i.) This is immediately evident $\forall t \in [-\alpha, T] \setminus \{0\}$. (Why?) As such, it is sufficient to confirm that there is no jump discontinuity at $t = 0$.

ii.) The right-side is of the form studied in Chapter 4, and is continuous. (Why? So what?)

iii.) No. Does $\exists \lim_{t \to 0^-} \mathbf{x}(t)$? How does this help?

iv.) Parts (i) and (ii) follow easily because α_i is assumed to be continuous. (Why?) The extension argument should be adjusted. (Tell how.)

8.2.4. This is a simple modification.

8.2.5. i.) This follows directly from the continuity of $\mathbf{x}$. Check using Def. 1.8.4(i).

ii.) Since $\mathbf{x}$ is continuous on $[-\beta, T]$, it must be uniformly continuous on $[-\beta, T]$. (Why? So what?)

8.2.6. Refer to Exer. 8.2.4.

8.3.1. Note that $\forall 0 \le t \le T$,

$$\left\| (x_\varphi)_t - (x_\psi)_t \right\|_{\mathbb{C}_\beta} = \sup_{-\beta \le \theta \le 0} \left\| (x_\varphi)_t (\theta) - (x_\psi)_t (\theta) \right\|_{\mathscr{X}}$$

$$= \sup_{-\beta \le \theta \le 0} \left\| x_\varphi(\theta + t) - x_\psi(\theta + t) \right\|_{\mathscr{X}}.$$

(Now what?)

8.3.2. What is true about the operator $A + B$? Can you handle the delay in (8.37) as a special case of $F(t, u_t)$ in (8.23)? (How? So what?)

8.3.3. Does $A + B$ generate a C_0-semigroup? What kind of operator is $\sum_{j=1}^{n} B_j$? (Now what?)

8.3.4. Reformulate (8.1) abstractly by combining the approaches used in Exer. 5.5.26 and Exer. 5.5.28 and identifying the forcing term correctly. What is the operator B ?

8.3.5. Use Exer. 5.5.30, in the "one wave" case, to formulate (8.4) abstractly, along with identifying the forcing term as one with multiple delay (like (8.19)). What conditions can be imposed to ensure that the forcing term is sufficiently regular so that Cor. 8.3.3 can be applied?

8.3.6. Use Exer. 6.5.23 and Exer. 6.5.27 to help you reformulate (8.5) abstractly.

8.3.7. Use Exer. 6.5.13. This time the delay is of Volterra-type.

8.3.8. As usual, subtract the variation of parameters formulae for (8.40) and (8.41). Let $\varepsilon > 0$. Show $\exists N \in \mathbb{N}$ such that

$$n \geq N \implies \|z_n(t) - y(t)\|_{\mathscr{X}} \leq \frac{\varepsilon}{e^{M_1^\star T}} + M_1^\star \int_0^t \|z_n(s) - y(s)\|_{\mathscr{X}}\, ds.$$

(Now what?)

8.3.9. Replace $(z_n)_t$ by $z_n(\rho(t))$, assuming that $\rho : [0,T] \to \mathbb{R}$ is a continuous function such that $-\beta \leq \rho(t) \leq t$, $\forall t$. The approach is essentially the same otherwise.

8.3.10. The main differences from earlier convergence results are:

(i) The functions arising in the IC are defined on $[-\beta, 0]$.

(ii) The semi-linear terms are now $f_n(t, (u_n)_t)$ and $f(t, u_t)$. How does this affect the type of convergence you have?

8.3.11. A moment's thought suggests that (8.42) can be viewed as a special case of (8.23). (So what?)

8.3.12. i.) Apply Cor. 8.3.4.

ii.) Use Exer. 7.1.1 to aid in the abstract formulation.

8.3.13. Consider the variation of parameters formula on $[-\beta, T]$ and show that

$$\|u(t)\|_{\mathscr{X}} \leq e^{-\omega t} K(t), \forall t \in [-\beta, T],$$

for an appropriate function K. (So what?)

8.3.14. i.) See Exer. 8.3.2 to help you decide which growth condition to impose.

ii.) Assume that ω_{ij} $(i, j = 1, \ldots, M)$ are as in earlier versions of the model and that g_j $(j = 1, \ldots, M)$ are globally Lipschitz. What else do you need?

8.4.2. The changes are not difficult to implement. Just use (8.35) and proceed step-by-step through the details in the proof of Thrm. 5.6.3.

8.4.3. We will need to adjust **(H8.5)(c)** and **(H8.6)(c)**. (How?) How about **(H8.7)**?

8.4.4. Show that (8.7) can be reformulated abstractly as (8.45). Modify Exer. 8.3.7. (How?)

8.5.1. Mimic the proofs of Thrm. 8.3.2 and Cor. 8.3.3 using the new variation of parameters formula.

8.5.2. i.) Reformulate (8.49) as (8.46) in the same space used for (7.5).

ii.) This is routine. Subtract the variation of parameters formulae and estimate.

8.5.3. i.) Rewrite the PDE in (8.50) as

$$\frac{\partial}{\partial t}\left[I + \beta \frac{\partial^2}{\partial x^2}\right] w(x,t) = \alpha \frac{\partial^2 w}{\partial x^2}(x,t) + g(t) \int_{-\sigma}^{0} a(\tau) w_\tau(x, \cdot)\, d\tau.$$

Now, proceed to reformulate (8.50) as (8.46).

ii.) Argue as in Exer. 8.5.2(ii). Refer to the hints provided for that exercise.

8.5.4. For every $n \in \mathbb{N}$, consider

$$\begin{cases} (Bu_n)'(t) = Au_n(t) + f(t, u_n(t)) + F_n(t, (u_n)_t), 0 \le t \le T, \\ u_n(t) = \varphi(t), -\beta \le t \le 0, \end{cases} \tag{8.87}$$

and

$$\begin{cases} (Bu)'(t) = Au(t) + f(t, u(t)), 0 \le t \le T, \\ u(0) = \varphi(0). \end{cases} \tag{8.88}$$

Make certain that the IC belongs to the appropriate space. What are the other hypotheses?

8.5.5. You cannot directly apply Exer. 8.5.4. (Why?)

8.5.6. This is similar to (7.6) with delay, except for the presence of the term $\alpha z(x, y, t)$. How do you handle this? Also, the delay here is variable. Does this affect the abstract formulation?

8.6.1. Adapt the proof of Thrm. 8.3.2. Note that $\sup \{ e^{\lambda t} | -\infty \le t \le T \} = e^{\lambda T}$. (So what?)

8.6.2. Modify **(H8.8)** to account for the new space. Is $\sup \left\{ \frac{\|\phi(s)\|_{\mathscr{X}}}{h(s)} | -\infty \le s \le T \right\} < \infty$?

8.6.3. The only significant change occurs in the growth conditions imposed in **(H5.7)** and **(H5.8)**. What is this change? What does it affect in the proof?

8.6.4. In the case of variable delay, replace the subscript t in x_t and $\|x\|_t$ by $\rho(t)$, and assume that ρ is increasing and continuous.

8.7.1. i.) Modify Exer. 6.5.8 to account for the delay term. Can you reformulate (8.62) as (8.42) by appropriately defining g and k?

ii.) Once you have reformulated the IBVP abstractly as (8.42), adapt (8.25) accordingly.

8.7.2. i.) Use (8.43).

ii.) Once you have reformulated the IBVP abstractly, the continuous dependence estimate is easily obtained (as in Thrm. 8.3.2).

8.7.3. i.) It remains to verify that the forcing term is globally Lipschitz. However, when doing so, it becomes evident that the space to which the mild solution should belong is $C([0, T]; \mathscr{B}_{\mathscr{X}})$, where $\mathscr{B}_{\mathscr{X}} = \{ z \in \mathscr{X} \, | \, \|z\|_{\mathscr{X}} \le R \}$. To show that the forcing term is globally Lipschitz, let $\begin{bmatrix} S_1 \\ B_1 \end{bmatrix}, \begin{bmatrix} S_2 \\ B_2 \end{bmatrix} \in (\mathbb{L}^2(\Omega))^2$. Observe that

$$\left\| F \begin{bmatrix} S_1 \\ B_1 \end{bmatrix} - F \begin{bmatrix} S_2 \\ B_2 \end{bmatrix} \right\|_{\mathscr{X}} = \|G(S_1)B_1 - G(S_2)B_2\|_{\mathbb{L}^2(\Omega)} + \left\| \frac{1}{M} [-G(S_1)B_1 - G(S_2)B_2] \right\|_{\mathbb{L}^2(\Omega)}.$$

Square both sides, use the definition of the $\mathbb{L}^2(\Omega)$−norm, and make hefty use of the triangle inequality to get the desired estimate.

ii.) Appeal to Def. 8.2.3(ii), as extended to the Banach space setting. Is the forcing term Frechet differentiable? Does it matter to which spaces ϕ_1 and ϕ_2 belong?

iii.) The abstract form (8.39) should be modified accordingly.

8.7.4. The approach is similar to the one used in Exer. 5.5.34 and Exer. 6.5.18.

8.7.5. Reformulate (8.69) as (8.23) in $\mathscr{X} = \mathbb{L}^2(\Omega)$. What is true about the forcing term?

8.7.6. Consider a variant of (8.69), where the parameters h, α, β, and K are replaced by $\overline{h}, \overline{\alpha}, \overline{\beta}$, and $\overline{K}$, respectively. Subtract the corresponding variation of parameters formulae and estimate.

8.7.7. i.) Rewrite (8.70) as a system in a manner similar to how we handled (2.6).

ii.) Consider the abstract formulation of (8.70) and use its variation of parameters formula as in Exer. 8.7.6, where the parameters of interest are a and b.

iii.) Refer to the variation of parameters formula.

8.7.8. Reformulate (8.75) as a system of the form (8.72). How are the functions $c_i(x)$ in (8.74) chosen?

8.7.9. Making the suggested substitution in (8.76) yields

$$\frac{\partial z}{\partial x}\left(\gamma\frac{\partial^2 z}{\partial x^2} - \frac{\partial z}{\partial t}\right) - z\frac{\partial}{\partial x}\left(\gamma\frac{\partial^2 z}{\partial x^2} - \frac{\partial z}{\partial t}\right) = 0.$$

(Now what?)

8.7.10. i.) Reformulate (8.78) as a system of PDEs:

$$\begin{cases} \frac{\partial \psi_1}{\partial t}(\mathbf{x},t) + \triangle\psi_2(\mathbf{x},t) = \int_0^t a(t-s)\psi_1(\mathbf{x},s)ds, \ \mathbf{x} \in \Omega, t > 0, \\ \frac{\partial \psi_2}{\partial t}(\mathbf{x},t) - \triangle\psi_1(\mathbf{x},t) = \int_0^t a(t-s)\psi_2(\mathbf{x},s)ds, \ \mathbf{x} \in \Omega, t > 0, \\ \psi_1(\mathbf{x},t) = \psi_2(\mathbf{x},t) = 0, \ \mathbf{x} \in \partial\Omega, t > 0, \\ \psi_1(\mathbf{x},0) = \Psi_0^1(\mathbf{x}), \ \psi_2(\mathbf{x},0) = \Psi_0^2(\mathbf{x}), \ \mathbf{x} \in \Omega. \end{cases} \quad (8.89)$$

(Now what?)

ii.) This is a special case of the semilinear evolution equation (5.10); so, the continuous dependence estimate has already been established. Interpret the abstract result using the particular functions in (8.89).

iii.) & iv.) I am leaving this as an open exercise. Refer to the discussion of the Schrodinger equation in **[19, 91, 279]**.

8.7.11. Similar, but now the form is (8.45).

8.7.12. The form of (8.82) is (5.10), so if we can guarantee that the operator $A : \text{dom}(A) \subset \left(\mathbb{L}^2(\Omega)\right)^2 \to \left(\mathbb{L}^2(\Omega)\right)^2$ defined by

$$A\begin{bmatrix} u \\ v \end{bmatrix} = \begin{bmatrix} 0 & I \\ \beta\triangle & -\alpha I \end{bmatrix}\begin{bmatrix} u \\ v \end{bmatrix}$$

generates a C_0−semigroup on $\left(\mathbb{L}^2(\Omega)\right)^2$, then the result will follow from Thrm. 5.5.2 because the mapping $(t,u) \mapsto \begin{bmatrix} 0 \\ F(u) \end{bmatrix}$ is globally Lipschitz.

8.7.13. i.) Let $u = u_1 + iu_2$. Then, (8.83) can be viewed as the system

$$\begin{cases} \frac{\partial^2 u_1}{\partial t^2}(\mathbf{x},t) - \alpha \triangle u_1(\mathbf{x},t) + \beta u_1(\mathbf{x},t) = 0, \ \mathbf{x} \in \Omega, t > 0, \\ \frac{\partial^2 u_2}{\partial t^2}(\mathbf{x},t) - \alpha \triangle u_2(\mathbf{x},t) + \beta u_2(\mathbf{x},t) = 0, \ \mathbf{x} \in \Omega, t > 0, \\ u_1(\mathbf{x},t) = u_2(\mathbf{x},t) = 0, \ \mathbf{x} \in \partial \Omega, t > 0, \\ u_1(\mathbf{x},0) = u_0^1(\mathbf{x}), \ u_2(\mathbf{x},0) = u_0^2(\mathbf{x}), \ \mathbf{x} \in \Omega, \\ \frac{\partial u_1}{\partial t}(\mathbf{x},0) = u_1^1(\mathbf{x}), \ \frac{\partial u_2}{\partial t}(\mathbf{x},0) = u_1^2(\mathbf{x}), \ \mathbf{x} \in \Omega. \end{cases} \qquad (8.90)$$

ii.) Once you have incorporated the forcing term into (8.90), the abstract form of the resulting system is (5.10). (Why?)

iii.) The abstract form is (8.23).

iv.) Adapt (8.45) accordingly.

8.10.2   Level 2: An Additional Thrust in a Right Direction

8.1.1. The main differences are:

(i) information about the state process is required at times prior to the fixed time $t = 0$ used in all earlier versions of the models, and

(ii) the state process is evaluated at times different from those appearing on the left-side of the equation.

This will require a change of interpretation and framework.

8.2.1. i.) The IC should be $x(t) = \varphi(t), t \in [\max \{\alpha_i | i = 1, \dots, n\}, 0]$.

ii.) Look ahead to (8.13) to get an idea about the variation of parameters formula. Details concerning regularity are developed throughout the section.

8.2.2. i.) How about $-\alpha \le \alpha_i(t) \le t, \forall t$.

ii.) Make certain to prescribe $x(t) = \varphi(t)$ on a time interval that accounts for all values of t for which $t - \alpha_i(t) \le 0, \forall i = 1, \dots, n$.

8.2.3. i.) The portion of (8.13) defined for $0 \le t \le T$, when evaluated at $t = 0$, is $\varphi(0)$.

ii.) Use Prop. 4.1.1.

iii.) Begin with $\mathbf{x}(t)$ defined on $[-\alpha, 0]$ since it is known. The continuity of φ guarantees the existence of $\lim_{t \to 0^-} \mathbf{x}(t)$, so that we can extend the domain of the solution to $[0, \alpha]$ (Why?), and then to $[\alpha, 2\alpha]$, etc. using a standard extension argument, until the solution is globally defined on $[-\alpha, T]$. This process can only generate one function.

iv.) You could proceed by computing

$$\max_{0 \le t \le T} \left\{ \max_{1 \le i \le n} \alpha_i(t) \right\} = \alpha^\star$$

and begin the extension on $[-\alpha^\star, 0]$. Does $\exists \lim_{t \to 0^-} \mathbf{x}(t)$? (Now what?)

8.2.4. Replace $\mathbf{x}_s(-\alpha_j)$ in (8.19) by $\mathbf{x}_s(-\alpha_j(s))$. Why does this make sense?

8.2.5. i.) Let $\theta \in [-\beta, 0]$ and consider any sequence $\{\theta_n\} \subset [-\beta, 0]$ such that $\theta_n \to \theta$. Observe that

$$\mathbf{x}_t(\theta_n) = \mathbf{x}(t + \theta_n) \to \mathbf{x}(t + \theta) = \mathbf{x}_t(\theta),$$

as needed.

ii.) Use Prop. 1.8.8. Let $\varepsilon > 0$. There exists $\delta > 0$ such that

$$|t - t^\star| < \delta \implies \|\mathbf{x}(t) - \mathbf{x}(t^\star)\|_{\mathbb{R}^N} < \varepsilon.$$

Now, conclude that $\forall \theta \in [-\beta, 0]$,

$$\|\mathbf{x}_t(\theta) - \mathbf{x}_{t^\star}(\theta)\|_{\mathbb{R}^N} = \|\mathbf{x}(t + \theta) - \mathbf{x}(t^\star + \theta)\|_{\mathbb{R}^N} < \varepsilon.$$

8.3.1. Compare this to (8.35). To obtain the estimate, use (8.33) and compute an appropriate sup.

8.3.2. $A + B$ generates a C_0–semigroup by Prop. 3.6.1. Use

$$F(t, u_t) = g(t, u_t(\beta_1), \dots, u_t(\beta_N)). \tag{8.91}$$

Show that this function satisfies **(H8.1)** and conclude that Cor. 8.3.3 follows as a consequence of Thrm. 8.3.2. The case of variable delay is handled by modifying (8.91) slightly (as in Exer. 8.2.4) and then verifying that **(H8.1)** holds. Assume that $\rho : [0, T] \to \mathbb{R}$ is increasing and $-\beta \le \rho(t) \le t$, $\forall t$. Impose a global Lipschitz-type growth condition of the form

$$\|\bar{g}(t, u(\rho(t))) - \bar{g}(t, v(\rho(t)))\|_{\mathscr{X}} \le M_{\bar{g}} \|u - v\|_{\rho(t)},$$

where $\|u\|_{\rho(t)} = \sup \{\|u(\theta)\|_{\mathscr{X}} \mid -\beta \le \theta \le \rho(t)\}$.

8.3.3. Since $A + B$ generates a C_0–semigroup on $\mathscr{X}$ by Thrm. 6.5.1 and $\sum_{j=1}^{n} B_j \in \mathbb{B}(\mathscr{X})$, it follows that $A + B + \sum_{j=1}^{n} B_j$ generates a C_0–semigroup on $\mathscr{X}$ by Prop. 3.6.1.

8.3.4. Let $B = 0$ and define the forcing term by $t \mapsto \begin{bmatrix} 0 \\ f(v(t - \sigma_1), w(t - \sigma_2)) \end{bmatrix}$, where $v = x, w = x'$. Identifying the solution of (8.1) as $\mathbf{u}(t) = \begin{bmatrix} v(t) \\ w(t) \end{bmatrix}$, we see that the mapping $t \mapsto g(t, \mathbf{u}(t - \sigma_1), \mathbf{u}(t - \sigma_2))$ can be identified as the forcing term. (Provide the details.) Assuming that f is globally Lipschitz, we can conclude from Cor. 8.3.3 that (8.1) has a unique mild solution.

8.3.5. Since $u(t) = z(\cdot, \cdots, t)$ and the forcing term is rewritten as

$$F(t, u_t) = \sum_{i=1}^{M} g_i(t, \cdot, \cdots, z_t(-\sigma_i)),$$

assuming that g_i $(i = 1, \dots, M)$ is continuous in all variables and Lipschitz in the fourth variable should work.

8.3.6. Incorporating the delay is now a matter of viewing the right-side of (8.5) as $F(t, u_t)$. (Tell how.)

8.3.7. Suitably apply the manner in which (8.20) was formulated as $F(s, x_s)$ to each component of the forcing term. In (8.7), note that the delay is variable. How do you handle this abstractly? Can you somehow use Exer. 8.2.4?

8.3.8. Use Gronwall's lemma to conclude that

$$n \ge N \implies \|z_n(t) - y(t)\|_{\mathscr{X}} \le \left(\frac{\varepsilon}{e^{M_1^\star T}} \right) e^{M_1^\star T} = \varepsilon, \ \forall 0 \le t \le T.$$

Now take sups to conclude.

8.3.10. Use $\|\varphi_n - \varphi\|_{\mathbb{C}_\beta} \to 0$ and $\|f_n(t,x) - f(t,x)\|_{\mathbb{C}_\beta} \to 0$ (uniformly in t). How do the estimates change? Can you still apply Gronwall's lemma?

8.3.11. Under the conditions of Prop. 8.3.6, it is not difficult to show that

$$f(t,u_t) = g\left(t, u_t, \int_0^t k(s,u_s)ds\right)$$

satisfies **(H8.1)**.

8.3.12. i.) $f(t,u_t) = f(t)\int_{-\beta}^0 a(\tau)u_\tau d\tau$ is globally Lipschitz because

$$\|f(t,u_t) - f(t,v_t)\|_{\mathscr{X}} \leq \sup_{0 \leq t \leq T} \|f(t)\|_{\mathscr{X}} \|a\|_{\mathbb{L}^1(-\beta,0)} \|u-v\|_{\mathbb{C}_\beta}, \forall u,v \in \mathbb{C}_\beta,$$

(uniformly in t). (Now what?)

8.3.13. Choose K so that Gronwall's lemma can be applied. Then, the result follows.

8.3.14. ii.) Remember that $e^{\mathbf{A}t}$ is contractive if all eigenvalues of $\mathbf{A}$ have negative real parts. What does this suggest that you should impose on a_i? (Why?) Now, carry out the details.

8.4.2. Note that $t \mapsto \|\cdot\|_t$ is increasing. (So what?)

8.4.3. Replace $t \mapsto \|\cdot\|_t$ by $t \mapsto \|\cdot\|_{\rho(t)}$. **(H8.7)** likely works, but trace through the details to be certain. Where do you use the fact that ρ is increasing and $\rho(t) \leq t$?

8.5.1. Look back at Exer. 7.1.15 for compatibility issues.

8.5.2. i.) Apply Prop. 8.5.2.

8.5.3. i.) Assume that g is continuous and $a \in \mathbb{L}^1(-\sigma,0)$. The argument is then similar to Exer. 8.5.2.

8.5.4. Assume **(H7.1)** - **(H7.6)**, **(H$_\varphi$)**, **(H8.3)**, and **(H8.4)**. The conclusion is that $\|u_n - u\|_{\mathbb{C}} \to 0$.

8.5.5. The presence of B_n prevents this. Try to use both Exer. 7.1.20 and Exer. 8.5.4 to form suitable hypotheses.

8.5.6. Treat αI has a bounded perturbation of $\left(\frac{\partial^2}{\partial x^2} + \frac{\partial^2}{\partial y^2}\right)$ and modify (7.25) to account for the delay. Treat the variable delay as you have in earlier exercises.

8.6.1. The approach is the same, but now the estimates (8.29) - (8.32) change slightly.

8.6.2. Yes, because of (8.57), together with the continuity of $\frac{\|\phi(\cdot)\|_{\mathscr{X}}}{h(\cdot)}$ on $[0,T]$. Now, the argument proceeds as in Exer. 8.6.1.

8.6.3. The growth condition in **(H5.7)** becomes

$$\|f(t,x_t)\|_{\mathscr{X}} \leq K_1\left(t, e^{\lambda t}\|x\|_t\right), \forall t \in [0,T], x \in \mathscr{X}$$

a similar change is made to **(H5.8)**. The constant α in Lemma 5.6.2 will change accordingly, but the proof is essentially unaffected.

8.7.1. i.) Yes. Notice that the integrand is the product of an $\mathbb{L}^1$−function and one that is Lipschitz. (Why?) Be certain to handle the multiple delay correctly; refer to Exer. 8.2.6.

8.7.2. i.) Look back at Exercises 6.5.10 - 6.5.13 to recall the properties of the logistic terms. Use this to show that the Volterra delay term behaves appropriately.

8.7.3. i.) Use Exer. 1.4.4(iii) with the Cauchy-Schwarz inequality to obtain

$$\left\| F\begin{bmatrix} S_1 \\ B_1 \end{bmatrix} - F\begin{bmatrix} S_2 \\ B_2 \end{bmatrix} \right\|_{\mathscr{X}}^2 \leq 2\left(1 + \frac{1}{M}\right) \int_{\Omega} \left[M_G^2 |S_1 - S_2|^2 |B_1^2| + \right.$$
$$\left. |G(S_2)|^2 |B_1 - B_2|^2 \right] dx$$
$$\leq \left(2\left(1 + \frac{1}{M}\right) M_G^2 \right) \left[\|(S_1 - S_2)B_1\|_{\mathbb{L}^2(\Omega)}^2 + \right.$$
$$\left. \|S_2(B_1 - B_2)\|_{\mathbb{L}^2(\Omega)}^2 + \|G(0)(B_1 - B_2)\|_{\mathbb{L}^2(\Omega)}^2 \right]$$
$$\vdots$$
$$\leq \left[2\left(R^2 + \|G(0)\|_{\mathbb{L}^2(\Omega)}^2 \right) \left(\left(1 + \frac{1}{M}\right) M_G^2 \right) \right] \cdot$$
$$\left\| \begin{bmatrix} S_1 \\ B_1 \end{bmatrix} - \begin{bmatrix} S_2 \\ B_2 \end{bmatrix} \right\|_{\mathscr{X}}^2.$$

ii.) Trace back through the results governing the existence of a classical solution in the semi-linear case. Modify them accordingly.

8.7.5. The forcing term

$$f(t, C_t) = (g(t) - h)C_t + \beta C_t \left(1 - \frac{C_t}{K}\right)$$

belongs to $\mathbb{C}^2((0,T); \mathscr{X})$. (Why? So what?)

8.7.6. The calculations resemble those from Thrm. 8.3.2.

8.7.7. i.) Show that the system form of (8.70) can be viewed abstractly as (8.22) and argue that the forcing term is differentiable.

ii.) Only the integral term remains. Compute this precisely using Prop. 2.2.3.

8.7.8. You will need to simplify the term $\frac{\partial}{\partial x}\left(p(x,t)\frac{\partial c}{\partial x}(x,t)\right)$ to aid in identifying the operator A in (8.74). Assume that g is globally Lipschitz in its spatial variables.

8.7.9. Note that if z satisfies $\gamma \frac{\partial^2 z}{\partial x^2} - \frac{\partial z}{\partial t} = 0$, then u given by the substitution satisfies (8.76). Moreover, z has a nice representation formula (see Section 3.2). This can be substituted into the given change of variable formula for u to obtain a solution formula for (8.76). Apply the BCs and ICs appropriately.

8.7.10. i.) This IBVP is of the form (5.10) in $\mathscr{X} = \left(\mathbb{L}^2(\Omega)\right)^2$, where

$$A\begin{bmatrix} \psi_1 \\ \psi_2 \end{bmatrix} = \begin{bmatrix} 0 & -\triangle \\ \triangle & 0 \end{bmatrix} \begin{bmatrix} \psi_1 \\ \psi_2 \end{bmatrix}.$$

Does this operator generate a C_0−semigroup on $\mathscr{X}$?

8.7.12. Refer to our discussion of the classical wave equation and the beam equation.

8.7.13. i.) The system can be reformulated abstractly as (3.132) in an appropriate space.

ii.) Now use Thrm. 5.5.2 and Thrm. 5.6.3 assuming appropriate restrictions on F.
iii.) Apply Thrm. 8.3.2 and Thrm. 8.4.1 assuming appropriate restrictions on F.

Chapter 9

Nonlinear Evolution Equations

Overview

Using a modest dose of linear semigroup theory, we were able to develop a rather rich existence theory that formed a theoretical basis for a formal mathematical study of vastly different phenomena. We introduced a significant amount of complexity into the IBVPs by way of time delays, perturbations, and complex forcing terms. But, as rich as the theory is, it all hinges on the crucial assumption that the operator $A : \mathrm{dom}(A) \subset \mathscr{X} \to \mathscr{X}$ is linear and generates a C_0-semigroup on $\mathscr{X}$. The problem is that these two assumptions do not hold for many phenomena, including the various improvements on the models discussed throughout the text. As such, the question is whether or not we can somehow argue analogously as we did when generalizing the setting of Chapter 2 to Chapter 3 to develop a theory in the so-called *nonlinear* case. The answer is a tentative yes, but the extension from the linear to the nonlinear setting takes place on a much grander scale than the generalization of the finite-dimensional to the infinite-dimensional setting in the linear case and requires a considerably higher degree of sophistication. We shall explore how this theory unfolds for some basic nonlinear models in this chapter.

9.1   A Wealth of New Models

We introduce some new models, including some nonlinear variants of previously-developed models. In each case, we allude to the identifications that enable us to express the IVP/IBVP as an abstract evolution equation of a form similar to those encountered earlier in the text. We postpone a formal analysis until later in the chapter.

Model XXIII.1 Nonlinear Mechanics

We have discussed increasingly more complex models of elementary spring-mass systems. A simple nonlinear variant of (2.6) arising in classical mechanics is

$$\begin{cases} x''(t) + \alpha x(t) |x(t)| = 0, \, t > 0, \\ x(0) = x_0, \, x'(0) = x_1, \end{cases} \tag{9.1}$$

where $\alpha > 0$. Proceeding as we did to arrive at (2.8), (9.1) can be viewed as the following equivalent system of first-order ODEs:

$$\begin{cases} \begin{bmatrix} w_1(t) \\ w_2(t) \end{bmatrix}' = \begin{bmatrix} 0 & 1 \\ -\alpha |w_1(t)| & 0 \end{bmatrix} \begin{bmatrix} w_1(t) \\ w_2(t) \end{bmatrix}, \, t > 0, \\ \begin{bmatrix} w_1(0) \\ w_2(0) \end{bmatrix} = \begin{bmatrix} x_0 \\ x_1 \end{bmatrix}. \end{cases} \tag{9.2}$$

Reformulating (9.2) abstractly as (2.10) requires that we identify the operator $\mathscr{A}$ as

$$\mathscr{A} \begin{bmatrix} w_1 \\ w_2 \end{bmatrix} = \begin{bmatrix} 0 & 1 \\ -\alpha |w_1| & 0 \end{bmatrix}, \tag{9.3}$$

which is no longer a constant matrix. As such, Def. 2.2.2 of the matrix exponential is inapplicable.

Exercise 9.1.1. Is $\mathscr{A}$ a linear operator? Explain.

Model XXIV.1 Nonlinear Diffusive Phenomena

We have encountered numerous phenomena whose mathematical description involved a diffusion term of the general form $\alpha \triangle$, where α is a positive constant. At its root, this operator arises from the premise that diffusion is governed by Fick's law (see [273]), which yields a natural, albeit linear, description of dispersion. But, more complicated phenomena are governed by more complex laws often resulting in nonlinear diffusivity. Indeed, it is often necessary to replace $\alpha \triangle u$ by a more general operator of the form $\triangle f(u)$, where $f : \mathbb{R} \to \mathbb{R}$ is a continuous, increasing function for which $f(0) = 0$. For instance, such a term with $f(u) = u |u|^{m-1}$, where $m > 1$, occurs in an IBVP arising in the study of porous media. Other variants arise in models in the context of differential geometry with Ricci flow (see [315]), a nonlinear model of brain tumor growth (see [153]), porous media (see [101]), and other diffusive processes (see [199, 200, 312, 326, 334]).

The most rudimentary IBVP involving this nonlinear diffusion operator is

$$\begin{cases} \frac{\partial u}{\partial t}(x,t) = \triangle f(u(x,t)), \, x \in \Omega, \, t > 0, \\ u(x,t) = 0, \, x \in \partial \Omega, \, t > 0, \\ u(x,0) = u_0(x), \, x \in \Omega. \end{cases} \tag{9.4}$$

Given that the IBVP is homogenous, it is natural to reformulate (9.4) as an abstract evolution equation of a form similar to (3.87). This requires us to identify the operator $\mathscr{A} : \text{dom}(\mathscr{A}) \subset \mathscr{X} \to \mathscr{X}$ (for an appropriate space $\mathscr{X}$) as

$$\mathscr{A} u = \triangle f(u)$$

which is not linear unless $f(u) = au + b$, the typical *linear* diffusion operator. As such, the theory in Chapter 3 and all subsequent results are inapplicable to (9.4).

Model XXV.1 Hydrology and Groundwater Flow

The PDE governing one-dimensional lateral groundwater flow, referred to as the *Boussinesq equation* (discussed in [44, 290]) is given by

$$\frac{1}{S}\frac{\partial}{\partial x}\left(Kh(x,t)\frac{\partial h}{\partial x}(x,t)\right) = \frac{\partial h}{\partial t}(x,t), \ 0 \leq x \leq L, t > 0, \tag{9.5}$$

where the aquifer is modeled as the interval $[0,L]$, h is the hydraulic head, K is the hydraulic conductivity, and S is the specific yield. Assuming that there is no replenishment of water via rainfall by seepage through the soil surrounding the aquifer, (9.5) can be coupled with the following BCs

$$h(0,t) = M(t), \ \frac{\partial h}{\partial x}(L,t) = 0, \tag{9.6}$$

so that the aquifer is replenished at the end $x = 0$ and experiences no change at $x = L$.

Exercise 9.1.2. Why can the IBVP (9.5)-(9.6) not be subsumed as a special case of one of the evolution equations studied thus far in the text, for any choice of IC?

Model XXVI.1 Nonlinear Waves

We have accounted for linear dissipation in the models of wave phenomena via the inclusion of the term $\alpha\frac{\partial z}{\partial t}$, where $\alpha > 0$, in the PDE, where z represents the unknown displacement function. What if the dissipation is governed by a *nonlinear* function of $\frac{\partial z}{\partial t}$? For instance, consider the IBVP

$$\begin{cases} \frac{\partial^2 z}{\partial t^2} + \alpha\frac{\partial z}{\partial t} + \beta\left(\frac{\partial z}{\partial t}\right)^3 - c^2\frac{\partial^2 z}{\partial x^2} \\ = \int_0^t a(t-s)\sin(z(x,s))ds, \ 0 < x < L, t > 0, \\ z(x,0) = z_0(x), \ \frac{\partial z}{\partial t}(x,0) = z_1(x), \ 0 < x < L, \\ z(0,t) = z(L,t) = 0, t > 0, \end{cases} \tag{9.7}$$

where $z_0 \in \mathbb{H}^2(0,L) \cap \mathbb{H}_0^1(0,L)$, $z_1 \in \mathbb{H}_0^1(0,L)$, and $a : [0,T] \to \mathbb{R}$ is continuous (see **[89, 90, 151, 213]**). Following the discussion in Section 4.3, the second-order PDE portion of (9.7) can be written as the following equivalent system of first-order PDEs:

$$\frac{\partial}{\partial t}\begin{bmatrix} v_1 \\ v_2 \end{bmatrix}(x,t) = \begin{bmatrix} 0 & I \\ -c^2\frac{\partial^2}{\partial x^2} & \alpha + \beta(v_2)^2 \end{bmatrix}\begin{bmatrix} v_1 \\ v_2 \end{bmatrix}(x,t) \tag{9.8}$$

$$+ \begin{bmatrix} 0 \\ \int_0^t a(t-s)\sin(v_1)ds \end{bmatrix}, \ 0 < x < L, t > 0.$$

It is natural to view (9.8) abstractly as (6.31) in $\mathscr{X} = \mathbb{H}_0^1(0,L) \times \mathbb{H}^0(0,L)$. Doing so requires that we define $\mathscr{A} : \text{dom}(\mathscr{A}) \subset \mathscr{X} \to \mathscr{X}$ by

$$\mathscr{A}\begin{bmatrix} v_1 \\ v_2 \end{bmatrix} = \begin{bmatrix} 0 & I \\ -c^2 \frac{\partial^2}{\partial x^2} & \alpha + \beta (v_2)^2 \end{bmatrix} \begin{bmatrix} v_1 \\ v_2 \end{bmatrix} = \begin{bmatrix} v_2 \\ -c^2 \frac{\partial^2 v_1}{\partial x^2} + \alpha v_2 + \beta (v_2)^3 \end{bmatrix}$$

$$\text{dom}(\mathscr{A}) = \left(\mathbb{H}^2(0,L) \cap \mathbb{H}_0^1(0,L)\right) \times \mathbb{H}_0^1(0,L). \tag{9.9}$$

Exercise 9.1.3.
i.) Is $\mathscr{A}$ a linear operator?
ii.) Prove that $\mathscr{A}$ is dissipative (i.e., $\langle \mathscr{A}x, x \rangle_{\mathscr{X}} \leq 0$, $\forall x \in \text{dom}(\mathscr{A})$). In fact, variational methods can be used to argue that $\mathscr{A}$ is m-dissipative.
iii.) Replace the term $\beta \left(\frac{\partial z}{\partial t}\right)^3$ in (9.7) by $\beta \left(\frac{\partial z}{\partial t}\right)^{2m+1}$, where $m \in \mathbb{N}$. Do your responses to (i) and (ii) remain the same? Prove your assertions.

Model XXVII.1 Nonlinear Beams

The model of the deflection of beams explored in Section 6.5 can also be improved by accounting for nonlinear dissipation. The following generalization of IBVP (6.80) is one such improvement:

$$\begin{cases} \frac{\partial^2 w}{\partial t^2} + \alpha \frac{\partial^4 w}{\partial z^4} + \beta w + \left(\frac{\partial w}{\partial t}\right)^{2m+1} \\ \quad = \int_0^t a(t,s) f\left(s, z, w, \frac{\partial^2 w}{\partial z^2}, \frac{\partial w}{\partial s}\right) ds, \ 0 < z < a, 0 < t < T, \\ w(z,0) = w_0(z), \ \frac{\partial w}{\partial t}(z,0) = w_1(z), \ 0 < z < a, \\ \frac{\partial^2 w}{\partial z^2}(0,t) = w(0,t) = 0 = \frac{\partial^2 w}{\partial z^2}(a,t) = w(a,t), \ t > 0, \end{cases} \tag{9.10}$$

where $m \in \mathbb{N}$. Note the similarity to the nonlinear wave equation (see **[320]**).

Exercise 9.1.4. Express (9.10) abstractly as (6.31) in a suitable space $\mathscr{X}$. Is the operator $\mathscr{A}$ that arises in doing so linear?

Model XXVIII.1 Combustion

An elementary description of gas burning in a rocket is given by the IBVP

$$\begin{cases} \frac{\partial z}{\partial t}(x,t) = \frac{\partial^2 z}{\partial x^2}(x,t) - 2\left(\frac{\partial z}{\partial x}(x,t)\right)^{5/3}, \ -a \leq x \leq a, t > 0, \\ z(a,t) = z(-a,t), t > 0, \\ z(x,0) = z_0(x), \ -a \leq x \leq a. \end{cases} \tag{9.11}$$

We can reformulate (9.11) abstractly as (3.87) in the space

$$\mathscr{X} = \mathbb{C}_p([-a,a];\mathbb{R}) = \{f : [-a,a] \to \mathbb{R} \,|\, f \text{ is continuous} \wedge f(a) = f(-a)\}$$

$$\|f\|_{\mathscr{X}} = \sup_{-a \leq x \leq a} |f(x)| \tag{9.12}$$

by identifying the operator $\mathscr{A} : \mathrm{dom}(\mathscr{A}) \subset \mathscr{X} \to \mathscr{X}$ as

$$\mathscr{A} z = \frac{\partial^2 z}{\partial x^2} - 2 \left(\frac{\partial z}{\partial x} \right)^{5/3} \tag{9.13}$$

$$\mathrm{dom}(\mathscr{A}) = \left\{ z \in \mathscr{X} \,\middle|\, z, \frac{\partial z}{\partial x}, \frac{\partial^2 z}{\partial x^2} \in \mathbb{C}_p \left([-a,a] ; \mathbb{R} \right) \right\}.$$

More generally, we can replace the term $2 \left(\frac{\partial z}{\partial x} \right)^{5/3}$ by $G \left(\frac{\partial z}{\partial x} \right)$, where $G : \mathbb{R} \to \mathbb{R}$ is a continuous function such that $G(0) = 0$. In such cases, it can be shown that the resulting operator $\mathscr{A}$ is m-accretive (see [**271**]).

Model XXIX.1 Nonlinear Advection

A nonlinear extension of the linear advection models (cf. (4.59), (6.40)) is given by

$$\begin{cases} \frac{\partial z}{\partial t}(x,t) = \frac{\partial}{\partial x} \left(G(z(x,t)) \right) = \int_0^t a(t-s) f\left(s, z(x,s) \right) ds, \, 0 < x < a, t > 0, \\ \frac{\partial z}{\partial x}(0,t) = \frac{\partial z}{\partial x}(a,t), t > 0, \\ z(x,0) = z_0(x), \, 0 < x < a, \end{cases} \tag{9.14}$$

where $G : \mathbb{R} \to \mathbb{R}$ is a continuous function such that $G(0) = 0$ and $f : [0,T] \times \mathbb{R} \to \mathbb{R}$ and $a : [0,\infty) \to (0,\infty)$ are given mappings (see [**49, 183, 204, 207, 249**]). This IBVP can be reformulated abstractly as (6.31) in $\mathscr{X} = \mathbb{L}^1(0,a)$. In doing so, the operator $\mathscr{A} : \mathrm{dom}(\mathscr{A}) \subset \mathscr{X} \to \mathscr{X}$ is defined as

$$\mathscr{A} z = -\frac{\partial}{\partial x} \left(G(z) \right). \tag{9.15}$$

Model XXX.1 Hamilton-Jacobi Equations

The so-called *Hamilton-Jacobi equations* arise in classical mechanics (see [**5**]). A nonlinear version of these equations is given by

$$\frac{\partial z}{\partial t}(\mathbf{x},t) + f\left(\nabla z(\mathbf{x},t) \right) = 0, \, \mathbf{x} \in \mathbb{R}^N, t > 0, \tag{9.16}$$

where $z : \mathbb{R}^N \times (0,\infty) \to \mathbb{R}$, $\mathbf{x} = \langle x_1, \dots, x_N \rangle$, $\nabla z = \left\langle \frac{\partial z}{\partial x_1}, \dots, \frac{\partial z}{\partial x_N} \right\rangle$, and $f : \mathbb{R}^N \to \mathbb{R}$ is convex. It is possible to reformulate (9.16) abstractly as (3.87) in the nonreflexive Banach space

$$\mathscr{X} = \mathbb{L}^\infty \left(\mathbb{R}^N ; \mathbb{R} \right), \, \|v\|_{\mathscr{X}} = \sup_{\mathbf{x} \in \mathbb{R}^N} |v(\mathbf{x})| \tag{9.17}$$

by identifying the operator $\mathscr{A} : \mathrm{dom}(\mathscr{A}) \subset \mathscr{X} \to \mathscr{X}$ as

$$\mathscr{A} z = f\left(\nabla z \right), \, \mathrm{dom}(\mathscr{A}) = \mathbb{W}_k^\infty(\mathbb{R}^N). \tag{9.18}$$

It is important to point out that dom($\mathscr{A}$) is not dense in $\mathscr{X}$, which presents a new obstacle beyond nonlinearity.

There are many other interesting nonlinear models, some of which are explored in [200, 296]. Certainly, the nonlinearity of the operator $\mathscr{A}$ is an obstacle, but this is just the beginning.

9.2 Comparison of the Linear and Nonlinear Settings

Loosely speaking, all IBVPs from Section 9.1 can be reformulated as an abstract evolution equation of the form

$$\begin{cases} u'(t) & = \mathscr{A}u(t) + \mathfrak{F}(u)(t), 0 < t < T, \\ u(0) & = u_0, \end{cases} \tag{9.19}$$

in a Banach space $\mathscr{X}$ (some for which T is taken to be ∞), just as in Chapters 2 - 8. As such, it might be tempting to apply the theory already established without hesitation. That would be fine provided that the new operators $\mathscr{A} : \text{dom}(\mathscr{A}) \subset \mathscr{X} \to \mathscr{X}$ satisfied the necessary hypotheses. However, we are immediately faced with the fact that each of these operators is NON-linear, which throws a huge wrench into the works. Indeed, as pointed out in Section 8.8, the assumption of linearity crept into all aspects of our development in both subtle and very apparent ways.

Can we can patch up the proof when we cannot invoke linearity, or does the extension to the nonlinear setting fall apart completely? By way of preparation for addressing this question, we recount some main elements of the early theory in the linear case and make a comparison to the current setting. This will form the basis of our discussion in this chapter.

For simplicity, consider (9.19) with $\mathfrak{F} = 0$.

	Linear setting	**Nonlinear Setting**
A	Linear	Nonlinear (!!)
	Must be single-valued **[340]**	Can be multi-valued
	$\mathrm{dom}(A)$ is dense in $\mathscr{X}$	$\mathrm{dom}(\mathscr{A})$ need not be dense in $\mathscr{X}$
Generator	Unique generator	Can be >1 generator **[39]**
	Must be m-dissipative	Need not be m-dissipative
Properties	e^{At} is a linear operator	$e^{\mathscr{A}t}$ is a NON-linear operator
	$Ae^{At} = e^{At}A$	A and $e^{\mathscr{A}t}$ need not commute **[39]**
	e^{At} is strongly continuous	PROPERTY REMAINS SAME
	$e^{A(t+s)} = e^{At}e^{As}$	PROPERTY REMAINS SAME
Cauchy Problem	$t \mapsto e^{At}u_0$ is differentiable when $u_0 \in \mathrm{dom}(A)$	$t \mapsto e^{\mathscr{A}t}u_0$ need not be differentiable even when $u_0 \in \mathrm{dom}(A)$

Perusing the right column of the above table suggests that the present setting is significantly different from the linear setting. Indeed, the development of the theory for a general Banach space is rather sophisticated and requires new tools from non-linear functional analysis. Since our treatment is introductory, we shall restrict our attention to nonlinear problems whose abstract formulation takes place in a Hilbert space to minimize the technical complications that arise.

9.3 The Crandall–Liggett Theory

Several volumes and articles (see **[39, 40, 49, 100, 207, 237, 249, 266, 287, 288, 295, 316, 317, 325]**) have been devoted to a formal development of the theory in the nonlinear setting. We focus on a small snapshot of this development that lies at the root of the theory, namely the *Crandall-Liggett theory*. Interestingly, the ideas and techniques involved are not much different, in principle, than what we have become accustomed to using, albeit some of the technical details involving the "range/resolvent condition" are further complicated by the lack of linearity and commutativity.

What follows is a cursory outline of the development of the analog of the Hille-Yosida theorem as extended to the nonlinear setting.

9.3.1 m-Dissipativity

A central generation result from linear semigroup theory (cf. Thrm. 3.5.13) is

Theorem 9.3.1. *A linear operator* $A : \mathrm{dom}(A) \subset \mathscr{H} \to \mathscr{H}$ *generates a* C_0-*semigroup of contractions* $\{e^{At} : t \geq 0\}$ *on* $\mathscr{H}$ *if and only if* A *is densely-defined and* m-*dissipative.*

Expecting an identical theorem when $\mathscr{A}$ is nonlinear would be overly ambitious. But, we will not abandon hope for a partial extension of some kind. The first step is to establish an analogous concept of dissipativity for nonlinear operators. To this end, recall that by Prop. 3.5.12, a linear operator $A : \mathrm{dom}(A) \subset \mathscr{H} \to \mathscr{H}$ is dissipative if

$$\|(I - \alpha A)(f - g)\|_{\mathscr{H}} \geq \|f - g\|_{\mathscr{H}}, \tag{9.20}$$

$\forall f, g \in \mathrm{dom}(A)$ and $\alpha > 0$. The linearity of A allows us to view (9.20) in the equivalent form

$$\|(f - g) - \alpha(Af - Ag)\|_{\mathscr{H}} \geq \|f - g\|_{\mathscr{H}}, \tag{9.21}$$

$\forall f, g \in \mathrm{dom}(A)$ and $\alpha > 0$. This latter formulation is a more useful inequality for nonlinear operators $\mathscr{A}$, and provides the motivation for the following definition.

Definition 9.3.2. A nonlinear operator $\mathscr{A} : \mathrm{dom}(\mathscr{A}) \subset \mathscr{H} \to \mathscr{H}$ is *dissipative* if (9.21) holds, $\forall f, g \in \mathrm{dom}(\mathscr{A})$ and $\alpha > 0$.

Remarks.
1. The phrase "$-\mathscr{A}$ is accretive" is often used synonymously with "$\mathscr{A}$ is dissipative."
2. For a Hilbert space $\mathscr{H}$, (9.21) is equivalent to

$$\langle f - g, \mathscr{A}f - \mathscr{A}g \rangle_{\mathscr{H}} \leq 0, \forall f, g \in \mathrm{dom}(\mathscr{A}). \tag{9.22}$$

Since a real-valued function $h : I \subset \mathbb{R} \to \mathbb{R}$ is increasing on I iff

$$(h(x) - h(y))(x - y) \geq 0, \ \forall x, y \in I,$$

(9.22) can be naively interpreted as $\mathscr{A}$ being increasing.

Exercise 9.3.1. Prove that (9.21) is equivalent to (9.22).

The procedure used to verify the dissipativity of the operators $\mathscr{A}$ from Section 9.1 is comparable to the approach used in the linear setting, albeit the spaces and the operators can be somewhat more difficult to work with. We consider several such operators below and provide references for some of the technical details.

Examples:
1. Nonlinear Mechanics
   Define $\mathscr{A} : (0, \infty) \to \mathbb{R}$ by

$$\mathscr{A}z = -\alpha z |z|, \ z > 0. \tag{9.23}$$

Exercise 9.3.2. Verify that $\mathscr{A}$ satisfies (9.22). (Note that the inner product on $\mathbb{R}$ is just multiplication.)

2. Nonlinear Diffusion Operators

Many variants of the familiar Laplacian operator $\triangle$ arise when modeling nonlinear diffusion processes. For instance, consider the operator $\mathscr{A} : \text{dom}(\mathscr{A}) \subset \mathbb{L}^1(0,a) \to \mathbb{L}^1(0,a)$ defined by

$$\mathscr{A}z(x) = -\frac{\partial^2}{\partial x^2}\left(\log(z(x)+1)\right) \tag{9.24}$$

$$\text{dom}(\mathscr{A}) = \left\{ z \in \mathbb{C}\left([0,a];\mathbb{R}\right) \left| \frac{\partial z}{\partial x}(0) = \frac{\partial z}{\partial x}(a) = 0 \wedge \right.\right.$$
$$\left. -\log(z+1), -\frac{\partial}{\partial x}\log(z+1) \text{ are AC on } [0,a] \right\}.$$

Another common example of an operator defined on the same domain is

$$\mathscr{A}z = -\frac{\partial^2}{\partial x^2}\left(z^m(x)\right), \text{ where } m \in \mathbb{N}. \tag{9.25}$$

Both of these operators are dissipative on $\mathbb{L}^1(0,a)$, which is a consequence of the following proposition proved in **[249]**:

Proposition 9.3.3. *The operator* $\mathscr{A} : \text{dom}(\mathscr{A}) \subset \mathbb{L}^1(0,a) \to \mathbb{L}^1(0,a)$ *defined by*

$$\mathscr{A}z(x) = -\frac{\partial^2}{\partial x^2}\left(\varphi(z)\right), \tag{9.26}$$

where $\text{dom}(\mathscr{A})$ *is specified in (9.24) and* $\varphi : \mathbb{R} \to \mathbb{R}$ *is an increasing, continuous function for which* $\varphi(0) = 0$, *is dissipative on* $\mathbb{L}^1(0,a)$.

This proposition can be extended to other spatial domains $\Omega \subset \mathbb{R}^N, N \in \mathbb{N} \setminus \{1\}$, but this requires the use of inequalities involving multiple Lebesgue integrals not discussed in this text. (See **[283]** for a detailed discussion.)

3. Nonlinear Wave Operators

Let $\mathscr{X} = \mathbb{H}_0^1(\Omega) \times \mathbb{H}^0(\Omega)$, where $\Omega \subset \mathbb{R}^2$ and define $\mathscr{A} : \text{dom}(\mathscr{A}) \subset \mathscr{X} \to \mathscr{X}$ by (9.9). The dissipativity of $\mathscr{A}$ is easily verified since

$$\left\langle \mathscr{A}\begin{bmatrix} w \\ v \end{bmatrix}, \begin{bmatrix} w \\ v \end{bmatrix} \right\rangle_{\mathscr{X}} = \iint_\Omega \left(\nabla v \cdot \nabla w + (\triangle w - \alpha v - \beta v^3) v \right) dxdy$$
$$= \iint_\Omega (\nabla v \cdot \nabla w) \, dxdy + \iint_\Omega (v \triangle w) \, dxdy \tag{9.27}$$
$$- \iint_\Omega (\alpha v^2 + \beta v^4) \, dxdy.$$

Observe that $\iint_\Omega (\nabla v \cdot \nabla w) \, dxdy = 0$ since $v = w = 0$ on $\partial\Omega$. (Why?) The second term in (9.27) is shown to equal zero using integration by parts. As such, since $\alpha v^2 + \beta v^4 \geq 0, \forall v \in \text{dom}(\mathscr{A})$, we conclude that the right-side of (9.27) is less than or equal to zero, as needed.

Exercise 9.3.3. If $j(0) = j(a) = 0$, show that $\int_0^a j(x)h''(x)dx = 0$ using integration by parts. (This is a special case of the computation suggested above. A general integration by parts formula can be found in **[39, 40]**.)

Exercise 9.3.4. Suppose that the term $\beta \left(\frac{\partial z}{\partial t} \right)^3$ in (9.7) is replaced by $\beta \left(\frac{\partial z}{\partial t} \right)^{2m+1}$, where $m \in \mathbb{N} \setminus \{1\}$. Verify that the resulting operator $\mathscr{A}$ defined on the same space $\mathscr{X}$ is dissipative.

Exercise 9.3.5. The nonlinear operator $\mathscr{A}$ arising in the beam model (9.10) is similar to (9.9). Verify that $\mathscr{A}$ is dissipative in the appropriate space $\mathscr{X}$.

4. Combustion Model

Let $\mathscr{X}$ be given by (9.12) and define the operator $\mathscr{A} : \text{dom}(\mathscr{A}) \subset \mathscr{X} \to \mathscr{X}$ by

$$\mathscr{A}z = \frac{\partial^2 z}{\partial x^2} - G\left(\frac{\partial z}{\partial x} \right), \qquad (9.28)$$

where $G : \mathbb{R} \to \mathbb{R}$ is an increasing, continuous function for which $G(0) = 0$. Verifying the dissipativity of $\mathscr{A}$ is tricky in this general case, but an easy-to-follow argument when $G(r) = \sin(r)$ can be found in **[49]**.

5. Nonlinear Advection

Let $\mathscr{X} = \mathbb{L}^1(0, a)$. The operator $\mathscr{A} : \text{dom}(\mathscr{A}) \subset \mathscr{X} \to \mathscr{X}$ defined by (9.15) is dissipative. The special case when $G(r) = r^m$ is argued in **[49]** and the general argument can be found in **[249]**.

Guided by the linear case, we also need to impose a so-called *range condition*. One way to further characterize dissipative operators is to consider the same condition as before, namely $\text{rng}\,(I - \alpha \mathscr{A}) = \mathscr{X}$, $\forall \alpha > 0$. But, not all dissipative operators satisfy this condition, even in the linear case. So, we define the following strengthening of dissipativity.

Definition 9.3.4. An operator $\mathscr{A} : \text{dom}(\mathscr{A}) \subset \mathscr{X} \to \mathscr{X}$ is *m-dissipative* if $\mathscr{A}$ is dissipative and

$$\text{rng}\,(I - \alpha \mathscr{A}) = \mathscr{X}, \forall \alpha > 0. \qquad (9.29)$$

The forward inclusion ($\subset$) in (9.29) holds trivially. (Why?) As such, showing that $\mathscr{A}$ satisfies Def. 9.3.4 boils down to showing:

$$\forall y \in \mathscr{X}, \exists x \in \text{dom}(\mathscr{A}) \text{ such that } (I - \alpha \mathscr{A})x = y. \qquad (9.30)$$

On the surface, this might seem to be straightforward, but "solving" (9.30) often involves tricky technical details even though the process does resemble that of the linear case for a few simple situations. The majority of the time, the argument involves the use of more sophisticated tools, like elements of the theory of elliptic PDEs and variational methods. A thorough account of this theory can be found in **[125]**.

9.3.2 Nonlinear Semigroups

Linear, densely-defined m-dissipative operators were nice enough to generate linear C_0−semigroups. Stripping away linearity, however, opens the door to much more complicated behavior, as pointed out in the chart in Section 9.2. As such, we should not necessarily expect a nonlinear m-dissipative operator to generate a nonlinear semigroup. Indeed, as Crandall and Liggett pointed out, we must further restrict this class of operators considerably.

Before we present the theorem, we define what is meant by a *nonlinear* semigroup.

Definition 9.3.5. Let $\mathscr{X}$ be a Banach space and $\mathscr{Y} \subset \mathscr{X}$ be closed.
i.) A *nonlinear semigroup on* $\mathscr{Y}$ is a collection of nonlinear operators $\left\{ e^{\mathscr{A}t} : t \geq 0 \right\}$ that satisfies
   a.) $e^{\mathscr{A}(t+s)}f = e^{\mathscr{A}t}e^{\mathscr{A}s}f, \forall f \in \mathscr{Y}$ and $s, t > 0$,
   b.) $\lim_{t \to 0^+} \left\| e^{\mathscr{A}t}f - f \right\|_{\mathscr{X}} = 0, \forall f \in \mathscr{Y}$.
If we further have
   c.) There exists $\omega > 0$ such that

$$\left\| e^{\mathscr{A}t}f - e^{\mathscr{A}s}g \right\|_{\mathscr{X}} \leq e^{\omega t} \|f - g\|_{\mathscr{X}}, \forall f, g \in \mathscr{Y}, \tag{9.31}$$

then $\left\{ e^{\mathscr{A}t} : t \geq 0 \right\}$ is called ω−*contractive*. If $\omega = 0$, then we simply say it is *contractive*.
ii.) $\mathscr{A} : \mathrm{dom}(\mathscr{A}) \subset \mathscr{X} \to \mathscr{X}$ is a *(strong) infinitesimal generator* of $\left\{ e^{\mathscr{A}t} : t \geq 0 \right\}$ if

$$\mathscr{A}x = \lim_{h \to 0^+} \frac{e^{\mathscr{A}h}x - x}{h}, \tag{9.32}$$

$\forall x \in \mathscr{Y}$ for which the limit exists.

Exercise 9.3.6.
i.) Compare Def. 9.3.5 to Def. 3.3.1 and Def. 3.3.5. Comment on their similarities and differences.
ii.) How does Def. 9.3.5(i)(c) simplify if $\left\{ e^{\mathscr{A}t} : t \geq 0 \right\}$ is a linear semigroup?

Exercise 9.3.7. Show that $\forall t \geq 0$,

$$\lim_{h \to 0^+} e^{\mathscr{A}(t+h)}f = e^{\mathscr{A}t}f, \forall f \in \mathscr{Y}.$$

Remark. If $\mathscr{A} : \mathrm{dom}(\mathscr{A}) \subset \mathscr{X} \to \mathscr{X}$ is a linear operator that generates a C_0-semigroup $\left\{ e^{\mathscr{A}t} : t \geq 0 \right\}$, then this semigroup satisfies Def. 9.3.5 with

$$\mathscr{Y} = \mathrm{cl}_{\mathscr{X}}(\mathrm{dom}(\mathscr{A})) = \mathscr{X}.$$

It becomes evident upon reviewing the proof of Thrm. 3.5.13 that the proof breaks down if the linearity assumption is removed at certain stages. (Where precisely?) But, we can still apply the same basic approach. Specifically, the resolvent operators

$$R_\lambda(\mathscr{A}) = \frac{1}{\lambda}\left(I - \frac{1}{\lambda}\mathscr{A} \right)^{-1}, \lambda > 0, \tag{9.33}$$

are well-defined, since $\mathscr{A}$ is dissipative. As such, consider the family of operators

$$\left\{ \frac{n}{t} R_{\frac{t}{n}}(\mathscr{A}) \mid t > 0, n \in \mathbb{N} \right\},$$

and observe that $\forall t > 0$,

$$\frac{n}{t} R_{\frac{t}{n}}(\mathscr{A}) = \left[\left(I - \frac{t}{n} \mathscr{A} \right)^{-1} \right]^n. \tag{9.34}$$

The strategy is to consider a sequence of abstract IVPs whose mild solutions are

$$u_n(t) = \frac{n}{t} R_{\frac{t}{n}}(\mathscr{A}) u_0, t > 0, \tag{9.35}$$

and show that

$$\lim_{n \to \infty} u_n(t) = e^{\mathscr{A}t} u_0, t > 0. \tag{9.36}$$

This is exactly what we would expect since $\lim_{n \to \infty} \left(1 - \frac{at}{n} \right)^n = e^{at}$. The limit function in (9.36) is the mild solution of (9.19) (with $\mathfrak{F} = 0$) that we seek.

Proving this is no easy task. Thankfully, many of the resolvent properties established in Prop. 3.5.3 remain valid in this setting, albeit verifying them is a bit more complicated (see **[39, 40, 249]** for the details). When the dust clears, we obtain the following partial extension of Thrm. 3.5.4 due to Crandall and Liggett:

Theorem 9.3.6. (Crandall-Liggett) *Let $\mathscr{X}$ be a Banach space and $\mathscr{A} : \mathrm{dom}(\mathscr{A}) \subset \mathscr{X} \to \mathscr{X}$ a dissipative operator. If $\mathrm{dom}(\mathscr{A}) = \mathrm{rng}(I - \alpha \mathscr{A}), \forall \alpha > 0$, then $\mathscr{A}$ generates a contractive nonlinear semigroup $\left\{ e^{\mathscr{A}t} : t \geq 0 \right\}$ on $\mathrm{cl}_{\mathscr{X}}(\mathrm{dom}(\mathscr{A}))$ given by*

$$e^{\mathscr{A}t}[f] = \lim_{n \to \infty} \left[\left(I - \frac{t}{n} \mathscr{A} \right)^{-1} \right]^n f, f \in \mathrm{cl}_{\mathscr{X}}(\mathrm{dom}(\mathscr{A})), \tag{9.37}$$

where the limit is uniform on compact interval of t. More generally, if instead $\exists \omega > 0$ such that $\mathscr{A} - \omega I$ is dissipative and $\exists \alpha^\star > 0$ such that $\mathrm{dom}(\mathscr{A}) \subset \mathrm{rng}(I - \alpha \mathscr{A})$, $\forall 0 < \alpha < \alpha^\star$, then $\mathscr{A}$ generates an $\omega-$contractive semigroup on $\mathrm{cl}_{\mathscr{X}}(\mathrm{dom}(\mathscr{A}))$ given by (9.37).

A detailed proof of these results can be found in **[100]**. The following corollary immediately follows from Thrm. 9.3.6:

Corollary 9.3.7. *If $\mathscr{A} : \mathrm{dom}(\mathscr{A}) \subset \mathscr{X} \to \mathscr{X}$ is m-dissipative, then $\mathscr{A}$ generates a contractive nonlinear semigroup $\left\{ e^{\mathscr{A}t} : t \geq 0 \right\}$ on $\mathrm{cl}_{\mathscr{X}}(\mathrm{dom}(\mathscr{A}))$ given by (9.37).*

9.3.3   The Associated Homogenous Cauchy Problem

As a consequence of Cor. 9.3.7, nonlinear semigroups are associated with each of the operators arising in the context of the IBVPs in Section 9.1. Great! But, with the examples in view, a very natural second question concerns the relationship between this semigroup and the "solution" of the abstract IVP (in $\mathscr{X}$).

$$\begin{cases} u'(t) = \mathscr{A} u(t), t > 0, \\ u(0) = u_0. \end{cases} \tag{9.38}$$

The link in the linear case was very nice. Indeed, as long as $u_0 \in \text{dom}(\mathscr{A})$, (9.38) was guaranteed to possess a unique classical solution given by $u(t) = e^{\mathscr{A} t} u_0$. Moreover, even when $u_0 \in \mathscr{X} \setminus \text{dom}(\mathscr{A})$, $u(t) = e^{\mathscr{A} t} u_0$ was at least a mild solution, even though the differentiability of this solution was no longer guaranteed. The current situation, however, is somewhat bleak by comparison. For instance, for IBVPs whose operators satisfy Thrm. 9.3.6 in $\mathbb{L}^1(\Omega)$, there is no *a priori* guarantee that the solution $u(t) = e^{\mathscr{A} t} u_0$ is differentiable for any $u_0 \in \text{cl}_{\mathscr{X}}(\text{dom}(\mathscr{A}))$, even when $\text{cl}_{\mathscr{X}}(\text{dom}(\mathscr{A})) = \mathscr{X}$. (The nonreflexivity of $\mathscr{X}$ is the culprit here - see **[39, 266]**.) We are faced with the following new questions:

1. What are sufficient conditions that guarantee (9.38) has a classical solution? For what set of IC does this hold?
2. In cases where there does not exist a classical solution, in what "weaker senses" can a meaningful solution be defined?

Both of these questions have received considerable attention in the literature. Solutions of a "weaker sense" can be interpreted in different ways, depending on the nature of the problem under investigation. The space in which a solution is sought and the type of convergence used (which is connected to the topology induced on the space, be it strong, weak, or weak-star) come into play when defining a solution. We shall focus on only two types of solutions, beginning with the more familiar notion of a classical solution:

Definition 9.3.8. Let $u_0 \in \text{cl}_{\mathscr{X}}(\text{dom}(\mathscr{A}))$. A *classical solution* of (9.38) is a function $u : [0,T] \to \mathscr{X}$ such that
a.) u is continuous on $[0,T]$ and absolutely continuous on compact subsets of $[0,T]$,
b.) u is differentiable a.e. on $(0,T)$,
c.) u satisfies (9.38) a.e. on $[0,T]$.

(Remember, "a.e." means "almost everywhere" in a measure-theoretic setting; cf. Section 1.8.4.) When does such a solution exist? A significant struggle in answering this question is symptomatic of the possibility that $e^{\mathscr{A} t} u_0$ might not be differentiable. Results guaranteeing differentiability (of various types) of the semigroup exist. For instance, we have the following from **[207]**:

Proposition 9.3.9. *If $\mathscr{X}$ is reflexive, $\mathscr{A} : \text{dom}(\mathscr{A}) \subset \mathscr{X} \to \mathscr{X}$ is m-dissipative, and $u_0 \in \text{cl}_{\mathscr{X}}(\text{dom}(\mathscr{A}))$, then $u(t) = e^{\mathscr{A} t} u_0$ is the unique classical solution of (9.38).*

Remark. Some common reflexive Banach spaces arising in applications include $\mathbb{L}^1(\Omega)$ and any Hilbert space. However, some Banach spaces, like $\mathbb{C}\left([a,b]; \mathbb{R}^N\right)$, are not reflexive. So, while Prop. 9.3.9 is nice, it is not a magic bullet.

The study of (9.38) simplifies in the Hilbert space setting, but an alternative is to seek a weaker type of solution as we did when we defined the notion of a mild solution in the linear setting. Proceeding in the same manner, however, presents us with some immediate challenges because of the nonlinearity and noncommutativity of $\mathscr{A}$ and $e^{\mathscr{A}t}$. A natural approach is to consider an approximating sequence of IVPs whose solutions are "nice" and tend toward a limit function (using an appropriate sense of convergence) that can be viewed as a solution (of some sort) to the original IVP. Typically, a discretized scheme (of a numerical approximation flavor) can be used to accomplish this task (see [266]). Regardless of the approach used, preserving uniqueness is important, so that the limit function should coincide with $u(t) = e^{\mathscr{A}t}u_0$. Now, *how* should we define such a solution?

The new form of a solution should possess the quality that any classical solution should satisfy its defining characteristics, but not conversely (since this would fail when $e^{\mathscr{A}t}u_0$ is not differentiable). To this end, we begin with a classical solution of (9.38) (when $\mathscr{A}$ satisfies Thrm. 9.3.6) and use the dissipativity of $\mathscr{A}$ to derive the desired form. Following the derivation in [317], we proceed loosely as follows.

Let u be the classical solution of (9.38). Then,

$$\frac{du}{dt} = -\mathscr{A}u(t), \text{ a.e. on } (0,T). \tag{9.39}$$

Since $\mathscr{A}$ is dissipative, it is also accretive, so that $\forall x \in \text{dom}(\mathscr{A}), s \in [0,T]$, and $\lambda > 0$,

$$\|(u(s) - x) + \lambda\left(\mathscr{A}u(s) - \mathscr{A}x\right)\|_{\mathscr{X}} \geq \|u(s) - x\|_{\mathscr{X}}. \tag{9.40}$$

Using (9.39) and (9.40), we see that $\forall h > 0$,

$$\lim_{h\to 0}\frac{1}{h}\left[\left\|(u(s) - x) + h\left(-\frac{du}{dt}(s) - \mathscr{A}x\right)\right\|_{\mathscr{X}} - \|u(s) - x\|_{\mathscr{X}}\right] \geq 0. \tag{9.41}$$

Using standard norm properties in (9.41) yields

$$\lim_{h\to 0}\frac{1}{h}\lceil u(s) - x, \mathscr{A}u(s)\rceil \equiv$$

$$\lim_{h\to 0}\frac{1}{h}\left[\left\|(u(s) - x) + h\left(\frac{du}{dt}(s)\right)\right\|_{\mathscr{X}} - \|u(s) - x\|_{\mathscr{X}}\right] \leq \tag{9.42}$$

$$\leq \lim_{h\to 0}\frac{1}{h}\left[\|(u(s) - x) + h\mathscr{A}x\|_{\mathscr{X}} - \|u(s) - x\|_{\mathscr{X}}\right].$$

The differentiability of the mapping $s \mapsto \|u(s) - x\|_{\mathscr{X}}$ then yields

$$\frac{d}{ds}\|u(s) - x\|_{\mathscr{X}} = \lim_{h\to 0}\frac{\|u(s+h) - x\|_{\mathscr{X}} - \|u(s) - x\|_{\mathscr{X}}}{h},$$

so that

$$\frac{d}{ds}\|u(s) - x\|_{\mathscr{X}} = \lim_{h\to 0}\frac{1}{h}\lceil u(s) - x, \mathscr{A}u(s)\rceil. \tag{9.43}$$

Integrating both sides of (9.43) over $[s,t] \subset [0,T]$ finally results in

$$\|u(t) - x\|_{\mathscr{X}} - \|u(s) - x\|_{\mathscr{X}} \le \int_s^t \lim_{h \to 0} \frac{1}{h} \lceil u(\tau) - x, \mathscr{A}u(\tau) \rceil \, d\tau. \tag{9.44}$$

While inequality (9.44) is not a representation formula for a solution of (9.38) in the sense that the variation of parameters formula (5.11) provides a formula for a mild solution of (5.10), it does play a similar role when showing the solution map used in the proof of the existence-uniqueness theorem possesses certain characteristics (e.g., contractivity, compactness, etc.) This prompts us to define the following weaker notion of a solution of (9.38).

Definition 9.3.10. Let $u_0 \in \mathrm{cl}_{\mathscr{X}} (\mathrm{dom}(\mathscr{A}))$. An *integral solution* of (9.38) is a function $u : [0,T] \to \mathscr{X}$ such that
a.) $u(0) = u_0$,
b.) u is continuous,
c.) (9.44) holds, $\forall [s,t] \subset [0,T]$ and $x \in \mathrm{dom}(\mathscr{A})$.

Remarks.
1. Inequality (9.44) is due to Benilan.
2. We can make the substitution $u(t) = e^{\mathscr{A}t} u_0$ and $u(s) = e^{\mathscr{A}s} u_0$ to obtain an inequality involving $e^{\mathscr{A}t}$. Further, note that if $u(t) = e^{\mathscr{A}t} u_0$ is not differentiable, then the steps in the above derivation cannot be reversed. (Why?)
3. Since the operators in Section 9.3.1 satisfy the hypotheses of Thrm. 9.3.6, each of the corresponding IBVPs in Section 9.1 possesses a unique integral solution. Further, for those IBVPs for which the space is reflexive, the integral solution is actually a classical solution.

9.3.4 The Nonhomogenous Cauchy Problem

We now turn our attention to the nonhomogenous Cauchy problem

$$\begin{cases} u'(t) = \mathscr{A}u(t) + f(t), t > 0, \\ u(0) = u_0, \end{cases} \tag{9.45}$$

in $\mathscr{X}$, where $u_0 \in \mathrm{cl}_{\mathscr{X}} (\mathrm{dom}(\mathscr{A}))$. It is not surprising that the development of a variation of parameters formula for (9.45) in the spirit of (4.7) fails miserably when $\mathscr{A}$ is a nonlinear operator.

Exercise 9.3.8. Work through the development of (4.7) and point out exactly where it breaks down when $\mathscr{A}$ is nonlinear.

A moment's reflection on the derivation of (9.44) suggests that we can define a similar formula for (9.45). Indeed, the following is a natural extension of Def. 9.3.10 for (9.45):

Definition 9.3.11. Let $u_0 \in \text{cl}_{\mathscr{X}}(\text{dom}(\mathscr{A}))$. An *integral solution* of (9.45) is a function $u : [0,T] \to \mathscr{X}$ that satisfies Def. 9.3.10 in which the term $\mathscr{A}u(s)$ in the integrand of (9.44) is replaced by $f(s) - \mathscr{A}u(s)$.

The following result is the nonlinear analog of Prop. 4.2.2 due to Barbu and Pavel (see **[39, 266]** for a proof).

Proposition 9.3.12. *Assume that $\mathscr{A} : \text{dom}(\mathscr{A}) \subset \mathscr{X} \to \mathscr{X}$ satisfies the hypotheses of Thrm. 9.3.6. If $f \in \mathbb{L}^1_{loc}(0,\infty;\mathscr{X})$, then $\forall u_0 \in \text{cl}_{\mathscr{X}}(\text{dom}(\mathscr{A}))$, (9.45) has a unique integral solution. Moreover, if $g \in \mathbb{L}^1_{loc}(0,\infty;\mathscr{X})$ and v is the unique integral solution of the IVP*

$$\begin{cases} v'(t) = \mathscr{A}v(t) + g(t), t > 0, \\ v(0) = v_0 \end{cases} \tag{9.46}$$

in $\mathscr{X}$, then

$$\|u(t) - v(t)\|_{\mathscr{X}} - \|u(s) - v(s)\|_{\mathscr{X}} \le \int_s^t \|f(\tau) - g(\tau)\|_{\mathscr{X}} \, d\tau. \tag{9.47}$$

Exercise 9.3.9. Modify the IBVPs in Section 9.1 by incorporating an appropriate external forcing term $f \in \mathbb{L}^1_{loc}(0,\infty;\mathscr{X})$. Then, verify the existence and uniqueness of an integral solution of each resulting IBVP.

9.3.5   Nonlinear Functional Evolution Equations

We conclude this section with a discussion of a class of nonlinear functional evolution equations of the form

$$\begin{cases} u'(t) = \mathscr{A}u(t) + \mathfrak{F}(u)(t), 0 < t < T, \\ u(0) = u_0, \end{cases} \tag{9.48}$$

in $\mathscr{X}$, where $\mathfrak{F} : \mathbb{C}([0,T];\text{cl}_{\mathscr{X}}(\text{dom}(\mathscr{A}))) \to \mathbb{L}^1(0,T;\mathscr{X})$. Consider the following special case of the results in **[316]**.

Proposition 9.3.13. *Assume that $\mathscr{A} : \text{dom}(\mathscr{A}) \subset \mathscr{X} \to \mathscr{X}$ is m-dissipative and $u_0 \in \text{cl}_{\mathscr{X}}(\text{dom}(\mathscr{A}))$. Then,*
i.) *If $\mathfrak{F}$ is globally Lipschitz with $M_{\mathfrak{F}} < 1$, then (9.48) has a unique integral solution on $[0,T]$.*
ii.) *If $\mathfrak{F}$ is continuous and $\{e^{\mathscr{A}t} : t > 0\}$ is compact, then (9.48) has at least one integral solution on $[0,T]$.*

Outline of Proof. We use the standard fixed-point approach to prove both statements. For (i), let $v \in \mathbb{C} = \mathbb{C}([0,T];\text{cl}_{\mathscr{X}}(\text{dom}(\mathscr{A})))$ and consider the IVP

$$\begin{cases} u'(t) = \mathscr{A}u(t) + \mathfrak{F}(v)(t), 0 < t < T, \\ u(0) = u_0. \end{cases} \tag{9.49}$$

There exists a unique integral solution of (9.49); call it u_v. (Why?) As such, the solution map $\Phi : \mathbb{C} \to \mathbb{C}$ given by

$$\Phi v = u_v \qquad (9.50)$$

is well-defined. (Why?) It can be shown that Φ is a strict contraction using (9.47). (Tell how.) Hence, we conclude from the Contraction Mapping Theorem that (9.48) has a unique integral solution.

For (ii), apply Schauder's fixed point theorem and suitably modify the proof of Thrm. 5.7.3. $\qquad\qquad\square$

Exercise 9.3.10. Provide the details in the proof of Prop. 9.3.13(ii).

Exercise 9.3.11. Argue that IBVP (9.7) has a unique integral solution on $[0,T]$.

Exercise 9.3.12. Argue that IBVP (9.14) has a unique integral solution on $[0,T]$.

Some standard references for this type of equation are **[37, 291, 296]**.

9.4   A Quick Look at Nonlinear Evolution Inclusions

The last stop on this brief tour of the nonlinear theory is an encounter with a different type of complexity that can arise, namely the notion of a *multi-valued* (or *set-valued*) operator. Basically, we now consider functions for which the image of an input can be an entire set of values. A prototypical example is given by

$$f(x) = \begin{cases} \{-1\}, x < 0, \\ [-1,1], x = 0, \\ \{1\}, x > 0. \end{cases} \qquad (9.51)$$

Such functions can occur in the BCs of an IBVP when describing phenomena governed by nonlinear multi-valued constitutive laws or nonmonotone operators. This directly affects the structure of $\mathrm{dom}(\mathscr{A})$ (and thus, the nature of $\mathscr{A}$), when viewing the associated IBVP abstractly. We provide some examples of such IBVPs below to give you a flavor of the types of models in which this can arise. Some standard references include **[12, 110, 176, 247, 337]**.

9.4.1   Some Models

Model XXVIII.2 Combustion Model

The following IBVP is a generalization of (9.11) that accounts for more complicated BCs:

$$\begin{cases} \frac{\partial z}{\partial t}(x,t) = \frac{\partial^2 z}{\partial x^2}(x,t) - 2\left(\frac{\partial z}{\partial x}(x,t)\right)^{5/3}, \ -a < x < a, t > 0, \\ \frac{\partial z}{\partial x}(a,t) \in \beta_0\left(z(a,t)\right), t > 0, \\ -\frac{\partial z}{\partial x}(-a,t) \in \beta_1\left(z(-a,t)\right), t > 0, \\ z(x,0) = z_0(x), \ -a < x < a, \end{cases} \qquad (9.52)$$

where β_i $(i = 0,1)$ is a maximal monotone operator (see [61, 114]).

A typical β_i could be defined by (9.51) (with 1 replaced by a). Another commonly-used operator is

$$\beta_i(s) = \begin{cases} \varnothing, s = 0, \\ [0,\infty), \ s \neq 0. \end{cases} \qquad (9.53)$$

Taking into account this more general multi-valued BC when reformulating (9.52) abstractly requires that we treat the operator $\mathscr{A}$ a bit differently. It is reasonable to view (9.52) as an abstract evolution *inclusion* in $\mathscr{X} = \mathbb{C}\left([-a,a];\mathbb{R}\right)$ by making the identifications

$$\frac{\partial^2 z}{\partial x^2}(x,\cdot) - 2\left(\frac{\partial z}{\partial x}(x,\cdot)\right)^{5/3} \in \mathscr{A}z \qquad (9.54)$$

$$\mathrm{dom}(\mathscr{A}) = \left\{ z \in \mathbb{C}^2\left((-a,a);\mathbb{R}\right) \Big| \frac{\partial z}{\partial x}(a,\cdot) \in \beta_0\left(z(a,\cdot)\right) \wedge \right.$$

$$\left. -\frac{\partial z}{\partial x}(-a,\cdot) \in \beta_1\left(z(-a,\cdot)\right) \right\}.$$

Then, (9.52) can be written in the form

$$\begin{cases} u'(t) \in \mathscr{A}u(t), t > 0, \\ u(0) = u_0. \end{cases} \qquad (9.55)$$

Model XXVI.2 Waves with Multi-valued Dissipation

An IBVP need not become multi-valued solely through coupling a PDE with multi-valued BCs. Indeed, the equation itself can be multi-valued. Consider, for instance, the following IBVP related to (9.7) in a bounded domain $\Omega \subset \mathbb{R}^2$ with smooth boundary $\partial\Omega$:

$$\begin{cases} \frac{\partial^2 z}{\partial t^2} + \beta\left(\frac{\partial z}{\partial t}\right) - c^2 \triangle z \ni \int_0^t a(t-s)\sin(z(x,y,s))ds, \ 0 < t < T, \\ z(x,y,0) = z_0(x,y), \ \frac{\partial z}{\partial t}(x,,y,0) = z_1(x,y), \ (x,y) \in \Omega, \\ z(0,t) = 0, \ (x,y) \in \partial\Omega, \ 0 < t < T, \end{cases} \qquad (9.56)$$

where β is a maximal monotone operator with $0 \in \beta(0)$.

Exercise 9.4.1. Using the same space $\mathscr{X}$ as for (9.7), try to define the operator $\mathscr{A} : \mathrm{dom}(\mathscr{A}) \subset \mathscr{X} \to \mathscr{X}$ necessary to reformulate (9.56) as the abstract functional evolution inclusion

$$\begin{cases} u'(t) \in \mathscr{A} u(t) + \mathfrak{F}(u)(t), 0 < t < T, \\ u(0) = u_0, \end{cases} \tag{9.57}$$

in $\mathscr{X}$, where $\mathfrak{F} : \mathbb{C}([0,T]; \mathrm{cl}_{\mathscr{X}}(\mathrm{dom}(\mathscr{A}))) \to \mathbb{L}^1(0,T; \mathscr{X})$. Compare this to the identifications (9.8) and (9.9).

Model XXIV.2 Nonlinear Multi-valued Diffusion

The following model studied in **[334]** describes diffusion within a heterogenous material described by a bounded domain $\Omega \subset \mathbb{R}^3$ with smooth boundary $\partial \Omega$ comprised of two components undergoing phase changes.

$$\begin{cases} \frac{\partial}{\partial t} a(z) - \triangle \alpha_1(z) + \beta(z-w) \ni \int_0^t k_1(t-s) f(s,\mathbf{x}) ds, & \mathbf{x} \in \Omega, t > 0, \\ \frac{\partial}{\partial t} b(w) - \triangle \alpha_2(w) - \beta(z-w) \ni \int_0^t k_2(t-s) g(s,\mathbf{x}) ds, & \mathbf{x} \in \Omega, t > 0, \\ 0 \in \alpha_1(z(\mathbf{x},t)), 0 \in \alpha_2(w(\mathbf{x},t)), \mathbf{x} \in \partial\Omega, t > 0, \\ w(\mathbf{x},0) = w_0(\mathbf{x}), z(\mathbf{x},0) = z_0(\mathbf{x}), \mathbf{x} \in \Omega, \end{cases} \tag{9.58}$$

where $\alpha_1, \alpha_2, \beta : \mathbb{R} \to \mathbb{R}$ are maximal monotone operators such that $0 \in \alpha_i(0)$ $(i = 1,2), 0 \in \beta(0)$. The IBVP (9.58) can be reformulated abstractly as (9.57) where $\mathscr{X}$ is an appropriate subspace of $\mathbb{L}^1(\Omega) \times \mathbb{L}^1(\Omega)$ and $\mathscr{A} : \mathrm{dom}(\mathscr{A}) \subset \mathscr{X} \to \mathscr{X}$ is given by

$$\mathscr{A} \begin{bmatrix} z \\ w \end{bmatrix} = \begin{bmatrix} -\triangle \alpha_1(z) + \beta(z-w) \\ -\triangle \alpha_2(w) - \beta(z-w) \end{bmatrix}. \tag{9.59}$$

9.4.2 Evolution Inclusions

We mentioned that while the IBVPs in Section 9.4.1 resembled models studied earlier in the text, incorporating maximal monotone graphs (which can be multi-valued) into the PDE and/or BCs affects the nature of the operator $\mathscr{A} : \mathrm{dom}(\mathscr{A}) \subset \mathscr{X} \to \mathscr{X}$ in such a way as to render it multi-valued. We need to make precise what is meant by a multi-valued operator. Once this is done, the extension of the theory presented in Sections 9.2 and 9.3 to the multi-valued case is not difficult.

We begin with some notation.

Definition 9.4.1. Let $\mathscr{X}$ be a Banach space.
i.) The *power set* of $\mathscr{X}$, denoted by $2^{\mathscr{X}}$, is the set of all subsets of $\mathscr{X}$.
ii.) An operator $\mathscr{A} : \mathrm{dom}(\mathscr{A}) \subset \mathscr{X} \to 2^{\mathscr{X}}$ is called *multi-valued*.
  a.) For $x \in \mathscr{X}$, let $\mathscr{A} = \{ y \in \mathscr{X} | y \in \mathscr{A} x \}$. We say (x,y) is on the *graph of* $\mathscr{A}$.
  b.) The *domain of* $\mathscr{A}$ is given by $\mathrm{dom}(A) = \{ x \in \mathscr{X} | Ax \neq \emptyset \}$.

It is customary to think of A as a subset of $\mathscr{X} \times \mathscr{X}$, meaning that we identify the operator $\mathscr{A}$ with its graph. The following definition of dissipativity for a multi-valued operator might be what you would have naturally guessed.

Definition 9.4.2. $A : \operatorname{dom}(A) \subset \mathscr{X} \to 2^{\mathscr{X}}$ is *dissipative* if

$$\|x_1 - x_2\|_{\mathscr{X}} \le \|(x_1 - x_2) - \lambda (y_1 - y_2)\|_{\mathscr{X}}, \tag{9.60}$$

$\forall \lambda > 0, (x_i, y_i) \in A.$

Note that if A is single-valued, then Def. 9.4.2 coincides with (9.21).

Exercise 9.4.2. Using similar reasoning, adapt the definition of an integral solution to the multi-valued setting.

In essence, the theorems we have presented can be extended to this multi-valued setting by showing the desired estimates hold $\forall (x, y) \in A$ (which is done exactly as before since fixing such a member of A renders the estimate single-valued) and then quantifying over all such members of A. Despite the apparent complexity that a multi-valued operator presents, the details for numerous IBVPs carry over without significant issue. (See **[176]** for a more thorough discussion.) As such, under appropriate growth conditions on the forcing terms, we can conclude that the IBVPs in Section 9.4.1 have unique integral solutions.

9.5   Some Final Comments

We have merely provided an introduction to the notion of a nonlinear operator and some very basic theory of nonlinear evolution equations. Entire volumes have been written on the foundations of the theory, and the study of nonlinear evolution equations continues to be a very active area of research. Nonlinear evolution equations with time delays, and nonlinear variants of various classes of implicit evolution equations have been studied in **[295]**. We encourage you to continue your studies in this direction. There are numerous open problems waiting to be solved!

9.6   Guidance for Exercises

9.6.1   Level 1: A Nudge in a Right Direction

9.1.1. No. Try to simplify $\mathscr{A}(w_1 + w_2)$. What gets in the way?

9.1.2. Compare the term $\frac{\partial}{\partial x}\left(Kh(x,t)\frac{\partial h}{\partial x}(x,t)\right)$ to $K\frac{\partial^2 h}{\partial x^2}$. What is the critical difference?

9.1.3. i.) No. Compute $\mathscr{A}\left(\begin{bmatrix} v_1 \\ v_2 \end{bmatrix} + \begin{bmatrix} v_1^\star \\ v_2^\star \end{bmatrix}\right)$. What part of $\mathscr{A}$ renders it nonlinear?

ii.) Make certain to use the inner product on the given space $\mathscr{X}$. Refer to the aeroelasticity model for related computations.

iii.) Yes, they are the same. (Why?)

9.1.4. Using the same space $\mathscr{X}$ as (6.80), identify the operator $\mathscr{A}$ by

$$\mathscr{A} w = \alpha \frac{\partial^4 w}{\partial z^4} + \beta w + \left(\frac{\partial w}{\partial t}\right)^{2m+1},$$

or in an equivalent matrix form if you have converted the IBVP to a system. Is $\mathscr{A}$ linear?

9.3.1. Refer back to (3.128) to get started. The argument is similar, but do NOT make use of linearity anywhere.

9.3.2. Argue in two separate cases, namely when $x \le y$ and $y < x$.

9.3.3. Use Prop. 1.8.14(i) with $u = j(x)$, $dv = h''(x)dx$.

9.3.4. Refer to Exer. 9.1.3(iii).

9.3.5. Refer to Exer. 9.1.3 for similar calculations.

9.3.6. (i) (a) is comparable to Def. 3.3.1(iii), but the former is only guaranteed to work on a closed subspace $\mathscr{Y}$.

(i)(b) is comparable to Def. 3.3.5 with the same caveat.

(i)(c) is different. (How?)

9.3.7. Observe that $\forall f \in \mathscr{Y}$,

$$\left\| e^{\mathscr{A}(t+h)} f - e^{\mathscr{A}t} f \right\|_{\mathscr{X}} = \left\| e^{\mathscr{A}t} \left[e^{\mathscr{A}h} f \right] - e^{\mathscr{A}t} f \right\|_{\mathscr{X}} \le \underline{\;\;?\;\;}, \forall h > 0.$$

9.3.8. Neither $\mathscr{A}$ nor $e^{\mathscr{A}t}$ is linear, so that the final step in (4.7) is no longer valid. (What else?)

9.3.9. This process of conversion to an abstract evolution equation, imposing restrictions on the data, and verifying the hypotheses of a known theorem (here 9.3.12) has hopefully become standard. Use the information about the operators that we have acquired through the section to ensure the underlying hypotheses hold.

9.3.10. Define an appropriate closed ball and then investigate the solution map.

9.3.11. You argued that the forcing term generated a Lipschitz functional in (6.29). Use that with the abstract formulation of (9.7). (So what?)

9.3.12. Refer to the hint for Exer. 9.3.11.

9.4.1. As in [317], define $A : \operatorname{dom}(A) \subset \mathscr{X} \to 2^{\mathscr{X}}$ by

$$A \begin{pmatrix} u \\ v \end{pmatrix} = \left\{ \begin{pmatrix} -v \\ -\Delta u + w \end{pmatrix} \middle| w \in \mathbb{L}^2(\Omega) \wedge w(x) \in \beta(v(x)) \text{ a.e. on } \Omega \right\}$$

$$\operatorname{dom}(A) = \left\{ \begin{pmatrix} u \\ v \end{pmatrix} \in \mathbb{H}^2(\Omega) \times \mathbb{H}^1_0(\Omega) \,\middle|\, \exists w \in \mathbb{L}^2(\Omega) \text{ such that } \right.$$

$$w(x) \in \beta(v(x)) \text{ a.e. on } \Omega \}.$$

9.4.2. The same inequality works, but it needs to now hold $\forall (x,y) \in A$.

9.6.2   Level 2: An Additional Thrust in a Right Direction

9.1.1. $|w_1 + w_2| \neq |w_1| + |w_2|$, in general.

9.1.2. The operator $\frac{\partial}{\partial x}\left(Kh(x,t)\frac{\partial h}{\partial x}(x,t)\right)$ is not linear. Observe that

$$\frac{\partial}{\partial x}\left(K(h_1 + h_2)(x,t)\frac{\partial(h_1 + h_2)}{\partial x}(x,t)\right) =$$

$$\frac{\partial}{\partial x}\left(Kh_1(x,t)\frac{\partial h_1}{\partial x}(x,t)\right) + \frac{\partial}{\partial x}\left(Kh_2(x,t)\frac{\partial h_2}{\partial x}(x,t)\right) +$$

$$\underbrace{\frac{\partial}{\partial x}\left(Kh_2(x,t)\frac{\partial h_1}{\partial x}(x,t) + Kh_1(x,t)\frac{\partial h_2}{\partial x}(x,t)+\right)}_{\text{This is what makes the operator nonlinear.}}$$

9.1.3. i.) v_2^2 makes $\mathscr{A}$ nonlinear.

    ii.) & iii.) Look ahead to (9.27).

9.1.4. The operator $\mathscr{A}$ is nonlinear because of the term $\left(\frac{\partial w}{\partial t}\right)^{2m+1}$. (Explain why.)

9.3.2. Note that

$$\langle x - y, -\alpha x |x| + \alpha y |y|\rangle_{\mathbb{R}} = -\alpha(x - y)(x|x| - y|y|).$$

(So what?)

9.3.3. Observe that $du = j(x)$, $v = h'(x)$. So,

$$\int_0^a j(x)h''(x)dx = j(x)h'(x)\Big|_0^a - \int_0^a j'(x)h'(x)dx$$

$$= j(a)h'(a) - j(0)h'(0) - j(x)h(x)\,|_0^a$$

$$= 0 \text{ (Why?)}.$$

9.3.6. (i) (c) is included so that estimates can be obtained in the nonlinear setting, whereas the linearity of the semigroup helped with this in the linear setting without needing to include such an additional component in the definition.

9.3.7. Fill in the blank with $\left\|e^{\mathscr{A}h}f - f\right\|_{\mathscr{X}}$. As $h \to 0^+$, where does this go? Does this help with the squeeze theorem?

9.3.8. Also, $Ae^{\mathscr{A}t} \neq e^{\mathscr{A}t}\mathscr{A}$, in general, so that we can no longer go from the second to the third line in (4.7). (Why?)

9.3.10. Refer to **[317]**.

9.3.11. Show that the conditions of Prop. 9.3.13 are satisfied.

Chapter 10

Nonlocal Evolution Equations

Overview

We focus on a very particular type of generalization of a wide sampling of all of the IBVPs developed throughout the text in which the standard initial conditions are replaced by those which are state-dependent. The development of this theory requires only a small number of modifications to the general theory as presented, and the corresponding models offer a more accurate description of the underlying phenomena.

10.1   Introductory Remarks

The final topic of interest as we end our journey through deterministic evolution equations intersects an active area of research. Before launching in, it is beneficial to take a look back and recap the main course we have taken. We began with an exploration of some rudimentary notions and questions in the finite-dimensional setting in an attempt to establish a foundation for our study. The theory developed was used as a springboard into the study of linear PDEs via the use of C_0-semigroups. We developed a core theory for abstract ODEs in a Banach space that mimicked the finite-dimensional case, but which provided a unified framework within which to study many different PDEs arising from disparate applications under a common umbrella. Then, chapter by chapter, we considered successive modifications of the form of the abstract evolution equation that arose naturally in the mathematical modeling of various phenomena (by changing the nature of the forcing term, incorporating time delays, etc.). Each different type of complexity introduced gave rise to an evolution equation of a slightly different form that did not quite fall under the parlance of the theory developed up to that point; but, it *almost* did. We identified the nature of the modification each time and introduced a suitable change to the previously-studied abstract evolution equations that correctly accounted for this change. We then developed the abstract theory for the new class of evolution equation and finally, applied it to the IBVPs that launched the chapter. This process was easier to implement at certain junctures than others. Indeed, arguably the biggest leap was made when transitioning to Chapter 9 from the linear theory, while the second notable leap occurred between Chapters 2 and 3.

The purpose, pedagogically, of the present chapter is to consider a common extension of all types of models considered thus far by accounting for an additional complication occurring in the initial condition. Armed with the tools and strategies developed in the text, the idea is to formulate a new class of evolution equations and extend the core theoretical results to this setting. This will involve a short survey of the relevant literature, as well as the positing of open problems.

10.2 Motivation by Models

The IVP/IBVP used in the description of any application considered in this text was equipped with a condition that described precisely the state of the unknown at the beginning of the experiment, which was taken to be $t = 0$. But, the description of this initial state requires one either to perform some concrete measurement or to provide a theoretical guess, both of which are prone to error. There are continuous dependence estimates that help quantify and control for this error in the context of a model, but there are other ways of minimizing this error internal to the IVP/IBVP itself.

One way to minimize the negative effects incurred by bad measurements is to reformulate the entire IVP/IBVP within a probabilistic framework. This involves formally accounting for "noise" directly in the PDE in various ways. This approach is the topic of Volume 2. A more readily-applicable approach that can be used directly for the deterministic setting suggested by Byszewski **[68 - 80]** and others **[87, 88]** is to enhance the IC in the IVP/IBVP by incorporating more information about the system (e.g., its state at later times) into it. Since this by its very nature requires that we use values of the unknown at times other than $t = 0$, the IC is no longer a fixed condition but rather is now represented as a functional of the unknown u, say $u(0) = g(u)$. We refer to this as a *nonlocal initial condition*.

The study of evolution equations with nonlocal initial conditions was proposed by various researchers in the context of different problems in the early 1980s; some of the initial applications discussed in this context include elasticity and viscoelasticity in materials with memory. The techniques suggested are not limited to these models, as well shall see below. There has been considerable theoretical research on this type of problem during the past three decades; a small representative sample is: **[3, 6, 8, 10, 11, 21, 24, 25, 27, 28, 33, 68 - 80, 87, 88, 112, 162, 167, 182, 94, 214, 216, 217, 221, 257, 262, 313, 322, 323, 336, 338]**. It is impossible to recount every result, but we provide an overview of this work, focusing primarily on functional evolution equations with nonlocal initial conditions.

In preparation for our theoretical development, we present several IVPs and IB-VPs equipped with different types of initial conditions. We impose no conditions on the data at the moment and we refer you to the previous versions of the models for specific information regarding the mappings involved. Work through the rest of this section in its entirety and complete the following instructive exercise along the way:

Exercise 10.2.1. Compare each of the IVPs and IBVPs listed below to its corresponding sequence of predecessors to get a feel for how it progressively arrived at this stage. Pay particular attention to what has now changed. Write each one as an abstract evolution equation in an appropriate Banach space $\mathscr{X}$ and try to deduce the theme common to all of them.

Model II.6 Nonlocal Pharmacokinetics

Let $n \in \mathbb{N}$ and $0 < t_1 < t_2 < \ldots < t_n < T$ be fixed times. Consider the following nonlocal variant of IVP (5.1):

$$
\begin{cases}
\frac{dy}{dt} = -ay(t) + D(t, y(t), z(t)), \, 0 < t < T, \\
\frac{dz}{dt} = ay(t) - bz(t) + \overline{D}(t, y(t), z(t)), \, 0 < t < T, \\
y(0) = \sum_{i=1}^{n} \beta_i y(t_i), \, z(0) = \sum_{i=1}^{n} \beta_i^{\star} z(t_i),
\end{cases}
\tag{10.1}
$$

where $\beta_i, \beta_i^{\star} \, (i = 1, \ldots, n)$ are positive real numbers.

Model IX.3 Nonlocal Neural Networks

Consider the following nonlocal variant of IVP (5.5):

$$
\begin{cases}
\frac{d}{dt} x_1(t) = a_1 x_1(t) + \sum_{j=1}^{M} \omega_{1j}(t) g_j(x_j(t)), \, 0 < t < T, \\
\quad \vdots \\
\frac{d}{dt} x_M(t) = a_M x_M(t) + \sum_{j=1}^{M} \omega_{Mj}(t) g_j(x_j(t)), \, 0 < t < T, \\
x_1(0) = \int_0^T \alpha_1(s) f(s, x_1(s, \cdot)) ds, \\
\quad \vdots \\
x_M(0) = \int_0^T \alpha_M(s) f(s, x_M(s, \cdot)) ds,
\end{cases}
\tag{10.2}
$$

where $\alpha_i \in \mathbb{L}^2(0, T; \mathbb{R}) \, (i = 1, \ldots, M)$.

Model XI.3 Nonlocal Pollution Model

Let Ω be a bounded region in $\mathbb{R}^N$ with smooth boundary $\partial \Omega$, and let $z(\mathbf{x}, t)$ denote the pollution concentration at position $\mathbf{x} \in \Omega$ and time $t > 0$. Consider the following nonlocal variant of IBVP (6.47):

$$
\begin{cases}
\frac{\partial z}{\partial t}(\mathbf{x}, t) = k \triangle z(\mathbf{x}, t) + \overrightarrow{\alpha} \cdot \nabla z(\mathbf{x}, t) \\
\quad + \int_0^t a(t - s) g(s, z(\mathbf{x}, s)) ds, \, \mathbf{x} \in \Omega, \, 0 < t < T, \\
z(\mathbf{x}, 0) = \int_0^T a(t, s) g\left(s, z(s, \mathbf{x}), \int_0^s k(s, \tau, z(\tau, \mathbf{x})) d\tau\right) ds \, \mathbf{x} \in \Omega, \\
\frac{\partial z}{\partial \mathbf{n}}(\mathbf{x}, t) = 0, \, \mathbf{x} \in \partial \Omega, \, 0 < t < T,
\end{cases}
\tag{10.3}
$$

where **(H6.5)** - **(H6.7)** hold.

Model XIV.3 Nonlinear Nonlocal Functional Extensible Beam Model

Consider the following nonlocal variant of IBVP (9.10):

$$
\begin{cases}
\frac{\partial^2 w}{\partial t^2} + \alpha \frac{\partial^4 w}{\partial z^4} + \beta w + \left(\frac{\partial w}{\partial t}\right)^{2m+1} \\
\quad = \int_0^t a(t,s) f\left(s,z,w,\frac{\partial^2 w}{\partial z^2},\frac{\partial w}{\partial s}\right) ds,\ 0 \leq z \leq a, 0 < t < T, \\
w(z,0) = \int_0^T k_1(s) f_1\left(s,w(z,s),\frac{\partial w}{\partial s}(z,s)\right) ds,\ 0 \leq z \leq a, \\
\frac{\partial w}{\partial t}(z,0) = \int_0^T k_2(s) f_2\left(s,w(z,s),\frac{\partial w}{\partial s}(z,s)\right) ds,\ 0 \leq z \leq a, \\
\frac{\partial^2 w}{\partial z^2}(0,t) = w(0,t) = 0 = \frac{\partial^2 w}{\partial z^2}(a,t) = w(a,t),\ 0 < t < T,
\end{cases}
\tag{10.4}
$$

where $k_i \in C([0,T];\mathbb{R})$ $(i=1,2)$ and $f_i : [0,T] \times \mathbb{R} \times \mathbb{R} \to \mathbb{R}$ $(i=1,2)$.

Model XIII.3 Nonlocal Functional Aeroelasticity Model

Let $0 < \varepsilon < T$ and consider the following nonlocal variant of IBVP (6.52), where $0 \leq z \leq a$ and $0 < t < T$:

$$
\begin{cases}
\frac{\partial^2 w}{\partial t^2} + \beta_1 \frac{\partial w}{\partial t} + \beta_2 \frac{\partial}{\partial t}\left(\frac{\partial^4 w}{\partial z^4}\right) + \frac{\partial^4 w}{\partial z^4} - \beta_3 \frac{\partial^2 w}{\partial z^2} + \beta_4 \frac{\partial w}{\partial z} \\
\quad = g\left(t,z,\int_0^t a_1(t,s)w(s,z)ds, \int_0^t a_{21}(t,s)\frac{\partial w}{\partial s}(s,z)ds\right), \\
w(0,t) = w(a,t) = 0, t > 0, \\
\frac{\partial^2 w}{\partial z^2}(0,t) + \beta_2 \frac{\partial}{\partial t}\left(\frac{\partial^2 w}{\partial z^2}\right)(0,t) = 0, 0 < t < T, \\
\frac{\partial^2 w}{\partial z^2}(a,t) + \beta_2 \frac{\partial}{\partial t}\left(\frac{\partial^2 w}{\partial z^2}\right)(a,t) = 0, 0 < t < T, \\
w(z,0) = \int_{T-\varepsilon}^T [1+\sqrt{s}]\log(1+w(s,z))ds,\ 0 \leq z \leq a, \\
\frac{\partial w}{\partial t}(z,0) = \int_{T-\varepsilon}^T [1+\sqrt{s}]\log\left(1+\frac{\partial w}{\partial s}(s,z)\right) ds,\ 0 \leq z \leq a.
\end{cases}
\tag{10.5}
$$

Model XII.3 Nonlocal Epidemiological Model

Consider the following nonlocal variant of (6.48):

$$
\begin{cases}
\frac{\partial}{\partial t}\begin{bmatrix} P_H \\ P_V \end{bmatrix} = \begin{bmatrix} \alpha_H \triangle + \beta_H & 0 \\ 0 & \alpha_V \triangle - \gamma_V \end{bmatrix}\begin{bmatrix} P_H \\ P_V \end{bmatrix} \\
\quad + \begin{bmatrix} -\int_0^t \frac{\eta P_H(x,y,s)P_V(x,y,s)}{(1+P_H(x,y,s)+P_V(x,y,s))^2}ds \\ \int_0^t \frac{\eta P_H(x,y,s)P_V(x,y,s)}{(1+P_H(x,y,s)+P_V(x,y,s))^2}ds \end{bmatrix},\ (x,y) \in \Omega, 0 < t < T, \\
\begin{bmatrix} P_H \\ P_V \end{bmatrix}(x,y,0) = \begin{bmatrix} \int_0^T \alpha_1(s) f_1(s,P_H(x,y,s),P_V(x,y,s))ds \\ \int_0^T \alpha_2(s) f_2(s,P_H(x,y,s),P_V(x,y,s))ds \end{bmatrix},\ (x,y) \in \Omega, \\
\frac{\partial}{\partial \mathbf{n}}\begin{bmatrix} P_H \\ P_V \end{bmatrix}(x,y,t) = 0,\ (x,y) \in \partial\Omega, 0 < t < T,
\end{cases}
$$

$$\tag{10.6}$$

where $\alpha_i \in \mathbb{L}^2(0,T;\mathbb{R})$, $i = 1,2$.

Model XV.3 Nonlocal Sobolev Model of Soil Mechanics

Let $n \in \mathbb{N}$ and $0 < t_1 < t_2 < \ldots < t_n < T$ be fixed times. Consider the following nonlocal variant of (7.6):

$$
\begin{cases}
\frac{\partial}{\partial t}\left(z(x,y,t) - \frac{\partial^2}{\partial x^2}z(x,y,t) - \frac{\partial^2}{\partial y^2}z(x,y,t)\right) + \frac{\partial^2}{\partial x^2}z(x,y,t) + \frac{\partial^2}{\partial y^2}z(x,y,t) \\
\quad = \int_0^{a_2}\int_0^{a_1} k(t,x,y)f(t,z(x,y,t))\,dxdy,\ 0 < x < a_1, 0 < y < a_2, 0 < t < T, \\
z(x,y,0) = \sum_{i=1}^{n}\beta_i z(x,y,t_i),\ 0 < x < a_1, 0 < y < a_2, \\
z(0,y,t) = z(a_1,y,t) = 0,\ 0 < x < a_1, 0 < t < T, \\
z(x,0,t) = z(x,a_2,t) = 0,\ 0 < y < a_2, 0 < t < T,
\end{cases}
$$

$$(10.7)$$

where β_i $(i = 1, \ldots, n)$ are positive real numbers.

Model XXVIII.2 Nonlocal Combustion

Let $n \in \mathbb{N}$ and $0 < t_1 < t_2 < \ldots < t_n < T$ be fixed times. Consider the following nonlocal variant of (9.11):

$$
\begin{cases}
\frac{\partial z}{\partial t}(x,t) = \frac{\partial^2 z}{\partial x^2}(x,t) - 2\left(\frac{\partial z}{\partial x}(x,t)\right)^{5/3},\ -a \leq x \leq a, 0 < t < T, \\
z(a,t) = z(-a,t),\ 0 < t < T, \\
z(x,0) = \sum_{i=1}^{n}\beta_i(\cdot)x(t_i,\cdot) + \int_0^T \alpha(s)f(s,x(s,\cdot))\,ds,\ -a \leq x \leq a,
\end{cases}
$$

$$(10.8)$$

where **(H6.1)** - **(H6.3)** hold and β_i $(i = 1, \ldots, n)$ are positive real numbers.

Model XXVI.2 Nonlinear Nonlocal Waves

Consider the following nonlocal variant of (9.7):

$$
\begin{cases}
\frac{\partial^2 z}{\partial t^2} + \alpha\frac{\partial z}{\partial t} + \beta\left(\frac{\partial z}{\partial t}\right)^3 - c^2\frac{\partial^2 z}{\partial x^2} \\
\quad = \int_0^t a(t-s)\sin(z(x,s))ds,\ 0 < x < L, 0 < t < T, \\
z(x,0) = \int_{T-\varepsilon_1}^T [1 + \sqrt{s}]\log(1 + z(x,s))ds,\ 0 < x < L, \\
\frac{\partial z}{\partial t}(x,0) = \int_{T-\varepsilon_2}^T [1 + \sqrt{s}]\log\left(1 + \frac{\partial z}{\partial s}(x,s)\right)ds,\ 0 < x < L, \\
z(0,t) = z(L,t) = 0,\ 0 < t < T.
\end{cases}
$$

$$(10.9)$$

Model XXII.2 Nonlocal Semi-linear Sine-Gordon Equation

Consider the following nonlocal variant of (8.80):

$$
\begin{cases}
\frac{\partial^2 u}{\partial t^2}(\mathbf{x},t) + \alpha\frac{\partial u}{\partial t}(\mathbf{x},t) - \beta\triangle u(\mathbf{x},t) = F(\mathbf{x},t,u(\mathbf{x},t)),\ \mathbf{x} \in \Omega, 0 < t < T, \\
u(\mathbf{x},t) = 0,\ \mathbf{x} \in \partial\Omega, 0 < t < T, \\
u(\mathbf{x},0) = \int_0^T k_1(s)f_1\left(s,u(\mathbf{x},s),\frac{\partial u}{\partial s}(\mathbf{x},s)\right)ds,\ \mathbf{x} \in \Omega, \\
\frac{\partial u}{\partial t}(\mathbf{x},0) = \int_0^T k_2(s)f_2\left(s,u(\mathbf{x},s),\frac{\partial u}{\partial s}(\mathbf{x},s)\right)ds,\ \mathbf{x} \in \Omega,
\end{cases}
$$

$$(10.10)$$

where $k_i \in C([0,T];\mathbb{R})$ and $f_i : [0,T] \times \mathbb{R} \times \mathbb{R} \to \mathbb{R}\, (i = 1,2)$.

Common Theme: The common feature of all of the above nonlocal models is that the IC now depends on the unknown state rather than being a prescribed function independent of the state process (as has been the case up to this point). As such, the abstract formulation of these models must incorporate this change by replacing the old IC $u(0) = u_0$ by the new IC $u(0) = g(u)$, where g is an appropriate functional.

10.3   Some Abstract Theory

The key observation to make when reformulating the IVP and IBVPs in Section 10.2 as an abstract evolution equation is that regardless of whether or not the equation includes a functional forcing term, or if it is of Sobolev-type or if it is of second-order etc., the IC of each IVP/IBVP contains a functional term $g(u)$, where u stands for the state. In fact, as a preemptive measure we introduced each of the functionals arising within the ICs in Chapter 6, each defined on a variety of function spaces. But, how do we account for this when extending our core of theoretical results to this type of problem?

10.3.1   The Semi-Linear Case

We begin with the semi-linear nonlocal evolution equation

$$\begin{cases} u'(t) = Au(t) + f(t, u(t)), \ 0 < t < T, \\ u(0) = g(u), \end{cases} \tag{10.11}$$

in a Banach space $\mathscr{X}$, where $A : \mathrm{dom}(A) \subset \mathscr{X} \to \mathscr{X}$ is a linear (possibly unbounded) operator, $f : [0,T] \times \mathscr{X} \to \mathscr{X}$, and $g : C([0,T];\mathscr{X}) \to \mathscr{X}$. The definitions of mild and classical solutions of (10.11) are natural extensions of Def. 5.5.1. Precisely, we have

Definition 10.3.1. A function $u : [0,T] \to \mathscr{X}$ is a
i.) *classical solution* of (10.11) on $[0,T]$ if u satisfies Def. 5.5.1(i) with the IC replaced by $u(0) = g(u)$, where $\mathrm{rng}(g) \subset \mathrm{dom}(A)$;
ii.) *mild solution* of (10.11) on $[0,T]$ if u satisfies Def. 5.5.1(ii) with the IC replaced by $u(0) = g(u)$.

In both cases, the variation of parameters formula for the solution is given by

$$u(t) = e^{At}g(u) + \int_0^t e^{A(t-s)} f(s, u(s)) ds. \tag{10.12}$$

The general strategy employed for (10.11) naturally resembles the approach used in Chapter 6. (Why?) Indeed, the main steps are:

Step 1. Choose the function space $\mathcal{Y}$ to which you want the mild solution of (10.11) to belong. (It is necessary for $\mathcal{Y} \subset \text{dom}(g)$, but they need not be equal. (Why?))
Step 2. Define the solution map in the spirit of (5.65).
Step 3. Apply the appropriate fixed-point theorem or other method (as in the case of non-Lipschitz forcing terms) to establish the existence of a fixed-point which coincides with a global mild solution of (10.11) on $[0,T]$.
Step 4. (Classical solution) Depending on the regularity of the forcing term and the range of the functions used in defining the IC, you might not be able to conclude this level of regularity.

We wish to establish the standard results for (10.11) regarding existence and uniqueness of mild (and classical) solutions, continuous dependence on inital data and parameters, and related convergence schemes. We want to make certain to preserve uniqueness of the mild (and classical) solution in those cases where it would have been assured if $g(u)$ were constant, thereby resembling the fixed IC case. That is, the results we are extending should be subsumed as corollaries to the theorems governing (10.11).

Exercise 10.3.1. If we want to apply the above strategy, what seem to be natural conditions to impose on f and g to enable us to form such an extension of the theory?

Proposition 10.3.2. *Consider (10.11).*
i.) **(Lipschitz case)** *Assume* $(\mathbf{H_A})$ *and*
(H10.1) $g : \mathbb{C}([0,T];\mathcal{X}) \to \mathcal{X}$ *is a continuous mapping for which* $\exists M_g > 0$ *such that*
$$\|g(x) - g(y)\|_{\mathcal{X}} \le M_g \|x - y\|_{\mathbb{C}([0,T];\mathcal{X})}, \tag{10.13}$$
$\forall x, y \in \mathbb{C}([0,T];\mathcal{X})$.
    a.) *If f satisfies (5.55) globally on $\mathcal{X}$ and* $\overline{M_A}\left(M_g + TM_f\right) < 1$*, then (10.11) has a unique mild solution on* $[0,T]$.
    b.) *If* **(H5.1)** *holds,* $\text{rng}(g) \subset \text{dom}(A)$*, and* $\overline{M_A}\left(M_g + TM_f\right) < 1$*, then the mild solution from (a) is classical.*
ii.) **(Compactness case)** *Assume* $\left(\mathbf{H_A^\star}\right)$ *and*
(H10.2) $g : \mathbb{C}([0,T];\mathcal{X}) \to \mathcal{X}$ *is a continuous, compact mapping for which there exist positive constants c_1, c_2 such that*
$$\|g(x)\|_{\mathcal{X}} \le c_1 \|x\|_{\mathbb{C}([0,T];\mathcal{X})} + c_2,$$
$\forall x \in \mathbb{C}([0,T];\mathcal{X})$.
    If f satisfies **(H5.10)** *and* $\overline{M_A}(c_1 + c_2 T) < 1$*, then (10.11) has at least one mild solution on* $[0,T]$.

Outline of Proof: Let $v \in \mathbb{C}([0,T];\mathcal{X})$ and consider the IVP
$$\begin{cases} u_v'(t) = Au_v(t) + f(t, v(t)), \ 0 < t < T, \\ u_v(0) = g(v). \end{cases} \tag{10.14}$$

Observe that (10.14) has a unique mild solution u_v on $[0,T]$. (Why?) Define the solution map $\Phi : \mathbb{C}([0,T];\mathscr{X}) \to \mathbb{C}([0,T];\mathscr{X})$ by

$$(\Phi v)(t) = u_v(t) = e^{At} g(v) + \int_0^t e^{A(t-s)} f(s,v(s)) ds, \; 0 < t < T. \quad (10.15)$$

Φ is a well-defined, continuous mapping in each of the three separate cases in this theorem. (Verify this.) To verify (i)(a), we apply the Contraction Mapping Theorem. Observe that $\forall v_1, v_2 \in \mathbb{C}([0,T];\mathscr{X})$,

$$\|(\Phi v_1)(t) - (\Phi v_2)(t)\|_{\mathscr{X}} \le \left\| e^{At} \right\|_{\mathbb{B}(\mathscr{X})} \|g(v_1) - g(v_2)\|_{\mathscr{X}} +$$

$$\int_0^t \left\| e^{A(t-s)} \right\|_{\mathbb{B}(\mathscr{X})} \|f(s,v_1(s)) - f(s,v_2(s))\|_{\mathscr{X}} ds$$

$$\le \overline{M_A} \left[M_g \|v_1 - v_2\|_{\mathbb{C}([0,T];\mathscr{X})} + \int_0^t M_f \|v_1(s) - v_2(s)\|_{\mathscr{X}} ds \right]$$

$$\le \overline{M_A} \left(M_g + T M_f \right) \|v_1 - v_2\|_{\mathbb{C}([0,T];\mathscr{X})} \quad \text{(Why?)}$$

$$< \|v_1 - v_2\|_{\mathbb{C}([0,T];\mathscr{X})} .$$

Hence, Φ has a unique fixed-point which satisfies Def. 10.3.1(ii), as desired.

Part (i)(b) is argued as in Thrm. 5.5.2 (see **[11]**, for instance).

Finally, Schaefer's fixed-point theorem can be used to establish (ii) in a manner similar to the proof of Thrm. 5.7.3. The main argument remains intact, but you need to use the fact that if K is a bounded set, then the set $\{g(v)|v \in K\}$ is precompact in $\mathscr{X}$. (Why?) How does this enter into the argument? $\square$

Exercise 10.3.2. Complete the proof of Prop. 10.3.2.

The following generic exercises describe some of the directions that various authors have taken in the past two decades in this line of research. You have acquired all of the tools required to formulate and establish these results on your own. Try doing so!

Exercise 10.3.3. Establish a continuous dependence result for (10.11) under the assumptions of Prop. 10.3.2(i).

Exercise 10.3.4. Impose conditions on the data in each of (10.1), (10.2), and (10.10) so that upon reformulating them as (10.11), the mappings f and g satisfy the hypotheses of Prop. 10.3.2(i), thereby guaranteeing the global existence of mild solutions.

Exercise 10.3.5.
i.) Formulate a result in the spirit of Prop. 6.3.4 for (10.11).
ii.) Apply (i) directly to the IBVPs mentioned in Section 10.2. In each case, impose precise conditions on the data and verify that the hypotheses of (i) hold.

Open Exercise 10.3.6. Can you formulate a result governing the existence and uniqueness of a global mild solution of (10.11) under non-Lipschitz conditions? What issues do you encounter? Can they be overcome?

10.3.2 The General Functional Case

More generally, we consider the abstract nonlocal functional evolution equation

$$\begin{cases} u'(t) &= Au(t) + \mathfrak{F}(u)(t), 0 < t < T, \\ u(0) &= g(u), \end{cases} \tag{10.16}$$

in a Banach space $\mathscr{X}$, where $\mathfrak{F} : \mathbb{C}\left([0,T];\mathscr{X}\right) \to \mathbb{L}^1\left(0,T;\mathscr{X}\right)$ and $g : \mathbb{C}\left([0,T];\mathscr{X}\right) \to \mathscr{X}$. Such problems have been studied in **[6, 11, 248]**. Just as in the semi-linear case, the definition of a mild solution of (10.16) is an amalgam of Def. 6.3.1 and Def. 10.3.1(ii). Precisely, we have

Definition 10.3.3. A function $u : [0,T] \to \mathscr{X}$ is a *mild solution* of (10.16) on $[0,T]$ if u satisfies Def. 10.3.1(ii) with $f(t,u(t))$ replaced by $\mathfrak{F}(u)(t)$.

The following nonlocal generalization of Thrm. 6.3.2.

Proposition 10.3.4. *Consider (10.16).*
i.) *If* **(HA)**, **(H10.1)**, *and* **(H6.8)** *hold and* $\overline{M_A}\left(M_g + M_f\right) < 1$, *then (10.16) has a unique mild solution on* $[0,T]$.
ii.) *If* **(H$_A^\star$)**, **(H10.2)**, *and* **(H6.16)** *hold and* $\overline{M_A}\left(d_1 + c_1\right) < 1$, *then (10.16) has at least one mild solution on* $[0,T]$.

Exercise 10.3.7. Prove Prop. 10.3.4 by adapting the proofs of Thrm. 6.3.3 and Thrm. 6.4.1 appropriately.

Exercise 10.3.8. Formulate and prove a convergence result for (10.16) in the spirit of Prop. 6.3.4.

Exercise 10.3.9. Explain what changes must be made if the range of the functional $\mathfrak{F}$ is now the smaller space $\mathbb{C}\left([0,T];\mathscr{X}\right)$.

Exercise 10.3.10. Show that Prop. 10.3.2(i)(a) and (ii) can be obtained as a corollary of Prop. 10.3.4.

Exercise 10.3.11. Impose conditions on the data in each of (10.3), (10.5), and (10.6) so that upon reformulating them as (10.11), the mappings $\mathfrak{F}$ and g satisfy the hypotheses of Prop. 10.3.4(i), thereby guaranteeing the global existence of a mild solutions.

Assuming **(H7.1)** - **(H7.6)**, suitably adapted, the theoretical results for (10.16) can be applied directly to abstract nonlocal Sobolev functional evolution equations of the form

$$\begin{cases} (Bu)'(t) = Au(t) + \mathfrak{F}(u)(t), \ 0 < t < T, \\ u(0) = g(u), \end{cases} \qquad (10.17)$$

in a Banach space $\mathscr{X}$, where $\mathfrak{F} : \mathbb{C}([0,T];\mathscr{X}) \to \mathbb{L}^1(0,T;\mathscr{X})$ and $g : \mathbb{C}([0,T];\mathscr{X}) \to \mathscr{X}$. In fact, Prop. 7.1.6 - Prop. 7.1.9, and all of the related exercises, can be generalized to account for a nonlocal IC. These results, in turn, can be applied to (10.7) under suitable hypotheses. Complete the following instructive, rather lengthy, exercise to help facilitate your understanding of nonlocal problems.

Exercise 10.3.12. Generalize Prop. 7.1.6 - Prop. 7.1.9, and all of the related exercises, to the present nonlocal setting.

Delay versions of the above results, and for related nonlocal evolution equations, have been explored by many authors using essentially the same approach in an appropriate Banach space. It would be instructive for you to try to formulate such results on your own, and then consult the mentioned papers.

10.3.3   The Nonlinear Case

The last class of nonlocal evolution equations that we consider is when the operator $\mathscr{A}$ is nonlinear (as in Chapter 9). Precisely, consider

$$\begin{cases} u'(t) & = \mathscr{A}u(t) + \mathfrak{F}(u)(t), \ 0 < t < T, \\ u(0) & = g(u), \end{cases} \qquad (10.18)$$

in a Banach space $\mathscr{X}$, where $\mathscr{A} : \mathrm{dom}(\mathscr{A}) \subset \mathscr{X} \to \mathscr{X}$ is a nonlinear, unbounded (possibly multi-valued) operator, $g : \mathbb{C}([0,T];\mathrm{cl}_{\mathscr{X}}(\mathrm{dom}(\mathscr{A}))) \to \mathrm{cl}_{\mathscr{X}}(\mathrm{dom}(\mathscr{A}))$ and $\mathfrak{F} : \mathbb{C}([0,T];\mathscr{X}) \to \mathbb{L}^1(0,T;\mathscr{X})$. Evolution equations of the form (10.18) have been studied in **[6, 11]** under different growth and regularity conditions on $\mathfrak{F}$ and g. For example, consider the following generalization of Prop. 9.3.13(i):

Proposition 10.3.5. *Assume that $\mathscr{A} : \mathrm{dom}(\mathscr{A}) \subset \mathscr{X} \to \mathscr{X}$ is m-dissipative and that* **(H6.8)** *and* **(H10.1)** *(with the spaces suitably modified) hold. If $M_{\mathfrak{F}} + M_g < 1$, then (10.18) has a unique integral solution on $[0,T]$.*

Exercise 10.3.13. Prove Prop. 10.3.5 by suitably modifying the proof of Prop. 9.3.13(i).

Exercise 10.3.14. Try to formulate and prove a generalization of Prop. 9.3.13(ii) using Schaefer's fixed-point theorem. (A more general multi-valued variant of this result was established in **[11]**. Related results under modified growth conditions were more recently established in **[8]**.)

Exercise 10.3.15. Impose conditions on the data in each of (10.4), (10.8), and (10.9) so that upon reformulating them as (10.11), the mappings $\mathfrak{F}$ and g satisfy the hypotheses of Prop. 10.3.5(i), thereby guaranteeing global existence of integral solutions.

10.4   Final Comments

This is an area of ongoing research. Some recent work in the linear setting established under a variety of different types of growth conditions can be found in [21, 25, 27]. The delay case has received significant attention, as well; refer to [2, 24, 115]. Authors have treated abstract second-order evolution equations using cosine and sine families (cf. Section 4.3); refer to [162, 167]. Such work offers a different manner in which to study nonlocal wave equations. Rather general results for classes of abstract multi-valued evolution inclusions have been established in [11, 12], including a discussion of some interesting examples. At the moment, it remains an open question as to whether or not nonlocal variants of the results established under non-Lipschitz conditions on the forcing terms can be obtained.

Chapter 11

Beyond Volume 1...

Overview

The material developed in this volume has provided an introductory look at evolution equations designed to acquaint you with some essential and foundational notions and techniques, as well as a wealth of applications to which they apply. But, this is just the beginning. Several more chapters could have been included in this text that explored other interesting classes of abstract evolution equations, similarly rooted in concrete applications, using a similar approach. And, volumes more could be written for the study of abstract evolution equations for which a nice variation of parameters formula is no longer available. As encouragment for you to continue the current line of study, we provide very short encounters with three different, yet related, classes of abstract evolution equations below, each of which can be studied using a similar approach to the one developed in this text. We then conclude with a short introduction to Volume 2 on stochastic evolution equations.

11.1 Three New Classes of Evolution Equations

11.1.1 Time-Dependent Evolution Equations

Model IX.4 Time-Dependent Nonlocal Neural Networks
Consider the following time-dependent variant of IVP (10.2):

$$\begin{cases} \frac{d}{dt}x_1(t) = a_1(t)x_1(t) + \sum_{j=1}^{M} \omega_{1j}(t)g_j\left(x_j(t)\right), \\ \vdots \\ \frac{d}{dt}x_M(t) = a_M(t)x_M(t) + \sum_{j=1}^{M} \omega_{Mj}(t)g_j\left(x_j(t)\right), \\ x(0) = \int_0^T \alpha_1(s)f\left(s, x_1(s, \cdot)\right)ds, \\ \vdots \\ x(0) = \int_0^T \alpha_M(s)f\left(s, x_M(s, \cdot)\right)ds, \end{cases} \tag{11.1}$$

where $\alpha_i \in \mathbb{L}^2(0,T;\mathbb{R})$, $(i=1,\ldots,M)$.

Model V.6 Time-Dependent Functional Diffusion-Advection Equation

Let $n \in \mathbb{N}$ and $0 < t_1 < t_2 < \ldots < t_n < T$ be fixed times. Consider the following generalization of IBVP (6.40) governing a diffusive-advective process with accumulative external force and more general time-dependent diffusion:

$$\begin{cases} \frac{\partial z}{\partial t} + \alpha^2 \frac{\partial}{\partial x}\left(a(t,x)\frac{\partial z}{\partial x}\right) + \gamma b(t,x)\frac{\partial z}{\partial x} + c(t,x) \\ = \sum_{i=1}^n \beta_i(x)z(x,t_i) + \int_0^T \alpha(s)f(s,z)\,ds, 0 < x < L, 0 < t < T, \\ \frac{\partial z}{\partial x}(0,t) = \frac{\partial z}{\partial x}(L,t) = 0, t > 0, \\ z(x,0) = z_0(x),\ 0 < x < L, \end{cases} \tag{11.2}$$

where a,b,c are sufficiently smooth functions.

We can consider similar generalizations of all models developed in the text. The main change is that the operator identified by A in each case is now time-dependent. As such, we can naively reformulate (11.1) and (11.2) as the abstract nonlocal time-dependent evolution equation

$$\begin{cases} u'(t) &= A(t)u(t) + \mathfrak{F}(u)(t), 0 < t < T, \\ u(0) &= g(u). \end{cases} \tag{11.3}$$

The theory is understandably more complicated from the very beginning due to the time dependence of the operator A. However, it is possible to develop a theory in both the linear and nonlinear settings that resembles our development when the operator is not time dependent. Indeed, knowing nothing else and ignoring all technical details, we might expect each member of the family of operators $\{A(t)|t \geq 0\}$ to generate a semigroup. Of course, there is no reason to expect the semigroup to remain the same for each value of t. Rather, we have the following loose association:

$$A \mapsto \{e^{As}|s \geq 0\} \tag{11.4}$$
$$A(t) \mapsto \{U(t,s)|0 \leq s \leq t < \infty\} \tag{11.5}$$

For each fixed t, the hope is that the family of operators in (11.5) somehow resembles a semigroup as in (11.4). Indeed, as developed in **[266]**, this interpretation can be made formal, and in the linear case, the variation of parameters formula for a mild solution for (11.3) is given by

$$u(t) = U(t,0)g(u) + \int_0^t U(t,s)\mathfrak{F}(u)(s)\,ds. \tag{11.6}$$

A formula for an integral solution of (11.3) comparable to (9.47) in the nonlinear setting also exists. (See **[9, 222, 230, 266, 289]**, for instance.)

11.1.2   Quasi-Linear Evolution Equations

Further generalizing the time-dependent case, we can incorporate state dependence into the operators $A(t)$ in (11.3). Such operators arise naturally when reformulating the IBVPs arising in environmental science and mathematical physics abstractly. For instance, consider the following examples.

Model XI.4 Quasi-linear Pollution Model

Let Ω be a bounded region in $\mathbb{R}^N$ with smooth boundary $\partial \Omega$, and let $w(\mathbf{x}, t)$ denote the pollution concentration at position $\mathbf{x} \in \Omega$ and time $t > 0$. Consider the following generalization of (6.47) which now accounts for a more elaborate wind trajectory:

$$
\begin{cases}
\frac{\partial z}{\partial t} = k\triangle z + \sum_{i=1}^{N} a_i(t, \mathbf{x}, z(\mathbf{x}, t)) \frac{\partial z}{\partial x_i} + \int_0^t a(t-s) g(s, z) ds, \ \mathbf{x} \in \Omega, t > 0, \\
z(\mathbf{x}, 0) = z_0(\mathbf{x}), \ \mathbf{x} \in \Omega, \\
\frac{\partial z}{\partial \mathbf{n}}(\mathbf{x}, t) = 0, \ \mathbf{x} \in \partial \Omega, t > 0,
\end{cases}
\tag{11.7}
$$

where $a_i \ (i = 1, \ldots, N)$ are sufficiently smooth.

Model XXXI.1 Quasi-linear KdV equation

An initial model of the behavior of shallow water waves (later referred to as *solitons*) was formulated in the late nineteenth century by Korteweg and de Vries, resulting in the so-called *Korteweg-de Vries (KdV) equation*. A general version of the equation portion of the IBVP arising in this model is given by

$$
\frac{\partial z}{\partial t} + \frac{\partial^3 z}{\partial x^3} + \alpha(z) \frac{\partial z}{\partial x} = 0, \ x \in \mathbb{R},
\tag{11.8}
$$

where $\alpha(\cdot)$ is sufficiently smooth. (See **[143, 191, 307]**.)

Model VII.6 Quasi-linear Wave Equations

Certain wave phenomena are too complicated as to admit a description using the classical wave equation. Some specific areas in which more general wave equations naturally arise include fluid dynamics (via the Navier-Stokes equations), magnetohydrodynamics, and forestation. (See **[191, 250]** for details.) An example of a typical quasi-linear wave equation is

$$
\frac{\partial^2 z}{\partial t^2} - \alpha(z) \frac{\partial^2 z}{\partial x^2} = f\left(t, z, \frac{\partial z}{\partial t}, \frac{\partial z}{\partial x}\right),
\tag{11.9}
$$

where $\alpha(\cdot)$ is sufficiently smooth.

The equation portion of IBVPs (11.7) - (11.9) can be viewed as an abstract quasi-linear evolution equation of the form

$$
u'(t) = A(t, u)u(t) + \mathfrak{F}(u)(t).
\tag{11.10}
$$

The theory is considerably more complicated to develop for (11.10), but many researchers have developed results under different assumptions on the family of operators $\{A(t, u) | t \geq 0\}$. While a convenient variation of parameters formula does not exist for the general equation (11.10), various elements of the approach used in this text are applicable in its investigation. Some standard references include **[26, 34, 111, 116, 235, 250, 301, 324]**.

11.1.3   Integro-Differential Evolution Equations

Sometimes, the abstract formulation of an IBVP is an evolution equation that involves an integral operator. For instance, this arises when formulating mathematical models of heat conduction in materials with memory **[219]** and in viscoelasticity **[15]**.

Model V.6 Heat Conduction in Materials with Memory
Let $u = u(x,t)$ denote the temperature at position $x \in [0,L]$ at time $t > 0$ on a cross-section of a material assumed to have memory. Heat conduction through such material is described by the IBVP

$$
\begin{cases}
\frac{\partial u}{\partial t}(x,t) = -\frac{\partial v}{\partial x}(x,t) + f(x,t), \ 0 < x < L, t > 0, \\
-v(x,t) = \frac{\partial u}{\partial x}(x,t) + \int_0^t a(t-s)\frac{\partial u}{\partial x}(x,s)ds, 0 < x < L, t > 0, \\
u(x,0) = u_0(x),
\end{cases}
\tag{11.11}
$$

which can be written equivalently as

$$
\begin{cases}
\frac{\partial u}{\partial t}(x,t) = \frac{\partial^2 u}{\partial x^2}(x,t) + \frac{\partial^2}{\partial x^2}\int_0^t a(t-s)u(x,s)ds + f(x,t), 0 < x < L, t > 0, \\
u(x,0) = u_0(x).
\end{cases}
$$
$$\tag{11.12}$$

The presence of the second term on the right-side of (11.12) is the new feature preventing the IBVP from being of type (4.21). It turns out that this IBVP can be formulated as the abstract evolution equation

$$
\begin{cases}
u'(t) = A\left[u(t) + \int_0^t a(t-s)u(s)ds\right] + f(t), t > 0, \\
u(0) = u_0.
\end{cases}
\tag{11.13}
$$

Such evolution equations can be transformed, in some sense, into our previous setting of Chapters 4 and 5 with the help of a so-called *resolvent family*. Once the framework is established (as in **[276]**), one can study functional, delay, and nonlocal variants of (11.13). Some such work can be found in **[7, 82, 83, 98, 105, 148, 275]**.

11.2   Next Stop... Stochastic Evolution Equations!: Preface to Volume 2

The presence of "noise" in nature obscures the description of the behavior of any phenomena, and hinders the theory of *deterministic* evolution equations from providing a complete picture of such behavior. If noise is taken into account in these models, by way of perturbations of the operators involved or via a Wiener process, then the evolution equations become *stochastic* in nature. The development of the general theory is similar to the deterministic case, but a considerable amount of additional machinery is needed in order to rigorously handle the addition of noise, and questions regarding the nature of the solutions (which are now viewed as stochastic processes rather than deterministic mappings) must be addressed due to the probabilistic nature of the equations.

Volume 2 of this book series is devoted to the formulation of the theory of stochastic evolution equations. The primary overarching objective of the two-volume project is to very deliberately facilitate the transition from deterministic evolution equations to their stochastic counterparts. To this end, the volumes are written in very close alignment to facilitate the extension from the deterministic setting to the stochastic for each class of abstract evolution equations under investigation. The same concrete models are discussed in both volumes so that you can clearly see the effect of accounting for noise in the model. I suggest that as you work through Volume 2, refer back to the present volume frequently to make comparisons for each class of equations being studied. This will help build your intuition about the subject.

Bibliographic Remarks

Much of the theory presented in this text is considered classical. Many researchers and authors have contributed to its development throughout the past few decades and have presented the material in different ways and in different venues, including journal articles, research monographs, and textbooks. I have made no deliberate attempt to trace results back to their foundations, but rather I have relied on a rich supply of standard references to guide me while developing my own presentation of the material. The bibliography consists of more than 340 references covering various aspects of the theory. It is my hope that this list will provide you with a sufficient starting point to help you broaden your understanding of this area of study.

The chapter-by-chapter commentary provided below is meant to specifically acknowledge those references used to formulate the statement and proof of specific classical definitions, theorems, and examples in this text. Given the now-standard nature of this material, it is probable that numerous sources not mentioned here also contain versions of these results. While it would be virtually impossible to track down and identify all such sources, I have attempted to provide a thorough listing of references to which you can refer to round out your knowledge of the subject. All specific citations for the collection of models discussed throughout the text are provided in the actual body of the text and will not be repeated below.

Chapter 1

The topics included in this brief account of elementary analysis can be found in virtually any text on the subject. The standard references used to compile parts of this chapter are listed in the overview of the chapter.

Chapter 2

The practical and theoretical aspects of finite-dimensional homogenous linear systems of ODEs are discussed to varying degrees in virtually every undergraduate text devoted to the study of ODEs. The theory in Section 2.2 is discussed in detail, with specific attention paid to stability, in **[172, 220, 319]**. Propositions 2.2.4 - 2.2.10 are adapted to the finite-dimensional case from **[267, 318]**. Theorem 2.3.2 and Cor.2.3.3 are from **[123, 142, 267, 318]**. Proposition 2.4.1 is a special case of Lemma 8.1 in **[142]**.

Chapter 3

The material on linear operators presented in Section 3.1 can be found in all functional analysis texts. The primary references used to develop this section are **[118, 146, 185, 202, 210, 340]**. The example following Def. 3.1.14 is adapted from **[205]** and Prop. 3.1.17 is from **[320]**. The theory in the remainder of the chapter on linear C_0-semigroups and their application to the study of the homogenous Cauchy problem is classical and is developed in the sources listed at the beginning of Section 3.3. More specifically, the proof of Thrm. 3.3.4 most closely follows **[267, 318]**, with modifications occurring towards the end of the proof. The verification of the properties listed in Thrm. 3.3.6 is standard and most closely follows **[123, 205, 246, 267, 318]**. The discussion of regularity (i.e., Def. 3.4.5 - Prop. 3.4.7) most closely resembles the presentation in **[341]**. The definition and properties of resolvents and Yosida approximations in Section 3.5 are adapted from **[107, 142, 205, 267, 318]**. The proof of the Hille-Yosida theorem is well-known; the version presented in this text combines the approaches used in **[107, 205, 318]**. The example illustrating this theorem is adapted from **[318]**. Lemma 3.5.6 - Prop. 3.5.8 are modified from **[103]**. The proofs of Prop. 3.5.12 and Thrm. 3.5.13 are adapted from **[49, 318]**. The outline of the proof of Prop. 3.7.2 is adapted from **[142, 267]**.

Chapter 4

The computational aspects of the theory in this chapter are based on the usual variation of parameters approach. The content of Section 4.1 is a special case of the results in Section 4.2. The proof of Prop. 4.2.3 is adapted from **[267, 341]**. A more elaborate discussion of Prop. 4.2.4 can be found in **[267]**. The treatment of the wave equation using sine and cosine families is adapted from **[140 - 142]**; other aspects of this application are discussed in **[320]**.

Chapter 5

The material in Section 5.2 can be found in any standard functional analysis text; the primary sources used here are **[146, 185, 202, 340]**. The discussion in Sections 5.3 and 5.4 is based on standard material found in virtually any ODE text. The theoretical development in Section 5.5 is classical and is adapted from **[86, 97, 237, 267, 278, 318, 341]**. Specifically, the proof of Thrm. 5.5.2(i) follows the discussion in **[267, 318, 341]**. The outline of the proof of Prop. 5.5.3 follows **[142]**. The theoretical discussion in Section 5.5.3 is adapted from **[142, 318]**, and the proofs of Props. 5.5.7 and 5.5.8 are adapted from **[142, 318]**. A more elaborate discussion of extendability issues can also be found in these sources. The theory for non-Lipschitz type forcing terms discussed in Section 5.6 is newer and is also discussed in **[38, 96, 122, 229, 302, 333]**. A more thorough discussion of the theory formed under compactness assumptions can be found in **[267, 318]**.

Chapter 6

Evolution equations for which the forcing term is of a more general form that is conveniently identified as a mapping between function spaces have been studied occasionally, primarily in journal articles and advanced texts. (See **[6, 10, 11, 190, 317]**.) The principles upon which the theoretical results discussed in this chapter are based are rooted in such references.

Chapter 7

Section 7.1 is focused on Sobolev-type equations of the form considered in **[22, 30, 31, 33, 41, 64, 124, 149, 177, 188, 197, 215, 251, 292, 293, 294, 297, 303, 308, 327, 330]**. The underlying approach leading to the main results in Section 7.1.3 was extracted from these sources, most notably **[64, 293, 309]**. Propositions 7.1.6 and 7.1.9 are special cases of those in **[193]**. The theory of neutral evolution equations is expansive. The approach in the finite-dimensional case is a natural extension of the approach from Chapter 5, while the study of the infinite-dimensional case is facilitated by the theory of analytic semigroups. (See **[246, 267, 318]**, for instance.)

Chapter 8

Delay evolution equations have received much attention. The standard reference [156] was a guiding light in the development of Section 8.2 and 8.3. The results formed under non-Lipschitz growth conditions are newer (see [144, 145]); Thrm. 8.4.1 is based on these results. Other forms of delay, including unbounded delay and state-dependent delay, are also prevalent in the literature. (See [20, 58, 157, 159, 166, 171, 181, 218].)

Chapter 9

The literature for nonlinear evolution equations is equally vast, and arguably richer, than the linear theory. (See [39, 40, 49, 100, 207, 237, 249, 266, 287, 288, 295, 316, 317, 325].) The highlights presented in this chapter mainly follow [207, 249, 266, 317, 320]. It is also worth mentioning that the presentation provided in [49] is reader-friendly and discusses some interesting applications of the theory in detail.

Chapter 10

The topic of this chapter is a relatively new direction in evolution equations research. A representative sampling of journal articles devoted to the study of this topic is [3, 6, 8, 10, 11, 21, 24, 25, 27, 28, 33, 68 - 80, 87, 88, 112, 162, 167, 182, 94, 214, 216, 217, 221, 257, 262, 313, 322, 323, 336, 338].

Bibliography

[1] R.A. Adams and J.J.F. Fournier. *Sobolev Spaces*. Academic Press New York, 1975.

[2] S. Agarwal and D. Bahuguna. Existence and uniqueness of strong solutions to nonlinear nonlocal functional differential equations. *Electronic Journal of Differential Equations*, 2004(52):1–9, 2004.

[3] N.U. Ahmed. Differential inclusions on Banach spaces with nonlocal state constraints. *Nonlinear Functional Analysis and Applications*, 6(3):395–409, 2001.

[4] N.U. Ahmed. *Dynamic Systems and Control with Applications*. World Scientific, Singapore, 2006.

[5] S. Aizawa. A semigroup treatment of the Hamilton'Jacobí equation in several space variables. *Hiroshima Mathematics Journal*, 6:15–30, 1976.

[6] S. Aizicovici and Y. Gao. Functional differential equations with nonlocal initial conditions. *Journal of Applied Mathematics and Stochastic Analysis*, 10(2):145–156, 1997.

[7] S. Aizicovici and K.B. Hannsgen. Local existence for abstract semilinear Volterra integrodifferential equations. *J. Integral Equations Appl*, 5(3):299–313, 1993.

[8] S. Aizicovici and H. Lee. Nonlinear nonlocal Cauchy problems in Banach spaces. *Applied Mathematics Letters*, 18(4):401–407, 2005.

[9] S. Aizicovici and H. Lee. Existence results for nonautonomous evolution equations with nonlocal initial conditions. *Communications in Applied Analysis*, 11(2):285–297, 2007.

[10] S. Aizicovici and M. McKibben. Semilinear Volterra integrodifferential equations with nonlocal initial conditions. *Abstract and Applied Analysis*, 4(2):127–139, 1999.

[11] S. Aizicovici and M. McKibben. Existence results for a class of abstract nonlocal Cauchy problems. *Nonlinear Analysis*, 39(5):649–668, 2000.

[12] S. Aizicovici and V. Staicu. Multivalued evolution equations with nonlocal initial conditions in Banach spaces. *NoDEA: Nonlinear Differential Equations and Applications*, 14(3):361–376, 2007.

[13] J.M. Alonso, J. Mawhin, and R. Ortega. Bounded solutions of second order semilinear evolution equations and applications to the telegraph equation. *Journal de Mathématiques Pures et Appliquées*, 78(1):49–63, 1999.

[14] H. Amann and G. Metzen. *Ordinary Differential Equations: An Introduction to Nonlinear Analysis*. Walter de Gruyter, 1990.

[15] D. Andrade. A note on solvability of the nonlinear abstract viscoelastic problem in Banach Spaces, J. *Partial Diff. Equations*, 12:337–344, 1999.

[16] T.M. Apostol. *Mathematical Analysis*. Addison-Wesley Reading, MA, 1974.

[17] O. Arino, R. Benkhalti, and K. Ezzinbi. Existence results for initial value problems for neutral functional differential equations. *Journal of Differential Equations*, 138(1):188–193, 1997.

[18] L. Arlotti and J. Banasiak. Strictly substochastic semigroups with application to conservative and shattering solutions to fragmentation equations with mass loss. *Journal of Mathematical Analysis and Applications*, 293(2):693–720, 2004.

[19] A. Ashyralyev and A. Sirma. Nonlocal boundary value problems for the Schrodinger equation. *Computers and Mathematics with Applications*, 55(3):392–407, 2008.

[20] S. Baghli and M. Benchohra. Perturbed functional and neutral functional evolution equations with infinite delay in Fréchet spaces. *Electronic Journal of Differential Equations*, 2008(69):1–19, 2008.

[21] D. Bahuguna and S. Agarwal. Approximations of solutions to neutral functional differential equations with nonlocal history conditions. *Journal of Mathematical Analysis and Applications*, 317(2):583–602, 2006.

[22] D. Bahuguna and R. Shukla. Approximations of solutions to nonlinear Sobolev type evolution equations. *Electron. J. Differential Equations*, 2003(31):1–16, 2003.

[23] D.D. Bainov and D.P. Mishev. *Oscillation theory for neutral differential equations with delay*. Inst of Physics Pub Inc, 1991.

[24] K. Balachandran and M. Chandrasekaran. Existence of solutions of a delay differential equation with nonlocal condition. *Indian Journal of Pure and Applied Mathematics*, 27(5):443–449, 1996.

[25] K. Balachandran and M. Chandrasekaran. Existence of solutions of nonlinear integrodifferential equation with nonlocal condition. *Journal of Applied Mathematics and Stochastic Analysis*, 10(3):279–288, 1997.

[26] K. Balachandran and M. Chandrasekaran. Nonlocal Cauchy problem for quasilinear integrodifferential equation in Banach spaces. *Dynamic Systems and Applications*, 8:35–44, 1999.

[27] K. Balachandran and J.P. Dauer. Existence of solutions for an integrodifferential equation with nonlocal conditions in Banach space. *Libertas Mathematica*, 16:133–143, 1996.

[28] K. Balachandran and S. Ilamaran. Existence and uniqueness of mild and strong solutions of a semilinear evolution equation with nonlocal conditions. *Indian Journal of Pure and Applied Mathematics*, 25(4):411–418, 1994.

[29] K. Balachandran, D.G. Park, and S.M. Anthoni. Existence of solutions of abstract nonlinear second-order neutral functional integrodifferential equations. *Computers and Mathematics with Applications*, 46(8-9):1313–1324, 2003.

[30] K. Balachandran and J.Y. Park. Nonlocal Cauchy problem for Sobolev type functional integrodifferential equation. *Bulletin of the Korean Mathematical Society*, 39(4):561–570, 2002.

[31] K. Balachandran, J.Y. Park, and M. Chandrasekaran. Nonlocal Cauchy problem for delay integrodifferential equations of Sobolev type in Banach spaces. *Applied Mathematics Letters*, 15(7):845–854, 2002.

[32] K. Balachandran, J.Y. Park, and I.H. Jung. Existence of solutions of nonlinear extensible beam equations. *Mathematical and Computer Modelling*, 36(7-8):747–754, 2002.

[33] K. Balachandran and K. Uchiyama. Existence of solutions of nonlinear integrodifferential equations of Sobolev type with nonlocal condition in Banach spaces. *Proceedings of Indian Academy of Mathematical Sciences*, 110(2):225–232, 2000.

[34] K. Balachandran and K. Uchiyama. Existence of solutions of quasilinear integrodifferential equations with nonlocal condition. *Tokyo Journal of Mathematics*, 23(1):203–210, 2000.

[35] A.V. Balakrishnan. On the (Non-numeric) Mathematical Foundations of Linear Aeroelasticity. In *Fourth International Conference on Nonlinear Problems in Aviation and Aerospace*, pages 179–194.

[36] J.M. Ball. Initial-boundary value problems for an extensible beam. *Journal of Mathematical Analysis and Applications*, 42(1):61–90, 1973.

[37] P. Baras. Compacité de lopérateur f->u solution dune équation non linéaire (du/dt)+ Au =f. *CR Acad. Sci. Paris Sér. AB*, 286:1113–1116, 1978.

[38] D. Barbu. Local and global existence for mild solutions of stochastic differential equations. *Portugaliae Mathematica*, 55(4):411–424, 1998.

[39] V. Barbu. *Nonlinear Semigroups and Differential Equations in Banach Spaces*. Editura Academiei Bucharest-Noordhoff, Leyden, 1976.

[40] V. Barbu. *Analysis and Control of Nonlinear Infinite Dimensional Systems*, volume 190. Academic Press, 1993.

[41] G.I. Barenblatt, Y.P. Zheltov, and I.N. Kochina. Basic concepts in the theory of seepage of homogeneous liquids in fissured rocks (strata). *PMM, J. Appl. Math. Mech.*, 24:1286–1303, 1961.

[42] E. Barone and A. Belleni-Morante. A nonlinear initial-value problem arising from kinetic theory of vehicular traffic. *Transport Theory and Statistical Physics*, 7(1):61–79, 1978.

[43] A. Batkai and S. Piazzera. *Semigroups for Delay Equations*. AK Peters, Ltd., 2005.

[44] J. Bear. *Hydraulics of Groundwater*. McGraw-Hill New York, 1979.

[45] C. Bellehumeur, P. Legendre, and D. Marcotte. Variance and spatial scales in a tropical rain forest: changing the size of sampling units. *Plant Ecology*, 130(1):89–98, 1997.

[46] A. Bellen and N. Guglielmi. Solving neutral delay differential equations with state-dependent delays. *Journal of Computational and Applied Mathematics*, 229(2):350–362, 2009.

[47] A. Bellen, N. Guglielmi, and A.E. Ruehli. Methods for linear systems of circuit delay differential equationsof neutral type. *IEEE Transactions on Circuits and Systems I: Fundamental Theory and Applications*, 46(1):212–215, 1999.

[48] A. Belleni-Morante. *A Concise Guide to Semigroups and Evolution Equations*, volume 19 of *Advances in Mathematics for Applied Sciences*. World Scientific, 1994.

[49] A. Belleni-Morante and A.C. McBride. *Applied Nonlinear Semigroups: An Introduction*. John Wiley & Sons Inc, 1998.

[50] M. Benchohra, E.P. Gatsori, and S.K. Ntouyas. Nonlocal quasilinear damped differential inclusions. *Electronic Journal of Differential Equations*, 2002(07):1–14, 2002.

[51] H.C. Berg. Chemotaxis in bacteria. *Annual Review of Biophysics and Bioengineering*, 4(1):119–136, 1975.

[52] R.L. Bisplinghoff, H. Ashley, and R.L. Halfman. *Aeroelasticity*. Dover Publications, 1996.

[53] J. Bochenek. An abstract nonlinear second order differential equation. *Annales Polonici Mathematici*, 54:155–166, 1991.

[54] J. Bochenek. Second order semilinear Volterra integrodifferential equation in Banach space. In *Annales Polonici Mathematici*, volume 57, pages 231–241, 1992.

[55] J. Bochenek. Existence of the fundamental solution of a second order evolution equation. *Ann. Polon. Math*, 66:15–35, 1997.

[56] A. Bonfoh and A. Miranville. On Cahn-Hilliard-Gurtin equations. *Nonlinear Analysis*, 47(5):3455–3466, 2001.

[57] M. Boudart. *Kinetics of Chemical Processes*. Prentice-Hall, 1968.

[58] H. Bouzahir. Semigroup approach to semilinear partial functional differential equations with infinite delay. *Journal of Inequalities and Applications*, 2007, 2007.

[59] F. Brauer and J.A. Nohel. *The Qualitative Theory of Ordinary Differential Equations: An Introduction*. Dover Publications, 1989.

[60] R. Brayton. Nonlinear oscillations in a distributed network. *Q. Appl. Math.*, 24:289–301, 1967.

[61] H. Brézis. *Operateurs Maximaux Monotones*. North-Holland Amsterdam, 1973.

[62] H. Brézis and F. Browder. Partial differential equations in the 20th century. *Advances in Mathematics*, 135(1):76–144, 1998.

[63] H. Brézis, P.G. Ciarlet, and J.L. Lions. *Analyse Fonctionnelle: Théorie et Applications*. Masson, Paris, 1983.

[64] H. Brill. A semilinear Sobolev evolution equation in a Banach space. *J. Differential Equations*, 24(3):412–425, 1977.

[65] T.A. Burton. *Stability by Fixed Point Theory for Functional Differential Equations*. Dover Publications, Mineola, NY, 2006.

[66] T.A. Burton and C. Kirk. A fixed point theorem of Krasnoselskii-Schaefer type. *Mathematische Nachrichten*, 189(1):23–31, 1998.

[67] S.N. Busenberg and C.C. Travis. On the use of reducible-functional-differential equations in biological models. *Journal of Mathematical Analysis and Applications*, 89:46–66, 1982.

[68] L. Byszewski. Differential and functional-differential problems with nonlocal conditions. *Cracow University of Thechnology Monograph*, 184.

[69] L. Byszewski. Existence and uniqueness of solutions of nonlocal problems for hyperbolic equation. *Journal of Applied Mathematics and Stochastic Analysis*, 3(3):163–168, 1990.

[70] L. Byszewski. Strong maximum principles for parabolic nonlinear problems with nonlocal inequalities together with integrals. *Journal of Applied Mathematics and Stochastic Analysis*, 3(1):65–79, 1990.

[71] L. Byszewski. Strong maximum principles for parabolic nonlinear problems with nonlocal inequalities together with arbitrary functionals. *Journal of Mathematical Analysis and Applications*, 156:457–470, 1991.

[72] L. Byszewski. Theorem about existence and uniqueness of continuous solution of nonlocal problem for nonlinear hyperbolic equation. *Applicable Analysis*, 40(2):173–180, 1991.

[73] L. Byszewski. Theorem about existence and uniqueness of continuous solution of nonlocal problem for nonlinear hyperbolic equation. *Applicable Analysis*, 40(2):173–180, 1991.

[74] L. Byszewski. Existence of approximate solution to abstract nonlocal Cauchy problem. *Journal of Applied Mathematics and Stochastic Analysis*, 5(4):363–373, 1992.

[75] L. Byszewski. Existence and uniqueness of solutions of semilinear evolution nonlocal Cauchy problem. *Zesz. Nauk. Pol. Rzes. Mat. Fiz*, 18:109–112, 1993.

[76] L. Byszewski. Uniqueness criterion for solution of abstract nonlocal Cauchy problem. *Journal of Applied Mathematics and Stochastic Analysis*, 6(1):49–54, 1993.

[77] L. Byszewski. Application of properties of the right-hand sides of evolution equations to an investigation of nonlocal evolution problems. *Nonlinear Analysis*, 33(5):413–426, 1998.

[78] L. Byszewski and H. Akca. On a mild solution of a semilinear functional-differential evolution nonlocal problem. *Journal of Applied Mathematics and Stochastic Analysis*, 10(3):265–271, 1997.

[79] L. Byszewski and H. Akca. Existence of solutions of a semilinear functional-differential evolution nonlocal problem. *Nonlinear Analysis*, 34(1):65–72, 1998.

[80] L. Byszewski and V. Lakshmikantham. Theorem about the existence and uniqueness of a solution of a nonlocal abstract Cauchy problem in a Banach space. *Applicable Analysis*, 40(1):11–19, 1991.

[81] C. Capellos and B.H.J. Bielski. *Kinetic Systems*. John Wiley and Sons New York, 1972.

[82] R.W. Carr and K.B. Hannsgen. A nonhomogeneous integrodifferential equation in Hilbert space. *SIAM Journal on Mathematical Analysis*, 10:961–984, 1979.

[83] R.W. Carr and K.B. Hannsgen. Resolvent formulas for a Volterra equation in Hilbert space. *SIAM Journal on Mathematical Analysis*, 13:459–483, 1982.

[84] A. Casal and A. Somolinos. Forced oscillations for the sunflower equation, Entrainment. *Nonlinear Analysis Theory, Methods & Applications*, 6(4):397–414, 1982.

[85] H. Caswell. *Matrix Population Models*. Sinauer Associates Sunderland (MA), 2001.

[86] T. Cazenave, A. Haraux, and Y. Martel. *An Introduction to Semilinear Evolution Equations*. Clarendon Press Oxford, 1998.

[87] J. Chabrowski. On nonlocal problems for parabolic equations. *Nagoya Math. J*, 93:109–131, 1984.

[88] J. Chabrowski. On the non-local problem with a functional for parabolic equation. *Funkcial. Ekvac*, 27:101–123, 1984.

[89] G. Chen. Control and stabilization for the wave equation in a bounded domain. *SIAM Journal on Control and Optimization*, 17:66–81, 1979.

[90] G. Chen. Control and stabilization for the wave equation, part II. *SIAM Journal on Control and Optimization*, 19:114–122, 1981.

[91] V.V. Chepyzhov and M.I. Vishik. *Attractors for Equations of Mathematical Physics*. American Mathematical Society, 2002.

[92] P.R. Chernoff. Perturbations of dissipative operators with relative bound one. *Proceedings of the American Mathematical Society*, 33(1):72–74, 1972.

[93] P. Clement. *One-Parameter Semigroups*. Elsevier Science Ltd, 1987.

[94] L. Cobb. Stochastic Differential Equations for the Social Sciences. *Mathematical Frontiers of the Social and Policy Sciences*, pages 37–68, 1981.

[95] E.A. Coddington and R. Carlson. *Linear Ordinary Differential Equations*. Society for Industrial and Applied Mathematics, 1997.

[96] E.A. Coddington and N. Levinson. *Theory of Ordinary Differential Equations*. Tata McGraw-Hill, 1972.

[97] C. Corduneanu. *Principles of Differential and Integral equations*. Chelsea Publishing Company, 1977.

[98] C. Corduneanu. *Integral Equations and Applications*. Cambridge University Press Cambridge, 1991.

[99] R.G. Cox. The motion of long slender bodies in a viscous fluid. Part 1. General theory. *Journal of Fluid Mechanics*, 44(part 3):790–810, 1970.

[100] M.G. Crandall and T.M. Liggett. Generation of semi-groups of nonlinear transformations on general Banach spaces. *American Journal of Mathematics*, 93(2):265–298, 1971.

[101] Z. Cui and Z. Yang. Roles of weight functions to a nonlinear porous medium equation with nonlocal source and nonlocal boundary condition. *Journal of Mathematical Analysis and Applications*, 342:559–570, 2007.

[102] R.V. Culshaw and S. Ruan. A delay-differential equation model of HIV infection of CD4+ T-cells. *Mathematical Biosciences*, 165(1):27–39, 2000.

[103] R.F. Curtain and H.J. Zwart. *An Introduction to Infinite-Dimensional Linear Systems Theory*. Springer, 1995.

[104] M. Cyrot. Ginzburg-Landau theory for superconductors. *Rep. Progress Physics*, 36(2):103–158, 1973.

[105] G. Da Prato and M. Iannelli. Linear integro-differential equations in Banach spaces. *Rendiconti del Seminario Matematico dellUniversita di Padova*, 62:207–219, 1980.

[106] J.P. Dauer and K. Balachandran. Existence of solutions of nonlinear neutral integrodifferential equations in Banach spaces. *Journal of Mathematical Analysis and Applications*, 251(1):93–105, 2000.

[107] R. Dautray and J.L. Lions. Evolution Problems I, volume 5 of Mathematical analysis and numerical methods for science and technology, 1992.

[108] E.B. Davies. *One-Parameter Semigroups*. Academic Press, 1980.

[109] L. Debnath. *Nonlinear Partial Differential Equations for Scientists and Engineers*. Birkhauser, 2005.

[110] K. Deimling. *Multivalued Differential Equations*. Walter de Gruyter, 1992.

[111] J.M. Delort, D. Fang, and R. Xue. Global existence of small solutions for quadratic quasilinear Klein–Gordon systems in two space dimensions. *Journal of Functional Analysis*, 211(2):288–323, 2004.

[112] K. Deng. Exponential decay of solutions of semilinear parabolic equations with nonlocal initial conditions. *Journal of Mathematical Analysis and Applications*, 179(2):630–637, 1993.

[113] W. Desch, R. Grimmer, and W. Schappacher. Wellposedness and wave propagation for a class of integrodifferential equations in Banach space. *Journal of Differential Equations*, 74(2):391–411, 1988.

[114] V. Dolezal. *Monotone Operators and Applications in Control and Network Theory*. Elsevier, 1979.

[115] Q. Dong, Z. Fan, and G. Li. Existence of solutions to nonlocal neutral functional differential and integrodifferential equations. *International Journal of Nonlinear Science*, 5(2):140–151, 2008.

[116] Q. Dong, G. Li, and J. Zhang. Quasilinear nonlocal integrodifferential equations in Banach spaces. *Electronic Journal of Differential Equations*, 2008(19):1–8, 2008.

[117] J. Duan, P. Holmes, and E.S. Titi. Global existence theory for a generalized Ginzburg-Landau equation. *Nonlinearity(Bristol. Print)*, 5(6):1303–1314, 1992.

[118] N. Dunford and J.T. Schwartz. *Linear Operators, Part I, Wiley Interscience.* 1958.

[119] J. Dyson, R. Villella-Bressan, and G.F. Webb. Asynchronous exponential growth in an age structured population of proliferating and quiescent cells. *Mathematical biosciences*, 177:73–83, 2002.

[120] L. Edelstein-Keshet. Mathematical Models in Biology. *Birkhauser mathematics series.*, 1988.

[121] R.E. Edwards. *Fourier Series: A Modern Introduction*. Holt, Rinehart and Winston, 1967.

[122] Y. El Boukfaoui and M. Erraoui. Remarks on the existence and approximation for semilinear stochastic differential equations in Hilbert spaces. *Stochastic Analysis and Applications*, 20(3):495–518, 2002.

[123] K.J. Engel and R. Nagel. *One-Parameter Semigroups for Linear Evolution Equations*. volume 194 of Graduate Texts in Mathematics. Springer-Verlag, New York, 2000.

[124] M.E. Erdogan and C.E. İmrak. On some unsteady flows of a non-Newtonian fluid. *Applied Mathematical Modelling*, 31(2):170–180, 2007.

[125] L.C. Evans. *Partial Differential Equations*. vol. 19 in Graduate Studies in Mathematics. American Mathematical Society, 1998.

[126] H. Eyring, S.H. Lin, and SM Lin. *Basic Chemical Kinetics*. John Wiley and Sons New York, 1980.

[127] K. Ezzinbi and X. Fu. Existence and regularity of solutions for some neutral partial differential equations with nonlocal conditions. *Nonlinear Analysis*, 57(7-8):1029–1041, 2004.

[128] H.O. Fattorini. *Second Order Linear Differential Equations in Banach Spaces*. North Holland, 1985.

[129] H.O. Fattorini. *Infinite Dimensional Optimization and Control Theory*. Cambridge University Press, 1999.

[130] H.O. Fattorini and A. Kerber. *The Cauchy Problem*. Cambridge University Press, 1984.

[131] Z. Feng, W. Huang, and C. Castillo-Chavez. Global behavior of a multi-group SIS epidemic model with age structure. *Journal of Differential Equations*, 218(2):292–324, 2005.

[132] W.E. Fitzgibbon. Global existence and boundedness of solutions to the extensible beam equation. *SIAM Journal on Mathematical Analysis*, 13(5):739–745, 1982.

[133] R. Fitzhugh. Impulses and physiological states in theoretical models of nerve membrane. *Biophysical Journal*, 1(6):445–466, 1961.

[134] W.H. Fleming. Diffusion processes in population biology. *Advances in Applied Probability*, 7:100–105, 1975.

[135] G.B. Folland. *Introduction to Partial Differential Equations*. Princeton University Press, 1995.

[136] Y.C. Fung. *An Introduction to the Theory of Aeroelasticity*. Dover Publications, 2002.

[137] M. Gitterman. *The Noisy Oscillator*. World Scientific, 2005.

[138] A. Glitzky and W. Merz. Single dopant diffusion in semiconductor technology. *Mathematical Methods in the Applied Sciences*, 27(2):133–154, 2004.

[139] M. Goland. The flutter of a uniform cantilever wing. *Journal of Applied Mechanics*, 12(4):197–208, 1945.

[140] J.A. Goldstein. Semigroups and second-order differential equations. *Journal of Functional Analysis*, 4:50–70, 1969.

[141] J.A. Goldstein. On a connection between first and second order differential equations in Banach spaces. *Journal of Mathematical Analysis and Applications*, 30:246–251, 1970.

[142] J.A. Goldstein. *Semigroups of Linear Operators and Applications*. Oxford University Press New York, 1985.

[143] J.A. Goldstein. The KdV equation via semigroups, in Theory and Applications of Nonlinear Operators of Accretive and Monotone Type, 107-114. *Lecture Notes in Pure and Applied Mathematics*, 178:107–114, 1996.

[144] T.E. Govindan. An existence result for the Cauchy problem for stochastic systems with heredity. *Differential and Integral Equations -Athens*, 15(1):103–114, 2002.

[145] T.E. Govindan. Stability of mild solutions of stochastic evolution equations with variable delay. *Stochastic Analysis and Applications*, 21(5):1059–1077, 2003.

[146] D.H. Griffel. *Applied Functional Analysis*. Dover Publications, 2002.

[147] L.J. Grimm. Existence and uniqueness for nonlinear neutral-differential equations. *American Mathematical Society*, 77(3):374–376, 1971.

[148] G. Gripenberg, S.O. Londen, and O.J. Staffans. *Volterra Integral and Functional Equations*. Cambridge University Press, 1990.

[149] G. Gudehus. On the onset of avalanches in flooded loose sand. *Philosophical Transactions: Mathematical, Physical and Engineering Sciences*, 356(1747):2747–2761, 1998.

[150] R.B. Guenther and J.W. Lee. *Partial Differential Equations of Mathematical Physics and Integral Equations*. Dover Publications, 1996.

[151] V.E. Gusev, W. Lauriks, and J. Thoen. Evolution equation for nonlinear Scholte waves. *IEEE Transactions on Ultrasonics, Ferroelectrics and Frequency Control*, 45(1):170–178, 1998.

[152] R. Haberman. *Mathematical Models: Mechanical Vibrations, Population Dynamics, and Traffic Flow: An Introduction to Applied Mathematics.* Society for Industrial and Applied Mathematics, 1998.

[153] S. Habib, C. Molina-Paris, and T.S. Deisboeck. Complex dynamics of tumors: modeling an emerging brain tumor system with coupled reaction–diffusion equations. *Physica A: Statistical Mechanics and its Applications*, 327(3-4):501–524, 2003.

[154] T. Hagen. On the semigroup of linearized forced elongation. *Applied Mathematics Letters*, 18(6):667–672, 2005.

[155] T.C. Hagen. *Elongational flows in polymer processing.* PhD thesis, 1998.

[156] J.K. Hale. Theory of Functional Differential Equations. *Applied Mathematical Sciences*, 3, 1977.

[157] J.K. Hale and J. Kato. Phase space for retarded equations with infinite delay. *Funkcial. Ekvac*, 21(1):11–41, 1978.

[158] M. He. Global existence and stability of solutions for reaction diffusion functional differential equations. *Journal of Mathematical Analysis and Applications*, 199(3):842–858, 1996.

[159] H.R. Henríquez. Regularity of solutions of abstract retarded functional differential equations with unbounded delay. *Nonlinear Analysis*, 28(3):513–531, 1997.

[160] H.R. Henríquez and C.H. Vásquez. Differentiability of solutions of the second order abstract Cauchy problem. In *Semigroup Forum*, volume 64, pages 472–488. Springer, 2002.

[161] E. Hernández. Existence results for a class of semi-linear evolution equations. *Electronic Journal of Differential Equations*, 2001(24):1–14, 2001.

[162] E. Hernández. Existence of solutions to a second order partial differential equation with nonlocal conditions. *Electronic Journal of Differential Equations*, 2003(51):1–10, 2003.

[163] E. Hernández, H.R. Henríquez, and M.A. McKibben. Existence of solutions for second order partial neutral functional differential equations. *Integral Equations and Operator Theory*, 62(2):191–217, 2008.

[164] E. Hernández and M.A. McKibben. Some comments on: Existence of solutions of abstract nonlinear second-order neutral functional integrodifferential equations. *Computers and Mathematics with Applications*, 50(5-6):655–669, 2005.

[165] E. Hernández and M.A. McKibben. On state-dependent delay partial neutral functional–differential equations. *Applied Mathematics and Computation*, 186(1):294–301, 2007.

[166] E. Hernández, M.A. McKibben, and H.R. Henríquez. Existence results for partial neutral functional differential equations with state-dependent delay. *Mathematical and Computer Modelling*, 49(5-6):1260–1267, 2009.

[167] E. Hernández and M. Pelicer. Existence results for a second-order abstract Cauchy problem with nonlocal conditions. *Electronic Journal of Differential Equations*, 2005(73):1–17, 2005.

[168] S.C. Hille. Local well-posedness of kinetic chemotaxis models. *Journal of Evolution Equations*, 8(3):423–448, 2008.

[169] T. Hillen and A. Potapov. The one-dimensional chemotaxis model: global existence and asymptotic profile. *Mathematical Methods in the Applied Sciences*, 27(15):1783–1801, 2004.

[170] E.J. Hinch. The distortion of a flexible inextensible thread in a shearing flow. *Journal of Fluid Mechanics*, 74(part 2):317–333, 1976.

[171] Y. Hino, S. Murakami, and T. Naito. *Functional Differential Equations with Infinite Delay*. Springer, 1991.

[172] M.W. Hirsch and S. Smale. *Differential Equations, Dynamical Systems, and Linear Algebra*. Academic Press Inc, 1974.

[173] A.L. Hodgkin and A.F. Huxley. A quantitative description of membrane current and its application to conduction and excitation in nerve. *Bulletin of Mathematical Biology*, 52(1):25–71, 1990.

[174] K. Hoffman. *Analysis in Euclidean Space*. Prentice Hall, 1975.

[175] S.S. Holland. *Applied Analysis by the Hilbert Space Method*. Dover Publications, 1990.

[176] S. Hu and N.S. Papageorgiou. Handbook of Multivalued Analysis: Volume I: Theory. *Mathematics and its Applications -Dordrecht*, 419, 1997.

[177] R. Huilgol. A second order fluid of the differential type. *International Journal of Nonlinear Mechanics*, 3:471–482, 1968.

[178] A. Ida, S. Oharu, and Y. Oharu. A mathematical approach to HIV infection dynamics. *Journal of Computational and Applied Mathematics*, 204(1):172–186, 2007.

[179] K. Ito and F. Kappel. *Evolution Equations and Approximations*. World Scientific Pub Co Inc, 2002.

[180] Y. Ito. A semi-group of operators as a model of the lateral inhibition process in the sensory system. *Advances in Applied Probability*, 10:104–110, 1978.

[181] F. Izsak. An existence theorem for Volterra integrodifferential equations with infinite delay. *Electronic Journal of Differential Equations*, 2003(4):1–9, 2003.

[182] D. Jackson. Existence and uniqueness of solutions to semilinear nonlocal parabolic equations. *Journal of Mathematical Analysis and Applications*, 172(1):256–265, 1993.

[183] K. Jinno and A. Kawamura. Application of Fourier series expansion to two-dimensional stochastic advection-dispersion equation for designing monitoring network and real-time prediction of concentration. *IAHS Publications-Series of Proceedings and Reports-Intern Assoc Hydrological Sciences*, 220:225–236, 1994.

[184] A. Johnsson and D. Israelsson. Application of a theory for circumnutations to geotropic movements. *Physiologia Plantarum*, 21(2):282–291, 1968.

[185] M.C. Joshi and R.K. Bose. *Some Topics in Nonlinear Functional Analysis*. John Wiley and Sons, 1985.

[186] D. Jou, J. Casas-Vazquez, and G. Lebon. Extended irreversible thermodynamics revisited (1988-98). *Reports on Progress in Physics*, 62(7):1035–1142, 1999.

[187] G. Kallianpur and J. Xiong. Stochastic models of environmental pollution. *Advances in Applied Probability*, 26(2):377–403, 1994.

[188] M. Kanakaraj and K. Balachandran. Existence of solutions of Sobolev-type semilinear mixed integrodifferential inclusions in Banach spaces. *Journal of Applied Mathematics and Stochastic Analysis*, 16(2):163–170, 2003.

[189] G. Karakostas. Effect of seasonal variations to the delay population equation. *Nonlinear Analysis: Theory, Methods and Applications*, 6(11):1143–1154, 1982.

[190] A.G. Kartsatos and K.Y. Shin. Solvability of functional evolutions via compactness methods in general Banach spaces. *Nonlinear Analysis: Theory, Methods and Applications*, 21(7):517–535, 1993.

[191] T. Katō. Quasi-linear equations of evolution, with applications to partial differential equations. *Spectral Theory and Differential Equations*, pages 25–70, 1975.

[192] T. Katō. *Perturbation Theory for Linear Operators*. Springer Verlag, 1995.

[193] D.N. Keck and M.A. McKibben. Functional integro-differential stochastic evolution equations in Hilbert space. *Journal of Applied Mathematics and Stochastic Analysis*, 16(2):141–161, 2003.

[194] A.A. Kerefov. Nonlocal boundary value problems for parabolic equations. *Differ-entsianye Uravnenija*, 15:52–55, 1979.

[195] J.R. Kirkwood. *An Introduction to Analysis*. PWS Publishing Company, 1995.

[196] K. Knopp. *Theory and Application of Infinite Series*. Courier Dover Publications, 1990.

[197] D. Kolymbas. An outline of hypoplasticity. *Archive of Applied Mechanics (Ingenieur Archiv)*, 61(3):143–151, 1991.

[198] H. Komatsu. Fractional powers of operators. *Pacific Journal of Mathematics*, 19(2):285–346, 1966.

[199] Y. Konishi. On ut=uxx-F(ux) and the differentiability of the nonlinear semi-group associated with it. *Proceedings of the Japan Academy*, 48(5):281–286, 1972.

[200] J. Kopfová. Nonlinear semigroup methods in problems with hysteresis. *Dynamical Systems*, pages 580–589, 2007.

[201] R. Kosloff, M.A. Ratner, and W.B. Davis. Dynamics and relaxation in interacting systems: Semigroup methods. *Journal of Chemical Physics*, 106(17):7036–7043, 1997.

[202] E. Kreyszig. *Introductory Functional Analysis with Applications*. John Wiley and Sons New York, 1978.

[203] Z.F. Kuang and I. Pázsit. A class of semi-linear evolution equations arising in neutron fluctuations. *Transport Theory and Statistical Physics*, 31:141–151, 2002.

[204] T.G. Kurtz. Convergence of sequences of semigroups of nonlinear operators with an application to gas kinetics. *Transactions of the American Mathematical Society*, 186:259–272, 1973.

[205] G.E. Ladas and V. Lakshmikantham. *Differential Equations in Abstract Spaces*. Academic Press, 1972.

[206] V. Lakshmikantham and S.G. Deo. *Method of Variation of Parameters for Dynamic Systems*. CRC Press, 1998.

[207] V. Lakshmikantham and S. Leela. *Nonlinear Differential Equations in Abstract Spaces*. Pergamon, 1981.

[208] D. Lauffenburger, R. Aris, and K. Keller. Effects of cell motility and chemotaxis on microbial population growth. *Biophysical Journal*, 40(3):209–219, 1982.

[209] J.R. Ledwell, A.J. Watson, and C.S. Law. Evidence for slow mixing across the pycnocline from an open-ocean tracer-release experiment. *Nature*, 364(6439):701–703, 1993.

[210] J.R. Leigh. *Functional Analysis and Linear Control Theory*. Academic Press, 1980.

[211] M.A. Lewis, G. Schmitz, P. Kareiva, and JT Trevors. Models to examine containment and spread of genetically engineered microbes. *Molecular Ecology*, 5(2):165–175, 1996.

[212] J. Li. Hopf bifurcation of the sunflower equation. *Nonlinear Analysis: Real World Applications*, 10(4):2574–2580, 2009.

[213] Y. Li and Y. Wu. Stability of travelling waves with noncritical speeds for double degenerate Fisher-type equations. *Discrete and Continuous Dynamic Systems Series B*, 10(1):149–170, 2008.

[214] J. Liang, J. Liu, and T.J. Xiao. Nonlocal Cauchy problems governed by compact operator families. *Nonlinear Analysis*, 57(2):183–189, 2004.

[215] J. Lightbourne and S.M. Rankin. Partial functional differential equation of Sobolev type. *Journal of Mathematical Analysis and Applications*, 93(2):328–337, 1983.

[216] Y. Lin. Analytical and numerical solutions for a class of nonlocal nonlinear parabolic differential equations. *SIAM Journal of Mathematical Analysis*, 25:1577–1596, 1994.

[217] Y. Lin and J.H. Liu. Semilinear integrodifferential equations with nonlocal Cauchy problem. *Nonlinear Analysis*, 26(5):1023–1033, 1996.

[218] J. Liu, T. Naito, and N. Van Minh. Bounded and periodic solutions of infinite delay evolution equations. *Journal of Mathematical Analysis and Applications*, 286(2):705–712, 2003.

[219] J.H. Liu. Integrodifferential equations with non-autonomous operators. *Dynamic Systems and Applications*, 7:427–440, 1998.

[220] J.H. Liu. *A First Course In The Qualitative Theory of Differential Equations*. Prentice Hall, 2003.

[221] J.H. Liu. A remark on the mild solutions of non-local evolution equations. In *Semigroup Forum*, volume 66, pages 63–67. Springer, 2008.

[222] J.H. Liu and K. Ezzinbi. Non-autonomous integrodifferential equations with non-local conditions. *Journal of Integral Equations and Applications*, 15(1):79–93.

[223] J.Y. Liu and W.T. Simpson. Solutions of diffusion equation with constant diffusion and surface emission coefficients. *Drying Technology*, 15(10):2459–2478, 1997.

[224] M. Lizana. Global analysis of the sunflower equation with small delay. *Nonlinear Analysis*, 36(6):697–706, 1999.

[225] B. Lods. A generation theorem for kinetic equations with non-contractive boundary operators. *Comptes rendus-Mathématique*, 335(7):655–660, 2002.

[226] P. Lu, H.P. Lee, C. Lu, and P.Q. Zhang. Application of nonlocal beam models for carbon nanotubes. *International Journal of Solids and Structures*, 44(16):5289–5300, 2007.

[227] C.R. MacCluer. *Boundary Value Problems and Orthogonal Expansions: Physical Problems from a Sobolev Viewpoint.* Dover Publications, 1994.

[228] N.I. Mahmudov. Existence and uniqueness results for neutral SDEs in Hilbert spaces. *Stochastic Analysis and Applications*, 24(1):79–95, 2006.

[229] N.I. Mahmudov and M.A. McKibben. McKean-Vlasov stochastic differential equations in Hilbert spaces under Caratheodory conditions. *Dynamic Systems and Applications*, 15(3/4):357, 2006.

[230] A. Maio and A.M. Monte. Nonautonomous semilinear second order evolution equation in a Hilbert space. *International Journal of Engineering Science*, 23(1):27–38, 1985.

[231] X. Mao. Approximate solutions for a class of stochastic evolution equations with variable delays. *Numerical Functional Analysis and Optimization*, 12(5-6):525–533, 1991.

[232] X. Mao. Approximate solutions for a class of stochastic evolution equations with variable delays. II. *Numerical Functional Analysis and Optimization*, 15(1-2):65–76, 1994.

[233] N. Marheineke. *Turbulent Fibers–On the Motion of Long, Flexible Fibers in Turbulent Flows.* PhD thesis, PhD thesis, Technische Universitat Kaiserslautern, 2005.

[234] N.G. Markley. *Principles of Differential Equations.* John Wiley and Sons, 2004.

[235] P. Markowich and M. Renardy. The numerical solution of a class of quasilinear parabolic Volterra equations arising in polymer rheology. *SIAM Journal on Numerical Analysis*, 20(5):890–908, 1983.

[236] J.E. Marsden and T.J.R. Hughes. *Mathematical Foundations of Elasticity.* Dover Publications, 1994.

[237] R.H. Martin. *Nonlinear Operators and Differential Equations in Banach Spaces.* Krieger Publishing Co., Inc. Melbourne, FL, USA, 1986.

[238] M.P. Matos and D.C. Pereira. On a hyperbolic equation with strong damping. *Funkcial. Ekvac*, 34:303–311, 1991.

[239] M.A. Matovich and J.R.A. Pearson. Spinning a molten threadline. Steady-state isothermal viscous flows. *Industrial & Engineering Chemistry Fundamentals*, 8(3):512–520, 1969.

[240] M.A. McKibben. Second-order damped functional stochastic evolution equations in Hilbert space. *Dynamic Systems and Applications*, 12(3/4):467–488, 2003.

[241] M.A. McKibben. Second-order neutral stochastic evolution equations with heredity. *Journal of Applied Mathematics and Stochastic Analysis*, 2004(2):177–192, 2004.

[242] R.C. McOwen. *Partial Differential Equations*. Prentice Hall, 1996.

[243] C.V.M. Mee and P.F. Zweifel. A Fokker-Planck equation for growing cell populations. *Journal of Mathematical Biology*, 25(1):61–72, 1987.

[244] H. Meinhardt and A. Gierer. Applications of a theory of biological pattern formation based on lateral inhibition. *Journal of Cell Science*, 15(2):321–346, 1974.

[245] I.V. Melnikova and A. Filinkov. *Abstract Cauchy Problems: Three Approaches*. CRC Press, 2001.

[246] M. Miklavcic. *Applied Functional Analysis and Partial Differential Equations*. World Scientific Publishing Company Inc, 1998.

[247] E. Mitidieri and I.I. Vrabie. Existence for nonlinear functional differential equations. *Hiroshima Mathematics Journal*, 17(3):627–649, 1987.

[248] E. Mitidieri and I.I. Vrabie. Differential inclusions governed by nonconvex perturbations of m-accretive operators operators. *Differential and Integral Equations*, 2:525–531, 1989.

[249] I. Miyadera. *Nonlinear Semigroups*, volume 109. American Mathematical Society, 1992.

[250] A. Moameni. On the existence of standing wave solutions to quasilinear Schrodinger equations. *Nonlinearity- London*, 19(4):937, 2006.

[251] M.A. Murad, L.S. Bennethum, and J.H. Cushman. A multi-scale theory of swelling porous media: I. Application to one-dimensional consolidation. *Transport in Porous Media*, 19(2):93–122, 1995.

[252] J.D. Murray. *Mathematical Biology*. Springer Verlag, 2003.

[253] K. Nagel, P. Wagner, and R. Woesler. Still flowing: Approaches to traffic flow and traffic jam modeling. *Operations Research*, 51(5):681–710, 2003.

[254] Y. Naito, T. Suzuki, and K. Yoshida. Self-similar solutions to a parabolic system modeling chemotaxis. *Journal of Differential Equations*, 184(2):386–421, 2002.

[255] S. Nakagiri and J.H. Ha. Coupled sine-Gordon equations as nonlinear second order evolution equations. *Taiwanese Journal of Mathematics*, 5(2):297–315, 2000.

[256] R. Narasimha. Non-linear vibration of an elastic string. *Journal of Sound Vibration*, 8(1):134–146, 1968.

[257] S.K. Ntouyas and P. Ames. Global existence for semilinear evolution equations with nonlocal conditions. *Journal of Mathematical Analysis and Applications*, 210(2):679–687, 1997.

[258] S.K. Ntouyas and P.C. Tsamatos. Global existence for semilinear evolution integrodifferential equations with delay and nonlocal conditions. *Applicable analysis (Print)*, 64(1-2):99–105, 1997.

[259] S.K. Ntouyas and P.C. Tsamatos. Global existence for second order functional semilinear integrodifferential equations. *Mathematica Slovaca*, 50(1):95–109, 2000.

[260] A. Okubo and S.A. Levin. *Diffusion and Ecological Problems: Modern Perspectives.* Springer Verlag, 2001.

[261] B.G. Pachpatte. Inequalities for Differential and Integral Equations. *Mathematics in Science and Engineering*, 1998.

[262] C.V. Pao. Reaction diffusion equations with nonlocal boundary and nonlocal initial conditions. *Journal of Mathematical Analysis and Applications*, 195(3):702–718, 1995.

[263] J. Park and J.J. Bae. On the existence of solutions of strongly damped nonlinear wave equations. *International Journal of Mathematics and Mathematical Sciences*, 23(6):369–382, 2000.

[264] J. Park and Y.C. Kwun. Existence, uniqueness and norm estimate of solution for the nonlinear delay integro-differential system. *Journal of Korea Society of Mathematical Education Series B Pure and Applied Mathematics*, 8:137–144, 2001.

[265] S.K. Patcheu. On a global solution and asymptotic behaviour for the generalized damped extensible beam equation. *Journal of Differential Equations*, 135(2):299–314, 1997.

[266] N.H. Pavel. *Nonlinear Evolution Operators and Semigroups: Applications to Partial Differential Equations.* Springer, 1987.

[267] A. Pazy. *Semigroups of Linear Operators and Applications to Partial Differential Equations.* Springer, 1983.

[268] J.R.A. Pearson and M.A. Matovich. Spinning a molten threadline. Stability. *Industrial & Engineering Chemistry Fundamentals*, 8(4):605–609, 1969.

[269] R.L. Pego and M.I. Weinstein. Asymptotic stability of solitary waves. *Communications in Mathematical Physics*, 164(2):305–349, 1994.

[270] L.C. Piccinini, G. Stampacchia, and G. Vidossich. *Ordinary Differential Equations in R_n : Problems and Methods.* Springer Verlag, 1984.

[271] D. Pierotti and M. Verri. A nonlinear parabolic problem from combustion theory: attractors and stability. *Journal of Differential Equations*, 218(1):47–68, 2005.

[272] H.M. Portella, A.S. Gomes, N. Amato, and R.H.C. Maldonado. Semigroup theory and diffusion of hadrons in the atmosphere. *Journal of Physics A: Mathematical and General*, 31(32):6861–6872, 1998.

[273] D.L. Powers. *Boundary Value Problems*. Academic Press, 2nd edition, 1999.

[274] J.C. Principe. Artificial Neural Network. *The Electrical Engineering Handbook, CRC Press, LLC*, 2000.

[275] J. Pruss. Positivity and regularity of hyperbolic Volterra equations in Banach spaces. *Mathematische Annalen*, 279(2):317–344, 1987.

[276] J. Pruss. *Evolutionary integral equations and applications*. Birkhauser, 1993.

[277] S.M. Rankin III. A remark on cosine families. *Proceedings of the American Mathematical Society*, 79(3):376–378, 1980.

[278] S.M. Rankin III. Semilinear evolution equations in Banach spaces with application to parabolic partial differential equations. *Transactions of the American Mathematical Society*, 336(2):523–535, 1993.

[279] M. Reed and B. Simon. *Methods of Modern Mathematical Physics*. Academic Press, 1972.

[280] K. Ritz, J.W. McNicol, N. Nunan, S. Grayston, P. Millard, D. Atkinson, A. Gollotte, D. Habeshaw, B. Boag, C.D. Clegg, et al. Spatial structure in soil chemical and microbiological properties in an upland grassland. *FEMS microbiology ecology*, 49(2):191–205, 2004.

[281] A.W. Roberts and D.E. Varberg. *Convex Functions*. Academic Press INC, 1973.

[282] A.E. Rodkina. On existence and uniqueness of solution of stochastic differential equations with heredity. *Stochastics*, 12(3-4):187–200, 1984.

[283] H.L. Royden. *Real Analysis*. Macmillan New York, 1968.

[284] D. Savić. Model of pattern formation in animal coatings. *Journal of Theoretical Biology*, 172(4):299–303, 1995.

[285] H. Schaefer. Uber die Methode der a priori-Schranken. *Mathematische Annalen*, 129(1):415–416, 1955.

[286] M.J. Schramm. *Introduction to Real Analysis*. Prentice Hall, 1995.

[287] I. Segal. Non-linear Semi-groups. *The Annals of Mathematics*, 78(2):339–364, 1963.

[288] H. Serizawa. A semigroup treatment of a one dimensional nonlinear parabolic equation. *Proceedings of the American Mathematical Society*, 106(1):187–192, 1989.

[289] H. Serizawa and M. Watanabe. Time-dependent perturbation for cosine families in Banach spaces. *Houston Journal of Mathematics*, 12(4):579–586, 1986.

[290] S.E. Serrano, S.R. Workman, K. Srivastava, and B. Miller-Van Cleave. Models of nonlinear stream aquifer transients. *Journal of Hydrology*, 336(1-2):199–205, 2007.

[291] N. Shioji. Local existence theorems for nonlinear differential equations and compactness of integral solutions in Lp (0, T; X). *Nonlinear Analysis*, 26(4):799–811, 1996.

[292] R.E. Showalter. Existence and representation theorems for a semilinear Sobolev equation in Banach space. *SIAM Journal of Mathematical Analysis*, 3(3):527–543, 1972.

[293] R.E. Showalter. A nonlinear parabolic-Sobolev equation. *Journal of Mathematical Analysis and Applications*, 50:183–190, 1975.

[294] R.E. Showalter. Nonlinear degenerate evolution equations and partial differential equations of mixed type. *SIAM Journal of Mathematical Analysis*, 6:25–42, 1975.

[295] R.E. Showalter. *Monotone Operators in Banach Space and Nonlinear Partial Differential Equations*. American Mathematical Society, 1997.

[296] R.E. Showalter and P. Shi. Plasticity models and nonlinear semigroups. *Journal of Mathematical Analysis and Applications*, 216(1):218–245, 1997.

[297] R.E. Showalter and T.W. Ting. Pseudoparabolic partial differential equations. *SIAM Journal of Mathematical Analysis*, 1(1):1–26, 1970.

[298] J.G. Skellam. Random dispersal in theoretical populations. *Bulletin of Mathematical Biology*, 53(1):135–165, 1991.

[299] D. Solow. *How to Read and Do Proofs*. John Wiley and Sons, 1990.

[300] H. Tanabe. *Equations of Evolution*. Pitman London, 1979.

[301] N. Tanaka. A class of abstract quasi-linear evolution equations of second order. *Journal of the London Mathematical Society*, 62(01):198–212, 2000.

[302] T. Taniguchi. Successive approximations to solutions of stochastic differential equations. *Journal of Differential Equations*, 96(1):152–169, 1992.

[303] D.W. Taylor. *Research on Consolidation of Clays*. Massachusetts Institute of Technology, 1942.

[304] J. Tchuenche. Asymptotic stability of an abstract delay functional-differential equation. *Nonlinear Analysis*, 11(1):79–93, 2006.

[305] J. Tchuenche. Abstract formulation of an age-physiology dependent population dynamics problem. *Matematički Vesnik*, 60(2):79–86, 2008.

[306] R. Temam. *Infinite-Dimensional Dynamical Systems in Mechanics and Physics*. Springer, 1997.

[307] L. Tian and X. Li. Well-posedness for a new completely integrable shallow water wave equation. *International Journal of Nonlinear Science*, 4(2):83–91, 2007.

[308] T.W. Ting. Certain non-steady flows of second-order fluids. *Archive for Rational Mechanics and Analysis*, 14(1):1–26, 1963.

[309] K. Tomomi and M. Tsutsumi. Cauchy problem for some degenerate abstract differential equations of Sobolev type. *Funkcialaj Ekvacioj*, 40:215–226, 1997.

[310] C.C. Travis and G.F. Webb. Compactness, regularity, and uniform continuity properties of strongly continuous cosine families. *Houston Journal of Mathematics*, 3(4):555–567, 1977.

[311] C.C. Travis and G.F. Webb. Cosine families and abstract nonlinear second order differential equations. *Acta Mathematica Hungarica*, 32(1):75–96, 1978.

[312] P. Troncoso, O. Fierro, S. Curilef, and A.R. Plastino. A family of evolution equations with nonlinear diffusion, Verhulst growth, and global regulation: Exact time-dependent solutions. *Physica A: Statistical Mechanics and its Applications*, 375(2):457–466, 2007.

[313] P.N. Vabishchevich. Nonlocal parabolic problems and the inverse heat conduction problem. *Differential Equations*, 17(7):761–765, 1982.

[314] J.A.K.P.P. Van der Smagt and B. Krose. An Introduction to Neural Networks, 1993.

[315] J.L. Vazquez. Perspectives in nonlinear diffusion: between analysis, physics and geometry. *Proceedings of the International Congress of Mathematicians*, pages 609–634, 2007.

[316] I.I. Vrabie. The nonlinear version of Pazys local existence theorem. *Israel Journal of Mathematics*, 32(2):221–235, 1979.

[317] I.I. Vrabie. *Compactness Methods for Nonlinear Evolutions*. Chapman & Hall/CRC, 1995.

[318] I.I. Vrabie. *C0-Semigroups and Applications*. North Holland, 2003.

[319] I.I. Vrabie. *Differential Equations: An Introduction to Basic Concepts, Results and Applications*. World Scientific Publishing Company Inc, 2004.

[320] J.A. Walker. *Dynamical Systems and Evolution Equations: Theory and Applications*. Plenum Pub Corp, 1980.

[321] P. Waltman. *A Second Course in Elementary Differential Equations*. Dover Publications, 2004.

[322] Y. Wang and C. Zhou. Contraction operators and nonlocal problems. *Communications in Applied Analysis*, 5(1):31–38, 2001.

[323] Y.D. Wang. Theorems about the existence of solutions to problems with non-local initial value. *Acta Mathematica Sinica*, 17(2):197–206, 2001.

[324] Y. Wanli. A class of the quasilinear parabolic systems arising in population dynamics. *Methods and Applications of Analysis*, 9(2):261–272, 2002.

[325] G.F. Webb. Nonlinear perturbations of linear accretive operators in Banach spaces. *Israel Journal of Mathematics*, 12(3):237–248, 1972.

[326] G.F. Webb. *Theory of Nonlinear Age-Dependent Population Dynamics*. CRC Press, 1985.

[327] T. Weifner and D. Kolymbas. A hypoplastic model for clay and sand. *Acta Geotechnica*, 2(2):103–112, 2007.

[328] D. Wrzosek. Global attractor for a chemotaxis model with prevention of over-crowding. *Nonlinear Analysis*, 59(8):1293–1310, 2004.

[329] J. Wu. *Theory and Applications of Partial Functional Differential Equations*. Springer Verlag, 1996.

[330] J. Wu and H. Xia. Self-sustained oscillations in a ring array of coupled lossless transmission lines. *Journal of Differential Equations*, 124(1):247, 1996.

[331] J. Wu and H. Xia. Rotating waves in neutral partial functional differential equations. *Journal of Dynamics and Differential Equations*, 11(2):209–238, 1999.

[332] J. Wu and X. Zou. Patterns of sustained oscillations in neural networks with delayed interactions. *Applied Mathematics and Computation*, 73(1):55–75, 1995.

[333] B. Xie. Stochastic differential equations with non-Lipschitz coefficients in Hilbert spaces. *Stochastic Analysis and Applications*, 26(2):408–433, 2008.

[334] X. Xu. Application of nonlinear semigroup theory to a system of PDEs governing diffusion processes in a heterogeneous medium. *Nonlinear Analysis: Theory, Methods & Applications*, 18(1):61–77, 1992.

[335] Y. Xu, C.M. Vest, and J.D. Murray. Holographic interferometry used to demonstrate a theory of pattern formation in animal coats. *Applied Optics*, 22(22):3479–3483, 1983.

[336] X. Xue. Existence of solutions for semilinear nonlocal Cauchy problems in Banach spaces. *Electronic Journal of Differential Equations*, 2005(64):1–7, 2005.

[337] A. Yagi. Generation theorems of semigroups for multivalued linear operators. *Osaka Journal of Mathematics*, 28(2):385–410, 1991.

[338] T. Yamaguchi. Nonlocal nonlinear systems of transport equations in weighted Ll spaces: An operator theoretic approach. *Hiroshima Mathematics Journal*, 29:529–577, 1999.

[339] P. Yan and S. Liu. SEIR epidemic model with delay. *Anziam Journal*, 48(1):119–134, 2006.

[340] E. Zeidler. Nonlinear Functional Analysis: Part I. Springer Verlag, 1985.

[341] S. Zheng. *Nonlinear Evolution Equations*. Chapman & Hall/CRC, 2004.

[342] J. Zhu, Z. Li, and Y. Liu. Stable time periodic solutions for damped Sine-Gordon equations. *Electronic Journal of Differential Equations*, 2006(99):1–10, 2006.

Index

Printed and bound by CPI Group (UK) Ltd, Croydon, CR0 4YY

28/10/2024

01780204-0003